Springer Series in Experimental Entomology

Thomas A. Miller, Editor

Springer Series in Experimental Entomology
Editor: T.A. Miller

Neuroanatomical Techniques
Edited by N.J. Strausfeld and T.A. Miller

Sampling Methods in Soybean Entomology
Edited by M. Kogan and D. Herzog

Neurohormonal Techniques in Insects
Edited by T.A. Miller

Cuticle Techniques in Arthropods
Edited by T.A. Miller

Functional Neuroanatomy
Edited by N.J. Strausfeld

Techniques in Pheromone Research
Edited by H.E. Hummel and T.A. Miller

Measurement of Ion Transport and Metabolic Rate in Insects
Edited by T.J. Bradley and T.A. Miller

Neurochemical Techniques in Insect Research
Edited by H. Breer and T.A. Miller

Methods for the Study of Pest *Diabrotica*
Edited by J.L. Krysan and T.A. Miller

Insect-Plant Interactions
Edited by J.R. Miller and T.A. Miller

Immunological Techniques in Insect Biology
Edited by L.I. Gilbert and T.A. Miller

Heliothis: **Research Methods and Prospects**
Edited by M.P. Zalucki

Rice Insects: Management Strategies
Edited by E.A. Heinrichs and T.A. Miller

Molecular Approaches to Fundamental and Applied Entomology
Edited by J. Oakeshott and M.J. Whitten

John Oakeshott Max J. Whitten
Editors

Molecular Approaches to Fundamental and Applied Entomology

With 59 Illustrations

Springer-Verlag
New York Berlin Heidelberg London Paris
Tokyo Hong Kong Barcelona Budapest

John Oakeshott
CSIRO
Division of Entomology
Molecular Biology and
Physiology Section
Black Mountain, Canberra ACT
Australia

Max J. Whitten
CSIRO
Division of Entomology
Molecular Biology and
Physiology Section
Black Mountain, Canberra ACT
Australia

Library of Congress Cataloging-in-Publication Data
Molecular approaches to fundamental and applied entomology / [edited by] John Oakeshott, Max Whitten.
p. cm.—(Springer series in experimental entomology)
Includes bibliographical references and index.
ISBN 0-387-97814-3 (U.S.).—ISBN 3-540-97814-3 (German)
1. Molecular entomology. 2. Molecular genetics. 3. Insect pests—Control—Molecular aspects. I. Oakeshott, John. II. Whitten, Max. III. Series.
QL493.5.M65 1992
595.7′088—dc20 92-5035

Printed on acid-free paper.

Production coordinated by Chernow Editorial Services, Inc., and managed by Christin R. Ciresi; manufacturing supervised by Jacqui Ashri.
Typeset by Best-set Typesetter Ltd., Hong Kong.
Printed and bound by Edwards Brothers, Inc., Ann Arbor, MI.
Printed in the United States of America.

9 8 7 6 5 4 3 2 1

ISBN 0-387-97814-3 Springer-Verlag New York Berlin Heidelberg
ISBN 3-540-97814-3 Springer-Verlag Berlin Heidelberg New York

Series Preface

Insects as a group occupy a middle ground in the biosphere between bacteria and viruses at one extreme, amphibians and mammals at the other. The size and general nature of insects present special problems to the study of entomology. For example, many commercially available instruments are geared to measure in grams, while the forces commonly encountered in studying insects are in the milligram range. Therefore, techniques developed in the study of insects or in those fields concerned with the control of insect pests are often unique.

Methods for measuring things are common to all sciences. Advances sometimes depend more on how something was done than on what was measured; indeed a given field often progresses from one technique to another as new methods are discovered, developed, and modified. Just as often, some of these techniques find their way into the classroom when the problems involved have been sufficiently ironed out to permit students to master the manipulations in a few laboratory periods.

Many specialized techniques are confined to one specific research laboratory. Although methods may be considered commonplace where they are used, in another context even the simplest procedures may save considerable time. It is the purpose of this series (1) to report new developments in methodology, (2) to reveal sources of groups who have dealt with and solved particular entomological problems, and (3) to describe experiments which may be applicable for use in biology laboratory courses.

Thomas A. Miller
Series Editor

Preface

The advent of molecular biology in the mid-70s prompted a rash of brave promises about its ability to solve basic issues in other disciplines. Resources and personnel were lavished on this field because of these promises, yet nearly two decades later many of the promises remain unfulfilled. It therefore behooves current protagonists to argue their case with particular caution. This volume argues such a case for molecular entomology. Molecular biology is not presented as a universal panacea and its limitations are addressed explicitly. Nevertheless, it is shown that when coupled with complementary approaches from the more traditional disciplines, it can contribute new and unique insights into enduring problems in entomology.

Much of the progress thus far in molecular entomology has been confined to *Drosophila*, because the first generation of molecular technologies was most easily applied to organisms with good classical genetics and cytology, and the molecular work on *Drosophila* has emerged as one area in which many of the brave early promises have been fulfilled. *D. melanogaster* has become the model for understanding the molecular regulation of embryonic development, and seminal molecular studies exploiting this species have emerged in areas as diverse as population genetics and behavior.

Nevertheless, there are two reasons for expecting that the scope of molecular entomology can now be extended to nondrosophilids. The first is that the very success of the *Drosophila* work has generated a superb paradigm to guide the work on other insects. The second is that the increased power and sophistication of the later generations of molecular technologies has to some extent overcome the need to restrict their use to model species with good classical genetics.

We therefore believe it is timely to devote a volume in the series on experimental entomology to detailed consideration of the prospects for molecular entomology. We have tried not to produce a technical manual; there are many such already available. Further, we have endeavored to minimize the use of technical jargon; the mere language of molecular biology, and for that matter *Drosophila* genetics, can bar access to the field by nonpractitioners. Rather, we have aimed to illustrate to nondrosophilid entomologists the areas of their field in which molecular biology can now make a contribution. Conversely, too, we hope we can alert the (*Drosophila*) molecular biologists to some of the fascinating problems and opportunities in other areas of entomology that their own discipline can now address.

We begin with a chapter by ffrench-Constant, Roush, and Cariño that brings *Drosophila* genetics and molecular biology to bear on what has been the major issue in applied entomology over the last 50 years, namely resistance to chemical pesticides. Genes involved in many of the major resistance mechanisms have been cloned recently from *Drosophila*, and ffrench-Constant and colleagues show how these clones can now be exploited to establish some generalities about the molecular bases of resistance. Beyond their considerable heuristic interest, these insights may throw up new strategies for managing resistance problems. This first chapter is recommended reading not only for its insights into resistance mechanisms but also for its simple explanations of the basic genetic and molecular technologies employed by drosophilists; it therefore serves as valuable background for several later chapters that will focus on *Drosophila* as a model for other phenomena.

Chapters Two through Four introduce some of the biotechnological alternatives to chemical insecticides that are made possible by molecular biology. Two such alternative approaches involve the genetic engineering of insecticidal genes, either into microbial pathogens of pest insects or into crop or pasture plants subject to pest attack. In Chapter Two, Binnington and Baule show that the *Bt* gene encoding the delta-endotoxin of *Bacillus thuringiensis* is but one of a wide array of insecticidal genes that could prove useful in this respect. While genetic engineering is now possible for a large and ever increasing number of plants, it is only possible for a very small proportion of the microbial pathogens. One group of pathogens that can be engineered at present is the baculoviruses, and Chapter Three by Vlak discusses the prospects and problems associated with engineering baculovirus insecticides. In Chapter Four, Christian and colleagues discuss the prospects for engineering several other groups of insect viruses. These authors also show how genes encoding the pathogenic determinants of some insect viruses could themselves be genetically engineered into plants.

The focus moves from applied to more fundamental issues in Chapters Five and Six. Chapter Five, by Crozier, reviews the experimental and analytical procedures available for using molecular characters in systematics. Chapter Six, by Aquadro, shows how population data on molecular characters are providing new insights into old questions concerning the amount and adaptive significance of genetic polymorphism. Only limited insect data were available to Crozier, while Aquadro deliberately confines his attention to *Drosophila*, where the volume of data enables a consolidated treatment of the various experimental and analytical procedures. Nevertheless, it will be clear that the procedures espoused in these two chapters can be translated readily to nondrosophilid insects.

Compared to those applicable to systematic and population issues, many of the molecular techniques used to address questions in insect physiology and cell biology are more sophisticated and not yet readily applied to insects other than *Drosophila*. Nevertheless, these are the areas in which *Drosophila* molecular biology has so far had its greatest impact. We present select reviews of the remarkable advances that *Drosophila* molecular biology has permitted in understanding insect differentiation (Chapter Seven, by Tearle and others), sex determination (Chapter Eight, by Belote), diurnal rhythms (Chapter Nine, by Kyriacou), and meiosis (Chapter Ten, by Lyttle, Wu, and Hawley). We do not cover embryonic development because there are many excellent and accessible reviews of that area already (for example Pankratz and Jäckle, 1990, Trends in Genetics 6:287–292, and Glover 1991, Trends in Genetics 7:125–132). In time, issues in cell biology will become more accessible to molecular approaches in other insects. In the meantime, however, the generality of many of the principles emerging may be tested in another invertebrate with good classical genetics, namely the Nematode *Caenorhabditis elegans*. We refer interested readers to Wood (ed), 1988, The Nematode *Caenorhabditis elegans*, Cold Spring Harbor Lab., NY, for an entrée to *C. elegans* molecular biology.

What then are the prospects for carrying out the more sophisticated molecular experiments on nondrosophilids? There have been perhaps two critical limitations to date. One has been the much greater difficulty of cloning genes from insects which lack good classical genetics. However, this limitation is rapidly diminishing. In Chapter Eleven Trowell and East summarize impressive recent progress in techniques in protein biochemistry that will greatly facilitate protein-based, rather than genetically-based, cloning strategies. On the other hand, the second limitation, the lack of a genetic engineering technology for insects other than *Drosophila*, may take longer to solve. An engineering technology is an important research tool because the biological effects of cloned DNA, or manipulated versions of it, can only be fully assessed if it can be engineered back into the organism from which it was isolated. In Chapter Twelve, O'Brochta and Handler outline the prospects for developing engineering capability for nondrosophilids. There is no reason to doubt that it is an achievable aim; there is no reason to think *Drosophila* is inherently more amenable to engineering than other insects. Indeed, the pioneering work of O'Brochta and his colleagues suggests that the aim will be achieved relatively soon for some other dipterans. In keeping with the tenor throughout the volume we make no promises, however there is certainly good cause for optimism.

We thank all the authors who have contributed to this volume, first for the high quality of their submissions, and second for their patience in adapting submitted material to achieve overall coherence across chapters. Finally, we thank Ms. Audra Ankers and Mr. Eric Hines for expert and tireless editorial assistance.

CSIRO
Division of Entomology

John Oakeshott
Max J. Whitten

Contents

Contributors

Charles F. Aquadro
Section of Genetics and Development, Biotechnology Building, Cornell University, Ithaca, NY 14853, USA

Valerie J. Baule
Molecular Biology and Physiology Section, CSIRO Division of Entomology, P.O. Box 1700, Canberra, ACT 2601, Australia

John M. Belote
Department of Biology, Syracuse University, Syracuse, NY 13244, USA

Keith C. Binnington
Molecular Biology and Physiology Section, CSIRO Division of Entomology, P.O. Box 1700, Canberra, ACT 2601, Australia

Flerida A. Cariño
Plant Protection, Department of Scientific and Industrial Research, Lincoln, Christchurch, New Zealand

Peter D. Christian
Molecular Biology and Physiology Section, CSIRO Division of Entomology, P.O. Box 1700, Canberra, ACT 2601, Australia

Ross H. Crozier
Department of Genetics and Human Variation, La Trobe University, Bundoora, VIC 3083, Australia

David J. Dall
Molecular Biology and Physiology Section, CSIRO Division of Entomology, P.O. Box 1700, Canberra, ACT 2601, Australia

Peter D. East
Molecular Biology and Physiology Section, CSIRO Division of Entomology, P.O. Box 1700, Canberra, ACT 2601, Australia

Richard H. ffrench-Constant
Department of Entomology, Russell Laboratories, University of Wisconsin, Madison, WI 53706, USA

Jeremy Garwood
Research School of Chemistry, Research School of Biological Sciences, Australian National University, G.P.O. Box 4, Canberra, ACT 2601, Australia

Karl H. Gordon
Molecular Biology and Physiology Section, CSIRO Division of Entomology, P.O. Box 1700, Canberra, ACT 2601, Australia

Alfred M. Handler
Insect Attractants, Behavior and Basic Biology Laboratory, USDA/ARS, Gainesville, FL 32608, USA

Terry N. Hanzlik
Molecular Biology and Physiology Section, CSIRO Division of Entomology, P.O. Box 1700, Canberra, ACT 2601, Australia

R. Scott Hawley
Department of Molecular Genetics, Albert Einstein College of Medicine, Bronx, NY 10461, USA

Wayne R. Knibb
Department of Biochemistry, University of Adelaide, G.P.O. Box 498, Adelaide, SA 5001, Australia

C.P. Kyriacou
Department of Genetics, University of Leicester, University Road, Leicester LE1 7RH, United Kingdom

Trevor J. Lockett
CSIRO Division of Biomolecular Engineering, Molecular Biology Laboratories, P.O. Box 184, North Ryde, NSW 2113, Australia

Terrence W. Lyttle
Department of Genetics and Molecular Biology, University of Hawaii, Honolulu, HI 96822, USA

David A. O'Brochta
Center for Agricultural Biotechnology, Department of Entomology, University of Maryland, College Park, MD 20742, USA

Richard T. Roush
Department of Entomology, Comstock Hall, Cornell University, Ithaca, NY 14853, USA

Robert B. Saint
Department of Biochemistry, University of Adelaide, G.P.O. Box 498, Adelaide, SA 5001, Australia

Rick G. Tearle
Department of Biochemistry, University of Adelaide, G.P.O. Box 498, Adelaide, SA 5001, Australia

Stephen C. Trowell
Molecular Biology and Physiology Section, CSIRO Division of Entomology, P.O. Box 1700, Canberra, ACT 2601, Australia

Justinus M. Vlak
Department of Virology, Wageningen Agricultural University, Binnenhaven 11 6709 PD, Wageningen, The Netherlands

Chung-I Wu
Department of Biology, University of Rochester, Rochester, NY 14627, USA

Chapter 1

Drosophila as a Tool for Investigating the Molecular Genetics of Insecticide Resistance

Richard H. ffrench-Constant, Richard T. Roush, and Flerida A. Cariño

1. Introduction

Molecular studies on insecticide resistance are important for both fundamental and applied research. Resistance has developed so rapidly that we can gather meaningful data on selective changes during evolution (Kimura 1968; Roush and McKenzie 1987). Elucidation of the molecular mechanisms involved may allow for a clearer understanding of the modes of action of insecticides and the pathways for their degradation. Such knowledge could facilitate the rational design of novel insecticidal compounds and help in the formulation of pest control strategies for the future.

Studies of insecticide resistance in more than 30 species indicate that only a limited number of independent mechanisms are involved (Oppenoorth 1985; Soderlund and Bloomquist 1990). These mechanisms can be broadly dichotomized into altered target sites in the insect nervous system and altered metabolic systems involved in the degradation of the insecticides. The limited number of mechanisms found is in part because many different classes of insecticide either attack the same target site, or are metabolized by the same degradative enzymes, or both. Despite this apparent conservatism in resistance mechanisms, and the continued design of many insecticides around targets in the insect nervous system, resistance is still poorly understood at the molecular level. This is at least partly because most of the toxicological and biochemical studies on resistance have been done in pest species where comprehensive genetic and molecular studies are difficult to perform.

The availability of clones from homologous systems in vertebrates might be expected to facilitate the isolation of resistance genes in insects via heterologous hybridization. However, the feasibility of this approach is limited to the few cases where there are strong homologies between these systems. This condition requires insect proteins so well characterized that their homologies can be readily discerned. Often, insect gene products responsible for resistance are largely unknown, complex in nature, or difficult to isolate. An alternative approach is therefore necessary if we are to understand molecular mechanisms of resistance.

To overcome the relative genetic intractability of most insects in which resistance is found, Wilson (1988) proposed *Drosophila melanogaster* as a model for insecticide resistance studies. *D. melanogaster* offers a number of crucial advantages for genetic and molecular work. These advantages include its relatively small genome (approximately 165,000 kilobase pairs), the availability of strains with a variety of mutations and chromosome rearrangements (Lindsley and Grell 1968), and its ability to undergo P-element-mediated transformation (Rubin and Spradling 1982). Since Wilson's proposal to use *D. melanogaster* as a model, a number of useful mutants have been isolated and employed in the study of resistance. This chapter is designed to update the reader on current progress in this field and, through the use of examples, to discuss the circumstances under which studies of *D. melanogaster* are likely to be especially fruitful.

The aims of this chapter are essentially threefold: first, to acquaint the reader with the wide range of genetic and molecular manipulations possible with *D. melanogaster*; second, to illustrate how these techniques can be used to advance the study of resistance genes; and finally, to discuss how work with *D. melanogaster* can facilitate cloning of resistance genes from insects that are less genetically accessible.

2. *D. melanogaster* as a Genetic and Molecular Model System

2.1. Chromosomes: Homology and Mapping

D. melanogaster has only three major chromosomes. One of these, the X-chromosome (denoted chromosome I), is acro- or telocentric, having a single chromosome arm. The other two are metacentric autosomes, and their arms are designated IIL, IIR, IIIL and IIIR. The only other chromosomes are the Y, and another, tiny autosome (designated IV), both of which contain very few genes. The small number of major chromosomes allows the rapid mapping of any new mutant and has encouraged the isolation of countless genetic markers useful for fine scale genetic mapping. Apart from over 3,000 single gene loci for which mutants have been mapped, several hundred carefully mapped chromosomal rearrangements are also available.

There is also an excellent cytogenetic map of the giant polytene chromosomes that occur in the larval salivary glands of *D. melanogaster*. This map partitions the five major chromosome arms into 20 divisions (numbered 1 to 100, starting at the tip of chromosome I and ending at the tip of IIIR). There are about six subdivisions (denoted A–F) within each division, and between 4 and 22 discrete chromosome bands (numbered within each letter) per subdivision (Lindsey and Grell 1968). The reader is referred to Ashburner (1989a; 1989b) for a comprehensive treatment of *D. melanogaster* genetics and cytogenetics.

Rough homologies of the chromosomes and genetic maps of *D. melanogaster* with those from some other Diptera have also been established (Foster et al. 1981). Chromosomal homologies are potentially very useful in making predictions about the likely location of homologous genes. From such knowledge it is possible to predict roughly where genes in *D. melanogaster* homologous to resistance genes in pest species might be located. Using the chromosomal homologies proposed by Foster et al. (1981), we offer in Table 1.1 several hypotheses about the homologies in resistance genes between *D. melanogaster*, the housefly and sheep blowfly. Besides their evolutionary implications, these homologies can be used to take advantage of specialised chromosome systems developed in *D. melanogaster* (such as balancer and attached X chromosomes, see Ashburner 1989a; 1989b) to improve the efficiency of schemes to find the *Drosophila* homologs of genes known in other species.

Once a resistance gene has been found in *D. melanogaster* it can be mapped by recombination with known gene loci to a specific chromosome, next to a specific region within the chromosome, and finally, using chromosome deficiencies, to a specific cytological location. By these means, the gene can generally be mapped quite quickly to a single subdivision on the polytene chromosome, which is a small enough distance to initiate molecular analysis (for example, a chromosomal walk; Section 2.2.2 below).

Once molecular clones in the region have been obtained, in situ hybridization analyses of these clones to polytene preparations of various chromosomal deficiencies or rearrangements with breakpoints in the region can map the cloned material still more exactly, often to a single polytene band. A precise correspondence between genetic, chromosomal and molecular maps is thus possible. Such in situ analysis using chromosomal deficiencies and rearrangements also provides a convenient mechanism of monitoring progress in movement along the chromosome during chromosome walking or jumping (Section 2.2.2 below). The power of these techniques for the study of resistance genes (and neurophysiology) is illustrated by the cloning of the gene for dieldrin susceptibility in *D.*

Table 1.1. Possible homologies between insecticide resistance genes among three species of Diptera*

Musca domestica[†]		*Lucilia cuprina*[†]		*Drosophila melanogaster*	
I	—	II	—	IIL	—
II	insensitive AChe (OPs, carbamates)	IV		IIIR	Ace (acetylcholinesterase)[1]
	glutathione-S-transferase (OPs)		—		glutathione-S-transferase[2]
	carboxylesterase (malathion)(?[3††])		carboxylesterase[4]		malathion carboxyl esterase[5]
	hydrolase/esterase (OPs)		Rop-1 (esterase)[6]		esterase 23[5]
	microsomal oxidase (DDT, OPs, carbamates)		—		microsomal oxidase[7]
III	*kdr* (insensitivity to DDT, pyrethroids)	III	—	X	para (?)[8]
	pen (reduced penetration)		—		—
IV	cyclodiene insensitivity (e.g., dieldrin)	V	Rdl (dieldrin)	IIIL	Rdl (dieldrin)[9]
V	microsomal oxidase	VI	Rop-2(?)	IIR	microsomal oxidase[7]
	cyromazine resistance[10]				cyromazine resistance (?)[11]

* Chromosomal homologies (Roman numerals) are from Foster et al. (1981).

† Chromosomal locations and mechanisms of resistance genes in *M. domestica* are from Oppenoorth (1985), except as noted. For discussion on the possible regulation of genes on chromosome II of the house fly, see Plapp (1984; 1986). Chromosomal locations of resistance genes in *L. cuprina* from Foster et al. (1981), except as noted. Other references are: [1] Bender et al. (1983). [2] Cochrane et al. (1990b). [3] Discussion in Tsukamoto (1983), Oppenoorth (1985), and Raftos and Hughes (1986). [4] Raftos and Hughes (1986). [5] ME Spackman, RJ Russell, K-A Smyth and JG Oakeshott pers comm and Parker et al. (1991). [6] Hughes and Raftos (1985). [7] Hãllstrõm (1985) and Waters and Nix (1988). [8] Hall and Kasbekar (1989). [9] ffrench-Constant et al. (1990). [10] Shen and Plapp (1990). [11] J McKenzie pers comm.

†† Question marks indicate cases where resistance mechanisms are not clearly identified.

melanogaster, as described in Section 4.1 (below). Not only has this led to the first sequencing of an insect $GABA_A$ receptor, it offers an excellent avenue to answer questions about the mechanism of resistance unsuccessfully addressed for decades.

The advantages of *D. melanogaster* as a model system stand in contrast to many pests that have large numbers of chromosomes and few genetic markers. For example, *Heliothis virescens* has 31 chromosomes. This is one of the few pest insects wherein extensive efforts have been made to locate genetic markers, including restriction fragment polymorphisms for each of the chromosomes. More than half of the chromosomes have now been marked, and this has recently enabled mapping of an acetylcholinesterase gene through a mutant with altered insecticide sensitivity (Brown and Bryson 1990; Heckel et al. 1990). However, as a simple function of statistical probability, finding markers for the remaining half of the chromosomes will become more difficult. While the first few markers were unlikely to be linked, finding other unlinked markers becomes increasingly less likely, especially for smaller chromosomes. Calculations indicate that more than a hundred variants would have to be mapped before all 31 chromosomes are likely to be marked.

2.2. Chromosomes: Rearrangements, Walking, Microcloning and Artificial Chromosomes

2.2.1. Chromosomal Rearrangements

Chromosome rearrangements can take various forms, including simple, generally small deficiencies, translocations (duplications or rearrangements involving different chromosomes) and inversions (material is rearranged within chromosomes). The deficiencies are generally the most straightforward to use for gene mapping and examination of the function of resistance genes. For example, it takes just a single generation of crosses to put a chromosome bearing a mutant gene of interest in heterozygous condition against various homologous chromosomes carrying overlapping deficiencies (i.e., having slightly different breakpoints) in the region expected to contain the gene. Where resistant homozygotes can be distinguished from heterozygotes by an appropriate insecticide dose (i.e., resistance can be made effectively recessive, or nearly so), it is then possible to determine between which breakpoints the gene lies, simply by screening the various heterozygotes for effects of dosage on the mutant phenotype.

Even if appropriate deficiencies do not exist, other rearrangements with breakpoints in the relevant region can be used in their stead. For example, by appropriate crosses involving two Y-autosome translocations with slightly displaced autosomal breakpoints, it is possible to produce zygotes heterozygously deficient for the region between these displaced breakpoints, as well as complementary zygotes carrying a duplication for precisely the same region (Lindsley et al. 1972). By analogy with the deficiency mapping above, one can then localize an effectively recessive gene of interest between breakpoints by examining the various genotypes for dosage effects on the mutant phenotype.

Chromosome rearrangements can also be used to study interactions between the dosage effects of different loci. For example, Stern et al. (1990) used translocations to investigate interactions between two neurological sodium channel genes with effects on pyrethroid resistance, namely, *para*$^+$ and *nap*^{ts}. They found that increased dosage of *para*$^+$ suppresses *nap*^{ts}, an observation consistent with the hypothesis that *nap*^{ts} down-regulates *para* (see also Section 4.3 below).

Even if few chromosomal rearrangements are available in the region of interest, once a resistance mutant has been discovered, new rearrangements uncovering the resistance gene can be quickly generated. Where a dose can be applied that makes resistance effectively recessive, gamma-irradiated susceptible males can be crossed to homozygously resistant females, and their progeny screened for new rearrangements uncovering the gene, as exemplified in Section 4.1. This approach can be extremely powerful and is capable of generating one new rearrangement in every 3,000 progeny tested (RH ffrench-Constant and RT Roush unpublished data). Where the heterozygotes cannot be readily distinguished from the resistant homozygotes but can be distinguished from susceptible homozygotes (i.e., resistance is nearly or completely dominant), it is still possible to recover rearrangements in the resistance allele (by irradiating resistant males, crossing them to susceptible females, and saving the siblings of flies that die on exposure to pesticide due to a dysfunctional break in the resistance allele), although this requires maintaining single female lines and would be more tedious. Coupled with the superb genetic and chromosomal maps of *D. melanogaster*, it can be seen that chromosomal rearrangements permit very precise mapping of the resistance gene.

2.2.2. Chromosomal Walking

D. melanogaster is a particularly useful system for cloning genes whose functions are unknown but for which chromosomal locations can be determined. As explained above, the detailed maps of large polytene chromosomes allow direct chromosomal localization of single gene sequences via in situ hybridizations. An extensive list of clones of known chromosomal location is available (Merriam 1988; Merriam et al. 1991), and it is now possible to access most of the genome by using available clones in positions adjacent to the gene of interest as probes. Once the gene has been localized to a single polytene subdivision, it becomes an ideal candidate for molecular cloning and characterization via a chromosomal "walk" (Bender et al. 1983). A "walk" is simply the process of isolating a set of clones containing overlapping DNA sequences by using the DNA at the end of one cloned sequence as a probe to isolate clones containing the next sequences further along.

Chromosomal inversions with breakpoints within a resistance gene are particularly useful in identifying the position of the gene within a chromosomal walk. In contrast to a deficiency, which may eliminate a relatively large piece of chromosome, the inversion breakpoints precisely localize the gene to one of two points. These inversions may also be used in the walking process to "jump" over large regions of chromosome in order to get to an area of interest from a distant starting point. The reader is referred to Scott et al. (1983) and Campuzano et al. (1985) for examples of chromosomal walks and jumps.

A chromosomal walk can be conducted in any insect species. However, the availability of so many minutely-mapped chromosomal rearrangements in *D. melanogaster* makes it uniquely suitable for monitoring the progress of a walk and for expediting the walk by chromosome jumping. In addition, for no other insect species is there such a large collection of clones available for initiating a walk near a gene of interest.

2.2.3. Microcloning and Yeast Artificial Chromosomes

Following a precise cytological localization of a gene on the polytene chromosome, two techniques can vastly reduce the number of clones to be screened in order to isolate the gene: microcloning and the use of yeast artificial chromosomes. Both these techniques can generate "minilibraries" containing a high proportion of the clones of interest.

Microcloning relies on the high degree of resolution in the banding patterns of polytene chromosomes and on the large numbers of copies of any sequence contained in a polytene band. A band containing a gene of interest can be dissected out using a micromanipulator under a compound microscope (Pirrotta et al. 1983). Although very small amounts of DNA are recovered, a primary minilibrary can be made in a very small volume (12 nanolitres) surrounded by paraffin oil to prevent evaporation. The primary minilibrary of clones can then be amplified up to usable copy numbers. The amplified minilibrary can be used directly for screening for specific genes or for isolating sequences (e.g., P elements; see Section 2.3.3 below) that may be inserted in or near the gene of interest. Clones from a microcloned minilibrary can also be used to initiate a chromosomal walk; in particular, microcloning makes it possible to walk simultaneously from several startpoints in a region of interest until overlap indicates that the entire region has been covered.

Another method for producing a minilibrary of a region of interest involves the use of yeast artificial chromosomes (YACs). These are large fragments of DNA cloned as "artificial chromosomes" in yeast (Burke et al. 1987). YAC libraries are not yet available from pest insects, but a YAC library has been constructed from *D. melanogaster* (Garza et al. 1989a; 1989b). The constituent YAC clones can span several hundred kilobases of DNA, corresponding to several subdivisions of the polytene chromosomes. Cloning into more conventional cosmid vectors gives an average insert size of 33 kbp, and screening three genome equivalents requires 10,000 clones (three genome equivalents give a 95% probability of finding a unique sequence of interest). In contrast, only 1,500 YAC clones would be required to screen three equivalents of the *D. melanogaster* genome (Garza et al. 1989a).

One present emphasis of work done with the *D. melanogaster* YAC library is to map large numbers of constituent clones to polytene chromosomes by in situ hybridization. Lists of chromosome locations of the YACs so far analyzed in this way are published (Garza et al. 1989a). A complementary initiative has been to map overlaps among YAC clones by molecular means using end-specific probes obtained by the use of the polymerase chain reaction (Ochman et al. 1988; Garza et al. 1989b; and see Section 2.3.6 below). These overlapping regions can be used to develop continuous molecular maps of the *D. melanogaster* genome and to isolate smaller cosmid or bacteriophage clones containing the zones of overlap (Kafatos et al. 1991).

Like the minilibraries produced by microcloning, the availability of individual YAC clones (or sets of overlapping YAC clones) of known cytological location will greatly facilitate the process of chromosome walking. Thus, given the known cytological location of any gene in *D. melanogaster* involved in resistance, cloned DNA encompassing that gene will soon be readily accessible.

2.3. Mutant Isolation and Transposon Tagging

The classic method of selecting for resistance in the laboratory tends to result in a polygenic response, due to the limited genetic diversity of isolated laboratory populations (Roush and McKenzie 1987; Roush and Daly 1990). Field collection, chemical or radiation mutagenesis, and P element mutagenesis ("transposon tagging") are preferred methods for obtaining major resistance genes for cloning purposes.

2.3.1. Field Collection

Wild populations of *D. melanogaster* are excellent reservoirs of resistant mutants arising from insecticide selection. A dominant gene on chromosome IIR that is associated with elevated P450 monooxygenase levels and confers resistance to several insecticides has been found in populations of this species widely distributed across the globe (Ogita 1961; Shepanski et al. 1977; Halstrom et al. 1984; RT Roush and RH ffrench-Constant unpublished data; see Section 3.1). Another gene on chromosome III that also affects P450 monooxygenase levels but confers resistance only to malathion and closely related organophosphorous (OP) insecticides has been derived from a laboratory-selected population originally collected from the field (Holwerda and Morton 1983). Genes conferring high levels of cyclodiene resistance have been isolated from field populations (ffrench-Constant et al. 1990), as have several low level insecticide insensitive alleles of the acetylcholinesterase gene (Pralavorio and Fournier 1992). Clearly, isolation of a range of mutants from field *D. melanogaster* populations is possible even though insecticides are seldom deliberately applied to such populations. Subsequent gene cloning should enable direct comparison of sequences between resistance genes isolated from field collections and those obtained from laboratory mutagenesis.

2.3.2. Radiation and Chemical Mutagenesis

Mutagenesis has been applied to investigate potential mechanisms of resistance to novel compounds and to generate mutations for comparison to alleles that have already appeared in field populations. Both X-rays and chemical mutagens have been used successfully.

In general terms, X-rays are expected to be less useful than chemical mutagenesis for generating synthetic mutants resistant to insecticides. This is because X-rays exert their mutagenic effect by generating chromosome breakages which are subsequently incorrectly repaired. Consequently, many of the mutations are deletions or rearrangements of the genetic material rather than single base pair substitutions. As such, they are more likely to disable the gene completely than to create a new functional capacity. Nevertheless, X-rays were successfully used on *D. melanogaster* by Kikkawa (1964) to generate an OP resistant mutant that was then mapped to the genetic (i.e., recombinational) map position 64 on chromosome IIR. This particular mutation was not characterized any further, but appears to have been an allele of a resistance locus that has been studied at the molecular level by other researchers (see Section 3.1). More recently, T Wilson (pers comm) has used X-rays to generate mutants resistant to the insect growth regulator methoprene, although the levels of resistance conferred by these mutants are lower than those generated by chemical mutagenesis.

Chemical mutagenesis can be achieved using a range of compounds, but the most widely used mutagen is ethylmethanesulfonate (EMS). EMS typically causes single base pair mutations, occasionally resulting in high (greater than tenfold) resistance levels. It has been used to generate mutants in a range of insects; for example, three diazinon and four dieldrin resistant mutants have been produced in this way in the Australian sheep blowfly. Each mutant appears to be allelic with, and toxicologically indistinguishable from, genes selected in the field (Smyth et al. 1992; J McKenzie pers comm).

In *D. melanogaster*, EMS has been used to induce low levels of resistance to malathion (Pluthero and Threlkeld 1984) and to eserine sulfate, a carbamate that inhibits acetylcholinesterase (Burnell and Wilkins 1988). This mutation apparently reduces affinity of the enzyme for eserine sulfate. Whether or not this mechanism is used by field populations remains to be established.

Mutagenesis has also been used on *D. melanogaster* to investigate potentially new resistance mechanisms to novel insect growth regulators. EMS has been used to produce several mutants of the *Met* gene, some of which show 60- to 100-fold resistance to methoprene, a juvenile hormone (JH) analog (Wilson and Fabian 1986; 1987). Resistance to JHs or their analogs is mediated by metabolic rather than target site change (Sparks and Hammock 1983) in several other insects, including *Tribolium castaneum* and the house fly, *Musca domestica* (Dyte 1972; Cerf and Georghiou 1972). However, detailed biochemical work in *D. melanogaster* indicates that *Met* is involved in the production of an insensitive target, a JH receptor (Shemshedini and Wilson 1990). *Met* has been mapped by recombination to position 35.4 cM on the recombinational map of the X-chromosome and by chromosome deficiencies to the region 10C2-10D4 on the polytene chromosome map (Wilson and Fabian 1986; 1987). Such precise genetic and chromosomal localization should allow cloning of the gene via techniques such as chromosomal walking.

EMS has also been used to produce strains of *D. melanogaster* resistant to the growth regulator cyromazine (J McKenzie pers comm). Cyromazine's precise mode of action is uncertain, but it is believed to disrupt cuticle formation (Hughes et al. 1989). Two EMS generated resistant strains have been isolated, each with a resistance mutation on a different chromosomal arm, IIR and IIIL. The gene for cyromazine resistance in field populations of house flies is on chromosome V (Shen and Plapp 1990), which is homologous to chromosome IIR in *D. melanogaster* (Table 1), so it is possible that the EMS induced mutation on IIR in *D. melanogaster* is homologous to the field derived gene on chromosome V in the house fly. However, because the IIR gene maps near *vestigial* (J McKenzie pers comm), which is at recombinational map position 67, cyromazine resistance may also be a function of the dominant P450 monooxygenase resistance gene in *D. melanogaster* (recombinational map position 64; see Section 3.1). Another possibility is that cyromazine resistance is due to a completely novel mechanism that has not yet been characterized.

2.3.3. P element Mutagenesis

Although mutants generated by either chemical mutagens or radiation may enhance our ability to characterize a gene, they may not directly facilitate cloning, even in *D. melanogaster*. In contrast, P element mutagenesis of this species can both localize a gene and facilitate cloning by a technique called "transposon tagging." This may be especially useful where chromosome walking is difficult because of a lack of neighboring clones or the presence of repetitive DNA.

P element mutagenesis relies on the insertion of a transposable element called a P element into or near the gene of interest, thereby causing a detectable phenotype, usually corresponding to a loss or alteration of gene function. A number of different systems are now available for P element mutagenesis (Spradling 1986). These systems all have two components. The first is a defective (nonautonomous) P element that is incapable of catalyzing its own transposition, and the second is a gene capable of producing the transposase that is necessary for the transposition of the defective P. Once mutations have been produced by the movement of the P element, the transposase-producing gene is then outcrossed away to ensure stability of the insertion. The systems range from the "jumpstarter" method, where a single nonautonomous transposon ("mutator") is mobilized by a single "jumpstarter" element encoding the P element transposase (Cooley et al. 1988), to systems that involve the mobilization of a large number of "ammunition" mutator elements (Robertson et al. 1988).

In the "single-shot" jumpstarter approach, the mutator element is marked (with, for example, *neomycin resistance* or the *white* eye color gene), facilitating the selection and cloning of insertion mutants. This system can be used to generate a large number of single P element insertion stocks that can subsequently be screened for mutations (Cooley et al. 1988).

The "multiple-shot" approach is more efficient in screening for a specific resistance mechanism. "Multiple-shot" methods have been used in the isolation and cloning of the sodium channel gene *para* (Loughney et al. 1989), which may bear homology to the knockdown resistance gene (*kdr*) that confers nerve insensitivity resistance to DDT and pyrethroids in house flies. The *para* gene is on the X-chromosome, so males carrying this mutation are hemizygous and express the phenotype. Thus, alleles of *para* can be isolated from a P element screen either in males or, by failure to complement existing *para* alleles, in females. Four P element mutants of *Met*, the gene conferring resistance to methoprene, have also been isolated by the multiple-shot approach (in addition to those generated by EMS and radiation discussed above). The presence of a P element in the region of *Met* (lOC-D on the X-chromosome) for one allele, *Met A3*, was confirmed by in situ hybridization and will be used in cloning this gene (T Wilson pers comm).

One of the advantages of generating P element mutants is that the phenotype of revertants can be examined as proof of the causal connection between the phenotype and the mutation. In the example of *para*, remobilizing the insert $para^{hd2}$ allowed the recovery of wild type revertants. Analysis of these revertants by in situ hybridization and genomic Southern blotting revealed that reversion was accompanied by complete or partial loss of the P element, thus proving that the alteration of the phenotype was caused by the insertion (Loughney et al. 1989). Similar studies are being carried out on the *Met* P element mutants, and wild-type and lethal revertants have been recovered (T Wilson pers comm).

Following "tagging" of a mutant allele with a P element, a library can be constructed from the resulting strain, and the region flanking the P element (assumed to be the gene of interest) can be cloned directly. For example, a genomic bacteriophage library has been constructed from the *Met A3* P element mutant and clones hybridizing to a P element probe recovered. As this *Met* allele resulted from the mobilization of a number of P elements, all still present in the library, positive clones were hybridized to chromosomes in situ until one of them hybridized to the 10C-D region on the polytene chromosome where *Met* was shown to map. This clone is being used to recover a Met^{+} allele from a wild-type library that can then be used to rescue the mutant phenotype by transformation (see Section 2.3.4 below), as proof of gene cloning.

2.3.4. *Transformation*

Cloned genes can be introduced into the *D. melanogaster* genome by a procedure known as P-element mediated germline transformation (Spradling 1986; O'Brochta and Handler this volume). Phenotypic analysis of the subsequent transformants allows functional analysis of the cloned gene. Sometimes the technique is also used to verify the identity of the cloned gene, as illustrated in Section 4.1 (below). In cases where the gene's identity is already established, specific mutations can be introduced into the cloned gene in vitro by site-directed mutagenesis (Botstein and Shortle 1985), and the mutant clone is then transformed into *D. melanogaster*. This allows very precise sequence-to-function assignments within the gene of interest (e.g., Yu et al. 1987).

P-mediated transformation is achieved by microinjection of embryos with cloned DNA inserted into specific plasmid vectors containing the P element termini. DNA integration events are mediated by P element insertion in the presence of transposase. Transposase can be co-injected as a separate plasmid or endogenously expressed in some microinjection stocks. Vectors for transformation generally carry a genetic marker to enable the detection of successful integration of microinjected DNA.

P-element transposase activity has not been found outside the genus *Drosophila* (O'Brochta and Handler 1988). Recently there have been many attempts to transform other insects or to find transposable elements that may act as suitable vectors (Morris et al. 1989; Monroe et al. 1990; Warren and Crampton 1990; O'Brochta and Handler this volume), but for all practical purposes, transformation is still a unique advantage of *D. melanogaster*.

2.3.5. Cell Lines

One further advantage of *D. melanogaster* for the study of resistance mechanisms is the wide availability of cell lines. One such line of ecdysone-sensitive Kc cells has been used to study the action of the steroid molting hormone 20-hydroxyecdysone, which induces molting and metamorphosis (Wing 1988). The insecticide RH 5849 (1,2-dibenzoyl-1-tert-butylhydrazine) acts by mimicking the action of 20-hydroxyecdysone and competes with radiolabeled ponasterone A for high affinity ecdysone receptor sites from extracts of Kc cells. Populations of Kc cells selected for resistance to RH 5849 have fewer measurable ecdysone receptors. Thus, insecticides such as RH 5849, which are more easily synthesized than ecdysteroids, can be used as probes to characterize these receptors (Wing 1988).

Cultured cells of *D. melanogaster* can readily be transfected with cloned DNA (DiNocera and Dawid 1983), thus providing an alternative to P mediated germline transformation for the functional analysis of cloned genes. This alternative strategy may be particularly useful for other species where germline transformation systems are not yet available. For example, cell lines of mosquitoes are already available and vectors and selectable markers are being developed for a transfection system for these cells (Lycett et al. 1990; Monroe et al. 1990).

Mosquito cell lines resistant to methotrexate have been selected already, and resistance appears to be associated with tetraploidy, chromosomal rearrangements and an apparent duplication of one chromosome (Shotkoski and Fallon 1990). Such a mode of inheritance has not been documented for resistance in whole organisms but has been shown for drug resistant mammalian cell lines. This raises an important caveat for the use of selected cell lines: the mechanisms of resistance they exhibit may differ from those displayed by whole insects.

Nevertheless, cell lines remain attractive tools for studying particular issues in resistance mechanisms, like the expression of target site receptors. RH5849 resistance is a case in point. Another instructive example concerns the α_1 and β_1 receptors for gamma-aminobutyric acid

(GABA) in bovine brain. Clones of genes encoding these receptors have been expressed in cultured hamster ovary cells (Moss et al. 1990). Expression of the receptors on the surface of these cells facilitated electrophysiological studies and ligand binding studies on large quantities of readily accessible cells. This approach could be adapted to study similar systems in established cell lines of *D. melanogaster*.

2.3.6. Heterologous Probing and the Polymerase Chain Reaction

It will be clear from previous sections that a very large number of genes of defined function have been cloned from *D. melanogaster*, making it an excellent source of heterologous probes for isolating conserved sequences of resistance genes from other insects. Until the advent of the polymerase chain reaction (PCR) (Erlich 1989), heterologous probing entailed the use of a clone of the gene of interest from one species as a probe for the corresponding gene in, say, the pest species. PCR obviates the use of the heterologous clones as probes, although it still requires knowledge of the sequence of the heterologous gene. Heterologous probing with PCR has been used to generate a probe for the $para^{+}$ gene homolog in the house fly, which may be the *kdr* pyrethroid resistance gene, as described in Section 4.3 (below).

PCR is a method of amplifying specific DNA sequences by using two oligonucleotide primers that hybridize to opposite strands and flank the region of interest. Controlled by temperature cycling, a typical reaction undergoes a series of template denaturation, primer annealing and primer extension by the enzyme Taq DNA polymerase. Each cycle doubles the quantity of the product, as newly synthesized strands from one cycle act as templates for the next. Twenty cycles can give up to a millionfold amplification of the region of interest.

The use of conserved regions of genes as primers for PCR bypasses heterologous probing with cloned genes and allows the direct manufacture of probes from the insect of interest. A modification of the technique called anchor PCR allows the use of only a single primer of known sequence to amplify a specific DNA sequence (Erlich 1989). PCR can also be used in the construction of minilibraries for verifying the overlap between YAC clones and for walking in a YAC library (Garza et al. 1989b). A full review of PCR and its applications can be found in Erlich (1989), and further examples of its use in molecular entomology are given by Trowell and East (this volume).

3. Cloning Metabolic Resistance Genes

Alterations in three classes of enzymes, namely the P450 monooxygenases, glutathione S-transferases and various esterases, are associated with increased metabolism or sequestration of insecticides. These enzymes are relatively well-characterized in various insects and thus provide gene products that are relatively accessible to molecular manipulations.

3.1. Cytochrome P450 Monooxygenases

Cytochrome P450 monooxygenases (P450s or MFOs) are microsomal enzymes that have been implicated in the oxidation of various endogenous and exogenous compounds. P450s catalyze the insertion of one atom of molecular oxygen into the substrate and the reduction of the other oxygen atom to water. NADPH:cytochrome P450 reductase (P450 reductase), another microsomal enzyme, is responsible for the transfer of electrons to the P450 in the form of NADPH. In mammals, cytochrome b_5 (cyt b_5) provides one of the two electrons for some P450 reactions, but participation of cyt b_5 in insect P450 reactions has not been unequivocally demonstrated (Black and Coon 1987; Feyereisen et al. 1990).

Enhanced P450-dependent oxidation of insecticides has been observed in various resistant insects, from house flies, *Musca domestica* (Scott and Georghiou 1986), to *Heliothis armigera* (Little et al. 1989), but the mechanism remains poorly understood. The diversity of P450 structure and function suggests that several members of this enzyme family are present in any one insect species (Feyereisen at al. 1990).

Few insect P450 genes have been cloned and characterized at the molecular level. A gene encoding a phenobarbital-induced P450 (*CYP6A1*) has been cloned and sequenced from the multi-resistant Rutger's strain of house flies (Feyereisen et al. 1989). The gene for a different P450, *CYP4C1*, which is inducible by the hypertrehalosemic hormone, has been cloned from the cockroach *Blaberus discoidalis* but has not been studied in relation to insecticide resistance (Lee et al. 1990). A genomic clone containing part of a P450 gene has also been isolated from the mosquito *Aedes aegypti* (Bonet et al. 1990). Northern blot analysis suggests overexpression of this gene in a pyrethroid-resistant strain. The identities of these three cloned P450s are confirmed by the presence of the invariant "fingerprint" peptide F**G***C*G, where * denotes any amino acid (Nebert et al. 1989). A P450 protein has also been purified from the Rutger's strain by Wheelock and Scott (1990), and N-terminal amino acid sequencing of this P450 shows that it is different from the *CYP6A1* gene product isolated by Feyereisen et al. (1989).

The house fly *CYP6A1* gene is present in roughly the same number of copies in the susceptible strain *sbo* (*stw;bwb;ocra*) and the multi-resistant Rutger's strain. Northern blot hybridization shows a 1.9 kb mRNA encoded by the gene in both strains. A survey of 13 other strains shows that although high levels of *CYP6A1* mRNA are observed in many resistant strains, *CYP6A1* expression does not necessarily correlate with resistance. In the Rutger's strain, high constitutive expression is controlled by a gene on chromosome II, but the chromosomal location of the structural gene for *CYP6A1* remains to be established (FA Cariño, JF Koener and R Feyereisen unpublished data).

The house fly P450 reductase gene has also been cloned from the Rutger's strain (Feyereisen et al. 1990). The P450 reductase gene is on chromosome III, and levels of constitutive mRNA expression are similar in both the susceptible *sbo* and resistant Rutger's strains, suggesting that the observed resistance is not conferred by differential P450 reductase expression (J Koener, FA Cariño and R Feyereisen unpublished data).

In contrast to the above results, biochemical and genetic studies by Scott et al. (1990) showed elevated levels of P450s, P450 reductase and cyt b5 proteins in several resistant house fly strains. Their results suggest that all three components play a role in the expression of insecticide resistance in different strains. These apparently contradictory results emphasize the complexity of the genetics of the microsomal oxidative system and its role in resistance.

This complexity emphasizes the importance of parallel studies of similar enzymes in *D. melanogaster*. Genetic and biochemical studies of *D. melanogaster* indicate that a single gene or a pair of closely linked genes near recombinational map position 64 on chromosome II plays a dominant regulatory role in insecticide resistance and also affects P450 levels (Kikkawa 1964; Hallstrom 1985; Waters and Nix 1988). Other genes around recombinational map positions 51 and 58 on chromosome III are also important in regulating the activity of the P450 system (Hallstrom 1985; Hallstrom and Blanck 1985; Houpt et al. 1988). The region on chromosome II has been correlated with resistance to a wide range of compounds (organochlorines, carbamates and OPs) in strains from various locations, including Japan, the USA and England (Hallstrom et al. 1984; Shepanski et al. 1977; RT Roush and RH ffrench-Constant unpublished data). Another gene, at map position 56 on chromosome II but of undescribed mechanism, controls DDT resistance in a laboratory-selected strain (Dapkus 1992).

The cloning of the P450 structural genes from *D. melanogaster* is in progress. Waters and Nix (1988) have purified two electrophoretically defined subsets of P450 proteins (P450-A and P450-B) from this species. Total P450-A is expressed in similar levels in all strains tested, whereas expression of total P450-B is 50–100 times higher in strains resistant to DDT. Monoclonal antibodies raised against specific components of each subset show that the level of an immunoreactive form of P450-B is 10–20 times higher in resistant strains than in susceptible strains (Sundseth et al. 1989). The chromosome II locus described above is required for enhanced P450-B expression in resistant strains, but loci on chromosome III were also required for maximum expression. The monoclonal antibodies raised against P450-A and B have been used to screen expression libraries. Several A-specific and B-specific cDNA clones have been obtained and are currently being sequenced (Waters et al. 1990).

In a separate set of experiments, a strain of *D. melanogaster* selected in the laboratory for malathion resistance was used to raise a polyclonal antiserum against a partially purified P450 protein (Houpt et al. 1988). Elevated expression of an immunoreactive P450 protein apparently mapped to chromosome III (Cochrane et al. 1990a). Immunological screening of an expression library produced a putative P450 clone which cross-hybridizes with the house fly P450 cDNA that Feyereisen et al. (1989) had sequenced. The putative P450 clone from *D. melanogaster* was subsequently used to isolate a 1.8 kb clone from a cDNA library for this species. Interestingly, the 1.8 kb clone hybridizes only to a single genomic sequence in *D. melanogaster*, but to two species of mRNAs of 1.8 and 5.0 kb. The shorter mRNA is detectable in all strains but the 5.0 kb mRNA was found only in the resistant strains (Cochrane et al. 1990a). A fuller interpretation of this intriguing result awaits verification that the cloned gene is indeed a P450.

Studies on P450 reductase and cyt b_5 have not been pursued with equal vigor in *D. melanogaster*. This is unfortunate, since isolation of clones for these components of the microsomal oxidative system in *D. melanogaster* would facilitate the study of their roles in P450-based reactions. Regulatory mechanisms that control expression of the microsomal enzymes interacting with P450s could also be elucidated once clones for these enzymes become available.

3.2. Glutathione-S-Transferases

The glutathione-S-transferases (GSTs) are a group of enzymes which catalyze the conjugation of electrophilic compounds with reduced glutathione. GSTs are important in the metabolism of OPs that contain cleavable aryl or alkyl substituents (Dauterman 1985). Elevated levels of GST activity have been correlated with insecticide resistance in several insect species (Clark and Dauterman 1982; Hayaoka and Dauterman 1983; Clark et al. 1984). Several forms of GST are present in the cytosol as well as in the microsomes, each isoform with varying molecular structure and catalytic competence (Clark et al. 1984). Several GSTs have been cloned and characterized from plants, vertebrates and invertebrates (reviews by Mannervik and Danielson 1988; Pickett and Lu 1989), but isolation of insect GST clones by heterologous hybridizations is difficult because of limited sequence similarities across different gene families of GSTs.

The *D. melanogaster* GST family is composed of at least three classes of subunits of about 23,400 da, 28,500 da and 35,000 da. The latter two of these subunits, at least, are immunologically cross-reactive and together constitute a GST isoform reported to account for more than 90% of the total activity for the model GST substrate 1-chloro-2,4-dinitrobenzene (CDNB) (Cochrane et al. 1990b). There have been two independent reports on the cloning of GST subunit genes in *D. melanogaster*.

In one of these studies, a cDNA clone of a gene encoding a 23,839 da GST subunit has been isolated and sequenced by Toung et al. (1990). This clone shows amino acid sequence similarity with a cloned maize GST subunit, and functional proteins with catalytic activity for the GST substrate CDNB have been expressed from the cloned gene in *E. coli*. Genomic Southern blots using this cDNA as a probe indicated the presence of a GST family in *D. melanogaster*, and two different but potentially overlapping genomic clones have been isolated (Toung et al. 1990).

In the other molecular analysis of GST in *D. melanogaster*, antisera raised against the major GST isoform were used to isolate a putative GST clone coding for at least the 28,500 da subunit. When used as probe, the clone hybridizes to multiple genomic sequences and to two mRNAs of 1.3 and 2.3 kb. In situ studies with polytene chromosomes shows strong hybridization of the putative GST cDNA clone to a single site at cytological location 87C,D on the right arm of chromosome III (Cochrane et al. 1990b). Although the clone has no homology with other reported GST sequences, investigation of homologous genes in house flies and other insects may help confirm the identity of this clone and its role in resistance.

A clone of a GST subunit has also now been isolated from the sheep blow fly *L. cuprina*. Antibodies raised against the major GST isozyme in the pupae of *L. cuprina* have been used to isolate a cDNA clone from this species (P Board, RJ Russell and JG Oakeshott pers comm). Encoding a GST subunit of 208 amino acids, the *L. cuprina* GST clone shows higher than 80% identity with the sequence of the *D. melanogaster* GST isolated by Toung et al. (1990). Tests for associations of constitutive mRNA levels of this gene with resistance in *L. cuprina* are in progress.

3.3. Esterases

Elevated esterase levels have been observed in a number of resistant pest insects. Working with OP resistant populations of *Myzus persicae*, Devonshire and Sawicki (1979) first proposed a simple model of tandem duplications of the structural gene for a resistance-associated esterase, E4. E4, and the closely related FE4 from nontranslocation strains, confers resistance via hydrolysis and sequestration of insecticide esters (Devonshire and Moores 1982). Hybrid-arrested translation of poly(A)$^+$ RNA followed by immunoprecipitation of translation products indicated elevated levels of E4- or FE4-encoding mRNAs in resistant lines (Devonshire et al. 1986) and led to the subsequent isolation of cDNA clones. The use of the cDNA clone for E4 in probing the aphid genome provided direct evidence for gene amplification in *M. persicae* (Field et al. 1988).

Mouchès et al. (1986; 1987) also gave evidence for the amplification of an esterase gene in OP resistant strains of the mosquito *Culex quinquefasciatus*. By cloning highly amplified sequences from resistant strains, structural genes for esterase B1 have been isolated and characterized. The gene is amplified up to 250-fold in resistant mosquitoes. In neither this case nor the aphid case above have the molecular mechanisms involved in the amplification process yet been characterized (see Devonshire and Field 1991 for review).

Homologs of the resistance-associated esterases of *Myzus* and *Culex* have not yet been cloned from *D. melanogaster*. However, a large amount of information regarding the evolution of esterase gene families has become available following the cloning of the esterase-6 and esterase-P genes, located at cytological position 69A1 on chromosome 3L of *D. melanogaster* (Oakeshott et al. 1987; Collet et al. 1990). Like the resistance-associated esterase genes in *Myzus* and *Culex*, these genes have arisen by tandem gene duplication. Understanding the factors that affect the differential expression of esterase-6 and esterase-P will allow insight into the characteristics and function of duplicated families of esterase genes (Oakeshott et al. 1990; Healy et al. 1991).

The value of the esterase 6/esterase P model system has recently been enhanced by the discovery of a cluster of three esterase genes at cytological location 84D on chromosome IIIR of *D. melanogaster* (ME Spackman, RJ Russell, K-A Smyth and JG Oakeshott pers comm). Two of these genes, one encoding esterase 23 (*Est 23*) and the other malathion carboxylesterase (*Mce*), are directly homologous to the esterases encoded by the *R-OP1* and *R-mal* resistance genes of *L. cuprina* (Parker et al. 1991). The *R-OP1* gene is thought to be homologous to the mutant aliesterase gene associated with OP resistance in some house flies, so current work to clone this gene cluster (RJ Russell, P Kostakos, D Hartl and JG Oakeshott pers comm) will be of particular interest. The cloning strategy here involves heterologous probing of YAC libraries with PCR products designed from knowledge of the *Est6* and *EstP* gene sequences (see Sections 2.3.3 and 2.3.6).

4. Cloning Target Site Resistance Genes

A great deal is known in a general way about neurological differences between the nervous systems of insecticide susceptible and resistant insects (Sattelle et al. 1988 for review). However, the specific products of resistance genes coding for insensitive target sites are often poorly understood or difficult to isolate. Moreover, the genetic inaccessibility of the pest species studied has hampered our understanding of these resistance genes. Genetic analysis of *D. melanogaster* has contributed significantly to our general understanding of neurobiology (Hall and Greenspan 1979 for review) and shows great potential for understanding the specific molecular basis of target site resistance mechanisms.

4.1. Cyclodiene Resistance

The small number of chromosomes in *D. melanogaster*, the availability of numerous genetic markers for three-point recombination maps, and the established chromosomal homologies with other dipterans have been used to full advantage in cloning the gene responsible for cyclodiene resistance (dieldrin resistance, *Rdl*) from field populations of *D. melanogaster* (ffrench-Constant et al. 1990; ffrench-Constant and Roush 1991). Resistance to cyclodienes accounts for over 60% of reported cases of resistance across a range of pests (Georghiou 1986) and seems similar in all cases studied (Oppenoorth 1985).

Dieldrin resistance (*Dld-R*) maps to chromosome 4 in *Musca domestica* and to chromosome 5 (*Rdl*) in *Lucilia cuprina*. Using the established homologies among dipteran chromosomes, the dieldrin resistance gene was predicted to be on chromosome IIIL in *D. melanogaster* (Table 1). Dieldrin resistance (*Rdl*) in this species was detected in field collections from agricultural areas (ffrench-Constant et al. 1990). Like the dieldrin resistance genes in a wide range of species (Brown 1967; Oppenoorth 1985; Yarbrough et al. 1986), *Rdl* in *D. melanogaster* is semidominant; a dose of 30 μg/vial discriminates between Rdl^R/Rdl^R (R/R) and Rdl^R/Rdl^S (R/S) flies, while a dose of 0.5 μg/vial separates R/S from Rdl^S/Rdl^S (S/S) flies, albeit less precisely (ffrench-Constant et al. 1990; ffrench-Constant and Roush 1991). This semidominance facilitated classical genetic mapping of *Rdl* against various marker genes, which localized the resistance gene to the vicinity of recombinational map unit 25 (or cytogenetic map location 66) on the left arm of chromosome III (ffrench-Constant et al. 1990).

Following this approximate localization, the gene was further mapped by chromosome deficiencies. Postulating that when made hemizygous against a deficiency (in the absence of any susceptible gene product), the resistance gene would maintain full levels of resistance equivalent to R/R, progeny from a number of deficiencies spanning the chromosomal location were screened. Survival data for the resulting R/Df and R/S flies at 30 μg dieldrin/vial are presented in Figure 1.1. Only one deficiency, *Df(3L) 29A6*, exposed the resistance gene as a hemizygote, whereas the

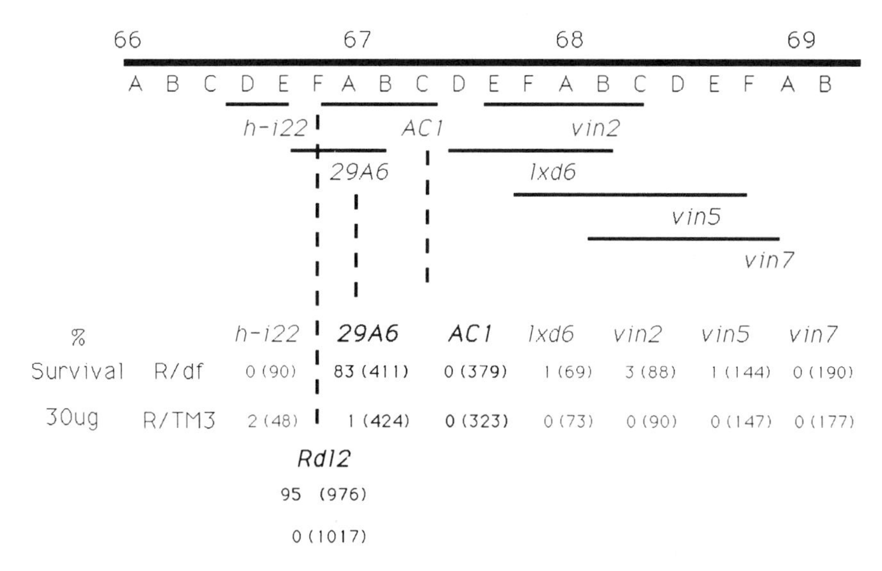

Figure 1.1. Chromosomal deficiencies near the cyclodiene resistance gene in *Drosophila melanogaster*. Cytological locations are represented by numbers (66–69) and letters (A–F) along the chromosome. The percentage survival when tested at the concentration (30 µg/vial) that kills all heterozygotes (with number of flies tested shown in parenthesis) is given for both R/Df and susceptible R/S (S from TM3 balancer) progeny for each deficiency (locations of which are represented by horizontal lines). Deficiency *Rdl* 2 was isolated by screening offspring of irradiated susceptible males (crossed to *Rdl/Rdl* females) for resistance, as described in the text (modified from ffrench-Constant and Roush 1990a).

overlapping deficiency *Df(3L) AC1* did not, thus localizing the resistance gene to a single subregion, 66F, of the polytene chromosome (ffrench-Constant and Roush 1991).

Following this isolation, new rearrangements uncovering the resistance gene were made (see Section 2.2.1), to give a more precise localization of the gene within 66F. Male *S/S* flies were gamma irradiated, crossed to *R/R* females, and their progeny screened at 30 µg dieldrin per vial. Only flies homozygous for resistance and a new rearrangement (i.e., resistance was uncovered by a new deficiency or inversion breakpoint) survived (ffrench-Constant et al. 1991). These rearrangements all failed to complement each other when crossed *inter se* (RH ffrench-Constant and RT Roush unpublished data), indicating that the product of the cyclodiene resistance gene is essential for viability.

Simultaneously with the isolation of the new rearrangements, a chromosomal walk was undertaken to clone the gene and identify the rearrangement breakpoints delimiting its position. The *Eco*R1 restriction map of the walk conducted across the chromosome region 66F is illustrated in Figure 1.2. The walk was initiated from the available phage clone (λ121) mapping to 66F1–2, corresponding to the border of the region of interest. The vector used for this walk was a modified CosPer cosmid (J Tamkun, University of California, Santa Cruz) containing 35–45 kb inserts and possessing P element ends, which allow it to be used in germline transformation. Once the walk spanned one visible cytological band, its orientation along the chromosome was confirmed by in situ hybridization. The progress of all subsequent steps in the walk was also checked by in situ hybridization to polytene chromosomes.

Inversions breaking within the resistance gene were particularly useful in identifying the position of the resistance gene within the walk. Several of the rearrangements generated to uncover resistance gave two new recombinant bands (suggesting an inversion or insertion) when fragments from the walk were used as probes in Southern analysis. A cluster of three such breakpoints (Figure 1.2) indicated the location of the resistance gene and suggested that transformation of cosmid 6 should rescue the susceptible phenotype. Transformation with cosmid 6, which carried a susceptible allele of the resistance gene, and appropriate crosses to a resistance strain converted resistant flies to heterozygotes susceptible to 30 μg/vial of dieldrin, confirming that the cloned DNA encompassed the resistance gene (ffrench-Constant et al. 1991).

Sequence analysis of the cloned region revealed strong similarities with GABA receptor genes from other species (ffrench-Constant et al. 1991), which is consistent with the pharmacological evidence that the target site for cyclodienes is a GABA receptor. The resistance allele will now be cloned and the amino acid sequence compared in order to elucidate the molecular nature of the cyclodiene resistance mutation in the GABA receptor gene.

4.2. Insensitive Acetylcholinesterase

Altered, insecticide-insensitive acetylcholinesterases confer resistance to carbamate and OP insecticides in a wide range of insects (Hama 1983). Although such mutants can be mapped by recombination in such insects as house flies, they have previously been inaccessible to molecular genetics. However, the cloning of the *Ace* gene in *D. melanogaster* (Hall and Spierer 1986) has provided an ideal starting point for molecular studies on insensitive acetylcholinesterases.

D. melanogaster itself is polymorphic for an *Ace* allele conferring low levels of insecticide resistance (Morton and Singh 1982; Burnell and Wilkins 1988). This resistance allele has been isolated and sequenced (Fournier and Berge 1990) and found to differ from the susceptible allele

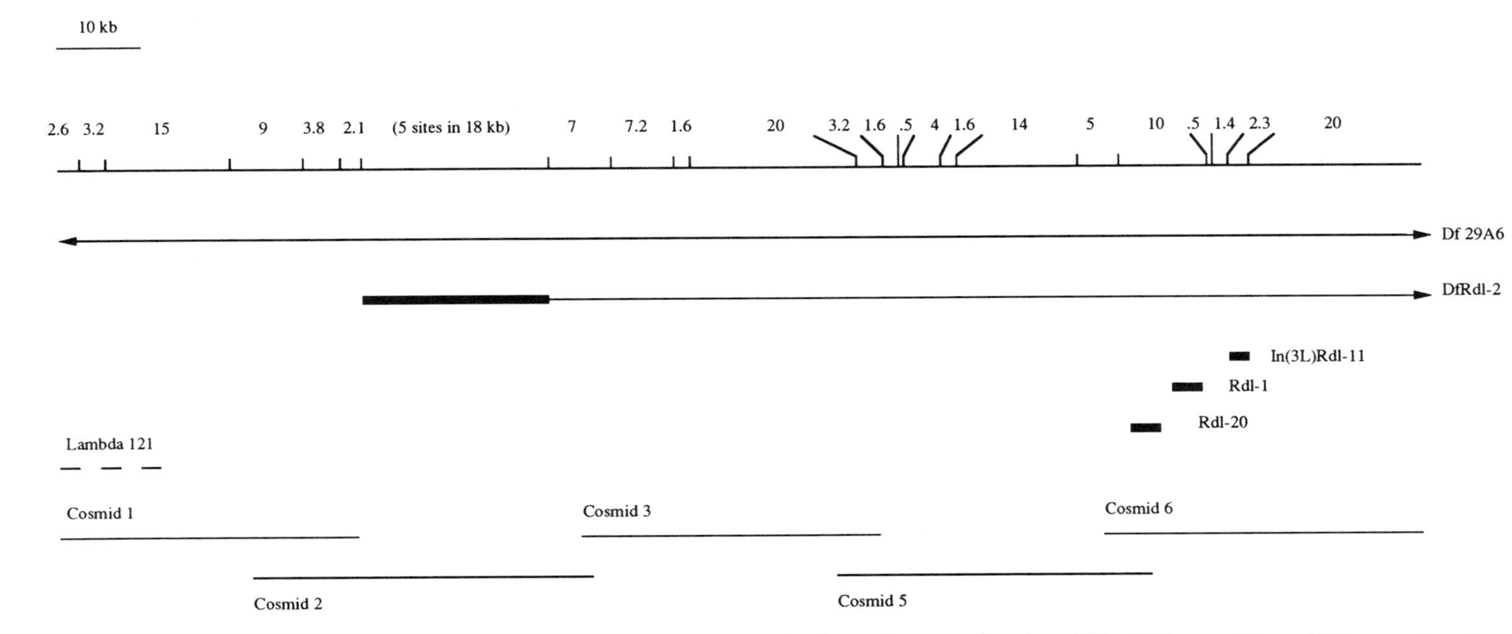

Figure 1.2. *Eco*R1 restriction map of a chromosomal walk through the polytene subregion 66F of *Drosophila*, which contains the cyclodience resistance gene. The positions of the cosmid "steps" of the walk are indicated below a restrictition map in kilobases. The positions of the rearrangement breakpoints marking the location of the gene are indicated by solid bars (see text).

only by the replacement of a single, highly conserved amino acid (a phenylalanine by a tyrosine).

However, of greater interest will be the use of *D. melanogaster Ace* probes to isolate resistant alleles from insects with more insensitive acetylcholinesterase, such as house flies (Devonshire and Moores 1984) and mosquitoes (ffrench-Constant and Bonning 1989). A cDNA clone from the *D. melanogaster Ace* gene has already been used to isolate a gene from the mosquito *Anopheles stephensi*, which shows 70% sequence similarity at the amino acid level and complete conservation at the active site (Malcolm and Hall 1990). Interestingly, intron-exon boundaries in the mosquito gene correspond exactly to those in *D. melanogaster*, but, due to the smaller introns of the mosquito gene, its genomic organization is far more compact than its counterpart in *D. melanogaster*.

4.3. Pyrethroid Resistance and *kdr*

Knock down resistance (*kdr*) was first described for DDT and pyrethroid resistance in house flies, but it now appears that similar mechanisms exist in many other insects (for review see Sawicki 1985). The voltage sensitive sodium channel of nerve membranes is thought to be the principal site of DDT and pyrethroid action (Grubs et al. 1988; Soderlund and Bloomquist 1990), and three biochemical mechanisms involving these channels have been proposed to explain *kdr* in house flies. These are: a 40–60% reduction in sodium channel density; alterations in the binding domain for DDT and pyrethroids on the sodium channel; and alterations in the lipid composition of *kdr* nerve membranes (Chiang and Devonshire 1982; Rossignol 1988; Sattelle et al. 1988; Pauron et al. 1989).

The mutant *nap*^{ts} (*no action potential, temperature sensitive*) of *D. melanogaster* shows nerve function failure at certain temperatures and has only half the sodium channel density of wild-type flies. *nap*^{ts} is an allele of the *maleless* (*mle*) locus on chromosome II (Kernan et al. 1989), which is involved in the regulation of genes on the X-chromosome. *maleless* is one of at least four loci needed for the twofold hyperactivation of X-linked gene expression in males, necessary to equal the output of the two X chromosomes in females (Lucchesi and Manning 1987). It has been proposed that *nap*^{ts} acts through down-regulation of the X-linked gene *para* (Hall and Kasbekar 1989; see below). Studies in several laboratories have confirmed that the *nap*^{ts} strain does have low level resistance to DDT and pyrethroids, but observed resistance levels do not reach the magnitude observed in *kdr* and *super-kdr* house flies (Hall et al. 1982; Jackson et al. 1984; Kasbekar and Hall 1988; Bloomquist et al. 1989). Most researchers in the area have therefore turned their attention to a more promising prospect, the *para* gene itself.

The gene *para* (*paralysis*) codes for a protein homologous to the alpha subunit of vertebrate sodium channels and is presumed to encode one of a number of classes of sodium channels in *D. melanogaster* (Loughney et al. 1989). *para*^{ts} is a temperature sensitive paralytic mutant with altered sodium channel properties. This and other alleles of *para* show either resistance or hypersensitivity to pyrethroids (Hall and Kasbekar 1989), suggesting that the study of the *para* alleles would identify suitable sodium channel mutants for characterization of *kdr* in *D. melanogaster*.

PCR has recently been employed to generate a conspecific probe for the homolog of the *para* gene in the house fly and a number of other insect species (Knipple et al. 1991). This probe was generated by using two degenerate primers separated by approximately 100 bp, this short distance between the primers allowing for most efficient PCR amplification. The precise priming sites were selected such that they did not flank known intronic regions but were within the amino acid sequences conserved in a rat sodium channel gene and *para* and not in the *DSCI* gene (a short sequence with sodium channel homology whose function is presently unclear). With the availability of the clones identified by these probes, similarities between the *D. melanogaster* and house fly model systems should eventually be established.

5. Applications

5.1. Cloning Resistance Genes from Pest Insects

In theory, resistance genes with well characterized products can be isolated and examined at the molecular level in virtually any insect. However, there are many cases, particularly involving target site resistance, where the resistance gene product is unknown or poorly understood. In such cases, the *D. melanogaster* model system is particularly useful, not only for studying resistance mechanisms in its own right, but also as a source of probes for isolating such genes from pest species.

Its potential in both these respects is illustrated by the case of cyclodiene resistance. Cyclodiene resistant strains of *D. melanogaster* show strong similarity to other insects in their cross-resistance to various cyclodienes and picrotoxin (ffrench-Constant and Roush 1990). Although cyclodienes are known to act on a GABA receptor (Eldefrawi and Eldefrawi 1987, and Section 4.1), the mechanism of resistance remains obscure. Resistance could arise from either insensitivity of a GABA receptor or changes in the total number of receptors (Tanaka 1987). The recent identification of GABA receptor cDNAs from the cyclodiene resistance locus of *D. melanogaster* (ffrench-Constant et al. 1991; and Section 4.1 above) should allow this issue to be resolved for *D. melanogaster* and at the same time allow cloning of homologous elements from other cyclodiene resistant insects, such as *L. cuprina*, where chromosomal walking is not yet feasible.

A similar approach could be used to clone other target site resistance genes from pest species. An important requisite here is that the gene of interest should be fairly well conserved across different taxa. This requirement is usually satisfied by target site resistance genes (Section 4) but may be less often satisfied for metabolic resistance genes (Section 3). The latter often belong to large multigene families, and homologies among different members of multigene families can obscure the relationships of specific genes across species.

On the other hand, *D. melanogaster* remains a useful model system for the study of metabolic resistance mechanisms in its own right. This is particularly true for the microsomal oxidative system, where the interaction of several components of an enzyme complex may determine the expression of insecticide resistance. Crosses could be designed to isolate each component, and subsequent use of chromosomal deficiencies and rearrangements would allow the study of one component in the presence of one or more of the other components. The well-characterized transformation system in *D. melanogaster* also allows in-depth studies on gene regulatory mechanisms that determine differential expression of insecticide metabolizing enzymes.

5.2. Understanding Gene Function

Aside from contributing to our understanding of resistance genes in insects of economic importance, the cloning of resistance genes enhances our understanding of basic insect physiology. Following gene cloning, a comparison of the sequences of susceptible and resistant alleles may identify the mutations that cause resistance and could allow identification of insecticide binding sites. For example, identification of the mutation conferring cyclodiene resistance is expected to elucidate the molecular biology of an insect GABA receptor and the binding site on the molecule for cyclodienes and picrotoxin (Section 4.1).

Functional regions of proteins can also be examined in great detail once the corresponding genes have been cloned. Following site-directed mutagenesis, the genes can be expressed *in vitro* or *in vivo* (by germ line transformation), and assays could measure the effect of the introduced mutation on the biochemical and/or structural characteristics of the expressed protein. To date, this approach has not been applied with specific reference to insecticide resistance, but it has been used to resolve other aspects of the substrate specificity of the acetylcholinesterase molecule that is the target for OPs and carbamates (Gibney et al. 1990).

5.3. Population Genetics

The impact of molecular studies of *D. melanogaster* on our understanding of population genetics has been reviewed by Aquadro this volume. Techniques such as PCR, whereby specific regions of DNA can be sequenced from individual insects, have obvious implications for our understanding of the population genetics of resistance genes. For example, the examination of variability in the sequences of resistance genes can establish the nature of the changes associated with resistance in field populations. It also examines the extent to which strong selection at a resistance locus influences variability in the surrounding sequences (linkage disequilibria). Such studies will enhance our knowledge of the frequency of resistance mutations, their geographical distributions, and the methods of their dispersal.

A recent study by Raymond et al. (1991) provides graphic evidence of the power of population genetic studies of resistance based on molecular markers of the resistance genes. OP resistance in *Culex pipiens* is attributed to overproduction of nonspecific esterase A and the highly amplified esterase B. Both classes of esterases are highly polymorphic but the geographical distribution of most variants is restricted. The exceptions are the A2 and B2 variants, which are associated with resistance and which are always found together. Restriction maps of esterase B2 from several strains collected from different regions in three continents show very high similarity. These results suggest that the worldwide occurrence of amplified esterase B2 reflects a single original mutation which has subsequently been dispersed by migration.

PCR is also being used to examine sequence variability in a number of *Ace* mutants of *D. melanogaster* (Arpagaus and Toutant 1990). These and other questions are central not only to our understanding of the population genetics of resistance but also to its modeling and management.

5.4. Resistance Management

Resistance management practices are commonly limited by the difficulty of monitoring the frequencies of resistance genes quickly but accurately in field populations. This limitation is at least partly due to reliance on bioassays, which can be prone to extreme variation, to score the resistance genes (ffrench-Constant and Roush 1990). Quick and accurate biochemical assays are available in some cases (Hughes and Raftos 1985; ffrench-Constant and Devonshire 1988; Field et al. 1989; Devonshire and Field 1991). However, molecular analyses such as those used to monitor the resistant esterase genes in the *Culex* populations, above, or E4 copy number in *Myzus persicae* (Field et al. 1988), clearly afford even more accurate detection of resistance genes and their frequencies in field populations. At present these molecular analyses are likely to utilize Southern blot, oligonucleotide probes, or PCR technology, which require significant expertise and facilities, and are more time-consuming than biochemical assays (ffrench-Constant and Roush 1990). One promising consequence of the cloning of resistance genes, however, is that it may facilitate the development of monoclonal antibody probes for resistance specific epitopes on the relevant proteins. Such antibodies have already been produced for a resistance-associated P450 from *D. melanogaster* (Sundseth et al. 1989; Waters et al. 1991; and Section 3.1 above). Corresponding antibodies from pest species could be incorporated into simple field kits that would allow rapid and unambiguous scoring of hundreds of individuals per person per day.

There is also a second route by which molecular analyses of resistance genes may support resistance management programs. Current management strategies often rely on the use of alternative compounds. A more detailed understanding of the genetic mutations associated with specific resistance mechanisms may lead to the biorational design of alternative compounds that can overcome those mechanisms. In principle, identification of highly conserved (and thus presumably essential) sequences in the mutant genes that contribute to resistance will help in the design of future pesticides or inhibitors which could synergize insecticide action.

6. Conclusions

We are well aware that generalizations on insecticide resistance mechanisms across different taxa would be of little value without research on other species as well as *D. melanogaster*. However, a greater understanding of the genetic and molecular mechanisms of resistance will be most rapidly achieved by taking advantage of the best features of each of several species. The extensive genetic and molecular information available for *D. melanogaster* makes it a good model system in its own right and provides a useful launching point for molecular studies of insecticide resistance in other species. We hope that this chapter encourages more researchers to exploit the advantages of the *D. melanogaster* system in studying resistance.

Acknowledgments. We thank D Soderlund, LC Waters, and TG Wilson for critical reviews of the manuscript. We are also grateful to P Kostakos for her editorial help. RH ffrench-Constant was supported at Cornell University by a postdoctoral fellowship from Merk Sharp and Dohme Research Laboratories and at the University of Wisconsin-Madison by PHS grant NS29623 during preparation of this manuscript.

References

Ahmad M, Gladwell RT, McCaffery AR (1989) Decreased nerve sensitivity is a mechanism of resistance in a pyrethroid resistant strain of *Heliothis armigera* from Thailand. Pestic Biochem Physiol 35:165–171

Arpagaus M, Toutant J-P (1990) *Drosophila* acetylcholinesterase: fine structure of molecular forms in wild type flies and mutants of the *Ace* locus. In Hagedorn H, Hildebrand HJ, Kidwell M, Law J (eds) Molecular Insect Science. Plenum Press, New York, p 276

Ashburner M (1989a) *Drosophila*: A Laboratory Handbook. Cold Spring Harbor Laboratory Press, Cold Spring Harbor, 1331 pp

Ashburner M (1989b) *Drosophila*: A Laboratory Manual. Cold Spring Harbor Laboratory Press, Cold Spring Harbor, 434 pp

Bender W, Spierer P, Hogness DS (1983) Chromosomal walking and jumping to isolate DNA from the *Ace* and *rosy* loci and the bithorax complex in *Drosophila melanogaster*. J Mol Biol 168:17–33

Black SD, Coon MJ (1987) P450 cytochromes: Structure and function. Adv Enzymol 60:35–87.

Bloomquist JR, Soderlund DM, Knipple DC (1989) Knockdown resistance to dichlorodiphenyl-trichloroethane and pyrethroid insecticides in the *nap*^{ts} mutant of *Drosophila melanogaster* is correlated with reduced neuronal sensitivity. Arch Insect Biochem Physiol 10:293–302

Bonet RG, Crampton J, Townson H (1990) Mosquito P-450 genes and pyrethroid insecticide resistance. In Hagedorn H, Hildebrand HJ, Kidwell M, Law J (eds) Molecular Insect Science. Plenum Press, New York, pp 280

Botstein D, Shortle D (1985) Strategies and applications of *in vitro* mutagenesis. Science 229:1193–1201

Brown AWA (1967) Genetics of insecticide resistance in insect vectors. In Wright JH, Pal R (eds) Genetics of Insect Vectors of Disease. Elsevier, Amsterdam, pp 505–552

Brown TM, Bryson PK (1990) Methyl paraoxon-resistant acetylcholinesterase controlled by a single gene in the tobacco budworm *Heliothis virescens*. In Hagedorn H, Hildebrand HJ, Kidwell M, Law J (eds) Molecular Insect Science. Plenum Press, New York, p 285

Burke DT, Carle GF, Olson MV (1987) Cloning of large segments of exogenous DNA into yeast by means of artificial chromosome vectors. Science 236: 800–812

Burnell AM, Wilkins NP (1988) An investigation of the *in vitro* inhibition of acetylcholinesterase by the carbamate inhibitor eserine sulphate in eserine resistant strains of *Drosophila melanogaster*. Comp Biochem Physiol 90C: 215–220

Campuzano S, Carromolino L, Cabrera CV, Ruiz-Gomez M, Villares R, Boronat A, Modolell J (1985) Molecular genetics of the *achaete-scute* gene complex of *Drosophila melanogaster*. Cell 40:327–338.

Cerf DC, Georghiou GP (1972) Evidence of cross-resistance to a juvenile hormone analogue in some insecticide-resistant houseflies. Nature 239:401–402

Chiang C, Devonshire AL (1982) Changes in membrane phospholipids, identified by Arrhenius plots of acetylcholinesterase and associated with pyrethroid resistance (*kdr*) in houseflies (*Musca domestica*). Pestic Sci 13:156–160

Clark AG, Dauterman WC (1982) The characterization by affinity chromatography of glutathione S-transferases from different strains of housefly. Pestic Biochem Physiol 17:307–314

Clark AG, Shamaan NA, Dauterman WC, Hayaoka T (1984) Characterization of multiple glutathione transferases from the house fly, *Musca domestica* L. Pestic Biochem Physiol 22:51–59

Cochrane BJ, Holtsberg F, Crocquet de Belligny P, Pursey J, Morton R (1990a) A cytochrome P-450 cDNA clone from *Drosophila melanogaster*. In Hagedorn H, Hildebrand HJ, Kidwell M, Law J (eds) Molecular Insect Science. Plenum Press, New York, pp 291–292

Cochrane BJ, Morrissey JJ, Phillips J (1990b) Isolation and characterization of a glutathione S-transferase cDNA from *Drosophila melanogaster*. In Hagedorn H, Hildebrand HJ, Kidwell M, Law J (eds) Molecular Insect Science. Plenum Press, New York, p 292

Cockburn AF (1990) Development of improved insect transformation techniques. In Hagedorn H, Hildebrand HJ, Kidwell M, Law J (eds) Molecular Insect Science. Plenum Press, New York, p 292

Collet C, Nielsen KM, Russell RJ, Karl M, Oakeshott JG, Richmond RC (1990) Molecular analysis of duplicated esterase genes in *Drosophila melanogaster*. Mol Biol Evol 7:525–546

Cooley L, Kelley R, Spradling A (1988) Insertional mutagenesis of the *Drosophila* genome with single P elements. Science 239:1121–1128

Dapkus D (1992) Genetic localization of DDT resistance in *Drosophila melanogasten* (Dipterai Drosophilidae). J Econ Entomol 85:340–347

Dauterman WC (1985) Insect metabolism: Extramicrosomal. In Kerkut GA, Gilbert LI (eds) Comprehensive Insect Physiology, Biochemistry and Pharmacology, Vol 12. Pergamon Press, Oxford, pp 713–730

Devonshire AL, Field LM (1991) Gene amplification and insecticide resistance. Ann Rev Entomol 36:1–23

Devonshire AL, Moore GD (1982) A carboxylesterase with broad substrate specificity causes organophosporus, carbamate and pyrethroid resistance in peach potato aphids (*Myzus persicae*). Pestic Biochem Physiol 18:235–246

Devonshire AL, Moore GD (1984) Different forms of insensitive acetylcholinesterase in insecticide-resistant house flies (*Musca domestica*) Pestic Biochem Physiol 21:336–340

Devonshire AL, Sawicki RM (1979) Insecticide-resistant *Myzus persicae* as an example of evolution by gene amplification. Nature 280:140–141

DiNocera P, Dawid IB (1983) Transient expression of genes introduced into cultured cells of *Drosophila*. Proc Natl Acad Sci USA 80:7095–7098

Dyte CE (1972) Resistance to synthetic juvenile hormone in a strain of flour beetle, *Tribolium castaneum*. Nature 238:48

Eldefrawi AT, Eldefrawi ME (1987) Receptors for γ-aminobutyric acid and voltage-dependent chloride channels as targets for drugs and toxicants. FASEB J 1:262–271

Erlich HA (1989) PCR Technology: Principles and Applications for DNA Amplification. Stockton Press, New York

Feyereisen R, Koener JF, Cariño FA, Daggett AS (1990) Biochemistry and molecular biology of cytochrome P450. In Hagedorn H, Hildebrand HJ, Kidwell M, Law J (eds) Molecular Insect Science. Plenum Press, New York, pp 263–272

Feyereisen R, Koener JF, Farnsworth DE, Nebert DW (1989) Isolation and sequence of cDNA encoding a cytochrome P-450 from an insecticide-resistant strain of the house fly, *Musca domestica*. Proc Natl Acad Sci USA 86: 1465–1469

ffrench-Constant RH, Bonning BC (1989) Rapid microtitre plate test distinguishes insecticide resistant acetylcholinesterase genotypes in the mosquitoes *Anopheles albimanus*, *An nigerrimus* and *Culex pipiens*. Med Vet Entomol 3:9–16

ffrench-Constant RH, Devonshire AL (1988) Monitoring frequencies of insecticide resistance in *Myzus persicae* (Sulzer) (Hemiptera: Aphididae) in England during 1985–86 by immunoassay. Bull Entomol Res 78:163–171

ffrench-Constant RH, Mortlock DP, Shaffer CD, MacIntyre RJ, Roush RT (1991) Molecular cloning and transformation of cyclodiene resistance in *Drosophila*: an invertebrate γ-aminobutyric acid subtype A receptor locus. Proc Natl Acad Sci USA 88:7209–7213

ffrench-Constant RH, Roush RT (1990) Resistance detection and documentation: the relative roles of pesticidal and biochemical assays. In Roush RT, Tabashnik BE (eds) Pesticide Resistance in Arthropods. Chapman and Hall, New York, pp 4–38

ffrench-Constant RH, Roush RT (1991) Gene mapping and cross resistance in cyclodiene insecticide resistant *Drosophila melanogaster* (Mg). Genet Res 57:17–21

ffrench-Constant RH, Roush RT, Mortlock D, Dively GP (1990) Isolation of dieldrin resistance from field populations of *Drosophila melanogaster* (Diptera: Drosophilidae). J Econ Entomol 83:1733–1737

Field LM, Devonshire AL, ffrench-Constant RH, Forde BG (1989) The combined use of immunoassay and a DNA diagnostic technique to identify insecticide resistant genotypes in the peach-potato aphid *Myzus persicae* (Sulz). Pestic Biochem Physiol 34:174–178

Field LM, Devonshire AL, Forde BG (1988) Molecular evidence that insecticide resistance in peach-potato aphids (*Myzus persicae* Sulz) results from amplification of an esterase gene. Biochem J 251:309–312

Foster GG, Whitten MJ, Konovalov C, Arnold JTA, Maffi G (1981) Autosomal genetic maps of the Australian sheep blowfly, *Lucilia cuprina* dorsalis R-D (Diptera: Calliphoridae), and possible correlations with the linkage maps of *Musca domestica L* and *Drosophila melanogaster* (Mg). Genet Res 37: 55–69

Fournier D, Berge J-B (1990) Resistance to insecticide: detection of a mutation in the acetylcholinesterase gene from *Drosophila melanogaster*. In Hagedorn H, Hildebrand HJ, Kidwell M, Law J (eds) Molecular Insect Science. Plenum Press, New York, p 302

Ganetzky B (1984) Genetic studies of membrane excitability in *Drosophila*: lethal interaction between two temperature-sensitive paralytic mutations. Genetics 108:897–911

Garza D, Ajioka JW, Burke DT, Hartl DL (1989a) Mapping the *Drosophila* genome with yeast artificial chromosomes. Science 246:641–646

Garza D, Ajioka JW, Burke DT, Hartl DL (1989b) Physical mapping of complex genomes. Nature 340:577–589

Georghiou GP (1986) The magnitude of the resistance problem. In Pesticide Resistance: Strategies and Tactics for Management. National Academy of Sciences, Washington DC, pp 14–43

Gibney G, Camp S, Dionne M, Macphee-Quigley K, Taylor P (1990) Mutagenesis of essential functional residues in acetylcholinesterase. Proc Natl Acad Sci USA 87:7546–7550

Grubs RE, Adams PM, Soderlund DM (1988) Binding of [^{3}H] saxitoxin to head membrane preparations from susceptible and knockdown-resistant house flies Pestic Biochem Physiol 32:217–223

Hall JC, Greenspan RJ (1979) Genetic analysis of *Drosophila* neurobiology. Ann Rev Genet 13:127–195

Hall LM, Kasbekar DP (1989) *Drosophila* sodium channel mutations affect pyrethroid sensitivity. In Narahashi TJ, Chambers H (eds) Insecticide Action from Molecule to Organism. Plenum, New York, pp 99–114

Hall LM, Spierer P (1986) The *Ace* locus of *Drosophila melanogaster*: structural gene for acetylcholinesterase with an unusual 5′ leader. EMBO J 5: 2949–2954

Hall LM, Wilson SD, Gitschier J, Martinez N, Strichartz GR (1982) Identification of a *Drosophila melanogaster* mutant that affects the saxitoxin receptor of the voltage-sensitive sodium channel. Ciba Foundation Symp 88:207–220

Hallstrom I (1985) Genetic regulation of the cytochrome P-450 system in *Drosophila melanogaster*. II. Localization of some genes regulating cytochrome P-450 activity. Chem-Biol Interact 56:173–184

Hallstrom I, Blanck A (1985) Genetic regulation of the cytochrome P-450 system in *Drosophila melanogaster*. I. Chromosomal determination of some cytochrome P-450 dependent reactions. Chem-Biol Interact 56:157–171

Hallstrom I, Blanck A, Atuma S (1984) Genetic variation in cytochrome P-450 and xenobiotic metabolism in *Drosophila melanogaster*. Biochem Pharmacol 33:13–20

Hama H (1983) Resistance to insecticides due to reduced sensitivity of acetylcholinesterase. In Georghiou GP, Saito T (eds) Pest Resistance to Pesticides. Plenum, New York, pp 299–332

Hayaoka T, Dauterman WC (1983) The effect of phenobarbital induction on glutathione conjugation of diazinon in susceptible and resistant flies house. Pestic Biochem Physiol 19:344–354

Healy MJ, Dumancic MM, Oakeshott JG (1991) Biochemical and physiological studies of soluble esterases from *Drosophila melanogaster*. Biochem Genet 29:365–388

Heckel DG, Bryson PK, Brown TM (1990) Genetic linkage analysis of a locus controlling acetylcholinesterase sensitivity to organophosphate insecticides in the tobacco budworm, Heliothis virescens. In Hagedorn H, Hildebrand HJ, Kidwell M, Law J (eds) Molecular Insect Science. Plenum Press, New York, p 312

Holwerda BC, Morton RA (1983) The in vitro degradation of [^{14}C]malathion by enzymatic extracts from resistance and susceptible strains of *Drosophila melanogaster* Pestic Biochem Physiol 20:151–160

Houpt DR, Pursey JC, Morton RA (1988) Genes controlling malathion resistance in a laboratory-selected population of *Drosophila melanogaster* Genome 30: 844–853

Hughes PB, Dauterman WC, Motoyama N (1989) Inhibition of growth and development of tobacco hornworm (Lepidoptera: Sphingidae) larvae by cyromazine. J Econ Entomol 82:45–51

Hughes PB, Raftos DA (1985) Genetics of an esterase associated with resistance to organophosphorus insecticides in the sheep blowfly, *Lucilia cuprina* (Weidemann) (Diptera: Calliphoridae). Bull Entomol Res 75:535–544

Jackson FR, Wilson SD, Strichartz GR, Hall LM (1984) Two types of mutants affecting voltage-sensitive sodium channels in *Drosophila melanogaster*. Nature 308:189–191

Kafatos FC, Louis C, Savakis C, Glover DM, Ashburner M, Link AJ, Siden-Kiamos I, Saunders RDC (1991) Integrated maps of the *Drosophila* genome: progress and prospects. Trends Genet 7:155–161

Kasbekar DP, Hall LM (1988) A *Drosophila* mutation that reduces sodium channel number confers resistance to pyrethroid insecticides. Pestic Biochem Physiol 32:135–145

Kernan M, Kuroda M, Baker BS, Ganetsky B (1989) *nap*^{ts} and *mle*; variations in one gene affecting sodium channel activity and transcription control in *Drosophila*. Soc Neurosci Abstr 15:196

Kikkawa H (1964) Genetical studies on the resistance to parathion in *Drosophila melanogaster* II Induction of a rèsistance gene from its susceptible allele. Botyu-Kagaku 29:37–42

Kimura M (1968) Evolutionary rate at the molecular level. Nature 217:624–626

Knipple DC, Payne LL, Soderlund DM (1991) PCR-generated conspecific sodium channel probe for the house fly. Arch Insect Biochem Physiol 16: 45–53

Lee YH, Keeley LL, Bradfield JY (1990) Hypertrehalosemic hormone regulated cytochrome P450 gene expression in the cockroach, *Blaberus discoidalis*. ESA Ann. Presentation 0191

Lindsley DL, Grell EH (1968) Genetic variations of *Drosophila melanogaster*. Carnegie Institution Washington Publication 627

Lindsley DL, Sandler L, Baker BS, Carpenter ATC, Denell R, Hall JC, Jacobs PA, Miklos GLG, Davis BK, Gethmann RC, Hardy RW, Hessler A, Miller SM, Nozawa H, Parry DM, Gould-Somero M (1972) Segmental aneuploidy and the genetic gross structure of the *Drosophila* genome. Genetics 71: 157–84

Little EJ, McCaffrey AR, Walker CH, Parker T (1989) Evidence for an enhanced metabolism of cypermethrin by a monooxygenase in a pyrethroid-resistant strain of the tobacco budworm (*Heliothis virescens* F). Pestic Biochem Physiol 34:58–68

Loughney K, Kreber R, Ganetzky B (1989) Molecular analysis of the *para* locus, a sodium channel gene in *Drosophila*. Cell 58:1143–1154

Lycett G, Eggleston P, Crampton J (1990) DNA transfection of an *Aedes aegypti* mosquito cell line. In Hagedorn H, Hildebrand HJ, Kidwell M, Law J (eds) Molecular Insect Science. Plenum Press, New York, pp 332

Malcolm CA, Hall LMC (1990) Cloning and characterization of mosquito acetylcholinesterase genes. In Hagedorn H, Hildebrand HJ, Kidwell M, Law J (eds) Molecular Insect Science. Plenum Press, New York, pp 57–66

Mannervik B, Danielson UH (1988) Glutathione transferases: structure and catalytic activity. CRC Crit Rev Biochem 23:283–337

Merriam J (1988) The *Drosophila* clone list by chromosome location. *Drosophila* Inform Serv 67:111–136

Merriam J, Ashburner M, Hartl DL, Kafatos FC (1991) Toward cloning and mapping the genome of *Drosophila* Science 254:221–225

Monroe TJ, Carlson JO, Clemens DL, Beaty BJ (1990) Selectable markers for transformation of mosquito and mammalian cells. In Hagedorn H, Hildebrand HJ, Kidwell M, Law J (eds) Molecular Insect Science. Plenum Press, New York, p 337

Morris AC, Eggleston P, Crampton JM (1989) Genetic transformation of the mosquito *Aedes aegypti* by micro-injection of DNA. Med Vet Entomol 3:1–7

Morton RA, Singh RS (1982) The association between malathion resistance and acetylcholinesterase in *Drosophila melanogaster*. Biochem Genet 20:179–198

Moss SJ, Smart TG, Porter NM, Nayeem M, Devin J, Stephenson FA, MacDonald RL, Barnard EA (1990) Cloned GABA receptors are maintained in a stable cell line: allosteric and channel properties. Eur J Pharmacol 189:77–88

Mouches C, Magnin M, Berge J-B, de Silvestri M, Beyssat V, Pasteur N, Georghiou GP (1987) Overproduction of detoxifying esterases in organophosphate-resistant Culex mosquitoes and their presence in other insects. Proc Natl Acad Sci USA 84:2113–2116

Mouches C, Pasteur N, Berge J-B, Hyrien O, Raymond M, De Saint Vincent BR, De Silvestri M, Georghiou GP (1986) Amplification of an esterase gene is responsible for insecticide resistance in a California *Culex* mosquito. Science 233:778–780

Nebert DW, Nelson DR, Feyereisen R (1989) Evolution of the cytochrome P450 genes. Xenobiotica 19:1149–1160

Oakeshott JG, Collet C, Phillis RW, Nielsen KM, Russell RJ, Chambers GK, Ross V, Richmond RC (1987) Molecular cloning and characterization of esterase-6, a serine hydrolase of *Drosophila*. Proc Natl Acad Sci USA 84: 3359–3363

Oakeshott JG, Healy MJ, Game AY (1990) Regulatory evolution of β-carboxyl esterases in *Drosophila*. In Barker JSF, Starmer WT, MacIntyre RJ (eds) Ecological and Evolutionary Genetics of *Drosophila*. Plenum Press, New York, pp 359–387

O'Brochta DA, Handler AM (1988) Mobility of P-elements in drosophilids and non-drosophilids. Proc Natl Acad Sci USA 85:6052–6056

Ochman H, Gerber AS, Hartl DL (1988) Genetic applications of an inverse polymerase chain reaction. Genetics 120:621–623

Ogita Z (1961) An attempt to reduce and increase insecticide-resistance in *D. melanogaster* by selection pressure. Botyu-Kagaku 26:7–30

Oppenoorth FJ (1985) Biochemistry and genetics of insecticide resistance. In Kerkut GA, Gilbert LI (eds) Comprehensive Insect Physiology, Biochemistry, and Pharmacology, Vol 12. Pergamon, New York, pp 731–773

Parker AG, Russell RJ, Delves AC, Oakeshott JG (1991) Biochemistry and physiology of esterases in organophosphate-susceptible and -resistant strains of the Australian sheep blowfly, *Lucilia cuprina*. Pestic Biochem Physiol 41:305–318

Pauron D, Barhanin J, Amichot M, Pralavorio M, Berge J-B, Lazdunski M (1989) Pyrethroid receptor in the insect Na^+ channel: alteration of its properties in pyrethroid-resistant flies. Biochemistry 28:1673–1677

Pickett CB, Lu AYH (1989) Glutathione S-transferses. Gene structure, regulation and bilological function. Ann Rev Biochem 58:743–764

Pirrotta V, Hadfield C, Pretorius GHJ (1983) Microdissection and cloning of the *white* locus and 3B1–3C2 region of the *Drosophila* X chromosome. EMBO J 2:927

Plapp FW Jr (1984) The genetic basis of insecticide resistance in the house fly: Evidence that a single locus plays a major role in metabolic resistance to insecticides. Pestic Biochem Physiol 22:194–201

Plapp FW Jr (1986) Genetics and biochemistry of insecticide resistance in arthropods: Prospects for the future. In Pesticide Resistance: Strategies and Tactics for Management. National Academy Press, Washington DC, pp 74–86

Pluthero FG, Threlkeld SFH (1984) Mutations in *Drosophila melanogaster* affecting physiological and behavioral response to malathion. Canad Entomol 116:411–418

Pongs O, Kecskemethy N, Müller R, Krah-Jentgens I, Baumann A, Kiltz HH, Canal I, Llamazares S, Ferrus A (1988) *Shaker* encodes a family of putative potassium channel proteins in the nervous system of *Drosophila*. EMBO J 7:1087–1096

Pralavorio M, Fournier D (1992) *Drosophila* acetylcholinesterase: Characterization of different mutants resistant to insecticides. Biochem Genet 30:77–83

Raftos DA, Hughes PB (1986) Genetic basis of a specific resistance to malathion in the Australian sheep blow fly, *Lucilia cuprina* (Diptera: Calliphoridae). J Econ Entomol 79:553–557

Raymond M, Callaghan A, Fort P, Pasteur N (1991) Worldwide migration of amplified insecticide resistance genes in mosquitoes. Nature 350:151–153

Robertson HM, Preston CR, Phillis RW, Johnson-Schlitz DM, Benz WK, Engels WR (1988) A stable genomic source of P element transposase in *Drosophila melanogaster*. Genetics 118:461–470

Rossignol DP (1988) Reduction in nerve membrane sodium channel content of pyrethroid resistant house flies. Pestic Biochem Physiol 32:146–152

Roush RT (1989) Designing resistance management programs: How can you choose? Pestic Sci 26:423–441

Roush RT, Daly JC (1990) The role of population genetics in resistance research and management. In Roush RT, Tabashnik BE (eds) Pesticide Resistance in Arthropods. Chapman and Hall, New York, pp 97–152

Roush RT, McKenzie JA (1987) Ecological genetics of insecticide and acaricide resistance. Ann Rev Entomol 32:361–380

Rubin GM, Spradling AC (1982) Genetic transformation of *Drosophila* with transposable element vectors. Science 218:348

Sattelle DB, Leech CA, Lummis SCR, Harrison BJ, Robinson HPC, Moores GD, Devonshire AL (1988) Ion channel properties of insects susceptible and resistant to insecticides. In Lunt GG (ed) Neurotox '88: Molecular Basis of Drug and Pesticide Action. Elsevier Biomedical Press, Amsterdam, pp 563–582

Sawicki RM (1985) Resistance to pyrethroid insecticides in arthropods. In Hutson DH, Roberts TR (eds) Insecticides. John Wiley, New York, pp 143–192

Scott JG, Georghiou GP (1986) Mechanisms responsible for high levels of permethrin resistance in the house fly. Pestic Sci 17:195–206

Scott JG, Lee SST, Shono T (1990) Biochemical changes in the cytochrome P450 monooxygenases of seven insecticide resistant house fly (*Musca domestica* L) strains. Pestic Biochem Physiol 36:127–134

Scott MP, Weiner AJ, Hazelrigg TI, Polisky BA, Pirrotta V, Scalenghe F, Kaufman TC (1983) The molecular organization of the *Antennapedia* locus of *Drosophila*. Cell 35:763–776.

Shanahan GJ (1961) Genetics of dieldrin resistance in *Lucilia cuprina* (Wied). Gent Agrar 14:307–321

Shemshedini L, Wilson TG (1990) Resistance to juvenile hormone and an insect growth regulator in *Drosophila* is associated with an altered cytosolic juvenile hormone-binding protein. Proc Natl Acad Sci USA 87:2072–2076

Shen J, Plapp Jr FW (1990) Cyromazine resistance in the house fly (Diptera: Muscidae): Genetics and resistance spectrum. J Econ Entomol 83:1689–1697

Shepanski MC, Glover TJ, Kuhr RJ (1977) Resistance of *Drosophila melanogaster* to DDT. J Econ Entomol 70:539–543

Shotkoski FA, Fallon AM (1990) Genetic changes in methotrexate resistant mosquito cells. Arch Insect Biochem Physiol 15:79–92

Smyth, K-A, Parker AG, Yen JL, McKenzie JA (1992) Selection of dieldrin-resistant strains of *Lucilia cuprina* (Diptera: Calliphoridae) after ethyl methanesulfonate mutagenesis of a susceptible strain. J Econ Entomol 85: 352–358

Soderlund DM, Bloomquist JR (1990) Molecular mechanisms of insecticide resistance. In Roush RT, Tabashnik B (eds) Pesticide Resistance in Arthropods. Chapman and Hall, New York, pp 58–96

Sparks TC, Hammock BD (1983) Insect growth regulators: Resistance and the future. In Georghiou GP, Saito T (eds) Pest Resistance to Pesticides. Plenum, New York, pp 615–668

Spradling AC (1986) P-element-mediated transformation. In Roberts DB (ed) *Drosophila*: A Practical Approach. IRL Press, Oxford, pp 175–197

Stern M, Kreber R, Ganetzky B (1990) Effects of a *Drosophila* sodium channel gene on behavior and axonal excitability. Genetics 124:133–143

Sundseth SS, Kennel SJ, Waters LC (1989) Monoclonal antibodies to resistance-related forms of cytochrome P450 in *Drosophila melanogaster*. Pestic Biochem Physiol 33:176–188

Tanaka K (1987) Mode of action of insecticidal compounds acting at inhibitory synapse. Pestic Sci 12:549–560

Toung Y-PS, Hsieh TS, Tu C-PD (1990) *Drosophila* glutathione S-transferase 1-1 shares a region of sequence homology with the maize glutathione S-transferase III. Proc Natl Acad Sci USA 87:31–35

Tsukamoto M (1983) Methods of genetic analysis of insecticide resistance In Georghiou GP, Saito T (eds) Pest Resistance to Pesticides. Plenum, New York, pp 71–98

Warren A, Crampton J (1990) Transposable genetic elements in the genome of the mosquito, *Aedes aegyptir*. In Hagedorn H, Hildebrand HJ, Kidwell M, Law J (eds) Molecular Insect Science. Plenum Press, New York, pp 379

Waters LC, Ch'ang L-Y, Kennel SJ (1990) Studies on the expression of insecticide resistance-associated cytochrome P450 in *Drosophila* using cloned DNA. Pestic Sci 30:456–458

Waters LC, Nix CE (1988) Regulation of insecticide resistance-related cytochrome P-450 expression in *Drosophila melanogaster*. Pestic Biochem Physiol 30:217–227

Wheelock GD, Scott JG (1990) Purification and initial characterization of the major cytochrome P-450 from a pyrethroid resistant house fly. In Hagedorn H, Hildebrand HJ, Kidwell M, Law J (eds) Molecular Insect Science. Plenum Press, New York, pp 381–382

Wilson TG (1988) *Drosophila melanogaster* (Diptera: Drosophilidae): A model insect for insecticide resistance studies. J Econ Entomol 81:22–27

Wilson TG, Fabian J (1986) A *Drosophila melanogaster* mutant resistant to a chemical analog of juvenile hormone. Devel Biol 118:190–201

Wilson TG, Fabian J (1987) Selection of methoprene-resistant mutants of *Drosophila melanogaster*. In Law JH (ed) Symposium on Molecular and Cellular Biology, New Series, Vol 49 Molecular Entomology; Monsanto-UCLA Symposium, Steamboat Springs, Colorado USA, April 6–13, 1986. AR Liss, New York, pp 179–188

Wing KD (1988) RH 5849, a nonsteroidal ecdysone agonist: effects on a *Drosophila* cell line. Science 241:467–469

Yarbrough JD, Roush RT, Bonner JC, Wise DA (1986) Monogenic inheritance of cyclodiene insecticide resistance in mosquitofish, *Gambusia affinis*. Experientia 42:851–853

Yu Q, Colot HV, Kyriacou CP, Hall JC, Rosbash M (1987) Behaviour modification by *in vitro* mutagenesis of a variable region within the *period* gene of *Drosophila*. Nature 326:765–769

Chapter 2

Naturally Occurring Insecticidal Molecules as Candidates for Genetic Engineering

Keith C. Binnington and Valerie J. Baule

1. Introduction

The use of persistent broad spectrum chemical insecticides has resulted in the development of high levels of insecticide resistance in hundreds of agriculturally and medically important insect pests (see reviews by Metcalf 1980; Georghiou 1986). Ever increasing doses of insecticide are needed for effective control, compounding problems of insecticide residues and environmental safety. Biological methods of control (which will be defined here as the use of living organisms to control insect pests), such as the use of fungi, viruses, bacteria, nematodes and autocidal control (e.g., sterile insect release), have many advantages over chemical control methods. Many are specific to a small number of pest species which gives them a high level of environmental safety. Since they are living organisms, some biological insecticides can be produced cheaply and, if established in the environment, can provide sustained control with little or no reapplication. Furthermore, many microbial insecticides are compatible with Integrated Pest Management (IPM).

On the other hand, biological methods of control also suffer many disadvantages when compared with chemical insecticides. Paradoxically, one of these is their narrow host range. Although high specificity is a positive attribute ecologically, and biological control agents can often provide effective insect pest control in stable ecological environments such as forests, this specificity also makes them inefficient and uneconomical against mixtures of pests. In addition to the problem of narrow host range, many biological insecticides also suffer disadvantages in respect of the slow rate of kill and the special environmental conditions that some require for growth, pathogenicity or storage. Some of these disadvantages may be alleviated using recombinant DNA (recDNA) technologies, for example, to increase host range or environmental stability or to enhance pathogenicity. We are interested in the latter aim, the use of genetic engineering to increase the efficacy of biological insecticides. The genetic engineering of plants to produce crops that are more resistant to insect pests or microbial pathogens that have enhanced insecticidal activity indicates that biological control agents can play a larger role in IPM and reduce our dependence upon chemical pesticides.

This chapter first presents an outline of how genetic engineering might be used to improve the pathogenicity of biological insecticides and then gives a brief resumé of some classes of genes which, when combined with a suitable delivery system, may constitute novel insecticidal agents.

2. Genetic Engineering of Biopesticides

2.1. Technical Criteria

The production of an effective genetically engineered insecticide is largely dependent upon meeting a number of technical criteria. Although an in-depth discussion of these criteria is outside the scope of this review, some of the more important ones are mentioned briefly below.

First, a suitable insecticidal molecule must be selected. For technical simplicity, this molecule would ideally be the product of a single gene. However, since some vectors (for example baculoviruses) are capable of accepting and expressing large amounts of foreign DNA, a product (not necessarily a protein) of a polygenic pathway may also be suitable. Thus we have concentrated our attention on insecticidal proteins encoded by single genes but have also considered some products of complex biosynthetic pathways. Although the nature of the insecticidal molecule chosen will vary with the pest species and the delivery or vector system to be used (discussed below), a sufficient understanding of the mode of action of the insecticidal agent as well as factors that may compromise its effectiveness is essential.

Second, a suitable organism that will act as the delivery vector for the insecticidal gene must be chosen. Current techniques dictate that the chosen vector must have a double stranded genome. For some such vectors direct manipulation of the genome may be technically feasible. However, for vectors with very large genomes (for example, baculoviruses and plants), a system to initiate recombination events may be the only means of introducing foreign DNA. In these cases an appropriate transfer vector must be constructed for insertion of the foreign gene into the delivery vector.

Third, the foreign gene must be expressed by the delivery vector. This will necessitate the use of appropriate 5′ and 3′ regulatory sequences to direct timing and quantity of transcription and, in some cases, 5′ untranslated regions and optimal codon usage for efficient translation of the foreign mRNA transcript. As all these parameters are properties of the host, some knowledge of the molecular biology of the host will facilitate expression of a foreign gene.

Fourth, the foreign gene product must be functional when expressed in or by the host organism. This feature may be dependent upon a number of factors such as correct post-translational modification and processing reactions, proper folding, transportation to intracellular compartments or secretion.

2.2. Potential Vectors

As alluded to above, the major factor to consider in the selection of an insecticidal gene is the biological vector used for the delivery or expression of the gene product. Factors that determine how the gene product is best presented to the pest include the ecology and feeding behavior of the insect and whether the gene product is lethal *per os*. Vectors currently under consideration for expression of insecticidal genes include entomopathogenic viruses and microorganisms, crop plants and the pest insect itself. Below is a short synopsis illustrating the capabilities and limitations of currently used entomopathogens that could serve as vectors for the delivery of insecticidal genes. Critical parameters of each vector are summarized in Table 2.1.

2.2.1. Genetic Engineering of Nematode/Bacteria Complexes

Many species of nematodes parasitize insects, including a large number of Coleoptera, Lepidoptera, Orthoptera, Hymenoptera and Diptera. Many of these nematodes exhibit high specificity to insects and have no harmful effects on either vertebrates or plants (Stoffolano 1973). Some of the most pathogenic nematodes are species of *Steinernema* and *Heterorhabditis* (Akhurst and Bedding 1986). This high pathogenicity is dependent on the presence of entomopathogenic bacteria (*Xenorhbadus*) that are symbiotically associated with the nematodes. The nematode penetrates the host insect through natural openings (mouth, spiracle, anus), invades the hemocoel and releases the bacterium. The bacterium then rapidly multiplies, and death of the insect generally occurs within 48 hours (Poinar and Thomas 1966; Petersen 1982).

Nematodes have the major advantage among biological insecticides in that they actively seek out their hosts. Unfortunately, this advantage is critically dependent on a moisture film for the nematodes to migrate in,

Table 2.1. Summary of important characteristics of biological control agents which are potential vectors of insecticidal genes*

Vector	Route of Infection	Specificity	Environmental Stability	Rate of Kill
Nematode	Natural openings	Broad	Poor, sensitive to dessicating environments, limited to soil dwelling or plant boring insects	Slow
Fungi	Surface contact	Narrow	Poor, sensitive to desiccating environments, require high humidity for sporulation	Slow
Bacteria	Ingestion	Broad to narrow	Generally good	Moderate
Virus	Ingested or injected by parasitic wasps but viral genes expressed intracellularly	Moderate to narrow	Many sensitive to UV light	Slow

* See text for details

restricting their effectiveness to soil dwelling or plant boring insect pests. Although molecular studies on the nonentomopathogenic nematode *Caenorhabditis elegans* are progressing rapidly, improvement of nematodes as insecticidal agents using recombinant DNA (recDNA) techniques is not currently practicable. Moreover, since the currently used species of nematodes are already highly pathogenic, the addition of an insecticidal gene would be of small consequence. The use of recDNA methods might produce a nematode with a higher tolerance to dessicating field environments; however, this will require a greater understanding of nematode physiology and such a change in the phenotype of the organism will almost certainly require the introduction of more than a single gene.

Although genetic engineering of nematodes may not be currently feasible, improving the bacteria symbiotically associated with them may be. New methods for mass production, storage and application of nematodes have greatly enhanced their potential as biological control agents (Bedding 1976). However, the symbiotic bacteria exhibit a phase variation that is detrimental to the production process and can limit the efficacy of the nematode/bacteria complex in the field. Analysis of the molecular mechanism of phase variation is currently under way with the goal of genetically engineering phase-stable strains (Bleakely and Nealson 1988; Frackman and Nealson 1990).

2.2.2. Genetic Engineering of Fungi

Over 400 species of entomogenous fungi have been identified (Ignoffo 1967), many of which play an important role in the natural control of pests found in diverse insect orders (Roberts and Humber 1981). Although many fungi possess the ecologically and environmentally positive attributes of high host specificity and infectivity by surface contact, little attention has been directed towards the commercialization of mycoinsecticides. This is due to their major disadvantages in respect to instability during large-scale production and formulation (reviewed by Weiser 1982; Bartlett and Jaronski 1988), and the need for appropriate humidity and temperature conditions in the field. Techniques involving classical genetics, parasexual recombination or somatic hybridization are expected to play a larger role than genetic engineering to improve the stability of fungi during production and to increase their tolerance to environmental stress (Heale 1988).

Fungal infection is generally initiated through contact and adhesion of spores with the insect surface. After germination of the spores, the hyphae secrete several cuticle softening enzymes such as hydrolase, proteases, esterases and chitinases to penetrate the host insect and invade the hemocoel. Death of the insect occurs through either the production of toxic metabolites or by general physiological disruption. Analysis of the molecular mechanisms of fungal infection and toxin production is progressing rapidly (Yoder et al. 1986; Heale 1988; Wraight et al. 1990). In addition, advances in fungal transformation systems have made the manipulation of specific genes possible (Hinnen et al. 1978; Case et al. 1979; Hynes 1986). It may be possible to enhance the pathogenicity of some entomopathogenic fungi by manipulating genes that are involved in host recognition, adhesion, lipase and protease activity and toxin production or by introducing novel insecticidal genes.

2.2.3. Genetic Engineering of Bacteria

To date, only bacteria in the family Bacillus have shown potential as microbial insecticides without requiring genetic engineering. Members of this family are obligate pathogens with the ability to infect and kill healthy insects. Major groups of the Bacilli form protoxins during sporulation that are activated to lethal toxins in the insect gut. The predominant species currently used commercially in insect control is *B. thuringiensis* (*B.t.*), although others, for example *B. sphaericus*, are also under development. Although many *Bacillus* strains are naturally highly pathogenic, the use of recDNA techniques to alter the toxin genes or to construct strains that express novel or multiple toxin genes may improve their efficacy as insecticides, guard against the development of resistance and broaden their host range (for example, see Bar et al. 1991).

The genetic engineering of many non-entomopathogenic bacteria is also under consideration (Kirschbaum 1985). Bacteria that colonize the surfaces and vascular systems of plants, or that inhabit specific feeding zones in aquatic environments, may be ideally suited as vectors for the delivery of insecticidal genes. However, the use of a bacterial vector has some limitations. Since infection will generally be dependent upon the ingestion of the bacteria, the insecticidal protein must either be a gut poison, and thus lethal *per os* (such as the *B.t.* toxin, Section 3.1.1) or small enough so that, if it is able to survive the environment of the insect gut, it can penetrate into the hemocoel (see Section 3.10).

2.2.4. Genetic Engineering of Viruses

Seven families of viruses are known to infect insects. Many of these viruses share similar properties to viruses that infect vertebrates and plants and in some cases can infect the same hosts (Kalmakoff and Longworth 1980). However, members of the baculovirus group, which includes nuclear polyhedrosis viruses (NPVs), granulosis viruses (GVs) and nonoccluded baculoviruses (NOVs), are predominantly restricted to invertebrates and exhibit a narrow host range (Ignoffo 1968; Summers et al. 1975; Kalmakoff and Longworth 1980). The major host species of baculoviruses are found in the Lepidoptera and Hymenoptera, but the viruses are also known to infect coleopteran and dipteran species to a lesser extent. Baculoviruses have been successful as biological control agents when used in stable environments such as forests (the NPV of the gypsy moth *Lymantria dispar*) and plantations (the NOV of the coconut palm rhinocerous beetle *Oryctes rhinocerous*). Although technically successful, the practical value of baculoviruses as insecticides for broad-acre crops is considered to be limited because they are generally not sufficiently pathogenic to give economic levels of control. In addition, the few baculoviruses that have been marketed (for example an NPV of *Heliothis* species, ELCAR) have not provided a suitable economic return to the commercial developer (reviewed by Huber 1986).

The major insecticidal limitation of many baculoviruses is their slow rate of kill and low pathogenicity after infecting larger larvae in later stages of development. The virus infects *per os* and generally initiates infection in the columnar cells of the insect midgut (Keddie et al. 1989). Although the virus replicates to high titers in these cells and spreads rapidly to other cells and organs, the time from infection to death may be as long as 2–3 weeks.

Baculoviruses have a number of characteristics that make them desirable as vectors of insecticidal genes. Since they have a double-stranded DNA genome, the use of genetic engineering to insert and express insecticidal genes is a promising strategy to increase their pathogenicity. A fundamental requirement for the engineering of insect viruses is a cell culture system in which the virus can be propagated and its genome manipulated. The development of the *Autographa californica* NPV system for the expression of foreign genes in a *Spodoptera frugiperda* cell line has contributed greatly as a model system for the genetic engineering of baculoviruses (for review, see Luckow and Summers 1988; Maeda 1989a). Although infection is *per os*, the virus replicates inside host cells. Since this obviates the need to transport foreign proteins across membranes, the range of insecticidal proteins that could be effective is wide.

Other potential viral vectors include viruses in the entomopox (EPV) group. Although similar to vaccinia, myxoma and other vertebrate viruses, the host specificity and environmental safety of the EPVs is considered comparable to that of the baculoviruses (Evans and Harrrap 1982). EPVs have been identified in several pests not susceptible to baculoviruses, including some in the orders Coleoptera and Orthoptera (Arif 1984). As with the NPVs, the EPVs suffer from low pathogenicity and slow rate of kill. However, since these viruses also have double-stranded DNA genomes, genetic engineering can be used to increase their desirability as insect pest control agents. The major limitation to the engineering of EPVs is the current paucity of orthopteran and coleopteran cell lines.

2.2.5. *Genetic Engineering of Plants*

Strain improvement of plants has traditionally been dependent on classical breeding and selection strategies. The application of genetic engineering to crop improvement offers the tremendous advantage of increased genetic variability and flexibility (reviewed by Meeusen and Warren 1989). Until recently, one drawback to this approach was the inability to transform and regenerate many important crop plants. However, there have been a number of recent advances in the transformation of corn, rice and other cereal crops (Cocking and Davey 1987; Rhodes et al. 1988; Datta et al. 1990) as well as important dicots (Mullins et al. 1990), and considerable effort is now being expended in the identification of genes that can be transferred into these plants to protect them from insect damage.

As with bacterial vectors, most insecticidal genes introduced into plants will have to encode a protein that is effective *per os*. In addition, genes inserted into food crops must be toxic only to insects and not compromise the nutritional value of the plant.

2.2.6. *Genetic Engineering of Pest Insects*

Gene transfer techniques in insects may also play a role in applied entomology. For example, genes conferring resistance to insecticides could be introduced to beneficial insects. Alternatively, the genetic engineering of pest species could be used as a means of spreading deleterious genes through populations.

The generation of insect strains with chromosomal rearrangements using classical genetics has provided evidence that autocidal control programs can be an effective means to control some insect pest species (for example, see Foster et al. 1988). Advances in the area of molecular entomology may increase the potential of autocidal control systems. The advent of *Drosophila melanogaster* transformation has allowed many previously inaccessible *Drosophila* genes to be cloned and analyzed. As a consequence, knowledge of the developmental and neurological mechanisms of this insect has expanded dramatically in the past seven years. *Drosophila* transformation is mediated by a particular transposable element called the P element. However, despite the spectacular success of the P element in transforming *Drosophila*, transformation of other insects is yet to be achieved. Genetic transformation of insects and potential applications in autocidal control programs are discussed by O'Brochta and Handler (this volume).

2.2.7. Novel Insecticide Production and Delivery Systems

Bacteria, viruses and fungi have all been developed as expression systems for large-scale production of foreign gene products. Many of these systems are capable of synthesizing, modifying and secreting a wide range of foreign proteins. The baculovirus expression system has been demonstrated to produce large amounts of foreign protein (up to 500 mg/ml, Luckow and Summers 1988). Furthermore, since the genome of the baculovirus is tolerant to insertions of large amounts of foreign DNA, it has been speculated that these expression systems could be used to produce insecticidal compounds that require multigenic biosynthetic pathways, for example pheromones, more efficiently than they could be synthesized in vitro (Miller et al. 1983).

In addition to the use of biological vectors to deliver insecticidal genes to pest insects, novel delivery systems may also be available in the future. Many of the powerful new therapeutic drugs being developed for human use are peptides or proteins. As such, novel drug delivery systems suitable for proteins such as encapsulation and the use of unique bioerodible polymers are receiving a great deal of attention (Saffran et al. 1986; Van Brunt 1989). This rapidly expanding field of proteinaceous drug delivery may uncover methods for the direct application of insecticidal proteins.

In conclusion, although a variety of biological vectors exist for the delivery of insecticidal genes, the development of many are at an early stage. The most severe problems in using nematodes and fungi as biological insecticides are their instability during production and formulation and their requirements for specific climatic conditions. The use of genetic engineering in conjunction with classical genetic selection may increase the pathogenicity of these organisms, as well as broadening the range of environmental conditions in which the pathogens can survive and infect their hosts.

As concluded by O'Brochta and Handler (this volume), genetic transformation of individual pest insect species will require a great deal of additional knowledge of genetic transposition and regulation of gene expression. Thus, the merits of procedures for genetically engineering pest insects for autocidal control will have to be judged in the future.

The most fully characterized and developed vectors for the delivery of insecticidal genes are plants, bacteria and viruses. The integration of insecticidal genes into plants has already proven to be an effective method to control pest insects (discussed in Section 3.1.1). In addition, laboratory studies of genetically engineered bacteria and viruses are also providing evidence that the insertion of insecticidal genes will enhance their efficacy as microbial insect control agents. The next section will focus on genes suitable for delivery by these vectors.

3. Candidate Insecticidal Genes

Although many proteins have been demonstrated to encode products with insecticidal properties, we have elected to review only those with proven or potential insect selectivity, as we feel that the issue of human health and environmental safety is non-negotiable. Candidate genes meeting this criterion that are discussed in this chapter include bacterial toxins, arthropod venoms, disruptors of juvenile hormone signaling, neuropeptides and antihormones, pheromones, viral proteins, plant defense proteins, genes that alter titers of specific mRNA populations and antibodies.

There are a number of ways that the above insecticidal genes can be categorized. Given that current technologies are best suited for the transfer of single genes, it is important to distinguish between single and polygenic products. The single gene products can then be further divided into those known to be insecticidal and those potentially insecticidal. Known insecticidal single gene products include: *Bacillus* toxins; the lectin arcelin; a trypsin inhibitor from cowpea; some peptides of scorpion venoms; neuropeptides and hormone esterases. Single gene products with potential but unproven insecticidal activity include chitinase, ribozymes and the heavy chain of antibody molecules.

Some polygenic candidates, such as secondary plant metabolites, are also worthy of consideration since genes for their production are already present in many plants and may be susceptible to manipulation. There are many natural insecticidal products that fall into this category, and decisions on which are most promising will depend on present and future knowledge of the genetic regulation of their synthesis and on their toxicity and effects on human health.

Another important distinction to make between the classes of insecticidal genes is their mode of action; particularly whether or not they are lethal *per os*. Proven and potential gut poisons include the *Bacillus* toxins, arcelin, protease inhibitors and antibodies. These genes have the potential to be expressed in plants and bacteria, whereas other genes such as neuropeptides and ribozymes will almost certainly require a delivery vector which replicates intracellularly.

Finally, even those proposals that appear outlandish at first may quickly be regarded as feasible. For example, when first suggested, the idea of using antibodies in plants sounded extremely radical. Now even this idea deserves merit in light of the recent finding that monoclonal antibody genes can be expressed in plants (Section 3.10).

Some of the more critical characteristics of the insecticidal genes that we have chosen to discuss are summarized in Table 2.2.

Table 2.2. Summary of characteristics of potential insecticidal proteins and peptides*

Gene	Known or Postulated Site of Action	Known or Expected Specificity
Bacterial Toxins	Gut	Moderate to high
Arthropod neurotoxins		
Scorpion	Na+ channels	High
Spider	Presynaptic ion channels, glutamate	Low to high
Hymenoptera	Various neural receptors, ion channels	High to moderate
Neuropeptides	Neuroendocrine system and its products present in the haemocoel	High
Plant products		
Lectins	Gut	Questionable
Enzymes and inhibitors	Gut	High
Viral proteins	Gut	Unknown
Antibodies	Gut, salivary enzymes, potentially enzymes and peptidases in both the gut and internal tissues	High
Pheromones	Sensory organs	High

* See text for details

3.1. Microbial Toxins

3.1.1. Bacillus thuringiensis *and Other Bacterial Toxins*

Bacteria produce a number of proteinaceous toxins such as hemolysins, necrotoxins and neurotoxins (reviewed by Lysenko 1967; Faust and Bulla 1982). Although many of these toxins are detrimental to higher vertebrates, some are either known or postulated to be specific towards insects. For this reason, as well as others mentioned in Section 2.2.3, the spore-forming Bacilli have received the most attention as biological control agents. Many members of this family produce proteinaceous insect-selective protoxins during sporulation. One member, *Bacillus thuringiensis* (*B.t.*) has been used as a microbial pesticide in the USA since the 1930s (reviewed by Luthy et al. 1982). The toxicology of *B.t.* is complex and potency against particular insects varies with the strain of *B.t.* used. For example, against lepidopteran larvae, strain HD-1, subspecies *kurstaki* is the most potent, whereas against dipteran larvae, the most potent *B.t.* strain is serotype H-14, subspecies *israelensis* (reviewed by Aronson et al. 1986). The biochemical and molecular basis of this specificity is unclear but is presumably due to a complex of proteins in the crystal product of the sporulating process. Moreover, *B.t.* can kill by virtue of its spores, vegetative cells, exoenzymes, exotoxins and parasporal crystal toxins (reviewed by Angus 1971). The parasporal crystal toxin (also known as δ-endotoxin) is a proteinaceous gut poison. As discussed below, the gene encoding this toxin has been used to genetically transform bacteria and plants and has been shown to convey insecticidal activity to the transgenic organisms.

The δ-endotoxin gene from *B.t.* var. *kurstaki* was first cloned in 1981 (Schnepf and Whiteley 1981), and since then many more endotoxin genes have been isolated. The classification of these genes on the basis of structure and specificity towards particular insect genera has been reviewed by Hofte and Whiteley (1989). At that time, 42 crystal protein genes had been sequenced and 14 distinct sequences could be identified from the data. Moreover, many strains contain more than one toxin in their crystal (Thomas and Ellar 1983). The actual role that each toxin peptide plays in the pathogenicity of the bacteria is unclear and synergism between individual peptides may also occur (Wu and Chang 1985; Chilcott and Ellar 1988).

The safety associated with the specific activity of the *B.t.* toxin and its record of use as a pesticide prompted the introduction of the toxin genes into other microorganisms as well as into plants. In one study, *B.t.* δ-endotoxin genes were cloned into a strain of *B.s.* with strong environmental persistence characteristics but with a limited target spectrum (Bar et al. 1991). Recombinant bacteria expressed *B.t.* toxin and demonstrated an increased host-range.

In another example, a *B.t.* toxin gene was integrated into the chromosome of two nonentomopathogenic bacteria (*Pseudomonas fluorescens* and *Agrobacterium radiobacter*) which colonize the roots of corn plants (Obukowicz et al. 1986). Transformed strains expressed the toxin at high levels (0.5%–1% of soluble cell protein) and, when added to an artificial diet, elicited 50% to 100% mortality to larvae of *Manduca sexta* within 4 days.

More recently, *B.t.* strains expressing novel δ-endotoxin combinations have been constructed (Crickmore et al. 1990). In this study, *B.t.* strains var. *kurstaki*, var. *israelensis* and var. *tenebrionis* were transformed with the cloned toxin genes of *B.t.* var. *tenebrionis*, var. *aizawai* IC1 and var. *sotto*. Stable transformants expressing both the native and introduced toxin genes indicated that the presence of novel toxin combinations may result in a synergistic toxic effect and perhaps even in changes in host specificity.

The use of *B.t.* toxin genes to increase the insecticidal efficacy of baculoviruses has also been considered (reviewed by Cameron et al. 1989). In a recent report, the δ-endotoxin gene from *B.t.* subspecies *kurstaki*

HD-73 was inserted into *Autographa californica* NPV (Merryweather et al. 1990). Although the toxin was expressed by cultured cells infected with the recombinant virus, its effect on *Trichoplusia ni* larvae could not be clearly established as the insects refused to consume food containing cell extracts. Larvae infected with recombinant virus were found to have *B.t.* toxin in their hemolymph; however, the pathogenicity of the recombinant virus was no higher than that of nonrecombinant virus.

The *B.t.* toxin gene has been successful as an insecticide when expressed in transgenic plants. Transfer of the *B.t.* endotoxin gene into tobacco demonstrated that expression and accumulation of the foreign gene product had the same insecticidal effects as the application of commercial *B.t.* products (Vaeck et al. 1987). Furthermore, endotoxin levels as low as 30 ng/gram leaf protein were sufficient to provide protection from *Manduca sexta* neonates. Introduction of the toxin gene into other plants such as tomato (Fischoff et al. 1987; Delannay et al. 1989) and cotton (Perlak et al. 1990) has also increased their resistance to attack from a range of lepidopteran pests in the field as well as in the laboratory. Recent developments in gene transfer systems for monocots (Klein et al. 1988; McCabe et al. 1988; Datta et al. 1990) have greatly expanded the potential for transfer of *B.t.* toxin to other economically important crop plants.

Although the *B.t.* toxins have been a success story as a method of biological control, it is clear that they are not a panacea. A recent report by Georghiou (1988) summarizes the results of a number of laboratory experiments selecting for insects resistant to *B.t.* Whereas most insects did not develop appreciable resistance when challenged with whole bacteria, laboratory strains of *Heliothis* selected with genetically engineered δ-endotoxin developed a 24-fold level of resistance within seven generations (Stone et al. 1989). In addition, several field populations of Indian meal moth *Plodia interpunctella* and almond moth *Codra cautella* have also demonstrated a measure of resistance to *B.t.* (McGaughey 1985a; 1985b; McGaughey and Beeman 1988). A recent report provided evidence that resistance to some *B.t.* insecticidal proteins is due to an alteration in binding of the toxin to membranes of midgut epithelium (Van Rie et al. 1990). Further analyses of the molecular mechanisms of toxicity and appropriate protein engineering of the toxin proteins, as well as the construction of bacteria which express a combination of toxins, may improve insecticidal activity and increase the effective life span and host range of *B.t.* toxins. In addition, screening of new *Bacillus* toxins and toxins with novel insecticidal activities from other entomopathogenic bacteria will play an important part in the design of new biological control strategies.

Other bacterial toxins amenable to recDNA techniques are those of *Bacillus sphaericus*. This bacteria is ubiquitous in soil and soil-aquatic systems and many strains are highly pathogenic to dipteran species. Several strains of *B. sphaericus* show a high specificity to a number of medically important mosquito species and are under active consideration as biocontrol agents (reviewed by Lacey and Undeen 1986). The majority of entomological and genetic research has been on three highly entomopathogenic strains; *B. sphaericus* 2363, 2297 and 1593. In each of these strains toxicity is linked to the production of two larvicidal proteins of 42 and 51 kD which are produced during sporulation and deposited in parasporal crystals. Similar to *B.t.* endotoxins, the parasporal toxins of *B. sphaericus* severely damage gut cells of infected larvae by binding to specific receptors (Davidson 1989). However, isolation and characterization of the genes encoding the toxins (Berry and Hindley 1987; Arapinis et al. 1988; Baumann et al. 1988; Berry et al. 1989) have demonstrated that the sequence of the *B. sphaericus* toxin genes are distinct from those of *B.t.* Moreover, it has not yet been clearly established whether the presence of both the *B. sphaericus* proteins is required for insecticidal activity (Baumann et al. 1988; de la Torre et al. 1989).

Bacillus cereus is known to invade and cause disease in insects in the orders Coleoptera, Hymenoptera and Lepidoptera. Although this bacterium produces a number of toxic compounds, its pathogenicity has been strongly correlated with the production of the enzyme Phospholipase C (Heimpel 1955a; 1955b). This enzyme has a molecular weight of approximately 24,000 and a pH optimum for activity at 7.0. Insects with an alkaline pH in the midgut are generally resistant to the bacterium (Heimpel 1955a). Although phospholipases of other bacterial species are highly toxic to man (e.g., the phospholipase C of *Clostridium perfringens*), that of *B. cereus* is not toxic to mammals (Johnson and Bonventre 1967).

Nonsporeforming entomopathogenic bacteria are also known to produce enzymes and proteinaceous toxins. Two genera considered to have potential for providing commercial biological insecticides are *Pseudomonas* and *Serratia*. Many proteinaceous toxins are produced by species of *Pseudomonas*. Although few of these proteins have been purified or studied in any detail, the pathogenicity of these bacteria and their toxins toward man and other mammals will probably preclude their use in the genetic engineering of insecticidal organisms.

Serratia marcesens is highly pathogenic to a wide range of insects. This pathogenicity is believed to be due to the production of highly toxic proteinases and a chitinase (Lysenko 1974; 1976). Although many *Serratia* species are pathogenic to humans (Farmer et al. 1976) and many of the proteinases are known or suspected vertebrate toxins, there is a strong but untested probability that the chitinase should only affect invertebrates and fungi (see Section 3.7.2).

Another species, *S. entomophila*, is a pathogen of the beetle *Costelytra zealandica*. The exact mechanism of pathogenicity is not well characterized but thought to occur in three stages: (1) adhesion of the bacterium to the gut wall of the insect, (2) cessation of feeding, and (3) bacterial invasion of the hemocoel. Analysis of the molecular mechanisms of pathogenicity of these and other *Serratia* species may uncover suitable insecticidal genes for the genetic engineering of nonentomopathogenic bacteria.

3.1.2. Fungal Toxins

3.1.2.1. Polyoxins and Nikkomycins. Some important insect growth regulators inhibit chitin formation and so prevent epidermal cells from secreting normal cuticle. Those currently used as insecticides are synthetic chemicals, but there are also natural products that inhibit chitin synthetase. Two such groups, the polyoxins and nikkomycins, which are antibiotic products of *Streptomyces*, have potential for controlling insects and mites (Mothes and Seitz 1982; Schluter 1982; Turnbull and Howells 1982; Binnington 1985). They are pyrimidine nucleosides comprised of a uridyl-ribose moiety to which a dipeptide is attached. Because chitin is absent from vertebrates, the polyoxins and nikkomycins are relatively specific, as well as having the added advantage of being toxic to both fungi and insects. However, a study of the biosynthesis of the pyrimidine and uronic acid moieties of the polyoxins of *Streptomyces cacaoi* (Isono and Suhadolnik 1976), has shown that these chemicals are the products of a complex pathway.

3.1.2.2. Destruxins. Destruxins are cyclodepsipeptides consisting of cyclically-linked amino acids and are produced by an insect fungal pathogen, *Metarhizium anisopliae*. Their effects on insects are complex and include tetany and paralysis due to a direct effect on muscle membrane (Samuels et al. 1986). They also inhibit DNA, RNA and protein synthesis (Quiot et al. 1985; Vey et al. 1986) and are thought to prevent normal defense reactions against *Metarhizium*. For example, when hemocytes (from *Periplaneta* and *Schistocerca*) were exposed to destruxins A and B, their ability to form nodules was reduced. Just how they do this appears uncertain, but it may be through inhibition of synthesis of phenol oxidases and/or actin (Huxham et al. 1989).

As pointed out by Heale et al. (1989), destruxins are products of complex, multigenic pathways and as such are not promising candidates for gene manipulation. Also, the knowledge that they inhibit nucleic acid synthesis in insect cell cultures (Quiot et al. 1985) and are related to the mammalian immunosuppressant cyclosporin A (Huxham et al. 1989) raises important safety questions.

3.1.2.3. Avermectins. The avermectins are a group of neurotoxic chemicals deployed with great success over the past decade for the control of arthropods and nematodes. Like the polyoxins, the avermectins are products of complex pathways of *Streptomyces* but differ in that they have a macrolide structure and are neurotoxins. Avermectins are thought to act at multiple sites, and the exact mode of toxicity has not been thoroughly elucidated. They are known to specifically increase membrane chloride ion permeability, perhaps by either binding to γ-aminobutyric acid (GABA) binding sites or by regulating the release of endogenous GABA (reviewed by Turner and Schaeffer 1989). Although avermectins are considered to be safe to vertebrates at low concentrations, they are not invertebrate selective. Avermectins create an increase in membrane chloride ion permeability in vertebrates and have been demonstrated to bind specifically to a site associated with, but not identical to, the GABA receptor in membranes of rat brain (Drexler and Sieghart 1984a; 1984b; 1984c; Soderlund et al. 1987).

3.2. Arthropod Neurotoxins

Most of the major classes of synthetic chemical insecticides in current use are neurotoxins. Many of these chemicals act on only a small number of target sites in the insect nervous system, namely voltage sensitive sodium channels, acetylcholinesterase and GABA receptor/chloride ion channel complexes. This has resulted in the development of resistance and cross-resistance between different insecticide classes. An additional problem with many of these chemicals is their lack of insect selectivity. Thus, a major goal in insecticide design is the identification of novel neurological target sites and affectors of these sites that are specific to the insect nervous system. Because many arthropods prey upon insects, venoms of predatory and parasitic species are a potential source of insect-selective neurotoxins.

Many of the arthropod toxins reported are nonproteinaceous molecules. However, some are peptides or proteins and are thus potential candidates for the genetic engineering of biological insecticides. A number of recent reviews on arthropod venoms can be found (Bettini 1978; Zlotkin 1985; Piek 1986; Olivera et al. 1990). Only toxins that are known to be, or may be, insect selective and are proteinaceous, and thus amenable to genetic engineering, will be covered here.

3.2.1. Scorpion Toxins

Perhaps the most widely studied of arthropod toxins are those from the scorpion. Using morphological and biochemical criteria, scorpions can be divided into two groups, the buthoids and the chactoids. There is only one family in the buthoids, the Buthidae, but this contains 40% of all known scorpion species and all the known medically important ones. The chactoids, which can be separated into five families, feed mainly on insects and have no known medical significance.

Over one hundred toxins from a number of scorpions have been isolated and characterized. The buthoid toxins form a family of small, basic neurotoxins of 60–70 amino acid residues and have the potential to form up to four disulfide bridges. The primary amino acid sequences suggest that extensive folding could result in a series of conformational variants. The toxins are all thought to be active on potential-dependent sodium channels of excitable membranes and show selectivity for distinct neurotoxin receptor sites on voltage-sensitive sodium channels (reviewed in Catterall 1980; 1984). Although there is significant amino acid sequence similarity between the toxins, many can differentiate between insect, crustacean, or mammalian sodium channels (Zlotkin 1983). Buthoid scorpions that have been demonstrated to possess venoms with one or more insect-selective toxins include *Androctonus australis* (Aa), *Buthotus judaicus* (Bj), *Androctonus mauretanicus* (Am) and *Leiurus quinquestriatus* (Lq) (reviewed by Zlotkin 1987, but also see De Dianous et al. 1987). These toxins can be catagorized into two classes by their mode of action. Excitatory toxins, such as Aa insect toxin-1, Am insect toxin, Bj insect toxin-1 and Lq Insect toxin-1, cause a fast contractional paralysis brought on by repetitive firing of the action potential in neurons. Depressant toxins, such as Bj insect toxin-2 and Lq insect toxin-2, result in a slow flaccid paralysis caused by a progressive and irreversible depolarization of axonal membranes. Both excitatory and depressant toxins affect sodium ion conductance and display high affinities to similar binding sites on neuronal membranes.

The three-dimensional structures of various insect-nonselective (Fontecilla-Camps et al. 1980; 1988; Almassy et al. 1983) and insect-selective scorpion toxins (Fontecilla-Camps 1989) have been predicted using X-ray crystallography and computer modeling techniques. Comparisons of the toxins indicate that overall they are highly conserved. However, the insect-selective toxin features one atypical disulfide bridge and an extended carboxy-terminal region, believed to exist as a β-sheet strand (Fontecilla-Camps 1989; Loret et al. 1990), which may act to confer the insect-selectivity. In addition, specificity may also be conferred by the presence of unique toxin binding sites in the insect nervous system, as it has been reported that binding sites for Aa insect toxin-1 appear to be unique to insect neuronal membranes (reviewed by Zlotkin 1985).

The toxin of one chactoid scorpion has been reported, and this forms a third functional class of scorpion toxins, referred to as cooperative toxins (Lazarovici et al. 1982). In contrast to the buthoid insect toxins, the two insect toxins of *Scorpio maurus palmatus* (SmIT1 and SmIT2) act synergistically to reversibly block both sodium and potassium ion channels. The physical characteristics of chactoid toxins also differ by having smaller molecular weights (SmIT1, 3232 MW; SmIT2, 3963 MW) and the potential to form only 2–3 disulfide bridges.

Recently, a gene encoding an insect-selective toxin from the buthoid scorpion *Buthus eupeus* (Be) was integrated into the genome of the *Autographa californica* baculovirus NPV (AcNPV) with the aim of producing a more virulent viral insecticide (Carbonell et al. 1989). Since tissue of the Be scorpion was not available to construct a cDNA or genomic library, the amino acid sequence of Be Insectotoxin-1 was used to design a synthetic gene 112 base pairs (bp) in length. A construct of the synthetic toxin gene was placed behind the human interleukin-3 signal peptide and under the control of the polyhedrin promoter of the AcNPV, and recombinant virus was used to express the toxin in an infected insect cell line. Although immunodetectable levels of toxin were produced, the protein did not possess any paralytic activity. Since the levels of transcription of the toxin gene were similar to those of the polyhedrin gene of nonrecombinant virus, the authors concluded that the lack of toxin activity could be due to either low efficiency of translation, instability of the toxin or, since the amino acid sequence used to design the gene was for the mature toxin and not the protoxin, incorrect folding of the newly synthesized toxin.

3.2.2. Hymenopteran Toxins

The venoms of many Hymenoptera, such as the social wasps, ants and bees, can contain a potent cocktail of biologically active amines, pain-producing neuropeptides, allergens and neurotoxins that affect vertebrates and invertebrates. Many of the proteinaceous toxins identified in bee and wasp venom, such as apamin, mellitin and bradykinin-like peptides (reviewed by Habermann 1971; Piek 1984 and references therein), are amenable to genetic engineering techniques but will not be considered here due to their harmful effects on vertebrates.

Although less familiar than social wasps, many species of solitary wasps also produce venoms. Most solitary wasps are entomophagous, and many paralyze their prey. A recent survey of paralyzing activity of more than 500 solitary wasps revealed that over 90% of the species listed contain venom with at least some paralyzing activity, and that all terrestial and some aquatic insect groups were affected (Piek and Spanjer 1986). Although normally the venoms cause little pain in vertebrates, they are often very active in insects and spiders and generally cause locomotor paralysis.

Of the many wasps with paralyzing activity identified, few have been studied at the pharmacological or biological level. Venoms of solitary aculeate wasps have been found to contain a number of compounds including acetylcholine, bradykinin and histamine (reviewed in Piek and Spanjer 1986). The venom of the digger wasp *Philanthus triangulum* has been examined in detail and is known to contain a number of nonproteinaceous components that interfere with neural transmission in the central nervous system and at neuromuscular junctions (Piek et al. 1971; Piek and Njio 1975; Piek and Spanjer 1986). Studies on the biochemistry or pharmacology of venoms of the solitary terrebrant wasps is also greatly lacking. Of these wasps, members of the family Braconidae have been studied in greatest detail. Proteinaceous neurotoxins have been reported for the species *Microbracon brevicornis* (Tamashiro 1971; Lee 1971), *M. gelechiae* (Piek et al. 1974) and *M. hebetor* (Beard 1952) but have been examined in any detail only in *M. hebetor*. Although very toxic to some lepidopteran larvae, variation in host sensitivity between different species of the same genus has been recorded, and insects in other orders have been found to be up to 1,000-fold less sensitive to the venom than those in Lepidoptera (Piek et al. 1982). The toxin was found to have essentially no effect on the higher invertebrates or vertebrates that were examined.

Analysis of components of *M. hebetor* venom revealed two proteinaceous neurotoxins of high molecular weight (Visser et al. 1983). These neurotoxins, referred to as MTX-A (43,700 MW) and MTX-B (56,700 MW) have similar isoelectric points and amino acid compositions. Pharmacological studies of the toxins demonstrated that they display different dose-response curves and species specificity. Both toxins act presynaptically to block excitatory glutamatergic transmission at neuromuscular junctions, have no effect on cholinergic transmission and are thought to act by inhibiting the release of synaptic vesicles (Piek 1966; Walther and Reinecke 1983). Recently, DW Miller and CA Johnson (Genetics Institute, Cambridge, MA) used an end-point dilution assay to trace *M. hebetor* toxin activity through a purification schedule based on four column chromatography steps. The toxin purified as a complex of several distinct subunits of 32, 22, 18 and 17 kD in size and approximately 1 pg was required for permanent paralysis of last instar *Galleria* larvae. Protein sequence information on the subunit polypeptides was subsequently used for the isolation of corresponding cDNA clones from a venom gland cDNA library. Nucleic acid sequence of several of the cDNA clones exhibited similarity to phospholipases of other Hymenoptera (DW Miller pers comm).

In addition to neurotoxins, many endoparasitic wasps also produce factors that alter the normal development of their hosts (reviewed by Beckage 1985). Studies of host larvae parasitized by different endoparasitic hymenopteran species have revealed that the presence of the parasite results in a variety of hormonal disruptions ranging from "juvenilizing effects" (Iwantsch and Smilowitz 1975; Beckage 1985) to premature metamorphosis (Leluck and Jones 1989, see also Section 3.6). Although the agent(s) involved in these developmental disruptions have not yet been identified, their presence invites speculation on other classes of insecticidal proteins that may be produced by parasitic wasps.

3.2.3. Spider Toxins

Spider venoms have proven to be a rich source of diverse neurotoxins, including compounds that act at specific ion channels, receptor channels and voltage-sensitive channels. The major drawback of many of the proteinaceous spider toxins to their use in the genetic engineering of insecticides is the lack of evidence for their specificity towards insects. Some toxins described below may or may not be insect-selective but have been added to this review as gene products with insecticidal potential that require more research.

Venom of the black widow spider *Lactrodectus mactens* has neurotoxic effects on several neuromuscular preparations of vertebrates and invertebrates. The venom acts presynaptically and has been found to mediate the release of every neurotransmitter examined in vertebrates, including acetylcholine, noradrenalin and GABA, as well as the release of GABA and glutamate in neural tissue of invertebrates (reviewed by Zlotkin 1985). Fractionation using discontinuous polyacrylamide gel electrophoresis resulted in the separation of black widow spider venom into 11 fractions with toxic activity (Ornberg et al. 1976). Comparative tests of the different fractions on neural preparations of frog, mouse, housefly, and crayfish and cockroach heart identified some fractions that were selective to either vertebrates or invertebrates, as well as some that could distinguish between insect and crustacean tissue. The mechanism of action or specificity of these toxins is not clear. One possibility is that, due to the different morphology of nervous tissue in different animals, access of a toxin to the nerve endings may vary. Also, one of the vertebrate toxins, α-latrotoxin, binds with high affinity to an unidentified protein component of neural membranes (Ornberg et al. 1976). This raises the possibility that specificity may be related to the presence or absence of different toxin receptors in susceptible or immune animals.

Venom of the spider *Hololena curta* is known to contain postsynaptic toxins that block glutamate receptors in some insects and birds (Usherwood and Duce 1985; Jackson et al. 1985). Fractionation of the venom results in the separation of three active presynaptic neurotoxic components designated curtatoxin I, II and III (Stapleton et al. 1990). These toxins are sequence-related polypeptides 36 to 38 amino acid residues in length, are high in cysteine and are amidated at the carboxy-terminal. As with scorpion venoms, formation of intramolecular disulfide bonds is believed to be important in the structure, subsequent activity and perhaps even specificity of the toxins. The exact mechanisms of these neurotoxins and their host specificity are unknown.

Another proteinaceous neurotoxin, designated *Hololena* toxin, has also been isolated from *H. curta* venom and is thought to act by blockage of presynaptic calcium channels (Bowers et al. 1987). Specificity studies have shown that *Hololena* toxin blocks neuromuscular transmission in larvae of *Drosophila melanogaster* but not in late pupal flight muscle or in the cutaneous pectoris muscle of the frog *Pipis ranus*. Although additional comparative analysis is necessary, the demonstration by these initial studies of some degree of specificity is encouraging.

Venom of the *Agelenopsis*, or Funnel web, spiders has been found to contain three classes of neurotoxins referred to as α, μ and ω-agatoxins, which act as postsynaptic and presynaptic antagonists (α and ω) as well as presynaptic activists (μ) (Adams et al. 1989). The α-agatoxins are acylpolyamines but the other two classes are polypeptides. Although agatoxins are widely used as probes for the study of nervous systems of both vertebrates and invertebrates, we are unaware of any comparative toxicology assays with test animals other than insects and do not know whether any of the agatoxins may be insect-selective.

3.3. Disruptors of Juvenile Hormone Signalling

There are two major classes of hormones that control insect development: the ecdysteroids, which are chemically related to the vertebrate steroid hormones, and the juvenile hormones, which are known only in insects and crustaceans. Of these two classes of hormones, disruption of the titers of JHs has shown the greatest potential for insect control. The field success of the juvenoid insecticides (which are functional JH mimics) has prompted research into antijuvenile hormone agents (AJHAs) to protect plants that cannot tolerate an extended larval stage of pest insects. The design of AJHAs, such as the precocenes and fluoromevalonate, indicates that this approach is possible but, as yet, few are potent or effective enough for practical field application (Menn 1985; Staal 1986)

Juvenile hormones are a family of lipidic hormones produced by paired glands called corpora allata (CA). One function of the JHs is to maintain an insect in a juvenile stage through the early larval or nymphal molts. A decline in the titer of JH corresponds with the onset of the final molt and, in holometabolous insects, the onset of metamorphosis and cessation of feeding. This decline in the JH titer is postulated to result from both reduced rates of JH synthesis and increased enzymatic degradation (reviewed in Hammock 1985). In adults, JH is important for the development and function of male and female accessory glands as well as for the synthesis of yolk proteins and their uptake by developing eggs (Koeppe et al. 1985).

JHs are postulated to act in a manner analogous to steroid hormones. This analogy is supported by the structural similarities between JHs and retinoic acid (Evans 1988) and the identification of two classes of JH binding proteins; "carrier" hemolymph JH binding proteins (Van Mellaert et al. 1985; Rayne and Koeppe 1988, and references therein) and cellular JH binding proteins (Wisniewski et al. 1988, and references therein). By comparison with steroid hormones, the signaling pathway of JHs appears to contain a number of processes that could be disrupted by genetically engineered insecticides. These include synthesis of the JHs, receptor mechanisms of responsive tissues and control of JH titer through enzymatic degradation.

Production of JHs in the CA is thought to be modulated by neuropeptide hormones (reviewed in Tobe and Stay 1985). These include the allatostatins, which maintain low levels of synthesis in unstimulated glands, the allatohibins, which reduce the rate of JH synthesis and the allatotropins, which stimulate JH synthesis. As described in Section 3.4, overexpression of an inhibitory or stimulatory neuropeptide due to the introduction of the neuropeptide gene or, alternatively, decreased expression of a neuropeptide using antisense mRNA or ribozymes, could cause a detrimental neuroendocrine imbalance.

In order to mediate its effect, JH must be available to responsive tissues. As mentioned above, hemolymph JH binding proteins and cellular JH binding proteins have been identified. Overexpression of the cellular binding proteins in the hemolymph may limit the amount of JH available to bind to endogenous cellular receptors. Alternative strategies may include the use of in vitro mutagenesis to increase the binding affinity of the hemolymph binding protein or by making JH binding with these proteins irreversible.

Degradation of JHs occurs mainly by either ester hydrolysis, which is mediated by JH esterase, or epoxide hydration, catalyzed by epoxide hydrases (for reviews see de Kort and Granger 1981; Hammock 1985). Most research on the enzymatic degradation of JHs has been on JH esterase and relatively little known of the epoxide hydralases. Hammock et al. (1990) have recently reported the effects of precocious expression of JH esterase in insect larvae. In their study, a cDNA for JH esterase was isolated from a *Heliothis virescens* cDNA library, placed under the control of the polyhedrin promoter and incorporated into the genome of the *Autographa californica* NPV. An insect cell line infected with the recombinant NPV expressed biologically active JH esterase. When first instar larvae of *Trichoplusia ni* were infected with recombinant virus, they grew more slowly than uninfected larvae or larvae infected with nonrecombinant virus. Although the genetically engineered virus was effective only in first instar larvae, the authors speculated that in vivo protein stability, as well as earlier and amplified expression of the JH esterase gene, may produce a more effective viral insecticide.

In addition to genes encoding enzymes that degrade JHs, those encoding enzymes involved in the biosynthetic pathway of these hormones might also be targeted. This might be accomplished using antisense mRNA or ribozymes, as described in Section 3.9.

Finally, a number of JH analogs have been identified in plants (reviewed by Bowers 1982). Further understanding of the synthesis of these JH analogs and the regulation of their biosynthetic pathways is needed to see if their expression can be increased overall, or in particular tissues, to protect the plant against insects (see Section 3.8).

3.4. Neuropeptides and Antineuropeptides

Neuropeptides form a diverse group of biologically active peptides that function as neurotransmitters, neuromodulators and neurohormones. In insects, neuropeptides are known to regulate over two dozen critical physiological processes including metamorphosis, development and maintenance of homeostasis. Research on insect neuropeptides has been slow due to difficulties in isolating sufficient amounts of peptide for analysis and characterization. These problems have been substantially alleviated by recent advances in peptide purification and isolation and by the ability to determine amino acid sequences using picomole concentrations of peptide. To date, more than 40 unique insect neuropeptide structures have been described, and it has been speculated that this number will more than double during the early 1990s (Holman et al. 1990). A perspective of insect neuropeptides can be obtained from recent reviews on the insect endocrine system (Scharrer 1987; Holman et al. 1990; Kelly et al. 1990) and the relationship between invertebrate and vertebrate neurohormones (Greenberg and Price 1983; De Loof 1987; Veenstra 1988).

As illustrated in Table 2.3, neuropeptides can be classified into eight families on the basis of their function or sequence similarities (reviewed by Holman et al. 1990). The increase in the number of neuropeptides isolated and characterized, and the accompanying understanding of the critical and fundamental processes in which many neuropeptides are involved, has led to renewed interest in the design of insecticides that exploit neuroendocrine target sites (reviewed by O'Shea 1985; Keeley and Hayes 1987; Evans et al. 1989; Kelly et al. 1990). In addition, although amino acid sequence similarity and antibody cross-reactivity between vertebrate and invertebrate neuropeptides exist, the diversity of neuropeptides, and their functions suggests that many will be insect, or at least arthropod, selective. Until recently, neuropeptides were largely dismissed as potential insecticidal agents since their proteinaceous nature made them unsuitable to be delivered *per os*. This view was supported by a report by Quistad et al. (1984) which demonstrated that when added to an artificial diet and fed to *Manduca sexta*, proctolin was almost completely broken down into its constitutive amino acids in less than five hours after ingestion. In the same report, it was shown that when topically applied, less that 1% of the neuropeptide penetrated the larval cuticle.

Table 2.3. Families of insect neuropeptides and their known or postulated functions*

Family	Properties
AKH/RPCH	13 insect neuropeptides structurally related to the RPCH hormone of crustaceans. Family includes AKH I and II from locust as well as some myotropins
Myotropins	Family includes a series of 11 myotropic peptides selective for muscles of hindgut isolated from the cockroach *Leucophaea maderae*. Also includes proctolin
FMRFamide related peptides	Family of peptides structurally related to the FMRFamide neuropeptide of molluscs
Diuretic and antidiuretic hormones	Function in control of water balance, waste removal and ion balance. Isolated from locust, tobacco hornworm and house cricket
Eclosion hormone	Large (>6 MW) molecules that stimulate larval, pupal and adult ecdysis as well as other development changes
Steroidogenic hormones	Stimulate the synthesis of steroid hormones in target organs. Includes PTTH which stimulates the prothoracic glands and EDNH which stimulates the gonads of adults
Allatotropins and allatostatins	Control the secretion of Juvenile Hormone from the corpora allata
PBAN	Control the production of pheromone synthesis. Isolated from *Heliothis zea*

* See text for details and specific references

However, the design of nonproteinaceous antagonists that could be stable after ingestion (Evans et al. 1986), or formulation of peptides to withstand both environmental as well as insect gut degradation (Saffran et al. 1986), may circumvent such problems.

Advances in molecular and genetic engineering techniques have further broadened the potential for using neuropeptides in insect control. Using recDNA techniques, it should be possible to raise or lower neuropeptide titers, resulting in a detrimental endocrine imbalance. Possible strategies include 1) disruption of the regulation of synthesis, processing and secretion, 2) inhibition of degradatory proteases, and 3) inhibition or activation of neuropeptide receptors. Although all these routes may alter neuropeptide titers or action, the intricate nature of the neuroendocrine system and sensitive intraendocrine interactions (including antihormone responses) suggests that successful insect control will require a thorough understanding of the physiological mechanisms involved in neuropeptide synthesis and action.

Although our understanding of the molecular genetics of the insect neuroendocrine system is increasing, the small number of insect neuropeptide genes that have been isolated and characterized suggests the genomic organization of neuropeptide genes is complex. For example, the eclosion hormone gene of *M. sexta* is 7.8 kb in length and contains a coding region that is split into 30 exons. Transcription of this gene results in an 800 bp mRNA transcript that encodes an 88 amino acid precurser (Horodyski et al. 1989). A 26 amino acid signal peptide is then cleaved from this precursor to produce mature eclosion hormone. The structure of the FMRFamide (Phe-Met-Arg-Phe-NH_2)-related gene of *Drosophila melanogaster* is even more complex. This gene is composed of an initial exon of 106 bp (that encodes the signal peptide) followed by a 2.5 kb intron and a second exon of 1,352 bp. The primary RNA transcript is approximately 1,700 bases in length and the deduced protein encodes 13 putative FMRFamide-related peptides (Nambu et al. 1988; Schneider and Taghert 1988). The existence and possible functions of the putative neuropeptides are unknown.

Many neuropeptides characterized to date are synthesized as parts of larger precursor proteins which then undergo proteolytic processing. Some precursors (such as the FMRFamide and *Aplysia* egg-laying hormone precursor) contain multiple biologically active polypeptides (Herbert and Uhler 1982; Scheller et al. 1984). Analysis of the FMRFamide-related gene of *Drosophila* suggests that some insect neuropeptides are also synthesized as larger prohormones (Douglass et al. 1984; Loh et al. 1984). Processing of either vertebrate or invertebrate prohormones is not well understood. Endoproteolytic cleavage is postulated to be aided by an internal signal consisting of pairs of dibasic amino acids flanking the active neuropeptide sequence (Evans et al. 1989). However, cleavage of neuropeptides has also been demonstrated to occur at nonbasic amino acids (Hudson et al. 1981; Jornvall et al. 1981). One of the major questions concerning the mechanism and regulation of functional neuropeptide titers is whether there exists a large number of specific endopeptidases with high substrate specificity. Analysis of vertebrate cells transfected with prohormone genes, endopeptidase genes or both has demonstrated that cells expressing endogenous prohormones also have the ability to correctly process a wide range of heterologous neuropeptide precursors (Thim 1986; Thomas et al. 1986; 1988a; 1988b). This suggests processing may be regulated by a small number of endopeptidases with broad cleavage specificity and a wide range of possible substrates. Some of the prohormone activating enzymes may be accessible to recombinant DNA techniques. Introduction or altered expression of these enzymes could theoretically result in detrimental levels of active neuropeptide due to precocious or inhibited processing.

Neuropeptides function as circulating neurohormones and neuromodulators, or in some cases as neurotransmitters, and generally mediate their effects through specific receptors and second messenger signals. Antagonists that disrupt this interaction or agonists that increase the activation could create an endocrine imbalance. Although the design of neuropeptide analogs does not immediately fall in the realm of genetic engineering, there is a potential to clone genes encoding antihormones that mimic receptors. This comes about from an observation by Blalock and Smith (1984) who recognized that some DNA sequences coding for hydrophilic amino acids on the sense strand could often be found to encode hydrophobic amino acids on the antisense strand and *vice versa*. When the antisense strand of adrenocorticotropic hormone (ACTH) was translated into a peptide sequence, the synthesized "antipeptide" acted like an "antihormone" and selectively bound to ACTH (Bost et al. 1985). This pattern has also been confirmed between mRNAs for other vertebrate hormones and their receptors. If the pattern holds for insect systems, sequence information of the neuropeptide gene could aid in the design of "antihormone" genes. Cloning and introduction of these genes would result in the expression of "antihormones" which could compete with the hormone receptor and cause a physiological depletion of the neuropeptide.

It is currently believed that most neuropeptides are inactivated by enzymatic cleavage by aminopeptidases (Starratt and Steele 1984; Steel and Starratt 1985) or, in some cases, selective uptake by adjacent tissues (McKelvey and Blumberg 1986). Agents that inhibit the degradation of neuropeptides may also have insecticidal properties in a manner analogous to organophosphate and carbamate pesticides that inhibit acetylcholinesterase activity. Alternatively, stimulation of the degradatory process may result in a lethal decline in neuropeptide titers. Analysis of neuropeptide degradation in vertebrate systems indicates that inactivation is mediated by a small number of aminopeptidases with low specificity that are located on the plasma membrane (Turner et al. 1985). The degradation of proctolin in the cockroach *P. americana* seems to be catalyzed by an aminopeptidase(s) (Quistad et al. 1984; Starratt and Steele 1984; Steele and Starratt 1985). In addition, aminopeptidase and endopeptidase activity capable of degrading locust AKH and proctolin has been demonstrated in synaptosomal membranes of the locust *S. gregaria*. (Isaac 1988). More research on the characterization and isolation of these enzymes will be necessary to determine if their inhibition, or uncontrolled expression, could create a lethal or incapacitating neuroendocrine imbalance in insects.

In addition to disrupting the processing, action or degradation of neuropeptides, the titer of strategic peptides might also be disrupted by the introduction and overexpression of the neuropeptide gene. The feasibility of this approach was first substantiated by a study demonstrating large quantities of biologically active interferon in the hemolymph of larvae of *Bombyx mori* infected with recombinant baculovirus containing the human α-interferon gene (Maeda et al. 1985). Recently, the gene encoding the diuretic hormone of *Manduca sexta* (a peptide of 41 amino acids) was inserted into the *Bombyx mori* NPV (BmNPV) (Maeda 1989b). Recombinant virus was then used to infect silkworm larvae. Fat body mRNA isolated from infected larvae contained diuretic hormone mRNA of the correct size and hemolymph from infected larvae displayed over 100-fold higher levels of diuretic activity than hemolymph of uninfected larvae. However, the insecticidal effects of the increased expression of diuretic hormone were slight. Compared to larvae infected with non-recombinant NPV, larvae infected with the recombinant NPV grew slower, showed a decrease in hemolymph volume and died one day earlier. Although this study demonstrates the potential of using neuropeptide genes for insect control, it also highlights the importance of a detailed understanding of the physiology of the neuroendocrine system to facilitate the interpretation of experimental results and the design of more effective strategies.

3.5. Pheromones

Insect behavior is largely directed by the reaction of various sensory receptors to external chemical stimuli. Pheromones and pheromone mimics for more than 100 insects are currently known, and many synthetic pheromones are used in insect control management to bait insecticide traps, monitor pest populations and in confusion strategies to disrupt mating (reviewed by Ridgeway et al. 1990). One of the major obstacles to using pheromones for large-scale control programs is the high cost of production. Many of the synthetic pheromones are manufactured by chemical methods using biologically derived enzymes (Van Brunt 1987). While pheromones are products of complex biosynthetic pathways and as such are not currently suitable for the genetic engineering of biological control agents, genetic engineering may play a role in the inexpensive synthesis of large quantities of pheromones. As described in Section 2.2.4, baculovirus expression systems are capable of carrying and expressing large amounts of foreign DNA. It has been speculated that the genetic material for the entire pathway of a pheromone could be integrated into the virus for expression in cell cultures or infected insects (Miller et al. 1983). The use of these expression systems for economic production of large quantities of pheromones could stimulate an increased use of insect semiochemicals in pest control.

3.6. Viral Proteins

Some insect viruses encode and produce proteins which enhance either their infectivity or that of other viruses. For example, the *Trichoplusia ni* granulosis virus produces a proteinaceous viral enhancement factor (VEF) which damages the protective layer of the insect's gut (Derksen and Granados 1988; also cited in Beard 1989). Expression of VEF increases infectivity of the virus 25–100 fold. Other granulosis viruses contain a protein in the occlusion body matrix which enhances the infectivity of nuclearpolyhedrosis viruses (Tanada 1985, and references therein). This synergistic factor (SF) is a lipoprotein of approximately 126 MW. It appears to act by binding both to specific receptors of the midgut cell microvillus of susceptible insects and to the NPV envelope, aiding in attachment and subsequent infectivity of the virus (Uchima et al. 1988). Expression of these viral proteins in recombinant viruses, and perhaps even in plants and bacteria, may increase the efficacy of viral insecticides or act directly on feeding insects.

Another class of viral genes that may have insecticidal activity are those that encode proteins that disrupt the hormonal regulation of insect development and metamorphosis. Examples of these genes may be found in viruses in the family Polydnaviridae, which have been identified in the female reproductive tract of many endoparasitic wasps (reviewed by Stoltz and Vinson 1979; Stoltz et al. 1984). These viruses are injected into the host larvae or egg during oviposition and have been implicated in a wide range of physiological disturbances in the host, including suppression of the immune response (Edson et al. 1981), retarded growth (Vinson et al. 1979) and disruption of neuroendocrine-regulated development (Jones et al. 1986, and references therein).

Finally, a gene encoding an ecdysteroid UDP-glucosyl transferase was recently isolated from the *Autographa californica* NPV (O'Reilly and Miller 1989). This enzyme catalyzes the transfer of glucose from UDP-glucose to ecdysteroids and is believed to play an important role in the normal metabolism and disposal of ecdysteroids (Smith 1977). Expression of the UDP-glucosyl transferase gene in *Spodoptera frugiperda* larvae infected with this NPV demonstrated that the transferase inhibited host molting, probably due to increased inactivation of ecdysteroids.

3.7. Plant Defense Proteins

Plants have evolved a large chemical arsenal of secondary compounds to protect them from insect predators. Many are products of multigenic pathways and thus not easily amenable to genetic engineering. These are considered in Section 3.8. Others, such as lectins, enzymes and enzyme inhibitors, are proteinaceous and primary gene products. The potential of these molecules for use in the genetic engineering of plants and in some cases, bacteria, are discussed below.

3.7.1. Lectins

Lectins are proteins (usually glycoproteins) that bind to carbohydrate moieties of complex carbohydrates (reviewed by Etzler 1985). The carbohydrate-binding properties of lectins have made them useful as tools to probe the nature and distribution of membrane-bound, carbohydrate-containing structures (Goldstein and Hayes 1978). That this carbohydrate-binding property may also act to limit insect predation of plants, particularly of seeds, appears likely (Bell 1978). For example, in a recent study transgenic tobacco plants expressing the pea lectin (P-Lec) gene demonstrated a higher level of resistance to *Heliothis virescens* than non-recombinant control plants (Boulter et al. 1990).

As another example, Janzen et al. (1976) noted that although many species of bruchid beetles shared habitats with the black bean (*Phaseolus vulgaris*), which contains the lectin phytohemoaglutin (PHA), few attacked black bean seeds. Furthermore, beetles that did not feed on *P. vulgaris* succumbed when fed on seed meal incorporated with PHA. Although PHA appears to provide the black bean plants with a measure of resistance against some species of beetles, it is ineffective against two of the most important pests, the bean weevil *Acanthoscelides obtectus* and the Mexican bean weevil *Zabrotes subfasciatus*. Whereas most bean cultivars contain PHA, appreciable resistance is only seen in wild bean accessions (Schoonhoven and Cardona 1982; Schoonhoven and Valor 1983). This prompted Osborn et al. (1988a; 1988b) to screen wild accessions for resistance and to identify the mechanism behind that resistance. The result was the discovery of the major seed protein arcelin. Four variants of arcelin (arcelin 1–4) were identified, and it was determined that the presence of either arcelin 2, 3 or 4 gave resistance against the two major bean weevil pest species. Arcelin-1 had properties similar to PHA (Osborn et al. 1988b), and the cDNA for this protein was cloned. The arcelin-1 molecule is antigenically and genetically similar to PHA. In addition, a cDNA has been isolated from common beans that codes for a lectin-like protein different from both PHA and arcelin (Hoffman et al. 1982).

It has been proposed that, like PHA toxicity in mammals, arcelin might adversely affect the insect gut by binding to and disrupting epithelial cells (King et al. 1980; 1982; Gatehouse et al. 1984). So far it is not known if arcelin variants are intrinsically more toxic to weevils than PHA or whether seed resistance to weevils is due to the high expression of arcelin known to occur in seeds. Lectins are also present in bean tissues other than the seed but are generally at much lower levels (Etzler 1986).

Another lectin that has insecticidal potential is wheat germ agglutinin (WGA), a protein that binds to N-acetyl glucosamine residues and hence to chitin. Chitin is a relatively specific target for an insecticide as it is absent from vertebrates. It is not known if WGA protects any plant against insect attack but it is thought to defend some plants against fungal attack (reviewed by Mirelman et al. 1975; Etzler 1986). If ingested, WGA could bind to chitin in the peritrophic membrane—the thin cuticular inner tube of the insect gut through which nutrients must pass to enter the gut cells (reviewed by Spence 1991). WGA or closely similar molecules are widely distributed in species of the tribe Triticeae (includes wheat, barley and rye) and the family Graminaceae and are also found in rice (Etzler 1985).

Some plants possess multiple lectin genes, and the genetics of plant lectins is complex (reviewed by Etzler 1985). However, the large body of knowledge on lectin genetics will assist in any future genetic engineering of these molecules. Finally, while care would be needed to avoid the expression of proteins toxic to nontarget organisms, the transfer of insecticidal lectin genes into agricultural crop plants appears to have considerable potential.

3.7.2. Enzymes and Enzyme Inhibitors

It has been speculated that some enzymes in plants may have a role in defense against fungi (Kombrink et al. 1988) and that these enzymes may also be able to increase plant resistance to insect predators (Jaynes and Dodds 1987). One example is the degradative enzyme chitinase. The potential target of chitinase is chitin, a nitrogenous polysaccharide present both in fungi and in the lining of the insect gut. Other enzymes with insecticidal potential may also be present in plants, however, screening will be required to identify them.

A wide range of protease inhibitors (PIs) have been identified in plants (reviewed by Richardson 1977). These proteins can bind to, and thus inhibit, various proteolytic enzymes such as serine proteinases, sulphydryl proteinases, acidic proteinases and metalloproteinases. Although studied most extensively in the Leguminosea, Graminae and Solanacea, due to the importance of these plants in human nutrition, PIs are widespread throughout the plant kingdom. The majority of PIs are small proteins of approximately 70–90 amino acid residues. They are present in various plant tissues and particularly abundant in seeds and tubers. Leaves of plants in the families Solanaceae and Legumaceae also accumulate protease inhibitors, apparently in response to a signal or wound hormone released from the site of damage caused by insect attack (Ryan 1978), microbial infection (Peng and Black 1976) or mechanical damage (Cleveland and Black 1982).

PIs are toxic to a wide range of insects following ingestion (Gatehouse and Boulter 1983), probably because they inhibit proteinases found in the digestive tract of animals. One study on the larvae of *Heliothis zea* and *Spodoptera exigua* concluded that the binding of PIs to the proteases creates a demand for increased production of proteases and causes stress on the organs responsible for their synthesis (Broadway and Duffey 1986). Supplements of sulfur-containing amino acids to the larvae's diet alleviated the toxic effect, suggesting that shortages of such amino acids are responsible for the observed stress. In mammals, PIs also cause physiological damage including pancreatic hypertrophy (Gallaher and Schneeman 1984). As with the insect larvae, this is thought to be due to stress on the pancreas required to produce more proteinases with insufficient amino acids, especially those containing sulfur. The likelihood of a particular PI being harmful to man is dependant upon its ability to withstand the acidic conditions of the human gut (pH 2–3) and its resistance to the major human gut proteinase, pepsin.

Hilder et al. (1987; 1989) recently transferred a cowpea trypsin inhibitor gene (CpTI) into tobacco, with the aim of increasing the plant's resistance to insect attack. A cDNA encoding the protease inhibitor was placed under the control of the CaMV 35S gene promoter and transferred to tobacco. Compared with control plants (transformed with reverse constructs), regenerated transgenic plants, which accumulated CpTI in the leaves, showed increased resistance and less damage to leaves when infested with the tobacco budworm *H. virescens*. These results suggest that, following the proper manipulation of PI genes to get high levels of expression, PIs may act as insecticidal genes.

Finally, cross-breeding of transgenic tobacco plants expressing the pea lectin P-Lec with those expressing CpTI resulted in plants that produced both insecticidal proteins (Boulter et al. 1990). These plants demonstrated a higher level of insect resistance than either parent plant, suggesting that a plant's resistance to insect pests is likely to be enhanced by the incorporation of multiple insecticidal genes.

3.8. Plant Secondary Metabolites

About 10,000 plant secondary metabolites are known, and there may be as many as 100,000–400,000 more (Schoonhoven 1982). Some of these compounds, such as alkaloids, polyphenols and isoprenoids, naturally discourage insect predators (Bowers 1983; Bell 1987). The pyrethrums have been used as lead molecules for the design of effective insecticides.

It could be assumed that insect strains resistant to such compounds should already have evolved but there are a number of counter-arguments. For instance, the monocultural nature of most agriculture in the developed world is highly artificial and selection has often been for productivity regardless of insect resistance. Furthermore, pest populations are generally low in a natural ecosystem, so selection pressure for development of resistance may also be low (de Ponti 1982).

How can genetic engineering exploit secondary plant substances? For products that are effective, but absent from plants needing protection, it may be possible to transfer a suite of genes required for production of the substance into the crop plant. Alternatively, a substance may be produced in a crop plant but at ineffective concentrations or in the wrong tissues of the plant to confer protection. In such cases it may be sufficient to manipulate regulatory genes to evoke increased or tissue specific expression. It appears that a change in a single gene can raise the level of a product of a complex pathway. Evidence for this includes the observation that, through genetic variation, parts of individual plants can respond differently to herbivores, a phenomenon thought to result from newly arisen somatic mutations. For example, a species of *Eucalyptus* has higher levels of volatile oils (mainly terpenes) in branches resistant to predation by scarab beetles than in susceptible branches (Edwards et al. 1990).

3.8.1. Terpenes

This group of ubiquitous lipophilic biochemicals includes the insecticides pyrethrum and azadirachtin. Although terpenes are generally only moderately toxic to insects, these two chemicals are very potent (Brattsten 1983). Some terpenes, like ergosterol, are also highly toxic to mammals (Brattsten 1983). However, the significance of the terpenes goes beyond their direct toxic effects. Many also act as insect attractants and deterrents and their toxicity may be due to their ability to mimic insect hormones. Gossypol can act as an antifeedant for some lepidopteran species, yet it is a kairomone (a compound beneficial to the receiving organism) and a feeding stimulant for the boll weevil *Anthonomus grandis* (Seigler 1982).

Extracts of *Chrysanthemum* flowers contain a number of insecticidal chemicals known as pyrethrins. These chemicals have been used from ancient times for the control of insects. Synthetic pyrethroids based on the natural products have largely replaced the pyrethrins as insecticides.

The pyrethrins do not appeal as candidates for genetic engineering for one paramount reason—they are not effective when ingested because they are readily hydrolysed to nontoxic products (Matsumura 1975). A solution to this problem might be to induce the production of a synergist, such as sesamin, that would inhibit the metabolism of the toxins. It is interesting to note that the pyrethrum plant contains not only the pyrethrin toxins but also sesamin (Doskotch and El-Feraly 1969).

Azadirachtin can affect insects in a number of ways. As well as interfering with ecdysteroid synthesis, release or action (Sieber and Rembole 1983), it is also a potent antifeedant. When impregnated into filter paper, only 1 ng/cm^2 completely inhibits feeding by *Shistocerca gregaria* (Morgan and Thornton 1973). Another potent terpenoid antifeedant is warburganal; this chemical is effective against *Spodoptera exempta* at a concentration of 0.1 ppm.

3.8.2. Nicotine and Other Tobacco Alkaloids

Nicotine is the major toxic component of tobacco (*Nicotiana*), but other nicotinoids are present in lower amounts. Other plants that contain nicotinoids include *Asclepia syriaca* (milk weed), *Atropa belladonna* (deadly nightshade), *Equisetum arvense* (horse tails), *Lycopodium clavatum* (club moss) and an Australian plant, *Duboisia hopwoodi* (Jacobson 1971).

Although long used as an effective insecticide against a wide range of insects, nicotine is particularly useful against easily penetrated soft-bodied ones. If nicotine was expressed in plants or a baculovirus, its potency when ingested would be more relevant. The nicotinoids are alkaloids that bind to acetylcholine receptors and, because they cannot be degraded by acetylcholinesterase, produce uncontrolled nervous activity that results in death. Insects that feed on tobacco plants are generally capable of either detoxifying or excreting nicotine.

Even though the structure of nicotine is a simple one (from the postulated biosynthetic pathway in plants, nicotine is produced through condensation of a 3-carbon unit such as glyceraldehyde), it is clear that a number of enzymes are involved in its synthesis.

Plants other than tobacco also contain insecticidal alkaloids. Janzen et al. (1977) studied the toxicity of 11 alkaloids isolated from plants to larvae of the burchid beetle *Callosobruchus maculatus*. Nine, including colchicine, gramine, reserpine and atropine, were lethal at 0.1% concentration. Most of these chemicals are toxic to vertebrates as well as to insects and, to be of value, would need to be expressed in parts of plants that posed no threat to non-target organisms.

3.8.3. Phytoecdysones

Plant sterols have been shown to affect insect development in experiments using artificial or unnatural diets, and it is thought that they serve a limited but important protective function in plants (Kubo and Klocke 1983). Their effectiveness may be undermined by the insect's ability to metabolize and excrete the steroids. For example, although the bracken fern is high in ecdysone and 20-hydroxyecdysone, ingestion of the plant does not affect the development of desert locusts. This is thought to be due to dehydroxylation of the phytoecdysones in the locust midgut (Carlisle and Ellis 1968).

3.8.4. *Rotenone and the Rotenoids*

Rotenone is produced by members of the family Leguminosae, the genus *Derris* being a particularly good source. It has long been used as a poison for insects and fish and has a low toxicity for mammals. The rotenoids are toxic to insects after ingestion and act by inhibiting mitochondrial respiration.

3.8.5. *The Unsaturated Isobutylamides*

These compounds are unsaturated, aliphatic straight-chain, C_{10-18} acids and are found in plants of the families Compositae and Rutaceae. Jacobson (1971) observed that some are at least as toxic to insects as are the pyrethrins. Affinin, a compound present in *Heliopsis longipes*, is active against a wide range of insects including dipterans, lepidopterans and coleopterans. Although some unsaturated isobutylamides are used as spices, they can be dangerous to humans and would not be suitable for expression in food stuffs.

In addition to the classes of secondary plant metabolites discussed above, there is an impressive number of insecticidal or repellent chemicals that have been extracted from plants. A comprehensive review of this topic up to 1970 was provided by Crosby (1971). Included were four insecticides being used commercially—quassia, sabadilla, hellebore and ryania—as well as many others in limited and local use.

3.9. Alteration of mRNA Levels

The value of genes encoding degradatory or processing enzymes to alter the level of a physiologically important protein product was discussed in Sections 3.3 and 3.4. Another way to regulate the amount of a given protein is to modify the amount of functional mRNA available for translation. There are currently two techniques which could be used to accomplish this. The first involves the use of antisense mRNA and the second uses ribonucleic acid enzymes, or ribozymes. These two methods, along with their potential uses, will be discussed below.

The regulation of gene expression by antisense RNA occurs naturally in some prokaryotes and also in some eukaryotes (Kimelman and Kirschner 1989). The expression of an antisense RNA gene results in a transcript complementary to the transcript produced by the endogenous target gene. This antisense RNA can then hybridize to the sense RNA in the nucleus, interfering with processing, modification or splicing steps, or in the cytoplasm, where it would inhibit translation. With the advent of recDNA technologies, inhibition of gene expression with antisense RNA has been successfully demonstrated in a wide variety of engineered plants and animals. Advantages of this system include the ability to manipulate the antisense RNA *in vitro* for enhanced activity *in vivo* and the potential to control the location, level and timing of expression using appropriate regulatory sequences.

A number of RNA molecules that catalyze their own cleavage exist in nature (Cech 1987). These molecules contain "substrate" domains, which dictate where splicing is to occur, as well as a domain to catalyze the cleavage. The potential of these self-cleaving RNAs to cleave "nonself" RNAs was recognized by Haseloff and Gerlach (1988). The result of their work was a generic technology for the design of "ribozymes": RNA cleaving enzymes with customized substrate specificity.

Since expression of both antisense RNAs and ribozymes must occur within the cell, the only appropriate delivery systems currently available are viral vectors. However, the number of possible target genes and the high specificity that can be achieved offset this limitation. Theoretically, any identified gene transcript can be targeted. Expression of physiologically important neuropeptides, enzymes or other proteins can be decreased or completely blocked. Alternatively, an enzyme catalyzing one step in a complex pathway for the synthesis of a hormone, pheromone or cuticle component could be targeted, resulting in disruption of the entire pathway.

3.10. Antibodies

Antibodies are composed of two identical light polypeptide chains and two identical heavy polypeptide chains. The chains form a Y-shaped molecule and segments of both the light and heavy chain present in the arms of the Y constitute the antigen binding site. The ability to generate antibodies that display strong and specific binding to a selected antigen and with defined degradative catalytic activities (Tramontano et al. 1986a; 1986b) underlies the strength in using antibodies for such diverse applications as environmental monitoring and therapeutic medicine. The use of antibodies in many areas of molecular entomology is reviewed by Trowell and East (this volume).

A number of studies have demonstrated that proteins may penetrate the gut of some insects and enter the hemolymph and tissues intact (Wigglesworth 1943; Smith et al. 1969; Schlein et al. 1976). Furthermore, it has been shown that orally administered antibodies can also enter the hemolymph and tissues, retain their binding affinity and specificity with the antigen that they were made against (Schlein and Lewis 1976) and even interfere with the normal function of the endogenous protein (Nogge and Gianetti 1980). Thus, one potential use of antibodies in insect control is the production of vaccines to render livestock more resistant to insect pests. Recently, this approach was successful when cattle were immunized with gut proteins isolated from the tropical cattle tick *Boophilus microplus*. Antibodies produced in the vaccinated cattle combined with the gut proteins of attacking ticks resulting in damage to the gut cells and subsequent death of the tick (Kemp et al. 1989).

Antibodies are also a potentially rich source of genetic material for the engineering of insecticides. Whereas it is difficult to isolate genes encoding polyclonal antibodies, genes for monoclonal antibodies (mAbs) are more easily attainable. Traditionally, the production of mAbs has been expensive as well as labor- and time-intensive. Recent advances in mAb technology have greatly improved methods for the production and identification of mAbs in a form suitable for genetic manipulation. In particular, Huse et al. (1989) developed a novel bacteriophage vector system that will express a library of functional antibody fragments. Using this system, spleen DNA from mice immunized with the desired antigen can be used to make either a light or a heavy chain library. When these libraries are ligated, *E. coli* infected with the combinatorial library expresses antibodies consisting of both a light and heavy chain. It is predicted that the diversity of different mAbs obtained could be as great as that produced *in vivo*.

Ward et al. (1989) demonstrated that the antigen binding segment of the antibody heavy chain may alone be sufficient for strong, specific binding to an antigen. Although the production of single site antibodies may not be possible for every antigen, this technique would simplify the transfer of antibody genes to vector systems and provide the opportunity to design smaller molecules which would have a greater potential to penetrate tissues and react with intracellular and intercellular targets (Ward et al. 1989).

Utilizing the various strategies to produce mAbs, a variety of novel, biologically active substances may be developed. Cloned mAbs could then be delivered to insects using viral vector systems. There exists the potential to generate mAbs that bind to, and thus disrupt the activity of, a protein involved in regulation of metamorphosis, development, reproduction or physiological homeostasis. Receptors for different endocrine hormones, neurotransmitters, neuropeptides and growth factors are also potential targets.

Since the limiting factor to making antibodies can often be the isolation of sufficient amounts of antigen, a practical approach to the cloning of receptor-binding mAbs is the production of anti-idiotypic mAbs (antibodies that are generated to antibodies previously raised against a specific ligand). Anti-idiotypic antibodies raised against anti-hormone antibodies are capable of interacting with hormone receptors, mimicking the effects of the ligand and evoking physiological responses (Sege and Peterson 1978). This "molecular mimicry" has been used to make anti-idiotypic antibodies to the nicotinic acetylcholine receptor (nAChR) (Wasserman et al. 1982) as well as the insulin receptor (Sege and Peterson 1978), angiotensin II receptor (Couraud 1986) and β-adrenergic receptor (Chamat et al. 1984).

It would seem possible to transfer genes for antibodies that are detrimental to insects into plants. The "insecticide" could be made completely nontoxic to humans, highly specific to the pest species, and amenable to alteration should resistance arise. The feasibility of this strategy has been substantiated by a recent study where mouse genes for a monoclonal antibody were transferred into a tobacco plant (Hiatt et al. 1989). By crossing plants expressing single immunoglobulin chains, progeny were obtained that expressed both chains simultaneously.

4. Concluding Comments

The development of insecticide resistance in many pest insect species and the increasing awareness of the deleterious effects of chemical insecticides on nontarget species have emphasized the need for more effective and selective insecticides. Genetic engineering has proven to be a powerful tool in a variety of disciplines including human medicine and pharmacology. Now it appears that vast benefits to mankind may also be realised through the genetic engineering of crops and microorganisms to improve the control of insect pests. Field tests of genetically engineered plants and microorganisms expressing the *B.t.* toxin gene and laboratory trials of viral vectors containing genes for enzymes and neuropeptides have yielded many encouraging results. There is now a twofold need for research into the molecular biology of other potential biological vectors and into the identification of suitable insecticidal genes for these vectors to carry. Although most of the insecticidal proteins that have been identified and characterized are the results of extensive field and laboratory studies, the development of more sophisticated techniques to screen for insecticidal molecules (for example, in vitro assay systems) may alleviate this problem. Taking into account the wide repertoire of insecticidal molecules that are present in nature, the genetic engineering of biological control agents promises to yield an exciting and diverse new generation of insecticides.

Acknowledgments. We thank TJ Higgins, T Hanzlik and D Dall for reviewing earlier drafts of this manuscript and offering valuable comments and criticisms. We are also indebted to P Christian, P East, P Atkinson and S Trowell for many useful discussions and ideas.

References

Adams ME, Herold EE, Venema VJ (1989) Two classes of channel-specific toxins from funnel web spider venom. J Comp Physiol A 164:333–342

Akhurst RJ, Bedding RA (1986) Natural occurrence of insect pathogenic nematodes (*Steinernematidae* and *Heterorhabditidae*) in soil in Australia. J Aust Entomol Soc 25:241–244

Almassy RJ, Fontecilla-Camps JC, Suddath FL, Bugg CE (1983) Structure of variant-3 scorpion neurotoxin from *Centruoides sculpturatus* Ewing, refined at 1.8 A resolution. J Mol Biol 170:497–527

Angus TA (1971) *Bacillus thuringiensis* as a microbial insecticide. In Jacobson N, Crosby DG (eds) Naturally Occurring Insecticides. Marcel Dekker, New York, pp 463–497

Arapinis C, de la Torre F, Szulmajster J (1988) Nucleotide and deduced amino acid sequence of the *Bacillus sphaericus* 1593M gene encoding a 51.4 kD polypeptide which acts synergistically with the 42 kD protein for expression of the larvicidal toxin. Nucl Acids Res 15:7731

Arif, BM (1984) The entomopoxviruses. Adv Virus Res 29:195–213

Aronson AI, Beckman W, Dunn P (1986) *Bacillus thuringiensis* and related insect pathogens. Microbiol Rev 50:1–24

Bar E, Lieman-Hurwitz J, Rahamim E, Keynan A, Sandler N (1991) Cloning and expression of *Bacillus thuringiensis israelensis* δ-endotoxin DNA in *B. sphaericus*. J Invert Pathol 57:149–158

Bartlett MC, Jaronski ST (1988) Mass production of entomogenous fungi for biological control of insects. In Burge MN (ed) Fungi in Biological Control Systems. Manchester University Press, New York, pp 61–85

Baumann L, Broadwell AH, Baumann P (1988) Sequence analysis of the mosquitocidal toxin genes encoding 51.4- and 41.9-kilodalton proteins from *Bacillus sphaericus* 2362 and 2297. J Bacteriol 170:2045–2050

Beard J (1989) Viral protein knocks the guts out of caterpillars. New Scientist 23:21

Beard RL (1952) The toxicology of Habrobracon venom: a study of a natural insecticide. Conn Agric Exper Stn New Haven Bull 562:1–27

Beckage NE (1985) Endocrine interactions between endoparasitic insects and their hosts. Ann Rev Entomol 30:371–413

Bedding RA (1976) New methods increase the feasibility of using *Neoaplectana* spp. (Nematoda) for the control of insect pests. Proc First Int Colloq Invert Pathol. Kingston, Canada, pp 250–254

Bell EA (1978) Toxins in seeds. In Harbourne JB (ed) Biochemical Aspects of Plant and Animal Coevolution. Academic Press, New York, pp 141–163

Bell EA (1987) Secondary compounds and plant herbivores. In Labeyrie V, Gabres G, Lachaise D (eds) Insects-Plants. Proc Sixth Int Symp Insect-Plant Relationships. Dr W Junk Publishers, Dordrecht, pp 19–23

Berry C, Hindley J (1987) *Bacillus sphaericus* strain 2362: identification and nucleotide sequence of the 41.9 kDa toxin gene. Nucl Acids Res 15:5891

Berry C, Jackson-Yap J, Oei C, Hindley J (1989) Nucleotide sequence of two toxin genes from *Bacillus sphaericus* IAB59: sequence comparisons between five highly toxinogenic strains. Nucl Acids Res 17:7516

Bettini S (ed) (1978) Arthropod Venoms: Handbook of Experimental Pharmacology Vol 48. Springer-Verlag, New York, 977 pp

Binnington KC (1985) Ultrastructural changes in the cuticle of the sheep blowfly, *Lucilia*, induced by certain insecticides and biological inhibitors. Tissue Cell 17:131–140

Blalock JE, Smith EM (1984) Hydropathic anti-complementarity of amino acids based on the genetic code. Biochem Biophys Res Commun 121:203–207

Bleakely B, Nealson KH (1988) Characterization of primary and secondary forms of *Xenorhabdus luminescens* strain Hm. FEMS Microbiol Ecol 53:241–250

Bost KL, Smith EM, Blalock JE (1985) Similarity between the ACTH receptor and a peptide encoded by an RNA that is complementary to ACTH mRNA. Proc Natl Acad Sci USA 82:1372–1375

Boulter D, Edwards GA, Gatehouse AMR, Gatehouse JA, Hilder VA (1990) Additive protective effects of different plant-derived insect resistance genes in transgenic tobacco plants. Crop Protect 9:351–354

Bowers CW, Phillips HS, Lee P, Jan YN, Jan LY (1987) Identification and purification of an irreversible presynaptic neurotoxin from the venom of the spider *Hololena curta*. Proc Natl Acad Sci USA 84:3506–3510

Bowers WS (1982) Endocrine strategies for insect control. Entomol Exper Appl 31:3–14

Bowers WS (1983) Phytochemical action on insect morphogenesis, reproduction and behavior. In Whitehead DL, Bowers WS (eds) Natural Products for Innovative Pest Management. Pergamon Press, Oxford, pp 313–322

Brattsten LB (1983) Cytochrome P-450 involvement in the interactions between plant terpenes and insect herbivores. In Hedin PA (ed) Plant Resistance to Insects. American Chemical Society, Washington DC, pp 173–195

Broadway RM, Duffey SS (1986) Plant proteinase inhibitors: mechanism of action and effect on the growth and digestive physiology of larval *Heliothis zea* and *Spodoptera-exiqua*. J Insect Physiol 32:827–833

Cameron IR, Possee RD, Bishop DHL (1989) Insect cell culture technology in baculovirus expression systems. TIBTECH 7:66–71

Carbonell LF, Hodge MR, Tomalski MD, Miller LK (1989) Synthesis of a gene coding for and insect-specific scorpion neurotoxin and attempts to express it using baculovirus vectors. Gene 73:409–418

Carlisle DB, Ellis PE (1968) Bracken and locust ecdysones: Their effects on molting in the desert locust. Science 159:1472–1474

Case ME, Schweizer M, Kushner SR, Giles NH (1979) Efficient transformation of *Neurospora crassa* by utilizing hybrid plasmid DNA. Proc Natl Acad Sci USA 76:5259–5263

Catterall WA (1980) Neurotoxins that act on voltage-sensitive sodium channels in excitable membranes. Ann Rev Pharmacol Toxicol 20:15–43

Catterall WA (1984) The molecular basis of neuronal excitability. Science 223: 653–660

Cech TR (1987) The chemistry of self-splicing RNA and RNA enzymes. Science 236:1532–1539

Chamat S, Hoebeke J, Strosberg AD (1984) Monoclonal antibodies specific for β-adrenergic ligands. J Immunol 133:1547–1552

Chilcott CN, Ellar DJ (1988) Comparative toxicity of *Bacillus thuringiensis* var. *israelensis* crystal proteins in vivo and in vitro. J Gen Microbiol 134: 2551–2558

Cleveland TE, Black LL (1982) Partial purification of proteinase-inhibitors from wounded tomato plants. Plant Physiol 69:537–542

Cocking EC, Davey MR (1987) Gene transfer in cereals. Science 236:1259–1262

Couraud PO (1986) Structural analysis of the epitopes recognized by monoclonal antibodies to angiotensin II. J Immunol 136:3365–3370

Crickmore N, Nicholls C, Earp DJ, Hodgman TC, Ellar DJ (1990) The construction of *Bacillus thuringiensis* strains expressing novel entomocidal δ-endotoxin combinations. Biochem J 270:133–136

Crosby DG (1971) Minor insecticides of plant origin. In: Jacobson M, Crosby DG (eds) Naturally Occurring Insecticides. Marcel Dekker, New York, pp 177–239

Datta SK, Peterhans A, Datta K, Potrykus I (1990) Genetically engineered fertile indica-rice recovered from protoplasts. Bio/Technology 8:736–740

Davidson EW (1989) Variation in binding of *Bacillus sphaericus* toxin and wheat germ agglutinin to larval midgut cells of six species of mosquitoes. J Invert Pathol 53:251–259

De Dianous S, Hoarau F, Rochat H (1987) Re-examination of the specificity of the scorpion *Androctonus australis* Hector insect toxin towards arthropods. Toxicon 25:411–417

de Kort CAD, Granger NA (1981) Regulation of the juvenile hormone titer. Ann Rev Entomol 26:1–28

Delannay X, LaVallee BJ, Proksch RK, Fuchs RL, Sims SR, Greenplate JT, Marrone PG, Dodson RB, Augustine JJ, Layton JG, Fischhoff DA (1989) Field performance of transgenic tomato plants expressing the *Bacillus thuringiensis* var. *kurstaki* insect control protein. Bio/Technology 7: 1265–1269

de la Torre F, Bennardo T, Sebo P, Szulmajster J (1989) On the respective roles of the two proteins encoded by the *Bacillus sphaericus* 1593M toxin genes expressed in *Escherichia coli* and *Bacillus subtilis*. Biochem Biophys Res Commun 164:1417–1422

DeLoof A (1987) The impact of the discovery of vertebrate-type steroids and peptide hormone-like substances in insects. Entomol Exper Appl 45:105–113

de Ponti OMB (1982) Plant resistance to insects: a challenge to plant breeders and entomologists. In Visser JH, Minks AK (eds) Proc Fifth Int Symp Insect-Plant Relationships. Wageningen, pp 337–347

Derksen ACG, Granados RR (1988) Alteration of a lepidopteran peritrophic membrane by baculoviruses and enhancement of viral infectivity. Virology 167:242–250

Doskotch RW, El-Feraly FS (1969) Isolation and characterization of (+)-sesamin and -cyclopyrethrosin from pyrethrum flowers. Canad J Chem 47:1139–1142

Douglass J, Civelli O, Herbert E (1984) Polyprotein gene expression: generation of diversity of neuroendocrine peptides. Ann Rev Biochem 53:665–715.

Drexler G, Sieghart W (1984a) Properties of high affinity binding site for tritium-labeled avermectin B_{1a}. Eur J Pharmacol 99:269–277

Drexler G, Sieghart W (1984b) Evidence for association of a high affinity avermectin binding site with benzodiazepine receptor. Eur J Pharmacol 101: 201–207

Drexler G, Sieghart W (1984c) Sulfur-35-labeled tert-butyl-bicyclophosphorothionate and avermectin bind to different sites associated with the gamma-aminobutryic acid-benzodiazepine receptor complex. Neurosci Letters 50:273–277

Edson KM, Vinson SB, Stoltz DB, Summers MD (1981) Virus in a parasitoid wasp: suppression of the cellular immune response in the parasitoid's host. Science 211:582–583

Edwards PB, Wanjura WJ, Brown WV, Dearn JM (1990) Mosaic resistance in plants. Nature 347:434

Etzler ME (1985) Plant lectins: molecular and biological aspects. Ann Rev Plant Physiol 36:209–234

Etzler ME (1986) Distribution and function of plant lectins. In Liener IE, Sharon N, Goldstein IJ (eds) The Lectins: Properties, Functions, and Applications in Biology and Medicine. Academic Press, Orlando, pp 371–435

Evans BE, Bock MG, Rittle KE, DiPardo RM, Whitter WL, Veber DF, Anderson AP, Freidinger RM (1986) Design of potent, orally effective, nonpeptidal antagonists of the peptide hormone cholecystokinin. Proc Natl Acad Sci USA 83:4918–4921

Evans HF, Harrap KA (1982) Persistence of insect viruses. In Mahy BWJ, Minson AC, Darby GK (eds) Virus Persistence. Cambridge University Press, pp 57–96

Evans PD, Robb S, Cuthbert BA (1989) Insect neuropeptides—Identification, establishment of functional roles and novel target sites for pesticides. Pestic Sci 25:71–83

Evans RM (1988) The steroid and thyroid hormone receptor superfamily. Science 240:889–895

Farmer JJ, III, Davis BR, Hickman FH, Presley DB, Bodey GP, Negut MH, Bobo RA (1976) Detection of *Serratia* outbreaks in hospital. Lancet 2: 455–459

Faust RM, Bulla LA Jr (1982) Bacteria and their toxins as insecticides. In Kurstak E (ed) Microbial and Viral Pesticides. Marcel Dekker, New York, pp 75–208

Fischoff DA, Bowdisch KS, Perlak FJ, Marrone PG, McCormick SH, Niedermeyer JB, Dean DA, Kusano-Kretzmer K, Mayer EF, Rochester DE, Rogers SG, Fraley RT (1987) Insect tolerant transgenic tomato plants. Bio/Technology 5:807–813

Fontecilla-Camps JC (1989) Three-dimensional model of the insect-directed scorpion toxin from *Androctonus australis* Hector and its implication for the evolution of scorpion toxins in general. J Mol Evol 29:63–67

Fontecilla-Camps JC, Almassy RJ, Suddath FL, Watt DD, Bugg CE (1980) Three-dimensional structure of a protein from scorpion venom: a new structural class of neurotoxins. Proc Natl Acad Sci USA 77:6496–6500

Fontecilla-Camps JC, Habersetzer-Rochat C, Rochat H (1988) Orthorhombic crystals and three-dimensional structure of the potent toxin II from the scorpion *Androctonus australis* Hector. Proc Natl Acad Sci USA 85: 7443–7447

Foster GG, Vogt WG, Woodburn TL, Smith PH (1988) Computer simulation of genetic control. Comparison of sterile males and field-female killing systems. Theor Appl Genet 76:870–879

Frackman S, Nealson KH (1990) The molecular genetics of Xenorhabdus. In Gaugler R, Kaya HK (eds) Entomopathogenic Nematodes in Biological Control. CRC Press, Boca Raton, Florida, pp 285–300

Gallaher D, Schneeman BO (1984) Nutritional and metabolic response to plant inhibitors of digestive enzymes. In M Friedman (ed) Nutritional and Toxicological Aspects of Food Safety. Plenum Press, New York, pp 299–320

Gatehouse AMR, Boulter D (1983) Assessment of the antimetabolic effects of trypsin inhibitors from cowpea (*Vigna unguiculata*) and other legumes on development of the bruchid beetle *Callosobruchus maculatus*. J Sci Food Agric 34:345–350

Gatehouse AMR, Dewey FM, Dove J, Fenton KA, Pusztai A (1984) Effect of seed lectins from *Phaseolus vulgaris* on the development of larvae of *Callosobruchus maculatus*—mechanism of toxicity. J Sci Food Agric 35: 373–380

Georghiou GP (1986) The magnitude of the resistance problem. In Pesticide Resistance: Stratagies and Tactics for Management. National Academy Press, Washington DC, pp 14–43

Georghiou GP (1988) Implications of potential resistance to biopesticides. In Roberts DW, Granados RR (eds) Biotechnology, Biological Pesticides and Novel Plant-Pest Resistance for Insect Pest Management. Insect Pathology Resource Center, Boyce Thompson Institute for Plant Research at Cornell University, Ithaca, New York, pp 137–145

Goldstein IJ, Hayes CE (1978) The lectins: Carbohydrate-binding proteins of plants and animal. Adv Carbohydr Chem Biochem 35:127–340

Greenberg MJ, Price DA (1983) Invertebrate neuropeptides: native and naturalized. Ann Rev Physiol 45:271–288

Habermann E (1971) Chemistry, pharmacology and toxicology of bee, wasp and hornet venoms. In Bucherl W, Buckley EE (eds) Venomous Animals and Their Venoms. Academic Press, New York, pp 16–93

Hammock BD (1985) Regulation of juvenile hormone titer: degradation. In Kerkut GA, Gilbert LI (eds) Comprehensive Insect Physiology, Biochemistry and Pharmacology Vol 7. Pergamon Press, Oxford, pp 431–472

Hammock BD, Bonning BC, Possee RD, Hanzlik TN, Maeda S (1990) Expression and effects of the juvenile hormone esterase in a baculovirus vector. Nature 344:458–461

Haseloff J, Gerlach WL (1988) Simple RNA enzymes with new and highly specific endoribonuclease activities. Nature 334:585–591

Heale JB (1988) The potential impact of fungal genetics and molecular biology on biological control, with particular reference to entomopathogens. In Burge MN (ed) Fungi in Biological Control Systems. Manchester University Press, UK, pp 211–234

Heale JB, Isaac JE, Chandler D (1989) Prospects for strain improvement in entomopathogenic fungi. Pestic Sci 26:79–92

Heimpel AM (1955a) The pH in the gut and blood of the Larch Sawfly, *Pristiphora erichsonii* (Htg.) and other insects with reference to the pathogenicity of *Bacillus cereus* Fr. and Fr. Canad J Zool 33:99–106

Heimpel AM (1955b) Investigations of the mode of action of strains of *Bacillus cereus* Fr. and Fr. pathogenic for the Larch Sawfly, *Pristiphora erichsonii* (Htg.). Canad J Zool 33:311–326

Herbert E, Uhler M (1982) Biosynthesis of polyprotein precursors to regulatory peptides. Cell 30:1–2

Hiatt A, Cafferkey R, Bowdish K (1989) Production of antibodies in transgenic plants. Nature 342:76–78

Hilder VA, Gatehouse AMR, Boulter D (1989) Potential for exploiting plant genes to genetically engineer insect resistance, exemplified by the cowpea trypsin inhibitor gene. Pestic Sci 27:165–171

Hilder VA, Gatehouse AMR, Sheerman SE, Barker RF, Boulter D (1987) A novel mechanism of insect resistance engineered into tobacco. Nature 330: 160–163

Hinnen A, Hicks JB, Fink GR (1978) Transformation of yeast. Proc Natl Acad Sci USA 75:1929–1933

Hoffman LM, Ma Y, Barker RF, (1982) Molecular cloning of *Phaseolus vulgaris* lectin mRNA and use of cDNA as a probe to estimate lectin transcript levels in various tissues. Nucl Acids Res 10:7819–7828

Hofte H, Whiteley HR (1989) Insecticidal crystal proteins of *Bacillus thuringiensis*. Microbiol Rev 53:242–255

Holman GM, Nachman RJ, Wright MS (1990) Insect neuropeptides. Ann Rev Entomol 35:201–217

Horodyski FM, Riddiford LM, Truman W (1989) Molecular analysis of eclosion hormone in the tobacco hornworm, *Manduca sexta*. Proc Int Symp Molecular Insect Science. Tucson, Arizona

Huber J (1986) Use of baculoviruses in pest management programs. In Granados RR, Federici BA (eds) The Biology of Baculoviruses Vol II Practical Application for Insect Control. CRC Press, Boca Raton, Florida, pp 181–202

Hudson P, Haley J, Crank M, Shine J, Niall H (1981) Molecular cloning and characterization of cDNA sequences coding for rat relaxin. Nature 291: 127–131

Huse WD, Sastry L, Iverson SA, Kang AS, Alting-Mees M, Burton DR, Bendovic ST, Lerner RA (1989) Generation of a large combinatorial library of the immunoglobulin repertoire in phage lambda. Science 246:1275–1281

Huxham IM, Lackie AM, McCorkindale NJ (1989) Inhibitory effects of cyclodepsipeptides, destruxins, from the fungus *Metarhizium anisopliae*, on cellular immunity in insects. J Insect Physiol 35:97–105

Hynes MJ (1986) Transformation of filamentous fungi. Exper Mycol 10:1–8

Ignoffo CM (1967) In Van Der Laan PA (ed) Insect Pathology and Microbiological Control. North Holland, Amsterdam, pp 91–117

Ignoffo CM (1968) Specificity of insect viruses. Bull Entomol Soc Amer 14:265

Isaac RE (1988) The metabolism of neuropeptides by membrane peptidases from the locust, *Schistocerca gregaria*. Pestic Sci 24:259–260

Isono D, Suhadolnik RJ (1976) The biosynthesis of natural and unnatural polyoxins by *Streptomyces cacaoi*. Arch Biochem Biophys 173:141–153

Iwantsch GF, Smilowitz Z (1975) Relationships between the parasitoid *Hyposoter exiquae* and *Trichoplusia ni*: prevention of host pupation at the endocrine level. J Insect Physiol 2:1151–1157

Jackson H, Urnes M, Gray WR, Parks TN (1985) Soc Neurosci Abstr 11:107

Jacobson M (1971) The unsaturated isobutylamides. In Naturally Occurring Insecticides. Marcel Dekker, New York, pp 137–176

Janzen DH, Juster HB, Bell EA (1977) Toxicity of secondary compounds to the seed-eating larvae of the bruchid beetle *Callosobruchus maculatus*. Phytochem 16:223–227

Janzen DH, Juster HB, Liener IE (1976) Insecticidal action of the phytohemagglutinin in black beans on a bruchid beetle. Science 192:795–796

Jaynes J, Dodds J (1987) Synthetic genes make better potatoes. New Scientist 115:62–64

Johnson CE, Bonventre PF (1967) Lethal toxin of *Bacillus cereus*. I. Relationships and nature of toxin, hemolysin, and phospholipase. J Bacteriol 94: 306–316

Jones D, Sreekrishna S, Iwaya M, Yang JN, Eberely M (1986) Comparison of viral ultrastructure and DNA banding patterns from the reproductive tracts of eastern and western hemisphere *Chelonus* spp. (Braconidae:Hymenoptera). J Invert Pathol 47:105–115

Jornvall H, Carlquist M, Kwauk S, Otte SC, McIntosh CHS, Brown JC, Mutt V (1981) Amino acid sequence and heterogeneity of gastric inhibitory polypeptide (GIP). FEBS Letters 123:205–210

Kalmakoff J, JF Longworth (1980) Microbial Control of Insect Pests. New Zealand DSIR Bulletin 228, 102 pp

Keddie BA, Aponte GW, Volkman LE (1989) The pathway of infection of *Autographa californica* nuclear polyhedrosis virus in an insect host. Science 243:1728–1730

Keeley LL, Hayes TK (1987) Speculations on biotechnology applications for insect neuroendocrine research. Insect Biochem 17:639–651

Kelly TJ, Masler EP, Menn JJ (1990) Insect neuropeptides: new strategies for insect control. In Casida JE (ed) Pesticides and alternatives. Elsevier, Amsterdam, pp 283–297

Kemp DH, Pearson RD, Gough JM, Willadsen P (1989) Vaccination against *Boophilus microplus*: localization of antigens on tick gut cells and their interaction with the host immune system. Exper Appl Acarol 7: 43–58

Kimelman D, Kirschner MW (1989) An antisense mRNA directs the covalent modification of the transcript encoding fibroblast growth factor in *Xenopus* oocytes. Cell 59:687–696

King TP, Pusztai A, Clarke EMW (1980) Kidney bean (*Phaseolus vulgaris*) lectin-induced lesions in rat small intestine. 1. Light microscope studies. J Comp Pathol 90:585–595

King TP, Pusztai A, Clarke EMW (1982) Kidney bean (*Phaseolus vulgaris*) lectin-induced lesions in rat small intestine. 3. Ultrastructural studies. J Comp Pathol 92:357–373

Kirschbaum JB (1985) Potential implication of genetic engineering and other biotechnologies in insect control. Ann Rev Entomol 30:51–70

Klein TM, Gradziel T, Fromm ME, Sanford JC (1988) Factors influencing gene delivery into *Zea mays* cells by high-velocity microprojectiles. Bio/Technology 6:559–563

Koeppe JK, Fuchs M, Chen TT, Hunt LM, Kovalick GE, Briers T (1985) The role of juvenile hormone in reproduction. In Kerkut GA, Gilbert LI (eds) Comprehensive Insect Physiology, Biochemistry and Pharmacology Vol 8. Permagon Press, Oxford, pp 165–203

Kombrink E, Schroder M, Hahlbrock K (1988) Several "pathogenesis-related" proteins in potato are 1,3-β-glucanases and chitinases. Proc Natl Acad Sci USA 85:782–786

Kubo I, Klocke JA (1983) Isolation of phytoecdysones as insect ecdysis inhibitors and feeding deterrents. In Hedin PA (ed) Plant Resistance to Insects. American Chemical Society, Washington, pp 329–346

Lacey LA, Undeen AH (1986) Microbial control of blackflies and mosquitoes. Ann Rev Entomol 31:265–296

Lazarovici P, Yanai P, Pelhate M, Zlotkin E (1982) Insect toxic components from the venom of a chactoid scorpion, *Scorpio maurus palmatus* (Scorpionidae). J Biol Chem 257:8397–8404

Lee BLY (1971) An investigation on the biochemical properties of *Microbracon brevicornis* venom. Masters Thesis, Northeastern University, Boston

Leluck J, Jones D (1989) *Chelonus* sp. Near *Curvimaculatus* venom proteins: analysis of their potential role and processing during development of host *Trichoplusia ni*. Arch Insect Biochem Physiol 10:1–12

Loh YP, Brownstein MJ, Gainer H (1984) Proteolysis in neuropeptide processing and other neural functions. Ann Rev Neurosci 7:189–222

Loret EP, Mansuelle P, Rochat H, Granier C (1990) Neurotoxins active on insects: amino acid sequences, chemical modifications, and secondary structure esimation by circular dichroism of toxins from the scorpion *Androctonus australis* Hector. Biochemistry 29:1492–1501

Luckow VA, Summers MD (1988) Trends in the development of baculovirus expression vectors. Bio/Technology 6:47–55

Lüthy P, Cordier J-L, Fischer H-M (1982) *Bacillus thuringiensis* as a bacterial insecticide: basic considerations and application. In Kurstak E (ed) Microbial and Viral Pesticides. Marcel Dekker, New York, pp 35–74

Lysenko O (1967) Bacterial toxins. In Van Der Laan PA (ed) Insect Pathology and Microbial Control. North-Holland, Amsterdam, pp 219–237

Lysenko O (1974) Bacterial exoenzymes toxic for insects—proteinase and lecithinase. J Hyg Epidemiol Microbiol Immunol 18:347–352

Lysenko O (1976) Chitinase of *Serratia marcescens* and its toxicity to insects. J Invert Pathol 27:385–386

Maeda S (1989a) Expression of foreign genes in insects using baculovirus vectors. Ann Rev Entomol 34:351–372

Maeda S (1989b) Increased insecticidal effect by a recombinant baculovirus carrying a synthetic diuretic hormone gene. Biochem Biophys Res Comm 165: 1177–1183

Maeda S, Kawai T, Obinata M, Fujiwara H, Horiuchi T, Saeki Y, Sato Y, Furusawa M (1985) Production of human alpha-interferon in silkworm using a baculovirus vector. Nature 315:592–594

Matsumura F (1975) Toxicology of Insecticides. Plenum Press, New York, 503 pp

McCabe DE, Swain WF, Martinell BJ, Christou P (1988) Stable transformation of soybean (*Glycine max*) by particle acceleration. Bio/Technology 6:932–926

McGaughey WH (1985a) Insect resistance to the biological insecticide *Bacillus thuringiensis*. Science 229:193–195

McGaughey WH (1985b) Evaluation of *Bacillus thuringiensis* for controlling Indian meal moths (Lepidoptera: Pyralidae) in farm grain bins and elevator silos. J Econ Entomol 78:1089–1094

McGaughey WH, Beeman RW (1988) Resistance to *Bacillus thuringiensis* in colonies of Indian meal moth and almond moth (Lepidoptera: Pyralidae). J Econ Entomol 81:28–33

McKelvey JF, Blumberg S (1986) Inactivation and metabolism of neuropeptides. Ann Rev Neurosci 9:415–434

Meeusen RL, Warren G (1989) Insect control with genetically engineered crops. Ann Rev Entomol 34:373–38

Menn JJ (1985) Prospects of exploitation of insect antijuvenile hormones for selective insect control. In von Keyserlingk HC, Jager A, von Szczepanski Ch (eds) Approaches to New Leads for Insecticides. Springer-Verlag, Berlin, pp 37–46

Merryweather AT, Weyer U, Harris MPG, Hirst M, Booth T, Possee RD (1990) Construction of genetically engineered baculovirus insecticides containing the *Bacillus thuringiensis* subsp. *kurstaki* HD-73 delta endotoxin. J Gen Virol 71:1535–1544

Metcalf RL (1980) Changing role of insecticides in crop protection. Ann Rev Entomol 25:219–256

Miller LK, Lingg AJ, Bulla LA Jr (1983) Bacterial, viral and fungal insecticides. Science 219:715–721

Mirelman D, Galun E, Sharon N, Lotan R (1975) Inhibition of fungal growth by wheat germ agglutinin. Nature 256:414–416

Morgan ED, Thornton MD (1973) Azadirachtin in the fruit of *Melia azedarach*. Phytochemistry 12:391–392

Mothes U, Seitz K-A (1982) Action of the microbiol metabolite and chitin synthesis inhibitor nikkomycin on the mite *Tetranychus urticae*; an electron microscope study. Pestic Sci 13:426–441

Mullins MG, Tang FCA, Facciotti D (1990) *Agrobacterium*-mediated genetic transformation of grapevines: Transgenic plants of *Vitis rupestris* scheele and buds of *Vitis vinifera* L. Bio/Technology 8:1041–1045

Nambu JR, Murphy-Erdosh C, Andrews PC, Feistner GJ, Scheller RH (1988) Isolation and characterization of a *Drosophila* neuropeptide gene. Neuron 1:55–61

Nogge G, Gianetti M (1980) Specific antibodies: a potential insecticide. Science 209:1028–1029

Obukowicz MG, Perlak FJ, Kusano-Kretzmer K, Mayer EJ, Watrud LS (1986) Integration of the delta-endotoxin gene of *Bacillus thuringiensis* into the chromosome of root-colonizing strains of pseudomonads using Tn5. Gene 45:327–331

Olivera BM, Rivier J, Clark C, Ramilo CA, Corpuz GP, Abogadie FC, Mena EE, Woodward Sr, Hillyard DR, Cruz LJ (1990) Diversity of *Conus* neuropeptides. Science 249:257–263

O'Reilly DR, Miller LK (1989) A baculovirus blocks insect molting by producing ecdysteroid UDP-Glucosyl transferase. Science 245:1110–1112

Ornberg RL, Smyth T, Benton AW (1976) Isolation of a neurotoxin with presynaptic action from the venom of the black widow spider (*Latrodectus mactans*, Fabr.). Toxicon 14:329–333

Osborn TC, Alexander DC, Sun SSM, Cardona C, Bliss FA (1988a) Insecticidal activity and lectin homology of arcelin seed protein. Science 240:207–210

Osborn TC, Burrow M, Bliss FA (1988b) Purification and characterization of arcelin seed protein from common bean. Plant Physiol 86:399–405

O'Shea M (1985) Neuropeptides in insects: possible leads to new control methods. In von Keyserlingk HC, Jager A, von Szczepanski Ch (eds) Approaches to New Leads for Insecticides. Springer-Verlag, Berlin, pp 133–151

Peng JH, Black LL (1976) Increased proteinase-inhibitor activity in response to infection of resistant tomato plants by *Phytophthora infestans*. Phytopathology 66:958–963

Perlak FJ, Deaton RW, Armstrong TA, Fuchs RL, Sims SR, Greenplate JT, Fischhoff DA (1990) Insect resistant cotton plants. Bio/Technology 8: 939–943

Petersen JJ (1982) Current status of nematodes for the biological control of insects. Parasitology 84:177–204

Piek T (1966) Site of action of venom of *Microbracon hebetor* Say (Braconidae, Hymenoptera). J Insect Physiol 12:561–568

Piek T (1984) Pharmacology of Hymenoptera venoms. In Tu AT (ed) Handbook of Natural Toxins Vol 2 Insect Poisons, Allergens, and Other Invertebrate Venoms. Marcel Dekker, New York, pp 135–185

Piek T (1986) Venoms of the Hymenoptera: Biochemical, Pharmacological and Behavioral Aspects. Academic Press, London, 570 pp

Piek T, Mantel P, Engels E (1971) Neuromuscular block in insects caused by the venom of the digger wasp *Philanthus triangulum* F. Comp Gen Pharmacol 2:317–331

Piek T, Njio KD (1975) Neuromuscular block in honeybee by the venom of the beewolf wasp (*Philanthus triangulum*). Toxicon 13:199–201

Piek T, Spanjer W (1986) Chemistry and pharmacology of solitary wasp venoms. In Piek T (ed) Venoms of the Hymenoptera. Academic Press, London, pp 161–307

Piek T, Spanjer W, Njio KD, Veenendaal RL, Mantel P (1974) Paralysis caused by the venom of the wasp *Microbracon gelechiae*. J Insect Physiol 20: 2307–2319

Piek T, Veenendaal RL, Mantel P (1982) The pharmacology of *Microbracon* venom. Comp Biochem Physiol C 72:303–309

Poinar GO, Thomas GM (1966) Significance of *Achromobacter nematophilus* Poinar and Thomas (Achromobacteraceae: Eubacteriales) in the development of the nematode, DD136 (*Neoaplectana* sp.: Steinernematidae). Parasitology 56:385–390

Quiot JM, Vey A, Vago C (1985) Effects of mycotoxins on invertebrate cells in vitro, In Maramorosch K (ed) Advances in Cell Culture. Academic Press, New York, pp 199–212

Quistad GB, Adams ME, Scarborough RM, Carney RL, Schooley DA (1984) Metabolism of proctolin, a pentapeptide neurotransmitter in insects. Life Sci 34:569–576

Rayne RC, Koeppe JK (1988) Relationship of hemolymph juvenile hormone-binding protein to lipophorin in *Leucophaea maderae*. Insect Biochem 18: 667–673

Rhodes CA, Pierce DA, Mettler IJ, Mascarenhas D, Detmer JJ (1988) Genetically transformed maize plants from protoplasts. Science 240:204–207

Richardson M (1977) The proteinase inhibitors of plants and micro-organisms. Phytochemistry 16:159–169

Ridgeway RL, Silverstein RM, Inscoe MN (eds) (1990) Behavior Modifying Chemicals for Insect Management: Applications of Pheromones and Other Attractants. Marcel Dekker, New York, 761 pp

Roberts DW, Humber RA (1981) Entomogenous fungi. In Cole GT, Kendrick B (eds) Biology of Conidial Fungi. Academic Press, New York, pp 201–236

Ryan CA (1978) Proteinase-inhibitors in plant leaves—biochemical model for pest-induced natural plant protection. Trends Biochem Sci 5:148–150

Saffran M, Kumar GS, Savariar C, Burnham JC, Williams R, Neckers DC (1986) A new approach to the oral administration of insulin and other peptide drugs. Science 233:1081–1084

Samuels RI, Reynolds SE, Charnley AK (1986) Mode of action of destruxins. In Samson RA, Vlak JM, Peters D (eds) Fundamental and Applied Aspects of Invertebrate Pathology. Society of Invertebrate Pathology, The Netherlands, p 261

Scharrer B (1987) Insects as models in neuroendocrine research. Ann Rev Entomol 32:1–16

Scheller RH, Kaldany RR, Kreiner T, Mahon AC, Nambu JR, Schaefer M, Taussig R (1984) Neuropeptides: Mediators of behavior in *Aplysia*. Science 225:1300–1308

Schlein Y, Lewis CT (1976) Lesions in haematophagous flies after feeding on rabbits immunized with fly tissues. Physiol Entomol 1:55–59

Schlein Y, Spira DT, Jacobson RL (1976) The passage of serum immunoglobulins through the gut of *Sarcophaga falculata* Pand. Ann Trop Med Parasitol 70:227–230

Schluter U (1982) Ultrasructural evidence for inhibition of chitin synthesis by nikkomycin. Wilhelm Roux Arch Devel Biol 191:205–207

Schneider LE, Taghert PH (1988) Isolation and characterization of a *Drosophila* gene that encodes multiple neuropeptides related to Phe-Met-Arg-Phe-NH2 (FMRFamide). Proc Natl Acad Sci USA 85:1993–1997

Schnepf HE, Whiteley HR (1981) Cloning and expression of the *Bacillus thuringiensis* crystal protein gene in *Escherichia* coli. Proc Natl Acad Sci USA 78:2893–2897

Schoonhoven AV, Cardona C (1982) Low levels of resistance to the Mexican Bean Weevil in dry beans. J Econ Entomol 75:567–569

Schoonhoven AV, Valor J (1983) Resistance to the Bean Weevil and the Mexican Bean Weevil (Coleoptera: Bruchidae) in noncultivated common bean accessions. J Econ Entomol 76:1255–1259

Schoonhoven LM (1982) Biological aspects of antifeedants. Entomol Exper Appl 31:57–69

Sege K, Peterson PA (1978) Use of anti-idiotypic antibodies as cell-surface receptor probes. Proc Natl Acad Sci USA 75:2443–2447

Seigler DS (1982) Role of lipids in plant resistance to insects. In Hedin PA (ed) Plant Resistance to Insects. American Chemical Society, Washington, pp 303–327

Sieber KP, Rembole H (1983) The effects of azadirachtin on the endocrine control of molting in *Locusta migratoria*. J Insect Physiol 29:523–527

Smith DS, Compher K, Janners M, Lipton C, Wittle L (1969) Cellular organization and ferritin uptake in the mid-gut of a moth *Ephestia kuhniella*. J Morphol 127:41–72

Smith JN (1977) In Parke DV, Smith RL (eds) Drug Metabolism: from Microbe to Man. Taylor and Francis, London, pp 219–232

Soderlund DM, Adams PM, Bloomquist JR (1987) Differences in the action of avermectin B_{1a} on the $GABA_A$ receptor complex of mouse and rat. Biochem Biophys Res Commun 146:692–698

Spence KD (1991) Structure and physiology of the peritrophic membrane. In Binnington KC, Retnakaran A (eds) Physiology of Insect Epidermis. CSIRO Australia (in press)

Staal GB (1986) Anti juvenile hormone agents. Ann Rev Entomol 31:391–429

Stapleton A, Blankenship DT, Ackermann BL, Chen T, Gorder GW, Manley GD, Palfreyman MG, Coutant JE, Cardin AD (1990) Curtatoxins: neurotoxic insecticidal polypeptides isolated from the funnel-web spider *Hololena curta*. J Biol Chem 265:2054–2059

Starratt AN, Steele RW (1984) In vivo inactivation of the insect neuropeptide proctolin in *Periplaneta americana*. Insect Biochem 14:97–102

Steele RW, Starratt AN (1985) In vitro inactivation of the insect neuropeptide proctolin in haemolymph. Insect Biochem 15:511–519

Stoffolano JG Jr (1973) Host specificity of entomophilic nematodes—a review. Exper Parasitol 33:263–284

Stoltz DB, Krell P, Summers MD, Vinson SB (1984) Polydnaviridae—A proposed family of insect viruses with segmented, double stranded, circular DNA genomes. Intervirology 21:1–4

Stoltz DB, Vinson SB (1979) Viruses and parasitism in insects. Adv Virus Res 24:125–171

Stone TB, Sims SR, Marrone PG (1989) Selection of tobacco budworm for resistance to genetically engineered *Pseudomanas fluorescens* containing the delta-endotoxin of *Bacillus thuringiensis* subs. *kurstaki*. J Invert Pathol 53: 228–234

Summers MD, Engler R, Falcon LA, Vail PV (eds) (1975) Baculoviruses for Insect Pest Control: Safety considerations. American Society for Microbiology, Washington DC, pp 179–184

Tamashiro M (1971) A biological study of the venoms of two species of Bracon. Hawaii Agric Exper Stn Tech Bull 70:1–52

Tanada Y (1985) A synopsis of studies on the synergistic property of an insect baculovirus: A tribute to Edward A Steinhaus. J Invert Pathol 45:125–138

Thim L, Hansen MT, Norris K, Hoegh I, Boel E, Forstrom J, Ammerer G, Fils NP (1986) Secretion and processing of insulin precursors in yeast. Proc Natl Acad Sci USA 83:6766–6770

Thomas G, Herbert E, Hruby DE (1986) Expression and cell-type specific processing of human proenkaphalin with a vaccinia recombinant. Science 232:1641–1643

Thomas G, Thorne BA, Hruby DE (1988a) Gene transfer techniques to study neuropeptide processing. Ann Rev Physiol 50:323–332

Thomas G, Thorne BA, Thomas L, Allen RG, Hruby DE, Fuller R, Thorner J (1988b) Yeast KIX2 endopeptidase correctly cleaves a neuroendocrine prohormone in mammalian cells. Science 241:226–230

Thomas WE, Ellar DJ (1983) *Bacillus thuringiensis* var. *israelensis* crystal δ-endotoxin: effects on insect and mammalian cells in vitro and in vivo. J Cell Sci 60:181–197

Tobe SS, Stay B (1985) Structure and regulation of the corpus allatum. Adv Insect Physiol 18:305–432

Tramontano A, Janda KD, Lerner RA (1986a) Catalytic antibodies. Science 234:1566–1570

Tramontano A, Janda KD, Lerner RA (1986b) Chemical reactivity at an antibody binding site elicited by mechanistic design of a synthetic antigen. Proc Natl Acad Sci USA 83:6736–6740

Turnbull AJ, Howells IF (1982) Effects of several larvicidal compounds on chitin biosynthesis by isolated larval integuments of the sheep blowfly *Lucilia cuprina*. Aust J Biol Sci 35:491–503

Turner AJ, Matsas R, Kenny AJ (1985) Are there neuropeptide-specific peptidases? J Biochem Pharmacol 34:1347–1356

Turner MJ, Schaeffer JM (1989) Mode of Action of Ivermectin. In Campbell WC (ed) Ivermectin and Abamectin. Springer-Verlag, New York, pp 73–88

Uchima K, Harvey JP, Omi EM, Tanada Y (1988) Binding sites on the mid-gut cell membrane for the synergistic factor of a granulosis virus of the armyworm (*Pseudaletia-unipuncta*). Insect Biochem 18:645–650

Usherwood PNR, Duce IR (1985) Antagonism of glutamate receptor channel complexes by spider venom polypeptides. Neurotoxicology 6:239–250

Vaeck M, Reynaerts A, Hofte H, Jansens S, De Beuckeleer M, Dean C, Zabeau M, Van Montagu M, Leemans J (1987) Transgenic plants protected from insect attack. Nature 328:33–37

Van Brunt J (1987) Pheromones and neuropeptides for biorational insect control. Bio/Technology 5:31–36

Van Brunt J (1989) Novel drug delivery systems. Bio/Technology 7:127–130

Van Mellaert H, Theunis S, De Loof A (1985) Juvenile hormone binding proteins in *Sarcophaga bullata* haemolymph and vitellogenic ovaries. Insect Biochem 15:655–661

Van Rie J, McGaughey WH, Johnson DE, Barnett BD, Van Mellaert H (1990) Mechanism of insect resistance to the microbial insecticide *Bacilus thuringiensis*. Science 247:72–74

Veenstra JA (1988) Immunocytochemical demonstration of vertebrate peptides in invertebrates: the homology concept. Neuropeptides 12:49–54

Vey A, Quoit JM, Pais M (1986) Toxemie d'origine fongique chez les invertebres wer ses consequences cytotoxiques: etude sur l'infection a *Metarhizium anisopliae* (Hyphomycete, Moniliales) chez les Lepidopteres et les Coleopteres. C Soc Biol 180:105–122

Vinson SB, Edson KM, Stoltz DB (1979) Effect of a virus associated with the reproductive system of the parasitoid wasp, *Campoletis sonorensis*, on host weight gain. J Invert Pathol 34:133–137

Visser BJ, Labryere WT, Spanjer W, Piek T (1983) Characterization of two paralyzing toxins (A-MTX and B-MTX) isolated from a homogenate of the wasp *Microbracon hebetor* (SAY). Comp Biochem Physiol B 75:523–538

Walther C, Reinecke M (1983) Block of synaptic vesicle exocytosis without block of Ca^{2+} influx. An ultrastructural analysis of the paralysing action of *Habrobracon* venom on locust motor nerve terminals. Neuroscience 9: 213–224

Ward SE, Gussow D, Griffiths AD, Jones PT, Winter G (1989) Binding activities of a repertoire of single immunoglobulin variable domains secreted from *Escherichia coli*. Nature 341:544–546

Wassermann NH, Penn AS, Freimuth PI, Treptow N, Wentzel S, Cleveland WL, Erlanger BF (1982) Anti-idiotypic route to anti-acetylcholine receptor antibodies and experimental myasthenia gravis. Proc Natl Acad Sci USA 79:4810–4814

Weiser J (1982) Persistence of fungal insecticides: influence of environmental factors and present and future applications. In Kurstak E (ed) Microbial and Viral Pesticides. Marcel Dekker, New York, pp 531–557

Wigglesworth VB (1943) The fate of haemoglobin in *Rhodnius prolixus* (Hemiptera) and other blood sucking arthropods. Proc Roy Soc Lond Ser B 131:313–329

Wisniewski JR, Wawrzenczyk C, Prestwich GD, Kochman M (1988) Juvenile hormone binding proteins from the epidermis of *Galleria mellonella*. Insect Biochem 18:29–36

Wraight SP, Butt TM, Galaini-Wraight S, Allee LL, Soper RS, Roberts DW (1990) Germination and infection processes of the Entomophthoralean fungus *Erynia radicans* on the potato leafhopper, *Empoasca fabae*. J Invert Pathol 56:157–174

Wu D, Chang FM (1985) Synergism in mosquitocidal activity of 26 and 65 kDa proteins from *Bacillus thuringiensis* subsp. *israelensis* crystal. FEBS Letters 190:232–236

Yoder OC, Weltring K, Turgeon BG, Garber RC, Van Etten HD (1986) Technology for molecular cloning of fungal virulence genes. In Bailey J (ed) Biology and Molecular Biology of Plant-Pathogen Interactions. Plenum, New York, pp 371–384

Zlotkin E (1983) Insect selective toxins derived from scorpion venoms: an approach to insect neuropharmacology. Insect Biochem 13:219–236

Zlotkin E (1985) Toxins derived from arthropod venoms specifically affecting insects. In Kerkut GA, Gilbert LI (eds) Comprehensive Insect Physiology, Biochemistry and Pharmacology Vol 10. Pergamon Press, Oxford, pp 499–546

Zlotkin E (1987) Pharmacology of survival, insect selective neurotoxins derived from scorpion venom. Endeavour, New Series 11:168–174

Chapter 3

Genetic Engineering of Baculoviruses for Insect Control

Justinus M. Vlak

1. Introduction

About one-third of the loss of agricultural products in the world is caused by pests and diseases. Pest control is therefore of prime importance to meet the needs for sufficient food for human consumption and raw material for industrial processing. The use of chemical pesticides has been a worldwide strategy for many years to combat insect pests. However, the environmental hazards and public concern associated with the use of such chemicals have fostered a concerted search for alternatives, including the use of microbial agents such as baculoviruses.

The use of viral insecticides, in particular baculoviruses, has several advantages over chemical insecticides. Baculoviruses are naturally occurring pathogens that are specific for a single or a few related insect species. They are environmentally safe, produce no toxic residues and are harmless to non-target animals, including beneficial insects and vertebrates. In addition, they can give economic short-term and long-term control and can be effective at low dosage. Also, major resistance against baculovirus infections has not been observed. Finally, the technology involved in their production is relatively simple and can be carried out at the farm level. The biology of baculoviruses and their application as viral insecticides have been the subject of several reviews (Kurstak 1982; Sherman and Maramorosch 1985; Granados and Federici 1986a, 1986b; Maramorosch 1987).

There are also a number of limitations associated with the use of baculoviruses as insecticides. Because they have a limited host-range, their commercial potential is small compared to broad spectrum insecticides. Production of viruses is also labor-intensive and not easily amenable to large-scale commercial automation. Although most baculoviruses are reasonably virulent, the sensitivity of their hosts to infection decreases with the age of the larvae. The most important limitation of baculoviruses is the fact that they act slowly (Figure 3.1); it may take several days before insects die, whereas chemical insecticides act quickly. As a consequence, insects continue to feed after infection and are able to cause considerable damage to the crop before death (Benz 1986).

Genetic engineering of these pathogens may alleviate some of the above limitations by enhancing their pathogenicity and speed of action. Over the past few years information has mounted on the genetics and molecular biology of baculoviruses (Kelly 1982; Doerfler and Bohm 1986; Vlak and Rohrmann 1985; Granados and Federici 1986a; Blissard and Rohrmann 1990 for reviews), boosted by the wide interest in the use of baculoviruses as expression vectors for foreign genes (Luckow and Summers 1988a; Miller 1988b; Kang 1988; Maeda 1989; Fraser 1989; Miller 1989; Vlak and Keus 1990; Possee et al. 1990). This chapter will review the current understanding of baculovirus molecular biology, and the prospects of improving their insecticidal properties by genetic engineering.

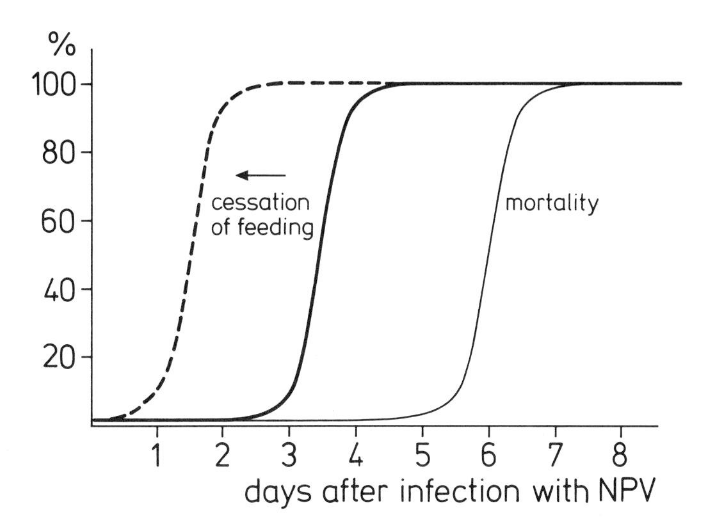

Figure 3.1. Schematic representation of the relation between larval mortality (thin line) and cessation of feeding (thick line) in percentage versus time after infection. The dotted line (cessation of feeding and/or mortality) represents the anticipated effect of a recombinant baculovirus with enhanced insecticidal activity.

Table 3.1. Occluded viruses of arthropods*

Order	Cytoplasmic Polyhedrosis Viruses	Entomopox Viruses	Baculoviruses
Class Insecta			
Lepidoptera	191	24	456
Diptera	37	9	27
Hymenoptera	6	3	30
Coleoptera	2	14	9
Neuroptera	2	—	—
Trichoptera	—	—	2
Orthoptera	—	9	
Class Crustacea			
Decapoda	—	—	2
Total	238	59	526

* Taken from data of Martignoni and Iwai (1986)

2. Baculovirus Biology

2.1. Systematics

Baculoviruses belong to the family *Baculoviridae* (Francki et al. 1991). Approximately 600 baculoviruses are known (Table 3.1) and the majority has been isolated from members of the orders Lepidoptera, Hymenoptera and Diptera (Martignoni and Iwai 1986). In contrast to other insect viruses, such as cytoplasmic polyhedrosis viruses or entomopoxviruses, baculoviruses are therefore restricted to arthropods and have no relatives in other kingdoms.

2.2. Structure

Baculovirus particles are characterized by the presence of rod-shaped (baculum = rod) nucleocapsids that are enveloped either singly or in bundles by a membrane (Figure 3.2). The virus particles themselves are usually embedded singly (granulosis viruses, or GVs) or in large numbers (nuclear polyhedrosis viruses, or NPVs) into large protein capsules, called occlusion bodies (OBs), granula or polyhedra. The OBs are surrounded by an envelope, consisting of protein and carbohydrate. OBs vary in size between 1–15 μm (NPVs) and 0.1–0.5 μm (GVs) in diameter and can be observed by light microscopy. A few baculoviruses are found exclusively in a nonoccluded form.

The major constituent of these OBs is a single protein (polyhedrin; granulin) with a subunit molecular weight of approximately 30,000 Dalton

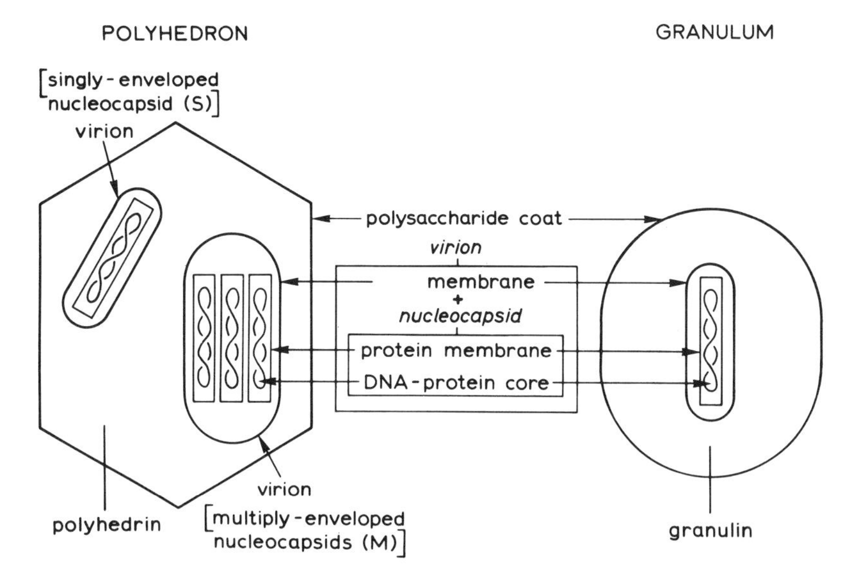

Figure 3.2. Schematic representation of the structure of baculovirus occlusion bodies showing a polyhedron from NPVs and a granulum from GVs.

(30 kDa). The protein sequence of granulins and polyhedrins is highly conserved among baculoviruses (Vlak and Rohrmann 1985; Rohrmann 1986). The OBs protect the virus against decay in the environment and probably also facilitate its retention on plant surfaces (Evans and Harrap 1982; Entwistle and Evans 1986).

Baculoviruses contain a double-stranded, circular DNA molecule varying in size between 80 and 160 kilobase pairs (kbp) (Burgess 1977; Smith and Summers 1978). The viral DNA contains sufficient information to code for at least 60 average-sized proteins. Physical maps of various baculovirus DNAs have been established, the most detailed one being of the 130 kbp type baculovirus, *Autographa californica* nuclear polyhedrosis virus (AcNPV; Vlak and Odink 1979; Vlak and Smith 1982). About 30 genes including the polyhedrin gene have been mapped on the genome (Figure 3.3). At present (July 1992) the complete AcNPV genome has been sequenced (RD Possee, pers. commun).

2.3. Replication

In nature OBs enter the larvae via ingestion of contaminated food. On entering the midgut the OBs are dissolved, as a consequence of the alkaline conditions and the presence of protease activities, releasing

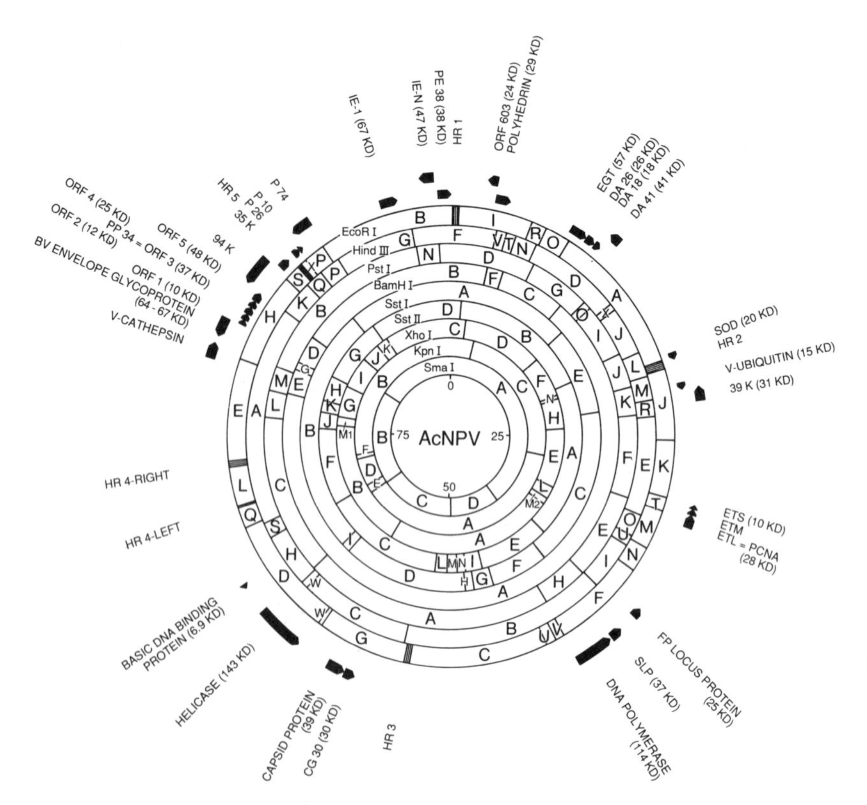

Figure 3.3. Physical map of AcNPV for nine restriction enzymes and location of various genes and transcripts. Numbers (in kDa) refer to appropriate molecular weights. Arrows indicate the approximate size and 5′-3′ orientation of the coding regions. The homologous repeat-regions (hr1-hr5) are indicated.

the occluded virions (Figure 3.4). The virions then infect midgut cells through a process of fusion of the viral envelope with the cell's microvilli. Replication occurs in the nucleus (Granados 1980; Granados and Lawler 1981; Granados and Williams 1986). In the case of the GV from *Trichoplusia ni*, at least, this process may be enhanced by a viral enhancement factor (VEF) present in the OBs (Granados and Corsaro 1990). Progeny virus particles (Extracellular Virus Particles, or ECVs) bud through the midgut cell membrane and are released into the insect hemolymph. These progeny ECVs are transported via the hemolymph to cells of other organs such as the fat body, which they enter by adsorptive endocytosis and where a second round of ECV replication then occurs (Volkman and Goldsmith 1985). In the later stages of the infection virus particles are not transported from the cell but retained in the nucleus and occluded into newly synthesized occlusion bodies. At the end of the infection, the cells lyse and release the OBs (Figure 3.4) from the insect corpses.

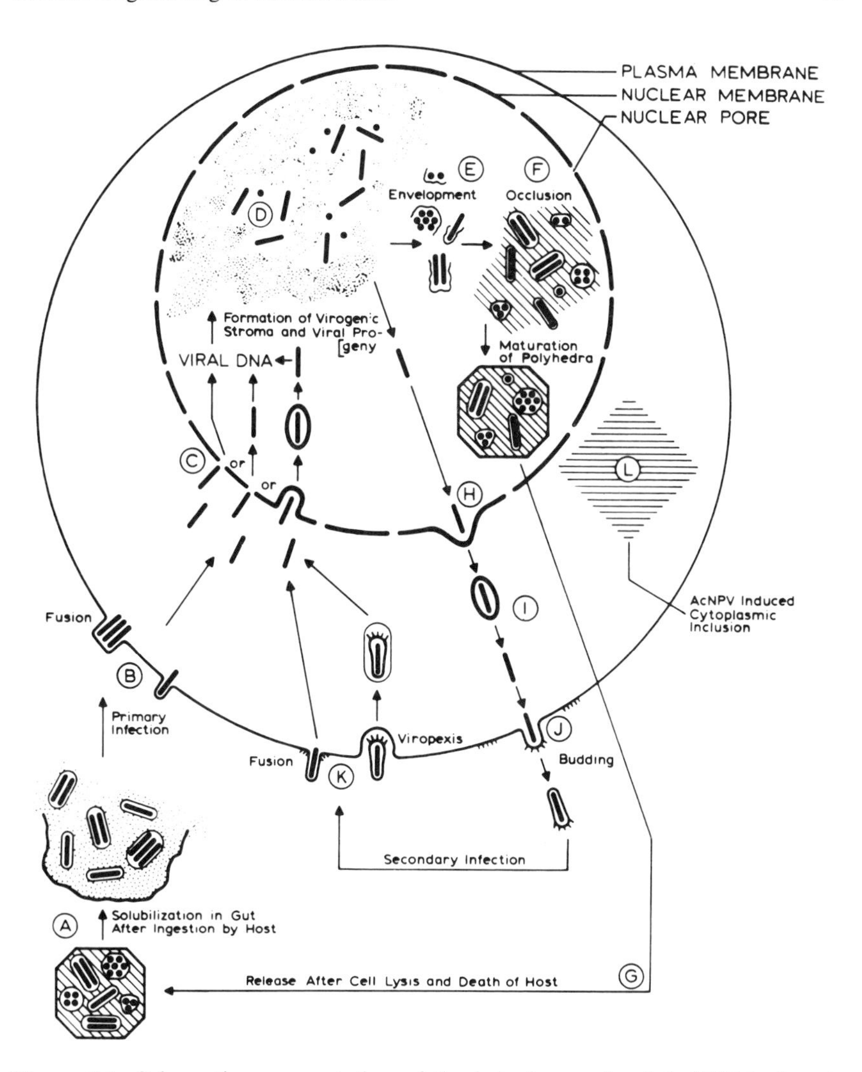

Figure 3.4. Schematic representation of the infection cycle of AcNPV in insects and insect cells. (**A**). Ingestion of polyhedra and solubilization by digestive juices in the insect gut. (**B**). Fusion of the viral envelope of the released virus with the plasma membrane of a midgut cell. (**C**). Entry of the nucleocapsid into the nucleus. (**D**). Formation of virogenic stroma where virus replication and assembly of progeny nucleocapsids occur. (**H**). Departure of nucleocapsids from the nucleus and formation of non-occluded virus particles (NOV) by the acquisition by budding of an envelope from the nuclear (**I**) or cellular (**J**) membrane. (**K**). Systemic infection of cells from other tissues by adsorptive endocytosis. (**E**). Envelopment of single or multiple nucleocapsids in membranes synthesized *de novo* in the nucleus. (**F**). Occlusion of singly and multiply enveloped virus particles into polyhedra. (**L**). Formation of cytoplasmic and nuclear inclusions (fibrillar structures) with unknown function. (**G**). Release of polyhedra from dead insect larvae.

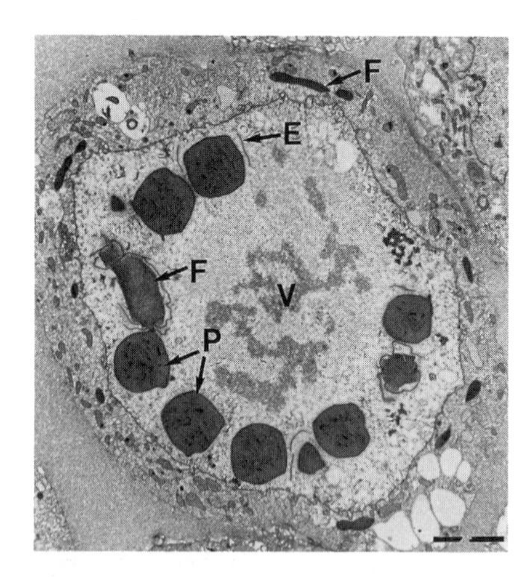

Figure 3.5. Thin section of *S. frugiperda* cells infected with wild-type AcNPV viewed in the electron microscope. P = polyhedron; V = virogenic stroma; F = fibrillar structure; E = electron-dense "spacers".

The replication of baculoviruses thus occurs in two distinct phases. In the first phase ECVs are produced, whereas in the second phase virus particles are occluded into OBs. This biphasic replication process can be mimicked in cultured insect cells, and this has allowed detailed cytopathological, molecular biological and genetic study of baculovirus replication. The replication of AcNPV in *Spodoptera frugiperda* (Sf) cells is the best studied example (Faulkner 1981; Granados and Lawler 1981; Kelly 1982). Figure 3.5 shows how the nuclei of such cells are enlarged and contain many polyhedra late in the infection cycle. Large fibrillar structures are present in the nucleus and cytoplasm of such cells.

2.4. Pathology

Depending on the virus isolate, the initial signs of infection appear about three days after virus ingestion. In the early stages of the disease, hemocytes show enlarged nuclei which are almost completely occupied by OBs, giving a whitish discoloration to the hemolymph. Usually, baculoviruses have a broad tissue tropism, including fat body, tracheal matrix and muscles. Differences in the rate of larval development are also apparent at this time, with infected larvae developing more slowly. Infected larvae also exhibit the classical signs of baculovirus infection, including loss of appetite and lethargy. In more advanced stages of disease, six to seven days after infection, the larval body usually develops a creamy yellow color due to accumulation of OBs in the infected tissues. Subsequently, the larvae become flaccid and hang by their prolegs from leaves or branches in a characteristic inverted position or lie detached from the plants on the soil surface. Larvae die three to eight days after the initial signs of infection, after which the cuticle ruptures, liberating about 10^9 OBs per larva. Even under optimal laboratory conditions, it may take three to five days before the insect dies (Figure 3.1). This slow speed of action and the late cessation of feeding are the major limitations in the use of these pathogens to control insect pests, particularly in crops with low damage thresholds (Benz 1986; Payne 1988).

Information about insects' defense mechanisms against virus infection is limited. The exoskeleton protects the insect against many pathogens but the entry of baculoviruses via the digestive tract bypasses this defense. The defense system of insects does not include an active immune response in the form of antibodies, as in vertebrates, and although antiviral proteins are observed, their role is unclear. The major form of resistance is of a genetic nature (Briese 1986), but larvae also become progressively less susceptible to virus infection with age. Hence, about 10^4 more OBs are required to achieve 50% mortality (LD_{50}) among fifth instar larvae than first instars (Table 3.2). Increased virulence must therefore be an important aim in the genetic improvement of baculoviruses.

3. Baculovirus Genetics

3.1. Gene Regulation

It was noted above that baculoviruses have a unique, temporally-regulated, biphasic replication cycle. In the first phase ECVs are pro-

Table 3.2. Mortality (LD_{50}) of *Spodoptera exigua* NPV, (numbers of OBs) against the five larval instars of beet armyworm, *Spodoptera exigua**†

Instar	LD_{50}	95% Fiducial Limits	
		Under	Upper
L1	4	3	6
L2	3	2	4
L3	39	21	94
L4	55	35	180
LS	11,637	6,210	32,544

* Data taken from Smits and Vlak (1988)
† Assay was performed using the peroral assay of Hughes and Wood (1981)

duced, and OBs are generated in the second (Figure 3.4). Regulation of this and other aspects of gene expression occurs at the transcriptional level (Friesen and Miller 1986; Guarino and Summers 1986; 1987). Upon infection of cells, viral genes of the **immediate early** class are transcribed by host factors, including RNA polymerases. Products of these genes, and possibly some host factors, turn on (*trans*-activate) an array of **delayed early** genes, including virus-encoded RNA and DNA polymerases, and turn off some host functions. Transcription of these delayed early genes obviously starts before the onset of DNA replication. Both immediate and delayed early gene expression may continue throughout the later stages of infection. **Late** genes are switched on concurrently with the onset of DNA replication, and their expression is promoted by **delayed early** and **immediate early** genes. The late genes code for structural proteins of the virus particles. Genes of each of the four classes are not clustered, but randomly distributed through the genome (Figure 3.3). The regulated expression of baculovirus genes is reflected by the synthesis of virus-specific proteins at different times after infection (Carstens et al. 1979; Dobos and Cochran 1980; Wood 1980; Kelly and Lescott 1981).

Some early genes are differentially regulated in time, and the promoter regions of some contain both immediate early and delayed early regulatory elements (Guarino and Summers 1987). The immediate early gene IE1 contains a promoter element that is recognized by host factors, and its product then *trans*-activates delayed early genes (Guarino and Summers 1986). The recognition of host factors by the promoters of immediate early genes and the availability of the appropriate host factors may be important determinants of the host specificity of baculoviruses. Baculovirus genomes also contain regions of repeat sequences (hr; Figure 3.3) (Cochran and Faulkner 1983) which can act in *cis* as "enhancers" for transcription of early genes (Guarino and Summers 1986; Guarino et al. 1986; Nissen and Friesen 1989). This type of enhancement presumably fine-tunes the transcriptional regulation facilitated by a virus encoded RNA polymerase or subunit of it (Fuchs et al. 1983). One or more hr regions may also serve as origins of DNA replication (Kool et al. 1992).

The hierarchical nature of baculovirus gene regulation is similar to that of other large DNA viruses, like adeno- or herpesviruses. Baculoviruses, however, have an additional, unique class of **very late** genes, which code for proteins that are involved in the late stages of virus infection and OB morphogenesis. Two of these very late genes, coding for polyhedrin and a protein of 10 kDa called p10, are hyperexpressed late after infection. Mutations or deletions in these very late genes do not affect ECV production (Smith et al. 1983b; Pennock et al. 1984; Gonnet and Devauchelle 1987; Vlak et al. 1988; Williams et al. 1989; Zuidema et al. 1989), and this forms the basis for the use of baculoviruses as vectors for the expression of foreign genes.

3.2. Gene Structure

The coding regions of baculovirus genes are contiguous and not spliced. Immediate and delayed early genes have promoter elements similar to those of eukaryotic organisms. A CAGT sequence, where initiation of transcription occurs, and an upstream TATA box are highly conserved. In AcNPV one of the IE1 mRNAs displays splicing in the leader sequences, but other baculovirus mRNA leaders are not known to be spliced. Transcripts of late are very late genes are initiated from a consensus promoter element, TAAG, which is conserved in all baculoviruses (Vlak and Rohrmann 1985; Rohrmann 1986; Rankin et al. 1988; Blissard and Rohrmann 1990). The very late genes polyhedrin and p10 are present in single copies and the TAAG motif is part of a conserved 12 nucleotide-long consensus sequence, AA*TAAG*TATTTT, which includes the ATAAG motif. In AcNPV this promoter element is located for polyhedrin around position−60 (Hooft van Iddekinge et al. 1983) and for p10 around position−70 (Kuzio et al. 1984) with respect to the translational start site (ATG) of these genes. The integrity of the leader sequence upstream from the translational start is absolutely essential for the high levels of expression from the polyhedrin (Matsuura et al. 1987; Rankin et al. 1988) and p10 (Weyer and Possee 1988; 1989) promoters.

Although there is only limited genetic relatedness among baculoviruses in terms of overall nucleotide sequence homology (Smith and Summers 1982), the order of genes in the genome appears to be conserved (Leisy et al. 1984; Blissard and Rohrmann 1990). While only limited comparative data are available as yet, information so far indicates that genes with similar functions and significant degrees of nucleotide and amino acid sequence similarity have similar physical locations in the genomes of different baculoviruses (Rohrmann 1986; Blissard and Rohrmann 1990).

3.3. Gene Function

The functions of a few baculovirus genes are known and their locations are shown on the physical map of AcNPV in Figure 3.3. Among these genes are three immediate early genes, IEl and IE-N (Guarino and Summers 1986) and PE-38 (Krappa and Knebel-Mörsdorf 1991), genes for the major nucleocapsid protein (Thiem and Miller 1989), the major ECV envelope glycoprotein (gp64; Whitford et al. 1989), DNA polymerase (Tomalski et al. 1988), a basic histone-like DNA-binding protein (Wilson et al. 1987), an ecdysteroid UDP-glucosyltransferase (O'Reilly and Miller 1989), the OB envelope gene (Zuidema et al. 1989), the polyhedrin (Vlak et al. 1981) and for p10 (Smith et al. 1982). Many more open reading frames (ORFs) have been detected and mapped on the genome, but their functions are not yet known. A similar but less detailed map has been constructed for *Orgyia pseudotsugata* NPV (Blissard and Rohrmann 1990).

Polyhedrin is the most prominent gene product and made very late after infection. In OBs it is arranged in a paracrystalline structure, which protects occluded virus particles against decay in the environment. Deletion of this gene reduces virulence, as the occluded virions are unprotected. P10 protein is located in fibrillar structures in the nucleus and cytoplasm of infected cells (Figure 3.5) (Van der Wilk et al. 1987; Croizier et al. 1987). These structures form a network in the cell, but the function of the network has not been fully determined. Deletion of the p10 gene does not impair OB formation, but the OBs are more sensitive to physical stress (Gonnet and Devauchelle 1987; Vlak et al. 1988) and are not easily lysed from the cell (Williams et al. 1989).

A few gene products have been implicated in the insecticidal activity of baculoviruses or in their infectivity or pathogenicity. The p74 gene, located downstream from the p10 gene (Figure 3.3), is involved in infectivity. Deletion of this gene abolishes infection of insects, whereas overproduction of p74 enhances infectivity fivefold (Carstens and Lu 1990). A protein termed the viral enhancement factor (Granados and Corsaro, 1990) found in a GV facilitates its infection of insects. The egt gene of AcNPV encodes an ecdysteroid UDP-glucose transferase which catalyzes the transfer of glucose to ecdysteroids (O'Reilly and Miller 1990). Ecdysteroids are essential for the induction of larval and pupal molts. Viral egt thus inactivates ecdysteroids by conjugating them with glucose. As a result the insect does not molt into the next larval instar and allows the virus to complete its replication in the host.

4. Genetic Manipulation of Baculoviruses

Four particular features of baculoviruses suit them for use as vectors for the high level expression of foreign genes in eukaryotic (insect) cells. First, the two very late genes, polyhedrin and p10, are hyperexpressed. The polyhedrin and p10 proteins contribute over 60% of the total cellular protein at the end of infection, both in dead insects and insect cells (Figures 3.5 and 3.6). Second, both these proteins are involved in the morphogenesis of OBs and are not essential for the replication of ECVs. The coding sequences for these proteins are therefore not essential and can be replaced by foreign genes. Their promoters are available to drive the expression of these foreign genes to high levels. Third, the size of the circular viral genome can be expanded to accommodate up to 25 kbp of additional DNA, while still being properly assembled in rod-shaped virus particles (Miller et al. 1986). Finally, post-translational modifications, such as glycosylation, phosphorylation and amidation, appear to be very

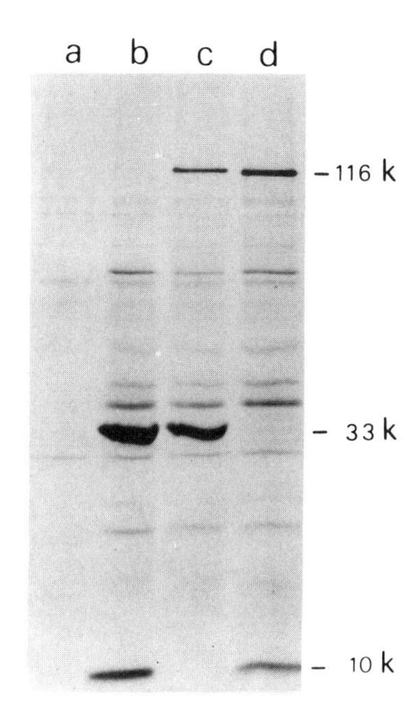

Figure 3.6. Polyacrylamide gel electrophoresis of uninfected *S. frugiperda* cells (a), cells infected with wild-type AcNPV (b) or with AcNPV recombinants expressing β-galactosidase under the control of the p10 (c) or polyhedrin (d) promoter. The molecular weight of β-galactosidase (116 kDa), polyhedrin (33 kDa) and p10 (10 kDa) are indicated.

similar, if not identical, in insect cells as in other, higher eukaryotic systems.

Given the above, the most commonly used strategies for engineering baculoviruses have exploited the polyhedrin or p10 promoters. In its simple form this is achieved by the allelic replacement of, say, the polyhedrin gene by foreign genes (Summers and Smith 1987; Figure 3.7). Since the baculovirus genome is too large (ca. 130 kbp) to be manipulated directly, special transfer vectors are designed. These vectors contain, in addition to a bacterial plasmid element, a baculovirus DNA segment with a cloning site downstream from the polyhedrin promoter, and 5′ and 3′ flanking sequences to facilitate recombination (Figure 3.7). The manipulations, which can be carried out with such a plasmid, range from introducing deletions or mutations of endogenous genes to the insertion of foreign genes.

For optimal expression of foreign genes in the recombinant, it is essential to maintain the complete polyhedrin or p10 promoter sequence (Possee 1986; Matsuura et al. 1987; Luckow and Summers 1988a; 1988b). One or more unique cloning sites are engineered behind the promoter as sites for the insertion of foreign genes. Recombinant transfer vectors usually contain the foreign gene's own ATG start codon and a short non-translated leader sequence. The foreign genes are transferred to the baculovirus genome by homologous recombination between the transfer

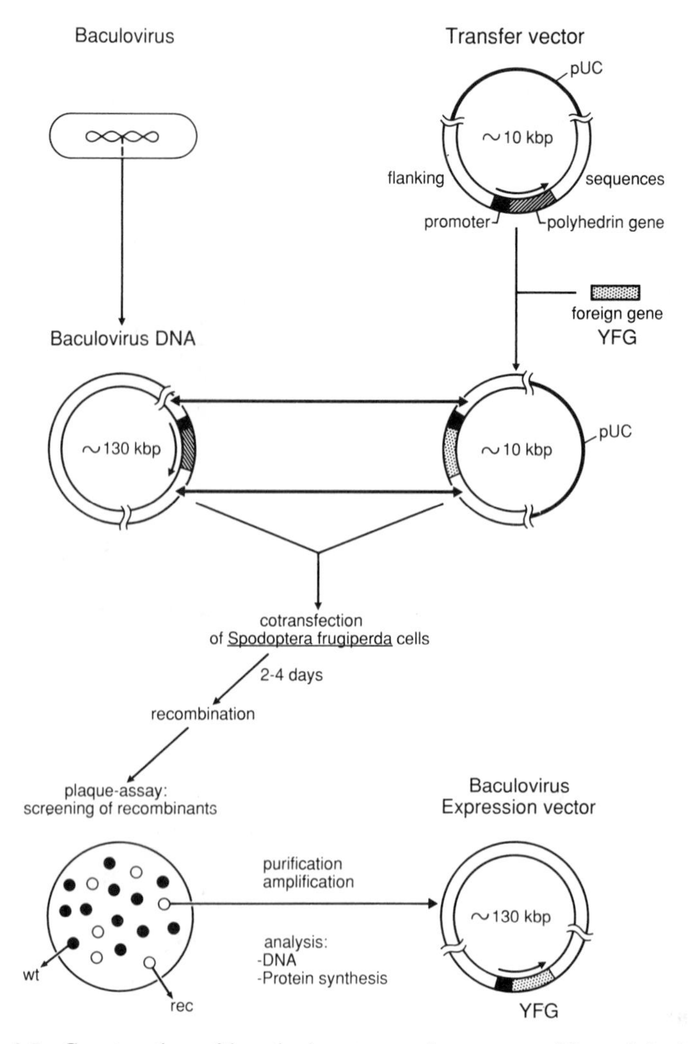

Figure 3.7. Construction of baculovirus expression vectors. The polyhedrin gene (dashed area) is replaced by "Your Favourite Gene" (stippled area). wt = wild-type virus (OB^+); rec = recombinant virus (OB^-); kbp = kilobase pairs.

vector and wild-type baculovirus DNA during replication in insect cells (Figure 3.7). The manipulated segment is targeted to the correct location on the viral genome by the homologous 5′ and 3′ flanking sequences.

In the case of the allelic replacement of the polyhedrin gene, recombinant viruses are usually recognized by the absence of OBs (using light microscopy; Smith et al. 1983b), or sometimes by the presence of the foreign gene (using Southern blot hybridization analysis), or

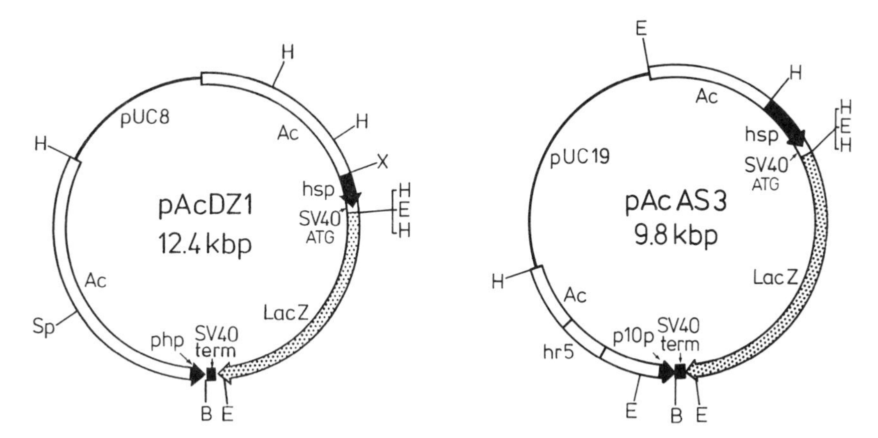

Figure 3.8. Structure of transfer vectors pAcDZl (polyhedrin vector) and pAcAS3 (p10 vector). E = *Eco*RI; H = *Hin*dIII; B = *Bam*HI; Sp = *Sph*I; hsp = heat shock promoter; php = polyhedrin promoter; p10p = p10 promoter; term = terminator; LacZ = β-galactosidase gene. The arrows indicate the direction of transcription.

recombinant protein (using immunofluorescence or immunoblot analysis). Alternative strategies involve the use of marker genes in the transfer vector, such as an additional polyhedrin gene (Emery and Bishop 1987) or the bacterial LacZ gene placed immediately downstream to the polyhedrin locus (Figure 3.8) (Vialard et al. 1990; Zuidema et al. 1990). These markers facilitate positive selection schemes for recombinants.

The insertion of foreign genes into the p10 locus is more complicated because phenotypic markers for the absence of p10 are lacking. A new generation of transfer vectors is available, which facilitates screening for recombinants using β-galactosidase produced by a LacZ gene as a marker (Figure 3.8) (Vlak et al. 1990).

When the recombination is successful, the polyhedrin or p10 promoters drive the expression of the foreign gene to levels equivalent to those of polyhedrin or p10 in wild-type virus, as exemplified by the expression of β-galactosidase (Figure 3.6). Hundreds of proteins from viral, bacterial, animal and plant origin have now been produced via such recombinant baculovirus expression vectors. The first recombinant vaccine against HIV was prepared from baculovirus-expressed gpl60 subunits (MicroGeneSys, New Haven, CT, USA). The potential of baculovirus expression vectors for the production of proteins for therapeutic, diagnostic and research purposes has been extensively reviewed elsewhere (Luckow and Summers 1988a; Miller 1988b; Kang 1988; Maeda 1989; Fraser 1989; Miller 1989; Vlak and Keus 1990; Possee et al. 1990).

5. Strategies for Engineering Baculovirus Insecticides

The three major aims in the engineering of baculovirus insecticides are: to increase the speed of action, in particular to induce early cessation of feeding; to enhance virulence, in particular to control older larvae at lower dosages; and to extend host specificity. A number of strategies to attain these goals will be discussed below. Some of these strategies result from detailed knowledge of baculovirus replication in insects and insect cells. Other strategies emerged when genes coding for insect toxins, hormones and enzymes became available. Introduction of these genes in particular may improve the insecticidal activity of baculoviruses. In many instances increased speed of action and enhanced virulence, albeit not an extended host range, can be achieved with a single strategy.

5.1. Alteration of Host Specificity

Studies on the basis for the host specificity of baculoviruses are scarce. It is not clear whether the barrier(s) in nonpermissive hosts are at the level of adsorption, penetration, transcription, DNA replication or virus morphogenesis. To extend the host range, either a targeted or a shotgun approach could be taken. The targeted approach, where specific genes are transferred to the recipient virus, requires a detailed understanding of the genetic and molecular basis of host specificity of baculoviruses. The shotgun approach involves testing of viruses in non-susceptible hosts and relies on stochastic processes (like mutation and non-homologous recombination) and selection to establish genotypes newly adapted for growth in that host.

An indication of the potential of the targeted approach comes from the observation that the expression of the immediate early genes IE1 and IE-N of AcNPV is under host control (Guarino and Summers 1986), whereas the expression of delayed early genes is activated by the immediate early genes. Thus, the host controls a key switch at the outset of the virus' transcription cascade but viral factors are necessary thereafter. This opens up the possibility that transfer of immediate early gene(s) between viruses could transfer at least some of their host-specificity properties.

As far as the shotgun approach is concerned, efforts to extend host range by adapting viruses to alternate insect hosts have not been very encouraging so far. Usually an endogenous virus is activated (Jurkovicova 1979). Recently Kondo and Maeda (1991) have been able to adapt *B. mori* NPV and *Spodoptera litura* NPV for growth in *S. litura* cells and *B. mori* cells, respectively. These infections are normally abortive. Recombinant viruses were obtained by co-infection of one cell type with the two viruses, followed by selection of progeny for growth in the other cell line. In this way a 3 kb fragment was identified in recombinant *B. mori* NPV that was responsible for the infectivity of these recombinants in *S. litura* cells.

5.2. Enhancement of Pathogenicity

One strategy for the improvement of baculovirus insecticides is the introduction of genes coding for proteinaceous insect toxins, hormones or metabolic enzymes (Table 3.3). Overexpression of these proteins may interfere quickly and effectively with the insect's metabolism and metamorphosis, which would clearly improve the insecticidal properties of the virus.

5.2.1. Toxin Genes

One candidate for the enhancement of insecticidal activity is the delta endotoxin of the bacterium *Bacillus thuringiensis* (*B.t.*; see Binnington and Baule this volume). *B.t.* preparations are widely used to control insects. The insecticidal activity is due to proteins, produced as crystalline inclusions during sporulation. When these inclusions are ingested by insects, they are dissolved and degraded by the action of gut proteases into toxic proteins (the delta endotoxins). These toxins bind to receptors on epithelial cells lining the midgut (Hofmann et al. 1988), possibly generating small holes in the cell membrane. This quickly destroys the regulation of ion transport and results in paralysis of the gut and mouth, followed by the cessation of feeding. An array of *B.t.* toxins with varying specificities is known (Whiteley and Schnepf 1986; Krieg 1986). The

Table 3.3. Candidate genes for introduction into baculoviruses and potential to enhance pathogenicity and insecticidal activity

Class	Protein	Potential
Toxins	Bt-toxin	+/−
	scorpion toxin	+++
	mite toxin	+++
	trypsin inhibitor	?
Hormones	eclosion hormone (EH)	+/−
	diuretic hormone (DH)	+
	prothoracicotropic hormone (PTTH)	+
	allatoropin	−
	allatostain	+
	proctolin	+
Enzymes	juvenile hormone esterase (JHE)	+

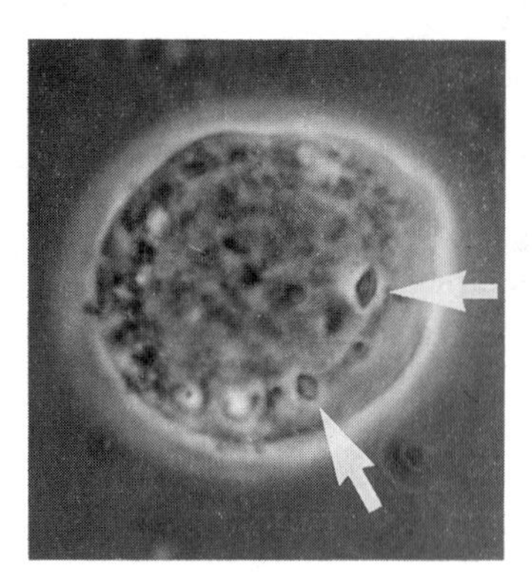

Figure 3.9. *S. frugiperda* cells infected with an AcNPV recombinant lacking the polyhedrin gene and expressing the *B.t.* protoxin under the control of the polyhedrin promoter. The arrows indicate the position of *B.t.* crystals.

toxins are encoded on extrachromosomal elements (plasmids) of the bacterium.

Successful use of the *B.t.* toxin gene engineered into baculoviruses depends on many variables, including the level, site and time of synthesis of the toxin, as well as the mode and site of its delivery, but most importantly on whether the toxin produced is biologically active. As an example, AcNPV recombinants have been constructed expressing the *B.t.*II protoxin gene of the strain *B.t. aizawai* 7.21 under the control of either the p10 or polyhedrin promoter (Martens et al. 1990), using the particular transfer vectors shown in Figure 3.8 (Zuidema et al. 1990; Vlak et al. 1990). These recombinants lacked the p10 and polyhedrin gene respectively. The recombinants expressed the protoxin in cultured *S. frugiperda* cells and formed bipyramidal-like crystals in the cytoplasm (Figure 3.9). These crystals are similar in structure to those observed after infection by the *B.t.* bacterium. The baculovirus-derived protoxin is lethal to susceptible insects (Table 3.4), suggesting that the protoxin is correctly processed by the insect. However, preliminary experiments suggested that the insecticidal activity of these AcNPV/*B.t.* recombinants (measured as LT_{50}) was not enhanced. Crystals of the protoxin were mainly produced in the fat body and not in the midgut cells or hemolymph, where the toxic effects of *B.t.* are normally produced.

The possibility of improving baculoviruses by introducing an insect-specific neurotoxin gene from the scorpion *Buthus eupeus* has also been investigated. The scorpion toxin gene was made by reverse genetics, i.e., by chemical synthesis of a gene that would encode the previously determined sequence of the protein. However, when this gene was expressed by baculovirus recombinants, it did not result in paralytic activity against target insects (Carbonell et al. 1988).

More recently, however, two other neurotoxin-encoding genes have been engineered into a baculovirus and increased its insecticidal activity. In the first case Stewart et al. (1991) and McCutchen et al. (1991) inserted

Table 3.4. A. Toxicity f BtII protoxin, produced by *E. coli* (as a control) and *Spodoptera frugiperda cells infected with various AcNPV strains, against second instar larvae of Pieris brassicae**

Source	Bt Toxin	Viral Genotype	Mortality
E. coli	—	na	0
E. coli 7.21A	+	na	100
Sf cells	—	na	3
Sf cells + AcNPV	—	$ph^+/p10^+$	0
Sf cells + AcNPV/DZ1	—	$ph^-/p10^+$	3
Sf cells + AcNPV/JM3	+	$ph^-/p10^+$	97
Sf cells + AcNPV/AS2	—	$ph^+/p10^-$	0
Sf cells + AcNPV/JM1	+	$ph^+/p10^-$	100

B. LC_{50} expressed as number of cells applied per cm^2 leaf disk and associated statistics

Source	LC_{50}	95% Fiducial Limits	
		Upper	Lower
E. coli + *Bt* 7.21A	2.7×10^4	1.7×10^4	4.6×10^4
Sf cells + AcNPV/JM3	3.4×10^2	1.9×10^2	5.9×10^2

* Mortality was scored two days after application of bacteria or (infected) insect cells on cabbage white leaf disks

† Genotypes are with respect to the presence or absence of polyhedrin (ph^+ or ph^-) and p10 ($p10^+$ or $p10^-$). na = not applicable

a cDNA copy from another scorpion, *Androctonus australis* toxin into AcNPV. The recombinant virus expressed the toxin protein, and its kill-time was 25% faster than wild type AcNPV. In the second case, the toxin gene from a predatory mite, *Pyemotes tritici*, was inserted into AcNPV and found to increase its speed of action by approximately 40% (Tomalski et al. 1991).

In both the latter cases the recombinant viruses had the toxin-encoding gene linked to a second copy of the polyhedrin promoter. The possibility therefore exists that rates of kill could be improved even further by having the expression of the toxin driven by an earlier promoter. This would, of course, also depend on the toxin being sufficiently active (in respect of its LD_{50}) to remain effective at the lower levels of expression associated with the earlier promoters. With respect to the latter point, Binnington and Baule (this volume) discuss other neurotoxins that might also be useful for enhancing the insecticidal potential of baculovirus insecticides.

5.2.2. Hormone Genes

There are three major categories of insect hormones: ecdysteroids, juvenile hormones and neurohormones. Neurohormones are mainly peptidergic in nature and are involved in the regulation of growth and development, reproduction and metabolism, and physiological homeostasis. They influence the secretion of ecdysteroids and juvenile hormones. Genes coding for neurohormones, either isolated directly from the insect or obtained by reverse genetics from the protein sequence, are therefore in principle attractive for engineering into baculoviruses. The number of known neurohormones is still very limited, although advances have recently been made in the characterization of neurohormones and their genes (Keeley and Hayes 1987; Menn and Borkovec 1989; Trowell and East this volume).

One neurohormone called eclosion hormone (EH) initiates ecdysis, the process leading to the shedding of old cuticle. The amino acid sequence of this hormone (60 aa) has been determined from *Manduca sexta* (Marti et al. 1987; Kataoka et al. 1987), and *Bombyx mori* (Kono et al. 1987), and the EH gene has been isolated from *M. sexta* and sequenced (Horodyski et al. 1989). Overproduction of EH by baculoviruses should stimulate the process of ecdysis and may cause premature death (Eldridge et al. 1991).

Prothoracicotropic hormones (PTTH) are involved in triggering the molting process. The PTTHs are 4 and 22 kDa peptides synthesized in the brain. They stimulate synthesis and release of ecdysteroids by the prothoracic gland, thereby initiating molting and metamorphosis. The genes for PTTH have recently been isolated from *B. mori* and sequenced (Adachi et al. 1989; Kawakami et al. 1989; Iwami et al. 1989). The engineering of baculovirus recombinants expressing these genes may promote premature molting and metamorphosis.

Other neurohormones that are important for the regulation of metamorphosis are allatostatins and allatotropins, which regulate the release of juvenile hormone (JH) from the *corpora allata* in the larval brain. The function of JH is explained in more detail in Section 5.2.3 below, but the salient feature here is that high levels of JH are required to maintain the larval stage. Allatotropins may have little value for use in the engineering of baculoviruses with enhanced pathogenicity because they increase JH titres. On the other hand allatostatins may be more useful because they reduce JH titres. Infection with a virus producing allatostatin may therefore induce precocious molts, which may well be lethal. The primary amino acid sequences of the allatostatins from *Manduca sexta* (Kataoka et al. 1989a) and *Diploptera punctata* (Woodhead et al. 1989) have been determined, and synthetic genes can now be designed by reverse genetics to study their effects when introduced into baculoviruses.

Other possible candidates for use in engineering baculovirus insecticides are diuretic hormone (DH) genes. Water balance and possibly blood pressure in insects are regulated by diuretic and antidiuretic hormones (Maddrell 1986). High levels of DH increase diuresis, resulting in cessation of feeding and excretion of water, followed by desiccation and early mortality. The diuretic hormone of *M. sexta* was recently isolated and sequenced (Kataoka et al. 1989b). A recombinant *B. mori* NPV containing a synthetic diuretic hormone gene has been used to express this *M. sexta* DH in *B. mori* larvae (Maeda 1990). A signal sequence was included to allow secretion DH into the hemolymph. Hemolymph volume of *B. mori* larvae infected with recombinants expressing DH was reduced by 25% as compared to insects infected with wild-type virus. The recombinant viruses killed infected larvae about 20% faster than wild-type virus, probably as a result of dehydration.

5.2.3. Enzyme Genes

Juvenile hormone plays a vital role in the control of insect morphogenesis and reproduction. A reduction in the titre of JH early in the last instar initiates metamorphosis and leads to cessation of feeding (De Kort and Granger 1981). This reduction in JH level is caused by the enzyme juvenile hormone esterase (JHE), which hydrolyzes JH into biologically inactive JH-acid (Hammock 1985). The production of high levels of JHE in early instars results in a decline of the JH titre, the cessation of feeding and in a molt. Recombinant baculoviruses expressing the JHE gene under the control of the polyhedrin promoter may thus reduce feeding in infected insects. The JHE gene from *Heliothis virescens* has been isolated and sequenced (Hanzlik et al. 1989), and Hammock et al. (1990) have constructed an AcNPV recombinant expressing this gene. The recombinant induced elevated levels of JHE in infected *Trichoplusia ni*, but a reduction of larval growth and feeding was only observed in first instar larvae. Similar results were obtained when *S. exigua* were infected with AcNPV recombinants expressing JHE under the control of the p10 promoter (PW Roelvink pers comm). In this case polyhedra could be used to monitor the course of infection. No effects on the viability or feeding behavior of second to fifth instar larvae were observed, despite elevated levels of JHE in the hemolymph. Explanations might be that the induced JHE level is too low to trigger the molt, or that JH biosynthesis is increased to compensate for the increased JHE. Alternatively, the JHE from *H. virescens* may be inactive in the hemolymph of heterologous hosts. The latter difficulty could be overcome by isolating homologous JHE genes.

5.2.4. Inhibitor and Activator Genes

Yet another strategy to improve the effectiveness of baculoviruses could be the expression of a trypsin-inhibitor (TI) (Binnington and Baule this volume). This inhibitor may block the action of gut proteases required for the digestion of food. Another possibility would be the expression of proctolin, a myoactive pentapeptide which increases the frequency of hindgut contractions, causing exhaustion (Holman et al. 1990). Although it is a neurotransmitter, proctolin has a severe effect on the gut when administered with the food. The insecticidal activities of both TI and proctolin thus depend on their expression or secretion into the midgut area, just as does the *B.t.* toxin. Attempts to use these molecules to engineer more effective baculovirus insecticides may therefore face the same difficulties as *B.t.*-containing baculoviruses (Section 5.3.1 above).

A protein, called a viral enhancement factor from *T. ni* GV, has been isolated that appears to attack the chitinous peritrophic membrane of lepidopteran larvae, facilitating the penetration of the occluded virions into the midgut cell membrane (RR Granados pers comm). The gene for this factor has been isolated (Granados and Corsaro 1990) and expressed in AcNPV recombinants (RR Gradados pers comm), thereby enhancing their biological activity. It is not clear whether such enhancement factors occur in all baculoviruses, but this product clearly holds some promise, also when expressed in transgenic plants.

5.2.5. Antisense RNA

It has been found that expression of bacterial and eukaryotic genes can be blocked by the production of antisense RNA. A copy of the coding sequence of the target gene is engineered in the opposite orientation downstream from a strong promoter. The antisense RNA thus produced is complementary to the "sense" transcript, and the two strands will anneal to form double-stranded RNA that cannot be translated into protein. This strategy has been successfully used to alter flower coloring (Van der Krol et al. 1988) and to suppress virus replication in plants (Powell et al. 1989). Important insect genes coding for metabolic or regulatory enzymes or hormones could be engineered in antisense orientation behind a baculovirus promoter. The expression of this antisense gene in infected insects could then block essential host functions. A baculovirus model system has been designed to explore the potential of this strategy (Roelvink et al. 1992).

5.3. Choice of Promoters and Sites of Insertion

Foreign genes can be inserted into those areas of the baculovirus genome that are not essential for virus replication in insects or insect cells. Most genes involved in the late stages of virus morphogenesis, for example polyhedrin or the polyhedron envelope protein (Figure 3.2b), are essential for OB formation but not for virus replication. It is therefore convenient to insert insecticidal genes in these non-essentical areas, preferably under the control of (duplicated) viral promoters.

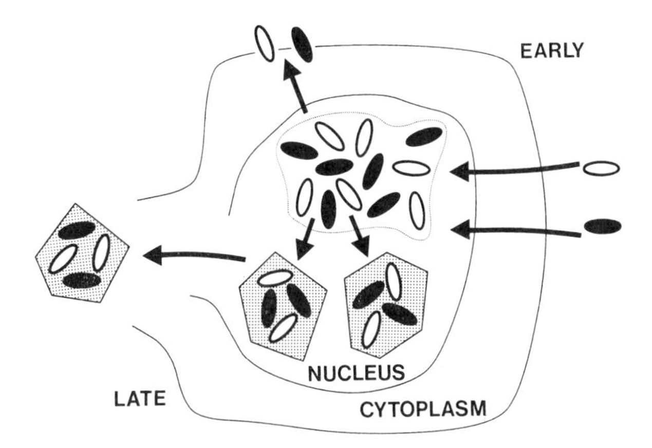

Figure 3.10. Co-occlusion of wild-type (open rods) and recombinant (closed rods) baculoviruses during co-infection of an insect cell (modified from HA Wood pers comm). Upper right-hand corner; early after infection: lower left-hand corner; late after infection.

High levels of the insecticidal protein are usually required to obtain recombinants with increased insecticidal activity. The polyhedrin and p10 promoters are therefore exploited. Insertion of insecticidal genes at the polyhedrin locus has, however, the disadvantage that the polyhedrin gene is lost and OBs are no longer produced, so virus transmission among insects is no longer possible. To alleviate this problem, recombinant viruses can be co-occluded with wild-type baculoviruses in an OB (Figure 3.10). Such an OB thus contains viruses with both wild-type and recombinant genotypes. The wild-type baculovirus provides the polyhedrin for the formation of the OB and complements the OB-negative phenotype of the recombinants. Upon infection of insects, the co-occluded recombinants can exert their insecticidal effect.

Alternatively, recombinants can be constructed where the polyhedrin gene is maintained, but its promoter duplicated and relocated to another site in the viral genome, e.g., proximal to its authentic location. The duplicated polyhedrin promoter is now able to drive the expression of the insecticidal gene. The transfer vectors hence contain both the polyhedrin gene in its original location and the foreign gene under the control of a duplicated polyhedrin promoter (Figure 3.7). This approach has the added advantage that recombinants can be conveniently retrieved after cotransfection with OB-negative viruses by the presence of OBs (Emery and Bishop 1987).

The locus of the p10 gene is ideal for the purpose of inserting insecticidal genes. It is not required for virus replication or OB formation, and its strong promoter can drive the expression of insecticidal genes, such as insect toxins, hormones and enzymes. In addition, deletion of the pio gene gives such recombinants a disadvantage in the field as compared to wild-type viruses (see below).

In most baculovirus infections polyhedrin and p10 are expressed at low level in the gut, whereas some of the insecticidal proteins, like *B.t.* toxin and proctolin, act on gut cells. In addition, the promoters for polyhedrin and p10 are expressed late after infection (Figure 3.2), which may result in delayed insecticidal action. In order to provoke an immediate insecticidal response, immediate early, delayed early or host promoters may therefore be more suitable than the major late polyhedrin or p10 promoters (Bishop et al. 1992). The insecticidal proteins would be expressed much earlier, albeit at a reduced level, because these promoters are weaker. However, it may not always be an advantage to kill the insect too quickly, as progeny OBs cannot be produced to perpetuate the recombinant virus.

6. Production of Recombinant Baculovirus Insecticides

The commercialization of baculovirus insecticides will require a suitable and economic production process. The two broad alternatives involve production in cultures of insect cells or in insect larvae. There are limitations and advantages to both systems. In either case it is essential that OBs are produced to allow infection of insects in the field.

6.1. In Insect Larvae

Insect larvae are optimal "bioreactors" for baculovirus production. Approximately $1-2 \times 10^9$ OBs are produced per larva, which amounts to 25–30% of their dry weight at death (Shapiro 1986). A single application of about 50 to 100 infected larval equivalents of OBs per hectare are required to obtain adequate control. Larvae are easy to rear, require simple media and, under some conditions, can therefore be obtained at relatively low costs. However, because of its specificity, each baculovirus requires a particular insect host. In addition, since insect larvae are sometimes cannibalistic, they cannot always be reared conveniently in large numbers. This adds to the labor costs and could make the commercial production of some baculovirus insecticides expensive.

Since OBs are required for infection of insects in the field, recombinants should have maintained a polyhedrin gene. As outlined in Section 5.3 above, this can be achieved by keeping the polyhedrin gene in its authentic location (e.g., by using p10 recombinants) (Vlak et al. 1990), or relocating a duplicate copy to another part of the genome (Emery and Bishop

1987). Another alternative outlined above entails coinfection of OB-minus recombinants and wild-type virus, resulting in co-occlusion of wild-type and recombinant viral genotypes in the same polyhedron (Figure 3.10). (Miller 1988a; Kuroda et al. 1989; Price et al. 1989). The wild-type genotype provides the polyhedrin gene for the formation of polyhedra and complements the OB-minus phenotype of the recombinants. In the field the OB-minus recombinant will have a selective disadvantage, as its persistence is greatly reduced (Entwistle and Evans 1986).

Yields of recombinant baculovirus insecticides from insect production systems may be reduced as a consequence of the accelerated death of the insects. When the insecticidal activity leads to almost instant mortality, an alternate insect host or cultured insect cells may be the only viable alternatives for the production of recombinant viral insecticides.

6.2. In Insect Cells

The advantages of insect cell cultures for the production of baculoviruses are convenient quality control, minimal contamination and convenient down-stream processing; the disadvantages are the costs of insect cell culture media and the difficulties in developing high volume bioreactor systems.

There are many insect cell lines, mostly derived from neonate larvae or from ovaries of adult females (Hink 1980). However, only a few cell lines replicate baculoviruses, mainly NPVs, in a fully permissive manner. The most popular cell lines are those derived from the fall armyworm *Spodoptera frugiperda* (Vaughn et al. 1977) and the cabbage looper *Trichoplusia ni* (Hink 1970). These cell lines are relatively easy to grow in larger volumes (Vaughn 1976; Miltenburger and David 1980; Hink 1982; Weiss and Vaughn 1986), the major limitations being sufficient oxygenation and the effect of hydrodynamic forces on cells. These forces can cause irreversible damage to the cells (Tramper et al. 1986). In addition, virus yields decline when cells are grown at high densities, since the replication of viral DNA is arrested (Wood et al. 1982).

Recently however, airlift bioreactors have been designed (Tramper et al. 1988; Tramper and Vlak 1988) and employed (Maiorella et al. 1988) for the production of recombinant baculoviruses. In these reactors rising air bubbles keep the cells in suspension and provide the oxygen required. The effect of the hydrodynamic forces can be reduced by the use of non-ionic surfactants such as the antifoam polymer Pluronic F-68 (Murhammer and Goochee 1988). Reactors with capacities of up to 400 liters are being used.

Fetal calf serum constitutes 80% of the media costs, which up to now has made the use of insects more economical (Tramper and Vlak 1986). Most cells grow on commercially available media such as TNM-FH

(Grace's medium modified with protein hydrolysates) or TC100 (Gibco, Flow), and require fetal calf serum for optimal growth and virus production. A major breakthrough is the availability of serum-free, low-protein media, such as SF-900 (Gibco) and EX-CELL 400 (JR Scientific), which may make the exploitation of insect cells for baculovirus production price-competitive with the use of insects in the near future. *S. frugiperda* cells, which are the most widely used (Vaughn et al. 1977), are usually grown in plastic flasks but will also grow happily in spinners and high volume (>20 liter) bioreactors.

With a yield of about 20 polyhedra per cell at a density of 10^6 cells per ml, a 10 liter bioreactor would produce 50–100 larval equivalents of OBs, sufficient to treat one hectare. Production of baculovirus insecticides will most likely occur in batch fermentation processes. However, continuous production systems are more economical and will be developed in the near future. A prototype of such a system has been designed and tested (Kompier et al. 1988; De Gooijer et al. 1989; Van Lier et al. 1990).

7. Risk Assessment

It is essential to assess the risks associated with the dissemination of genetically engineered viral insecticides in the environment before they are released. Some of these risks are similar to those associated with the deliberate release of wild-type baculoviruses, such as effects on nontarget hosts (Summers et al. 1975). The additional risks with genetically engineered viral insecticides may involve alteration of the host range, the spread of the engineered virus from the field site to other ecosystems, the instability of the viral genome, and the possible exchange of genetic information, in particular the insecticidal gene, with other organisms. Studies to assess these risks and to enhance their biosafety have just begun and will be essential when registration of genetically engineered viral insecticides is sought in the near future.

7.1. Alteration of Host Range

Host specificity is an important asset of recombinant baculoviruses in respect to their environmental safety. Normally, baculoviruses infect a single or a few related insect species. AcNPV has the widest host range of those NPVs studied to date, infecting 43 insect species (Payne 1986). Also important, baculoviruses do not adversely affect nontarget organisms, including vertebrates, plants and beneficial invertebrates (Doller 1985; Gröner 1986). These features provide baculoviruses with a high degree of inherent safety, and indeed, wild-type viruses have a long safety record as control agents of insect pests. So far, available data indicate that genetically engineered viral insecticides do not show altered host specificities (Bishop 1989).

7.2. Persistence

It is desirable to design genetically engineered baculovirus insecticides that have limited ability to survive or persist in the field. Preferably, they should also have less biological fitness than the wild-type without impinging on their insecticidal action. Biological fitness might be reduced by deleting genes from the baculovirus genome, following similar strategies to those used to insert foreign genes. These deletions might reduce virus persistence or progeny virus production, which would greatly limit their epizootic potential.

OB-minus recombinants are the most obvious examples of "crippled" viruses (Smith et al. 1983a; Bishop 1989). However, they maintain their full potential to kill host insects. The release of such OB-minus recombinants in the field has shown that they only persist for very short periods of time in soil, on vegetation or in caterpillar corpses (Bishop 1989). In order to obtain more insight into the field ecology of recombinant baculovirus, trial releases are now in progress with an OB-minus recombinant expressing the bacterial β-galactosidase gene as a marker enzyme. These releases have taken place in the United Kingdom (Institute of Virology in Oxford), the USA (Boyce Thompson Institute for Plant Research at Cornell, Ithaca, NY) and the German Federal Republic (Institüt für Biologische Schädlingsbekämpfung, Darmstadt).

In order to disseminate OB-minus recombinants containing an insecticidal gene into the environment, it was explained in Section 5 that they could be co-occluded into an OB of a wild-type baculovirus during a co-infection. The OB-minus genotypes should have a selective disadvantage compared to wild-type baculoviruses in the field, since only OBs are infectious and persistent. The OB-minus genotype should therefore gradually disappear from the genotype pool. A trial release of co-occluded OB-minus recombinant and wild-type AcNPVs is now in progress in the USA (Boyce Thompson Institute, Ithaca, NY).

The interaction between a virus and its host is the result of a long coevolutionary history. The deletion of viral genes will generally not be to the advantage of the virus as it would affect virulence, yield or persistence. However, the deletion of particular genes can lead to increased virulence and improved insecticidal action, for example, the removal of the OB envelope gene results in increased virulence due to more efficient dissolution of OBs in the alkaline environment of the gut (Zuidema et al. 1989). The removal of the egt gene enhances the speed of kill (O'Reilly and Miller 1991).

The deletion of the p10 gene of AcNPV does not affect the production of polyhedra (Vlak et al. 1988; Williams et al. 1989), but the processes of cell and insect lysis (Williams et al. 1989) are impaired. This would limit the dissemination of the virus in the environment, as the release of virus from larval corpses would be impaired.

The products of some viral genes are involved in redirecting the host cell machinery toward optimal production of the virus. One example detailed in the next section is the egt gene, which is found in baculoviruses and works to maintain the larval stage by preventing the molting process (O'Reilly and Miller 1989). Deletion of the egt gene is associated with precocious metamorphosis of infected larvae, which in turn is preceded by precocious cessation of feeding.

7.3. Yield Reduction

As prefaced above, a simple strategy for minimizing viral persistence is to minimize the yield of progeny virus from infected insects. The egt gene mentioned above encodes the enzyme UDP-glycosyl transferase (O'Reilly and Miller 1989). This enzyme glycosylates ecdysteroids, thus inactivating them and impeding molting and pupation of infected larvae. This allows further growth of the larvae and ultimately the production of large amounts of virus. Deletion of this egt gene from AcNPV prevents the inactivation of ecdysone in infected larvae and results in normal larval and pupal molts (O'Reilly and Miller 1990). When infected with egt-minus recombinants, larvae do not increase in size and most likely produce less OBs (O'Reilly and Miller 1990). Double recombinants, which lack the egt gene but carry and express JHE (Hammock et al. 1990; Section 5.2.3 above), might be particularly effective in inducing infected larvae to molt prematurely.

7.4. Gene Transfer and Heterologous Recombination

A major concern with the use of baculovirus recombinants is the possibility of transfer of the insecticidal gene to other baculoviruses with different host ranges that include important non-target hosts. Although many aspects of overall genome organization seem to be conserved among baculoviruses, levels of sequence similarity are nevertheless generally low (Smith and Summers 1982; Section 3.2). This suggests that the likelihood of heterologous recombination among baculoviruses will also be low. In fact, hybrid baculoviruses resulting from illegitimate recombination have never been reported from the field. Nevertheless, hybrids between two NPVs which do show strong DNA sequence similarity (96%) and overlapping host ranges have been recovered from mixed infections of compatible cell lines (AcNPV and *Rachiplusia ou* NPV in *Trichoplusia ni* or *Galleria mellonella* cells) (Summers et al. 1980; Croizier et al. 1988).

Other experiments were aimed to force recombination between different baculovirus species that do not normally infect the same cells. One experiment used AcNPV, which are infectious in *S. frugiperda* cells, and *Mamestra brassicae* NPV (MbNPV), which are not. Co-infection of these cells with MbNPV and an OB-minus mutant of AcNPV did not result in a single recombinant containing the polyhedrin gene (Roosien et al. 1986). Only when the copy number of the polyhedrin gene was substantially increased, by the addition of plasmids containing the MbNPV polyhedrin gene, were recombinants retrieved. The polyhedrin gene was not inserted into its normal location but at various sites along the AcNPV genome. Probably as a consequence of this, the infectivity of the recombinants was greatly diminished. Similar interspecies hybrid baculoviruses were obtained using AcNPV and plasmids containing the polyhedrin gene of *S. frugiperda* NPV (Gonzalez et al. 1989). These experiments indicate that transfer of genetic elements between baculovirus species can occur, but only under extreme conditions.

The transfer of baculovirus genes to non-target insects has not yet been observed. On the other hand, a low level of AcNPV replication was observed in *Aedes aegypti* cells (Sherman and McIntosh 1979), and transcriptional activity of baculovirus genes has been demonstrated in *Drosophila* cells (Carbonell et al. 1985), suggesting that baculovirus genes can be active in non-target hosts. However, deleterious effects of baculoviruses on non-target hosts have never been observed (Gröner 1986).

It is conceivable that transposon-mediated transfer of baculovirus sequences into the genome of a nonsusceptible host may occur after abortive infection (Fraser 1987). The possible involvement of baculoviruses in vectoring genetic information between insect species was suggested when transposon-like elements were discovered in baculoviruses (Miller and Miller 1982). However, evidence for this type of gene transfer from virus to insects has not been reported so far.

8. Application of Engineered Baculovirus Insecticides

There are two major strategies for the release of baculoviruses in the field. One strategy is the inoculative release of a baculovirus into the epicenter of the host pest's distribution. This virus might then cause an epizootic and provide natural, long-term control of the pest population. However, this strategy can only be successful in stable ecosystems (forests, grasslands), where also some damage by the pest can be tolerated and high densities of the pest occur. It also requires considerable persistence of the virus in the environment. Recombinant baculoviruses that are self-destructive (OB-minus), self-contained (pl0-minus) or unlikely to persist are not suitable for this strategy of control.

However, in many intense agro-ecosystems with frequent harvests, the damage threshold that can be tolerated is low and the population densities of the pest are insufficient to establish and maintain an epizootic. Under these circumstances inundative releases at frequent intervals are required for sufficient control. In order to obtain adequate short-term control, about 10^{11} OBs per hectare are required, equivalent to approximately 50 infected larvae.

Recombinant baculoviruses have not yet been released on a large scale. Only limited releases have been carried out, mainly for risk assessment studies (Bishop 1989). A major aim of these studies is to monitor the behavior and fate of recombinants in the environment and compare them to wild-type virus. Recombinants carrying benign marker genes, such as β-galactosidase, may be particularly useful for assessing virus persistence and the possibility of gene transfer to other organisms and viruses.

9. Conclusions

In this chapter the biology and molecular genetics of baculoviruses have been reviewed, and the possibilities and strategies for genetic engineering to improve the insecticidal activity of these pathogens have been outlined. The present knowledge about baculovirus gene structure, function and regulation allows the manipulation of the viral genome and the development of baculovirus expression vector and delivery systems. This technology can now be exploited and tailored for the construction of recombinants with novel insecticidal properties to combat insect pests.

The construction of baculovirus recombinants which cause quick cessation of feeding is now within reach. In addition, by deletion mutagenesis these recombinants can be made less persistent in the environment. The recombinants can be produced commercially in insects, provided that the insecticidal action does not cause immediate mortality. Otherwise, cell culture is an attractive alternative, since inexpensive media and large scale bioreactor systems are now available.

Development of additional strategies for improvement of baculoviruses requires a more detailed understanding of insect biochemistry and physiology, as well as a greater range of candidate insecticidal genes with which to tailor baculoviruses. An improved baculovirus insecticide ideally should have a broad but defined host range and cause rapid cessation of feeding without killing the host instantaneously. Only baculovirus insecticides with these specifications are commercially attractive to develop.

Baculovirus insecticides engineered with genes coding for insect-specific toxins, hormones or metabolic enzymes are not likely to impose additional environmental risks. These proteins are specific for insect species, and they are all natural elements of the insect biosphere. Recombinant baculovirus insecticides containing genes of this nature may therefore be considered as natural, insect-specific biocontrol agents producing biorational compounds. There is also the potential to modify the recombinants so as to reduce their persistence and host range, still further curtailing the risks associated with the deliberate release of genetically modified viral insecticides.

Acknowledgments. I thank the Uyttenboogaart-Eliasen Foundation for a travel grant in 1988 which stimulated the conception of this chapter. I acknowledge the assistance of FJJ von Planta and PJ Kostense for fine artwork and J Bakker for excellent photographic reproduction. I thank JWM van Lent for providing Figure 3.5 and M Kool for producing Figure 3.3. Studies performed by the author and reported herein were supported by grants from the "Stichting Technische Wetenschappen" of the Netherlands Organization of Science and from the Biomolecular Action Programme (Contracts No. BAP-0118-NL and BAP-0416-NL) of the Commission of the European Communities.

References

Adachi T, Takiya S, Suzuki Y, Iwami M, Kawakami A, Takahashi SY, Ishizaki H, Nagasawa H, Suzuki A (1989) cDNA structure and expression of bombyxin, an insulin-like brain secretory peptide of the silkworm *Bombyx mori*. J Biol Chem 264:7681–7685

Benz GA (1986) Introduction: historical perspectives. In Granados RR, Federici BA (eds) The Biology of Baculoviruses Vol I Biological Properties and Molecular Biology. CRC Press, Boca Raton, Florida, pp 1–35

Bishop DHL (1989) Genetically engineered viral insecticides—A progress report 1986–1989. Pestic Sci 27:173–189

Blissard GW, Rohrmann GF (1990) Baculovirus diversity and molecular biology. Ann Rev Entomol 53:127–155

Briese DT (1986) Insect resistance to baculoviruses. In Granados RR, Federici BA (eds) The Biology of Baculoviruses Vol II Practical Application for Insect Control. CRC Press, Boca Raton, Florida, pp 237–263

Burgess S (1977) Molecular weights of lepidopteran baculovirus DNAs: derivation by electron microscopy. J Gen Virol 37:501–510

Carbonell LF, Klowden MJ, Miller LK (1985) Baculovirus-mediated expression of bacterial genes in dipteran and mammalian cells. J Virol 56:153–160

Carbonell LF, Hodge MR, Tomalski MD, Miller LK (1988) Synthesis of a gene coding for an insect-specific scorpion neurotoxin and attempts to express it using baculovirus vectors. Gene 73:409–418

Carstens EB, Tjia ST, Doerfler W (1979) Infection of *Spodoptera frugiperda* cells with *Autographa californica* nuclear polyhedrosis virus. Virology 99:8–17

Carstens EB, Lu A (1990) Amplification of a region within the p74 gene of *Autographa californica* nuclear polyhedrosis virus enhances viral infectivity in insects. J Gen Virol 71:3035–3040

Cochran MA, Faulkner P (1983) Location of homologous DNA sequences interspersed at five regions in the baculovirus AcMNPV genome. J Virol 45:961–970

Croizier G, Gonnet P, Devauchelle G (1987) Localization cytologique de la protéine non-structurale p10 du baculovirus de la polyédrose nucléaire du Lépidoptère *Galleria mellonella* L. Compt Rend Acad Sci 305:677–681

Croizier G, Croizier L, Quiot JM, Lereclus D (1988) Recombination of *Autographa californica* and *Rachiplusia ou* nuclear polyhedrosis viruses in *Galleria mellonella* L. J Gen Virol 69:177–185

De Gooijer CD, Van Lier FLJ, Van den End EJ, Vlak JM, Tramper J (1989) A model for baculovirus production with continuous insect cell cultures. Appl Microbiol Letters 30:497–501

De Kort CAD, Granger N (1981) Regulation of the juvenile hormone titer. Ann Rev Entomol 26:1–28

Dobos P, Cochran MA (1980) Protein synthesis in cells infected by *Autographa californica* nuclear polyhedrosis virus (AcNPV): the effect of cytosine arabinoside. Virology 103:446–464

Doerfler W, Böhm P (eds) (1986) Current Topics in Microbial Immunology Vol 131 The Molecular Biology of Baculoviruses. Springer Verlag, Berlin

Döller G (1985) The safety of insect viruses as biological control agents. In Maramorosch K, Sherman KE (eds) Viral Insecticides for Biological Control. Academic Press, New York, pp 399–439

Eldridge R, Horodyski FM, Morton DB, O'Reilly DR, Truman JW, Riddiford LM, Miller LK (1991) Expression of an eclosion hormone gene in insect cells using a baculovirus vectors. Insect Biochem 21:341–351

Emery VC, Bishop DHL (1987) The development of multiple expression vectors for high level synthesis of eukaryotic proteins: expression of LCMV-N and AcNPV polyhedrin protein by recombinant baculovirus. Protein Engineering 1:359–366

Entwistle PF, Evans HF (1986) Viral control. In Kerkut GA, Gilbert LI (eds) Comprehensive Insect Physiology, Biochemistry and Pharmacology Vol 13. Pergamon Press, Oxford, pp 347–412

Evans HF, Harrap KA (1982) Persistence of insect viruses. In Minson AC, Darby GK (eds) Virus Persistence. SGM Symposium, Cambridge University Press, Cambridge, pp 57–96

Faulkner P (1981) Baculoviruses. In Davidson EA (ed) Pathogenesis of Invertebrate Microbial Diseases. Allanheld, Osmun & Co, Totowa, New Jersey, pp 3–37

Francki RIB, Fauquet CM, Knudson DL, Brown F (eds) (1991) Classification and Nomenclature of Viruses, Springer-Verlag, New York, 450 pp

Fraser MJ (1987) FP mutation of nuclear polyhedrosis viruses: a novel system for the study of transposon-mediated mutagenesis. In Maramorosch K (ed) Biotechnology in Invertebrate Pathology and Cell Culture. Academic Press, New York, pp 265–293

Fraser MJ (1989) The baculovirus infected cell as a eukaryotic expression system. In Muzyczka N (ed) Current Topics in Microbiology and Immunology (in press)

Friesen PD, Miller LK (1986) The regulation of baculovirus gene expression. In Doerfler W, Böhm P (eds) Current Topics in Microbial Immunology Vol 131 The Molecular Biology of Baculoviruses. Springer-Verlag, Berlin, pp 31–49

Fuchs YL, Woods MS, Weaver RF (1983) Viral transcription during *Autographa californica* nuclear polyhedrosis virus infection: a novel RNA polymerase induced in infected *Spodoptera frugiperda* cells. J Virol 48:641–646

Gonnet P, Devauchelle G (1987) Obtention par recombinaison dans le gène du polypeptide P10 d'un baculovirus exprimant le gène de résistance à la néomycine dans les cellules d'insecte. Compt Rend Acad Sci 305:111–114

Gonzalez MA, Smith GE, Summers MD (1989) Insertion of the SfMNPV polyhedrin gene into an AcMNPV polyhedrin deletion mutant during viral infection. Virology 170:160–175

Granados RR (1980) Infectivity and mode of action of baculoviruses. Biotech Bioengineering 22:1377–1405

Granados RR, Corsaro BG (1990) Baculovirus enhancing proteins and their implication for insect control. Proc Fifth Int Colloq Invert Pathol Microbial Control. Adelaide, Australia, pp 174–178

Granados RR, Federici BA (eds) (1986a) The Biology of Baculoviruses Vol I Biological Properties and Molecular Biology. CRC Press, Boca Raton, Florida

Granados RR, Federici BA (eds) (1986b) The Biology of Baculoviruses Vol II Practical Application for Insect Control. CRC Press, Boca Raton, Florida

Granados RR, Lawler KA (1981) In vivo pathway of *Autographa californica* baculovirus invasion and infection. Virology 108:297–308

Granados RR, Williams KA (1986) In vivo infection and replication of baculoviruses. In Granados RR, Federici BA (eds) The Biology of Baculoviruses Vol I Biological Properties and Molecular Biology. CRC Press, Boca Raton, Florida, pp 89–108

Gröner A (1986) Specificity and safety of baculoviruses. In Granados RR, Federici BA (eds) The Biology of Baculoviruses Vol I Biological Properties and Molecular Biology. CRC Press, Boca Raton, Florida, pp 177–202

Guarino LA, Summers MD (1986) Functional mapping of a trans-activating gene required for expression a baculovirus delayed early gene. J Virol 57:563–571

Guarino LA, Summers MD (1987) Nucleotide sequence and temporal expression of a baculovirus regulatory gene. J Virol 61:2091–2099

Guarino LA, Gonzalez MA, Summers MD (1986) Complete sequence and enhancer function of the homologous DNA regions of *Autographa californica* nuclear polyhedrosis virus. J Virol 60:224–229

Hammock BD (1985) Regulation of the juvenile hormone titer: degradation. In Kerkut GA, Gilbert LI (eds) Comprehensive Insect Physiology, Biochemistry, and Pharmacology Vol 7. Pergamon Press, New York, pp 431–472

Hammock BD, Bonning BC, Possee RD, Hanzlik TN, Maeda S (1990) Expression and effects of the juvenile hormone esterase in a baculovirus vector. Nature 344:458–461

Hanzlik TN, Abdel-Aal YAI, Harshman LG, Hammock BD (1989) Isolation and sequencing of cDNA clones coding for juvenile hormone esterase from *Heliothis virescens*. J Biol Chem 264:12419–12425

Hink WF (1970) Established cell line from the cabbage looper. Nature 226:466

Hink WF (1980) The 1979 compilation of invertebrate cell lines and culture media. In Kurstak E, Maramorosch K, Dübendorfer A (eds) Invertebrate Systems In Vitro. Elsevier/North-Holland Biomedical Press, Amsterdam, pp 553–578

Hink WF (1982) Production of *Autographa californica* nuclear polyhedrosis virus in cells from large suspension cultures. In Kurstak E (ed) Microbial and Viral Pesticides. Marcel Dekker, New York, pp 493–506

Hofmann C, Vanderbruggen H, Höfte H, Van Rie J, Jansens A, Van Mellaert H (1988) Specificity of *Bacillus thuringiensis* δ-endotoxins is correlated with the presence of high-affinity binding sites in the brush border membrane of target insect midguts. Proc Natl Acad Sci USA 85:7844–7848

Holman GM, Nachman RJ, Wright MS (1990) Insect neuropeptides. Ann Rev Entomol 35:201–217

Hooft van Iddekinge BJ, Smith GE, Summers MD (1983) Nucleotide sequence of the polyhedrin gene of *Autographa californica* nuclear polyhedrosis virus. Virology 131:561–565

Horodyski FM, Riddiford LM, Truman JW (1989) Isolation and expression of the eclosion hormone gene from the tobacco hornworm, *Manduca sexta*. Proc Natl Acad Sci USA 86:8123–8127

Hughes P, Wood HA (1981) A synchronous peroral technique for the bioassay of insect viruses. J Invert Pathol 37:154–159

Iwami M, Kawakami A, Ishizaki H, Takahashi SY, Adachi T, Suzuki Y, Nagasawa H, Suzuki A (1989) Cloning a gene encoding bombyxin, an insulin-like brain secretory peptide of the silkmoth *Bombyx mori* with prothoracicotropic activity. Devel Growth Diff 31:31–37

Jurkovicova M (1979) Activation of latent infections in larvae of *Adoxophyes orana* (Lepidoptera: Tortricidae) and *Barathra brassicae* (Lepidoptera: Noctuidae) by foreign polyhedra. J Invert Pathol 34:213–223

Kang CY (1988) Baculovirus vectors for expression of foreign genes. Adv Virus Res 35:177–192

Kataoka H, Troetschler RG, Kramer SJ, Cesarin BJ, Schooley DA (1987) Isolation and primary structure of the eclosion hormone of the tobacco hornworm, *Manduca sexta*. Biochem Biophys Res Commun 146:746–750

Kataoka H, Toschi A, Li JP, Carney RL, Schooly DA, Kramer SJ (1989a) Identification of an allatotropin from adult *Manduca sexta*. Science 243: 1481–1483

Kataoka H, Troetschler RG, Li JP, Kramer SJ, Carney RL, Schooley DA (1989b) Isolation and identification of a diuretic hormone from the tobacco hornworm, *Manduca sexta*. Proc Natl Acad Sci USA 86:2976

Kawakami A, Iwami, Nagasawa H, Suzuki A, Ishizaki H (1989) Structure and organization of four clustered genes that encode bombyxin, an insulin-like brain secretory peptide of the silkmoth *Bombyx mori*. Proc Natl Acad Sci USA 86:6843–6847

Keeley LL, Hayes TK (1987) Speculations on biotechnology applications for insect neuroendocrine research. Insect Biochem 17:639–651

Kelly DC (1982) Baculovirus replication. J Gen Virol 63:1–13

Kelly DC, Lescott T (1981) Baculovirus replication: protein synthesis in *Spodoptera frugiperda* cells infected with *Trichoplusia ni* nuclear polyhedrosis virus. Microbiologica 4:35–47

Kompier R, Tramper J, Vlak JM (1988) A continuous process for the production of baculovirus using insect-cell cultures. Biotech Letters 10:849–854

Kondo A, Maeda S (1991) Host range expansion by recombination of the baculoviruses *Bombyx mori* nuclear polyhedrosis virus and *Autographa californica* nuclear polyhedrosis virus. J Virol 65:3625–3632

Kono T, Nagasawa H, Isogai A, Fugo H, Suzuki A (1987) Amino acid sequence of eclosion hormone of the silkworm *Bombyx mori*. Agric Biol Chem 51: 2307–2308

Kool M, Van den Berg PMMM, Tramper J, Goldbach RW, Vlak JM (1992) Identification of two putative origins of DNA replication of *Autographa californica* nuclear polyhedrosis virus. Virology 191

Krappa R, Knebel-Mörsdorf D (1991) Identification of the very early transcribed baculovirus gene PE-38. J Virol 65:805–812

Krieg A (1986) *Bacillus thuringiensis*, ein mikrobielles Insectizid: Grundlage und Anwendung. Acta Phytomed 10:1–191

Kuroda K, Gröner A, Frese K, Drenckhahn D, Hauser Ch, Rott R, Doerfler W, Klenk H-D (1989) Synthesis of biologically active influenza virus hemagglutinin in insect larvae. J Virol 63:1677–1685

Kurstak E (ed) (1982) Microbial and Viral Pesticides. Marcel Dekker, New York

Kuzio JD, Rohel Z, Curry CJ, Krebs A, Carstens EB, Faulkner P (1984) Nucleotide sequence of the p10 polypeptide gene of the *Autographa californica* nuclear polyhedrosis virus. Virology 139:414–418

Leisy D, Rohrmann G, Beaudreau GS (1984) Conservation of the genome organization in two multicapsid nuclear polyhedrosis viruses. J Virol 52: 699–702

Luckow VE, Summers MD (1988a) Trends in the development of baculovirus expression vectors. Bio/Technology 6:47–55

Luckow VE, Summers MD (1988b) Signals important for high-level expression of foreign genes in *Autographa californica* nuclear polyhedrosis virus expression vectors. Virology 167:56–71

Maddrell SHP (1986) Hormonal control of diuresis in insects. In Borkovec AE, Gelman DB (eds) Insect Neurochemistry and Neurophysiology. Humana, Clifton, New Jersey, pp 79–90

Maeda S (1989) Expression of foreign genes in insects using baculovirus vectors. Ann Rev Entomol 34:351–372

Maeda S (1990) Increased insecticidal effect by a recombinant baculovirus carrying a synthetic diuretic hormone gene. Biochem Biophys Res Commun 165:1177–1183

Maiorella B, Inlow D, Shauger A, Harano D (1988) Large-scale insect cell-culture for recombinant protein production. Bio/Technology 6:1406–1410

Maramorosch K (ed) (1987) Biotechnology in Invertebrate Pathology and Cell Culture. Academic Press, New York

Martens JWM, Honée G, Zuidema D, Van Lent JWM, Visser B, Vlak JM (1990) Insecticidal activity of a bacterial crystal protein expressed by a baculovirus recombinant in insect cells. Appl Environm Microbiol 56:2769–2770

Marti T, Takio K, Walsh KA, Terzi G, Truman JW (1987) Microanalysis of the amino acid sequence of the eclosion hormone from the tobacco hornworm *Manduca sexta*. FEBS Letters 219:415–418

Martignoni ME, Iwai PJ (1986) A Catalog of Viral Diseases of Insects, Mites, and Ticks. USDA Forest Service PNW-195. Washington DC, USA

Matsuura Y, Possee RD, Overton HA, Bishop DHL (1987) Baculovirus expression vectors: the requirements for high level expression of proteins, including glycoproteins. J Gen Virol 68:1233–1250

McCutchen BF, Choudary PV, Crenshaw R, Maddox D, Kamita SG, Palekar N, Volrath S, Fowler E, Hammock BD, Maeda S (1991) Development of a recombinant baculovirus expressing an insect-selective neurotoxin—Potential for pest control. Bio/Technology 9:848–852

Menn JJ, Borkovec AB (1989) Insect neuropeptides: potential new insect control agents. J Agric Food Chem 37:271–278

Miller DW (1988) Method for Producing a Heterologous Protein in Insect Cells. International Patent Application No. W088/02030

Miller DW, Miller LK (1982) A virus mutant with an insertion of a copia-like transposable element. Nature 299:562–564

Miller DW, Safer P, Miller LK (1986) An insect baculovirus host-vector system for high-level expression of foreign genes. In Setlow JK, Hollaender A (eds) Genetic Engineering: Principles and Methods. Plenum Press, New York, pp 277–298

Miller LK (1988) Baculoviruses as gene expression vectors. Ann Rev Microbiol 42:177–199

Miller LK (1989) Insect baculoviruses: Powerful gene expression vectors. BioAssays 11:91–95.

Miltenburger HG, David P (1980) Mass production of insect cells in suspension. Devel Biol Stand 46:183–186

Murhammer DW, Goochee CF (1988) Scaleup of insect cells cultures: protective effects of pluronic F-68. Bio/Technology 6:1411–1418

Nissen MS, Friesen PD (1989) Molecular analysis of the transcriptional regulatory region of an early baculovirus gene. J Virol 63:493–503

O'Reilly DR, Miller LK (1989) A baculovirus blocks insect molting by producing ecdysteroid UDP-glucosyl transferase. Science 245:1110–1112

O'Reilly DR, Miller LK (1991) Improvement of a baculovirus pesticide by deletion of the egt gene. Bio/Technology 9:1086–1089

Payne CC (1986) Insect pathogenic viruses as pest control agents. Fortschr Zool 32:183–200

Payne CC (1988) Pathogens for the control of insects: where next? Phil Trans Royal Soc London B 318:225–248

Pennock GD, Shoemaker C, Miller LK (1984) Strong and regulated expression of *Escherichia coli* β-galactosidase in insect cells with a baculovirus vector. Mol Cell Biol 4:399–406

Possee RD (1986) Cell-surface expression of influenza hemagglutinin in insect cells using a baculovirus vector. Virus Res 5:43–59

Possee RD, Weyer U, King LA (1990) Recombinant antigen production using baculovirus expression vectors. In Dimmock NJ, Griffiths PD, Madeky CR (eds) Control of Virus Diseases. Soc Gen Microbiol Symp 45:53–76

Powell PA, Stark DM, Sanders PR, Beachy RN (1989) Protection against tobacco mosaic virus in transgenic plants that express tobacco mosaic virus antisense RNA. Proc Natl Acad Sci USA 86:6949–6952

Price PM, Reichelderfer CF, Johansson BE, Kilbourne ED, Acs G (1989) Complementation of recombinant baculoviruses by coinfection with wild-type virus facilitates production in insect larvae of antigenic proteins of hepatitis B virus and influenza. Proc Natl Acad Sci USA 86:1435–1456

Rankin C, Ooi BG, Miller LK (1988) Eight base pairs encompassing the transcriptional start point are the major determination for baculovirus polyhedrin gene expression. Gene 70:39–49

Roelvink PW, Van Meer MMM, De Kort CAD, Possee RD, Hammock BD, Vlak JM (1992) Temporal expression of *Autographa californica* nuclear polyhedrosis virus polyhedrin and p10. J Gen Virol 73:1481–1489

Rohrmann GF (1986) Polyhedrin structure. J Gen Virol 67:1499–1513

Roosien J, Usmany M, Klinge-Roode, Meijerink PHS, Vlak JM (1986) Heterologous recombination between baculoviruses. In Samson RA, Vlak JM, Peters D (eds) Fundamental and Applied Aspects of Invertebrate Pathology. Foundation Fourth Int Colloq Invert Pathol, Wageningen, The Netherlands, pp 389–392

Shapiro M (1986) In vivo production of baculoviruses. In Granados RR, Federici BA (eds) The Biology of Baculoviruses Vol II Practical Application for Insect Control. CRC Press, Boca Raton, Florida, pp 31–61

Sherman KE, Maramorosch K (1985) Viral and Microbial Insecticides. Academic Press, New York

Sherman KE, McIntosh AH (1979) Baculovirus replication in mosquito (dipteran) cell line. Infect Immun 26:232–234

Smith GE, Summers MD (1978) Analysis of baculovirus genomes with restriction endonucleases. Virology 89:517–527

Smith GE, Summers MD (1982) DNA homology among subgroup A, B, and C baculoviruses. Virology 123:393–406

Smith GE, Vlak JM, Summers MD (1982) In vitro translation of *Autographa californica* nuclear polyhedrosis virus early and late mRNAs. J Virol 44: 199–208

Smith GE, Fraser MJ, Summers MD (1983a) Molecular engineering of the *Autographa californica* nuclear polyhedrosis virus genome: deletion mutations within the polyhedrin gene. J Virol 46:584–593

Smith GE, Summers MD, Fraser MJ (1983b) Production of human B-interferon in insect cells infected with a baculovirus expression vector. Mol Cell Biol 3:2156–2165

Smits PH, Vlak JM (1988) Selection of nuclear polyhedrosis viruses as biological control agents of *Spodoptera exigua* (Lep.: Noctuidae). Entomophaga 33: 299–308

Stewart LMD, Hirst M, Lopez-Ferber M, Merryweather AT, Cayley PJ, Possee RD (1991) Construction of an improved baculovirus insecticide containing an insect-specific toxin gene. Nature 352:85–88.

Summers MD, Smith GE (1987) A manual of methods for baculovirus vectors and insect cell culture procedures. Texas Agr Exp Stat Bull No. 1555

Summers MD, Engler R, Falcon LA, Vail P (eds) (1975) Baculoviruses for Insect Pest Control: Safety Considerations. Amer Soc Microbiol, Washington

Summers MD, Smith GE, Knell JD, Burand JP (1980) Physical maps of *Autographa californica* and *Rachiplusia ou* nuclear polyhedrosis virus recombinants. J Virol 34:693–703

Thiem SM, Miller LK (1989) Identification, sequence, and transcriptional mapping of the major capsid protein gene of the baculovirus *Autographa californica* nuclear polyhedrosis virus. J Virol 63:2008–2018

Tomalski MD, Miller LK (1991) Insect paralysis by baculovirus-mediated expression of a mite neurotoxin gene. Nature 352:82–85.

Tomalski MD, Wu J, Miller LK (1988) The location, sequence, transcription, and regulation of a baculovirus DNA polymerase gene. Virology 167:591–600

Tramper J, Vlak JM (1986) Some engineering and economic aspects of continuous cultivation of insect cells and the production of baculoviruses. Ann NY Acad Sci 469:279–288

Tramper J, Vlak JM (1988) Bioreactor design for growth of shear sensitive mammalian and insect cells. In Mizrahi A (ed) Advances in Biotechnological Processes Vol 7 Upstream Processes: Equipment and Techniques. AR Liss, New York, pp 199–228

Tramper J, Williams JB, Joustra D, Vlak JM (1986) Shear sensitivity of insect cells in suspension. Enzyme Microbiol Technol 8:33–36

Tramper J, Smit D, Straatman J, Vlak JM (1988) Bubble-column design for growth of fragile insect cells. Bioproc Eng 3:37–41

Van der Krol AR, Mol JNM, Stuitje AR (1988) Modulation of eukaryotic gene expression by complementary RNA or DNA sequences. Bio Techniques 6:958–976

Van der Wilk F, Van Lent JWM, Vlak JM (1987) Immunogold detection of polyhedrin, p10 and virion antigens in *Autographa californica* nuclear polyhedrosis virus-infected *Spodoptera frugiperda* cells. J Gen Virol 68: 2615–2623

Van Lier FLJ, Van den End EJ, De Gooijer CD, Vlak JM, Tramper J (1990) Continuous production of baculovirus in a cascade of insect cell reactors. Appl Microbial Biotech 33:43–47

Vaughn JL (1976) Production of NPV in large scale cultures. J Invert Pathol 28:233–237

Vaughn JL, Goodwin RH, Tompkins GJ, McCawley P (1977) The establishment of two cell lines from the insect (Lepidoptera: Noctuidae). In vitro 13: 213–217

Vialard J, Lalumière M, Vernet T, Briedis D, Alkhatib G, Henning D, Levin D, Richardson C (1990) Synthesis of the membrane fusion and hemagglutinin proteins of measles virus using a novel baculovirus vector containing the β-galactosidase gene. J Virol 64:37–50

Vlak JM, Keus JAR (1990) Baculovirus expression vector system for production of viral vaccines. In Mizrahi A (ed) Advances in Biotechnological Processes Vol 14 Viral Vaccines. AR Liss, New York, pp 97–127

Vlak JM, Odink KG (1979) Characterization of *Autographa californica* nuclear polyhedrosis virus deoxyribonucleic acid. J Gen Virol 44:333–347

Vlak JM, Rohrmann GF (1985) The nature of polyhedrins. In Maramorosch K, Sherman KE (eds) Viral Insecticides for Biological Control. Academic Press, New York, pp 489–542

Vlak JM, Smith GE (1982) Orientation of the genome of *Autographa californica* nuclear polyhedrosis virus: a proposal. J Virol 41:1118–1121

Vlak JM, Smith GE, Summers MD (1981) Hybridization selection and *in vitro* translation of *Autographa californica* nuclear polyhedrosis virus mRNA. J Virol 40:762–771

Vlak JM, Klinkenberg FA, Zaal KJM, Usmany M, Klinge-Roode EC, Geervliet JBF, Roosien J, Van Lent JWM (1988) Functional studies on the p10 gene of *Autographa californica* nuclear polyhedrosis virus using a recombinant expressing a p10-beta-galactosidase fusion gene. J Gen Virol 69:765–776

Vlak JM, Schouten A, Usmany M, Belsham GJ, Klinge-Roode EC, Maule AJ, Van Lent JWM, Zuidema D (1990) Expression of cauliflower mosaic virus gene I using a baculovirus vector based upon the baculovirus p10 gene and a novel selection method. Virology 179:312–320

Volkman LE, Goldsmith PA (1985) Mechanisms of neutralization of budded *Autographa californica* nuclear polyhedrosis virus by monoclonal antibody: inhibition of the target antigen. Virology 143:185–195

Weiss SA, Vaughn JL (1986) Cell culture methods for large scale propagation of baculoviruses. In Granados RR, Federici BA (eds) The Biology of Baculoviruses Vol I Biological Properties and Molecular Biology. CRC Press, Boca Raton, Florida, pp 63–88

Weyer U, Possee RD (1988) Functional analysis of the p10 gene 5′ leader sequence of the *Autographa californica* nuclear polyhedrosis virus. Nucl Acids Res 16:3635–3653

Weyer U, Possee RD (1989) Analysis of the promoter of the *Autographa californica* nuclear polyhedrosis virus p10 gene. J Gen Virol 70:203–208

Whiteley HR, Schnepf HE (1986) The molecular biology of parasporal crystal body formation. Ann Rev Microbiol 40:549–576

Whitford M, Stewart S, Kuzio J, Faulkner P (1989) Identification and sequence analysis of a gene encoding gp67, an abundant envelope glycoprotein of the baculovirus, *Autographa californica* nuclear polyhedrosis virus. J Virol 63: 1393–1399

Williams GV, Rohel DZ, Kuzio J, Faulkner P (1989) A cytopathological investigation of *Autographa californica* nuclear polyhedrosis virus p10 gene function using insertion/deletion mutants. J Gen Virol 70:187–202

Wilson ME, Mainprize TH, Friesen PD, Miller LK (1987) Location, transcription, and sequence of a baculovirus gene encoding a small arginine-rich polypeptide. J Virol 61:661–666

Wood HA (1980) *Autographa californica* nuclear polyhedrosis virus-induced proteins in tissue culture. Virology 102:21–27

Wood HA, Johnston LB, Burand JP (1982) Inhibition of *Autographa californica* nuclear polyhedrosis virus replication in high-density *Trichoplusia ni* cell cultures. Virology 119:245–254

Woodhead AP, Stay B, Seidel SL, Khan MA, Tobe SS (1989) Primary structure of four allatostatins: neuropeptide inhibitors of juvenile hormone synthesis. Science 86:5997–6001

Zuidema D, Klinge-Roode EC, Van Lent JWM, Vlak JM (1989). Construction and analysis of an *Autographa californica* nuclear polyhedrosis virus mutant lacking the polyhedral envelope. Virology 173:98–108

Zuidema D, Schouten A, Usmany M, Maule AJ, Belsham GJ, Roosien J, Klinge-Roode EC, Van Lent JWM, Vlak JM (1990) Expression of cauliflower mosaic virus gene I in insect cells using a novel polyhedrin-based baculovirus expression vector. J Gen Virol 71:2201–2210

Chapter 4

Insect Viruses: New Strategies for Pest Control

Peter D. Christian, Terry N. Hanzlik, David J. Dall, and Karl H. Gordon

1. Introduction

For the last 50 years, insect control programs have relied heavily on the use of chemical insecticides. Today, however, awareness of the unintended biological consequences of the use of broad-spectrum insecticidal compounds and changing attitudes toward this ecological cost are now coupled with the ever-increasing problem of target-pest resistance. On the other hand, it now seems possible that advances in biotechnology will lead to novel uses of entomopathogens, and that such approaches might have the potential to circumvent a majority of these problems. In this chapter we outline the possibilities for integration of insect viruses, or their constitutive parts, into novel insect control agents and strategies.

The chapter begins with a brief description of the criteria for economically viable pest control strategies and compares chemical- and entomopathogen-based strategies against these criteria; in the latter category, particular emphasis is placed upon insect viruses (see Section 2). As described there, prior experience with insect viruses has shown that improvements in their efficacy, large-scale production and perceived safety will be needed if they are to play a major role in the control of insect pests. In Section 3 we consider the improvements in efficacy likely to derive from novel sources such as virally-based vector systems (the baculoviruses discussed by Vlak in Chapter 3 being precedents for a more general capability), the exploitation of viruses as novel sources of insecticidal agents (covered briefly by Binnington and Baule in Chapter 2), and the use of viruses as molecular tools in the design of insecticidal agents and associated control programs. Section 4 discusses the challenges implicit in proposed production and formulation processes, and in the use of these agents and strategies in the field.

This chapter focuses mainly on the use of insect viruses for the control of crop pests and makes little mention of the control of insect pests of humans and livestock. This is simply a reflection of our interests and expertise and does not imply that similar strategies would not prove valuable in the latter contexts.

2. Control of Insect Pests

2.1. The Ideal Insect Control Agent/Strategy

Control of insect pests has previously been based around three major strategies. As shown in Figure 4.1, these are the application of insecticides, the use of traditional biological control agents (e.g., diseases and predators, often as part of cultural control practices), and the use of resistant or tolerant varieties of plants and animals. Table 4.1 summarizes the relative merits of these three strategies with respect to the requirements described below.

In this chapter we reserve the term "insecticide" to denote an agent designed to produce rapid insect mortality following application in a conventional manner, e.g., by spraying, dusting or dipping. Agents with less conventional modes of activity will then be collectively described as "pest control agents." Regardless of form, however, each control mechanism should satisfy three rigorous criteria. Thus, each should be highly efficacious, safe both to the user and the environment, and easy to produce and implement at competitive cost from readily available

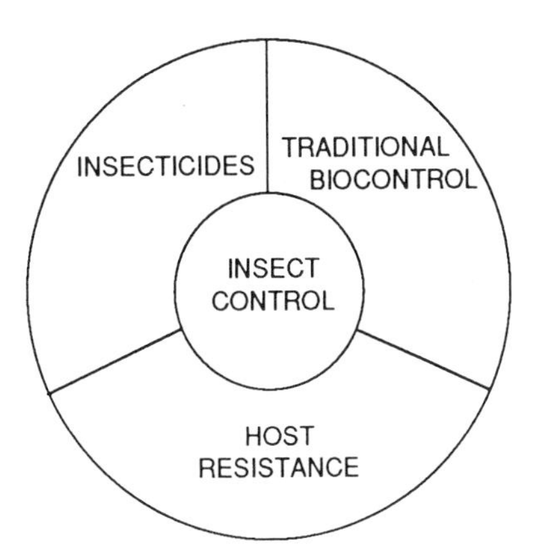

Figure 4.1. Major strategies used for control of insect pests.

Table 4.1. Summary of characteristics of insect pest control strategies

Characteristic	Biological Control	Host Mediated Resistance	Insecticides	
			Viral	Chemical
Efficacy				
Able to control target species	+	N/A	+	++
Free of resistance selection	+	+	?	−
Specific for target species	++	++	++	−
Acceptance by users	+	++	+	++
Safety				
Safe to produce & use	++	++	++	?
Biodegradable	++	++	++	?
Breakdown products are:				
non toxic	++	?	++	?
non mutagenic	++	?	++	?
Perceived as safe by public	++	++	+	+/−
Production				
Easy to discover & develop	+	−	+	−
Cheap & easy to make	−	N/A	+	++
Easy to formulate & apply	+	N/A	+	+

++ characteristics are ideal
+ " " acceptable
− " " deleterious
? " " highly variable or insufficient data
N/A " " not applicable

materials. As described below, few of the measures presently available for insect control satisfy these prerequisites.

2.2. Chemical and Biological Insecticides

Ancient forms of insect control were based around natural biological control, augmented by cultural practices and the use of resistant crop varieties. Since World War II, however, the use of chemical insecticides has become the mainstay of insect control in most of the world's agricultural systems. Nevertheless, and despite the current prominence of chemical insecticides, it is worth remembering that biological agents have long been used to control insects.

The first recorded description of an insect disease dates back to Aristotle (384–322 B.C.), and the first unequivocal demonstration of the parasitic and infectious nature of an insect pathogen (a fungus) was made by Agostino Bassi in 1834 (see Steinhaus 1956, for review of the early history of insect pathology). From this point insect pathogens were regarded as agents with potential for control of insect pests. Five major groups of entomopathogens are recognized (bacteria, fungi, nematodes, protozoa and viruses), and all have been used as control agents in various situations. In this text we use viral pathogens both to detail the development of biological insecticides and to describe the types of problem which hamper their commercial acceptance.

The potential for use of viruses as insecticides can be traced to 1856 when Cornalia and Maestri described and demonstrated the association between an insect virus and an identifiable disease (Steinhaus 1956). Nevertheless, it was not until the early part of the twentieth century that a deliberate attempt to control an insect population with a virus was successful. In this instance a nuclear polyhedrosis virus (NPV) was used to control an introduced pest, the European spruce sawfly (*Gilpinia hercyniae*), in Canada. Since this initial success, a number of milestones in the use of viral insecticides have been reached. These include aerial application of viruses (first used to control alfalfa caterpillar, *Colias philodice*, in California in 1949, and European pine sawfly, *Neodiprion sertifer*, in Canada in 1950), small-scale use of a number of viruses in the 1960s and early 1970s, registration of the first viral insecticide in 1975 (the NPV of *Helicoverpa zea*) and the subsequent large-scale commercialization of the latter under the tradename Elcar™.

In spite of the above developments, only nine viruses were being commercially produced in the late 1980s (Cunningham 1988), and furthermore, all were members of the family Baculoviridae. It should, in fact, be stated here that the only notable exception to the general commercial failure of entomopathogens remains that of the bacterium *Bacillus thuringiensis* (*B.t.*), which currently accounts for about 1% of the world insecticide market. Given this admission, it is important for the purposes of the chapter to determine why microbial insecticides in general, and viral insecticides in particular, have so far proven commercially unsuccessful.

For viral insecticides, at least, a substantial portion of the explanation relates to innate characteristics of insect viruses and includes factors such as their slow rate of kill, difficulties in mass producing them cheaply and easily, and unfavorable popular perceptions of their safety (Section 4.2). Conversely, the rise of chemical insecticides to their position of prominence derives from desirable characteristics such as high efficacy and ease of manufacture (Table 4.1). These traits led to wide initial acceptance of chemical insecticides, but by the 1960s considerable public anxiety was aroused by their nontarget toxicities and consequent environmental impact (Carson 1962). In addition, the evolution of target-pest resistance became an increasing concern, and by 1975 over 200 species of agriculturally important arthopods were known to be resistant to commonly used products. With the major groups of insecticides losing efficacy through the evolution of target pest resistance, and losing general acceptance through concerns about safety, pessimistic predictions were made about their continued efficacy (Forgash 1984).

With the introduction of pyrethroids in the late 1970s both environmental and target-resistance problems appeared to have been solved. Nevertheless, resistance problems to pyrethroids soon occurred, and by the mid 1980s it was apparent that while pyrethroids might offer respite in the battle to control insect pests, they were inevitably destined to go the same way as other groups of chemical insecticides (Georghiou and Lagunes 1988).

Thus, neither chemical nor biological insecticides have yet formed the basis of an ideal control agent or strategy (see Table 4.1). However, with recent rapid developments in genetic engineering and biotechnology the pest control industry is beginning to look further than continued development of chemical insecticides, and is now focusing on alternative agents and strategies which might facilitate sustainable control methodologies. The remainder of this chapter describes the manner in which insect viruses could play an important role in the development of these new agents and strategies.

2.3. Viruses and Insects

Insects are hosts to many viruses from 12 recognized families (Matthews 1982) as well as to a very large number of viruses whose affinities remain to be determined. In addition, they transmit many plant and vertebrate viruses. A summary of the major groups of viruses associated with insects of importance to agricultural crops is provided in Table 4.2.

Viruses can interact with insects in three fundamental ways. In one type of association, the insect acts as a vector for a virus which is unable to enter or replicate in insect cells, in a second type the insect is one of two alternating hosts, and in a third type the insect is the sole viral host. This range of interactions presents many opportunities for using viruses to design novel insect control agents and strategies. Before considering these, however, it is useful to outline the basics of viral replicative strategy.

As shown in Figure 4.2, a generalized viral replicative strategy can be divided into four phases. The first of these is entry into the uninfected

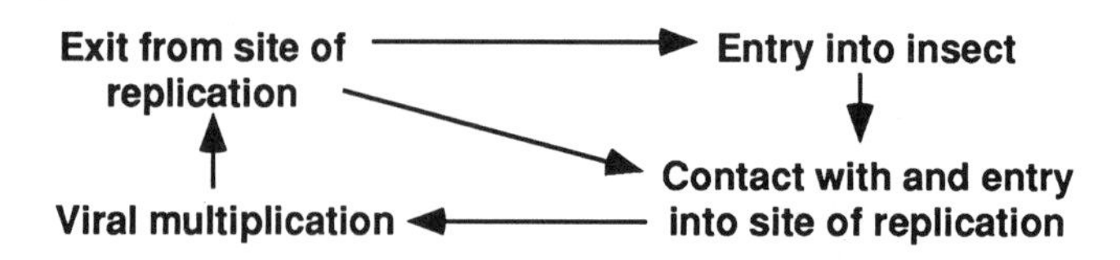

Figure 4.2. Generalised replicatative strategy of insect viruses.

Table 4.2. Virus families with members capable of entering insects or mites of agricultural importance

Virus Family or Group[a]	Genome Type[b]	Occlusion Body	Replication Status[c] in		Genera or Sub-groups Infecting Vertebrates	Insect Hosts or Vectors[d]	Used as Insecticide	References
			insects	plants				
Baculoviridae	1 ds DNA				–			
a. Nuclear polyhedrosis viruses (NPVs) (>500)		+	Host	–	N/A (e)	Diptera, Hymenoptera, Isoptera, Lepidoptera, Neuroptera, Orthoptera, Thysanura	+	Yearian and Young 1982; Granados and Federici 1986; Fazairy and Hassan 1988; Adams and Bonami 1991; Vlak, this volume
b. Granulosis viruses (GVs) (ca. 150)		+	Host	–	N/A	Lepidoptera	+	
c. Non-occluded viruses (NOVs) (3 + 15)		–			N/A	Coleoptera, Lepidoptera	+	
Poxviridae	1 ds DNA			–	+			
Entomopoxviruses (>30)		+	Host	–	N/A	Coleoptera, Diptera, Hymenoptera, Lepidoptera, Orthoptera	+	Arif and Kurstak 1991; Vlak, this volume
Iridoviridae	1 ds DNA				+			
insect iridescent viruses (>30)		–	Host	–	N/A	Coleoptera, Diptera, Hemiptera, Hymenoptera, Lepidoptera	–	Ward and Kalmakoff 1991
Polydnaviridae (>40)	~10–~30 ds DNA	–	Host	-?	–	Hymenoptera	–	Krell 1991
Ascoviridae (7)	1 ds DNA	–	Host	-?	–	Lepidoptera	–	Federici 1983

(continued)

Table 4.2. *Continued*

Virus Family or Group[a]	Genome Type[b]	Occlusion Body	Replication Status[c] in		Genera or Sub-groups Infecting Vertebrates	Insect Hosts or Vectors[d]	Used as Insecticide	References
			insects	plants				
Parvoviridae	1 ss DNA				+			
densonucleosis viruses (ca. 20)		–	Host	–	N/A	Diptera, Lepidoptera, Orthoptera	–	Kawase and Kurstak 1991
Reoviridae	10–12 ds RNA				+			
a. Cytoplasmic polyhedrosis viruses (CPVs) (ca. 250)		+	Host	-[f]	N/A	Coleoptera, Diptera, Hymenoptera, Lepidoptera	+	Martignoni and Iwai 1986; Belloncik 1989
b. Non-occluded insect reoviruses (ca. 14)		–	Host	?	N/A	Coleoptera, Diptera, Hemiptera, Thysanoptera	–	Adams and Bonami 1991
c. Phytoreo-viruses (3 + 1), Fijiviruses (5 + 2)		–	Propagative	Host	N/A	Hemiptera	N/A	Nault and Ammar 1989; Nuss and Dall 1990
Birnaviridae (2)	2 ds RNA	–	Host	–	+	Diptera	–	Miahle et al. 1983; Dobos et al. 1991
Rhabdoviridae	1 ss (–) RNA	–			+			
a. Drosophila σ virus		–	Host	–	N/A	Diptera	–	Brun and Plus 1980
b. Plant rhabdoviruses (>50 + >40)		–	Propagative	Host	N/A	Hemiptera	N/A	Nault and Ammar 1989

Picornaviridae (3 + ca. 30)	1 ss (+) RNA	–	Host	-[f]	+	Diptera, Lepidoptera, Orthoptera	–	Moore 1991b
Nodaviridae (6)	2 ss (+) RNA	–	Host	Replicates[g]	-[h]	Coleoptera, Diptera, Hymenoptera, Lepidoptera	–	Hendry 1991
Caliciviridae[i] (1)	1 ss (+) RNA	–	Host	–	+	Lepidoptera	–	Hillman et al. 1982
Tetraviridae (6)	1 ss (+) RNA	–	Host	-?	-[e]	Lepidoptera	+	Moore 1991a
Unclassified small RNA viruses	1-? ss (+?) RNA	–	Host	?	?	Acari, Coleoptera, Diptera, Hemiptera, Hymenoptera, Lepidoptera, Orthoptera	+	Hendry et al. 1985; Adams and Bonami 1991
Geminiviruses (24 + 10)	1–2 ss RNA	–	Circulative	Host	–	Hemiptera	N/A	Nault and Ammar 1989
Subterranean clover stunt virus group (1)	3–4 ss DNA	–	Circulative	Host	–	Hemiptera		Büchen-Osmond et al. 1988
Pea enation mosaic virus group (1)	2 ss (+) RNA	–	Circulative	Host	–	Hemiptera	N/A	Harris 1981
Marafiviruses (2 + 1)	1 ss (+) RNA	–	Propagative	Host	–	Hemiptera	N/A	Nault and Ammar 1989
Tenuiviruses (4 + 1)	4–5 ss (+/–) RNA	–	Propagative	Host	–	Hemiptera	N/A	Nault and Ammar 1989
"Umbraviruses"[j] (6)	1 ss (+) RNA	–	Circulative	Host	–	Hemiptera	N/A	Waterhouse and Murant 1983; Cockbain et al. 1986

(continued)

Table 4.2. *Continued*

Virus Family or Group[a]	Genome Type[b]	Occlusion Body	Replication Status[c] in		Genera or Sub-groups Infecting Vertebrates	Insect Hosts or Vectors[d]	Used as Insecticide	References
			insects	plants				
Luteoviruses (12)	1 ss (+) RNA	–	Circulative	Host	–	Hemiptera	N/A	Martin et al. 1990
Sobemoviruses[k] (10 + 4)	1 ss (+) RNA	–	Circulative	Host	–	Hemiptera	N/A	Gibb and Randles 1990
Tomato spotted wilt virus group (2)	4 ss (-?)[l] RNA	–	Circulative	Host	?	Thysanoptera	N/A	Büchen-Osmond et al. 1988

N/A = not applicable; ? = insufficient data available

[a] Insect-vectored plant viruses which do not enter the vector—the non-persistent or fore-gut-borne viruses—and insect-vectored vertebrate viruses are not included in this table. For the first 13 families (down to the Tetraviridae), nomenclature is according to the Atlas to Invertebrate Viruses (Adams and Bonami 1991); the other virus groups follow International Committee on Taxonomy of Viruses (ICTV) nomenclature (Matthews 1982 or Büchen-Osmond et al. 1988). The number of viruses or isolates so far described within each family or genus is given in parentheses; where appropriate, this is shown as (definite + possible) members. Possible members are those for which there are insufficient data to confidently place them in this family or group.

[b] The number and type of nucleic acid components comprising the virus genome are given. ss: single-stranded; ds: double stranded; for ss RNA genomes, + means that the genomic RNA is of the same sense as the mRNA(s), whereas – means that the genomic RNA is the strand complementary to the mRNA(s).

[c] – denotes neither a host, nor permits replication; *host* denotes the normal host in the wild, allowing viral growth and transmission, often accompanied by pathogenesis; the following terms refer to viruses traditionally identified as plant viruses: *circulative* means the virus enters the insect hemocoel from the gut, and is secreted via saliva, but without replication within the insect; *propagative* means that the virus is capable of replicating in its vector; *replicates* denotes that the virus is able to replicate under artificial conditions. See Harris (1981) and Nault and Ammar (1989) for a discussion of these terms, and Harrison (1987) for an overview of the proteins involved. See also Sylvester (1985) on the interactive transmission of multiply acquired viruses.

[d] Common names of host orders: Class Arachnida: Subclass Acari (mites); Class Insecta: Coleoptera (beetles); Diptera (flies); Hemiptera (bugs); Hymenoptera (wasps and bees); Isoptera (termites); Lepidoptera (moths); Neuroptera (lacewings); Orthoptera (grasshoppers); Thysanura (silverfish), Thysanoptera (thrips).

[e] Antibodies to several of these viruses have been detected in human sera; Doller et al. (1983).

[f] One virus in this group is also vectored by a plant (see Section 3.1.2, "RNA Viruses").

[g] Flock house virus has been shown to replicate in plants (Selling et al. 1990).

[h] Nodamura virus is able to replicate in mammalian cells (Hendry 1991).

[i] This group is represented by a single virus, *Amyelois transitella* chronic stunt virus. Although formally unclassified, it is very similar to the caliciviruses.

[j] These viruses, e.g. carrot mottle virus, require a helper virus for replication in the plant and circulation in the vector. The helper provides the capsid protein, since the "umbraviruses" encode no coat protein of their own. Five of the six members use luteoviruses; the other member, bean yellow vein-banding virus, uses pea enation mosaic virus. The "umbraviruses" (*Lat.* "umbra"—uninvited guests) were proposed as a new plant virus group by AF Murant at the 5th International Congress on Plant Pathology, Kyoto, 1988. The group and the name have not yet been formally approved by the ICTV.

[k] This refers to velvet tobacco mottle virus only; transmission by insects of other viruses in this group is not persistent (Büchen-Osmond et al. 1988).

[l] These viruses strongly resemble the bunyaviruses (which are insect-vectored) in a number of aspects, especially in the maturation of the enveloped virus particles. For this reason, and because their encapsidated genomic RNA is not infectious, the genomic RNA is tentatively descibed as negative-sense.

host and generally occurs by one of two routes. One of these is direct transmission from an infected parent *via* the germ-plasm (vertical transmission). Several insect viruses are known to use this route, for example, the *Drosophila* A and sigma viruses (Brun and Plus 1980) and aphid lethal paralysis virus (Hatfill et al. 1990). In many cases, and perhaps because the virus is so intimately tied to the host life-cycle, viruses transmitted in this manner have relatively few pathogenic effects. The second mode of transmission is by direct infection of the host (horizontal transmission), and in this strategy the virus generally enters *per os* when the host consumes virus-contaminated food. Most occluded viruses and small RNA viruses of insects can be transmitted in this manner (Adams and Bonami 1991).

Following entry into the host, the second phase in the viral life cycle involves contact with tissue permissive for replication; this contact is followed by entry of virus into cells of that tissue. Once inside a cell the third phase, viral multiplication, takes place. Viral strategies of multiplication are numerous and varied, and we do not intend to detail them here. In brief, however, they involve liberation of viral nucleic acid from its protein coat, expression of viral genes, replication of the viral nucleic acid and assembly of mature (infectious) virus particles. These particles then complete the fourth stage of the replication cycle, the exit from the cell and/or host, and, given favorable circumstance, they then establish a new round of infections. Timing of events in each of these replicative phases varies widely (see references listed in Table 4.2).

The design of novel control agents and strategies must utilize our knowledge of the different aspects of this replicative process. Thus, for instance, all viral gene expression is known to take place within host cells, so that it is possible to use engineered forms of viruses as agents to direct intracellular expression of foreign genes. Alternatively, since all virus particles contain elements which allow their entry into permissive cells, these "signals" have potential as "target location" agents to carry (cyto)toxic elements to particular cell types. In the following sections we discuss ways in which viruses could be used either as the bases of new control agents or as molecular sources or tools for developing other control strategies.

3. Viruses as the Basis for Designing New Insect Control Agents and Strategies

The problems with chemical insecticides described earlier do not preclude a large role for insecticides in future pest control programs. However, if they are to play such a role, they must fulfill as many as possible of the criteria listed in Table 4.1. One of the most important of those criteria is product safety, and in this respect virally-based insecticides would appear to offer an attractive alternative to those based on chemicals. Examination of other criteria, however, shows that additional problems must be addressed before viruses will provide viable alternative insecticides.

Seven of the viral groups listed in Table 4.2 have been used as insecticides, and as shown there, the majority of those are of the "occluded" type. The "occluded virus" group is an artificial and heterologous assemblage of viruses which incorporate infectious particles into a proteinaceous occlusion body (OB) during the latter stages of multiplication.

The relative success of occluded viruses when used as insecticides is closely related to production of the OB, since this protective structure confers stability to virus particles and thus allows lengthy retention of infectivity. The fact that most of these viruses produce large numbers of infectious progeny enhances their attraction as candidates for large scale production and distribution. A further advantage of occluded viruses is their wide distribution among the world's major insect pests.

Despite these factors, even the occluded viruses have not achieved real commercial success, probably as a result of two separate problems. One of these relates to the high cost of producing large quantities of the viruses and the other to their relatively slow rate of kill; in general these viruses take from five days to several weeks to kill an infected insect. While this rate of action can be acceptable in some circumstances, it is clearly not useful in, for example, high value crops with low damage thresholds, such as apples or cotton.

If the intrinsic target specificity of viruses (see Table 4.1) is to be translated into useful virally-based insecticides, then both of the problems mentioned above must be overcome. Thus, production systems must be developed which are both cheap and safe; factors which relate to production of viral insecticides are described in Section 4.3. On this point it is instructive to note that the most successful usage to date of a viral insecticide is that of the NPV of soybean looper, *Anticarsia gemmatalis*, in Brazil (Moscardi 1988). Much of the virus used in that program is produced by a government agency under field conditions, and the program includes a scheme in which farmers return dead insects from the field to assist in the production of more virus. While very successful in this instance, it is clear that such a program depends heavily on access to relatively cheap sources of labor.

In addition to improvement of production capabilities, but of equal importance, the viruses must be made more efficacious, perhaps through manipulation to increase their speed of action; it is this aspect of the design of virally-based control agents which we now address.

3.1. Viruses as Vectors of Insecticidal Agents

A number of methods have been proposed for generating faster acting viral insecticides, but greatest potential appears to lie in the development of viral vector systems. These systems will allow biologically-derived insecticidal molecules to be delivered quickly and efficiently to the target insect through viral infection. In Chapter 2 Binnington and Baule dealt with naturally occurring insecticidal molecules which might be used in conjunction with viral vector systems. In this section we deal with the virus groups which might be used to develop such systems.

3.1.1. DNA Viruses

Until recently most efforts to develop viral vector/expression systems concentrated on viruses with double-stranded (ds) DNA genomes. In part this was because most field control of insects had been achieved with viruses of this type, and in part because existing expression systems had largely been based on a small number of mammalian dsDNA viruses, such as the adenovirus SV40 (Solnick 1981), the bovine papillomavirus (Sarver et al. 1981) and vaccinia virus (Mackett et al. 1982). Using these systems as models, it was realised at an early stage that baculoviruses, with their dsDNA genomes, would also be readily amenable to genetic engineering (Smith et al. 1983; Smith and Summers 1988).

The first baculovirus expression system was based on the *Autographa californica* nuclear polyhedrosis virus (AcMNPV), and this system is discussed in detail by Vlak (Chapter 3). As reported therein, high level expression of many eukaryotic, prokaryotic and viral genes has now been achieved.

Vlak describes a number of attempts to increase the insecticidal activity of baculoviruses through insertion of foreign genes. Until recently most attempts had resulted in only slight improvements in their speed of action (Maeda 1989). The first reports of viruses which give significantly improved levels of control have now been published. In each instance a gene encoding an insect-specific toxin was introduced into the baculovirus genome; in one case the gene encoded a neurotoxin from the mite *Pyemotes tritici* (Tomalski and Miller 1991), and in two other cases a component of the venom of the North African scorpion *Androctonus australis* (Stewart et al. 1991; Maeda et al. 1991). In all cases virally-driven expression of the toxin gene substantially reduced the time taken to paralyze or kill the insect host.

Such work demonstrates that molecular techniques could be used to produce baculovirus insecticides with substantially increased field efficacies. Given the wide range of important insect pests from which baculoviruses have been isolated (Yearian and Young 1982), future prospects for developing engineered and highly efficient baculovirus insecticides are great. Demonstration of the potential of baculovirus systems has also led to examination of the possibilities for developing other dsDNA insect virus systems.

Engineering of entomopoxviruses (EPVs) has thus become a topic of considerable interest, particularly since the viruses in this group have potential as vectors of insecticidal agents to a range of pests which show little susceptibility to baculoviruses. Foremost among these target groups are various scarabeid beetles and grasshopper and locust species; for example, EPVs have been isolated from the agriculturally important beetle genera *Dermolepida* and *Aphodius* (Goodwin and Filshie 1975), and from orthopterans such as *Melanoplus sanguinipes* (Henry and Jutila 1966) and *Locusta migratoria* (Purrini and Rohde 1988).

While there has been, as yet, no published report of the successful engineering of a member of this group, recent cloning and sequencing of EPV genes which encode highly expressed viral proteins such as spheroidin (Yuen et al. 1990; Hall and Moyer 1991; DJD unpublished data), suggest that this goal will shortly be achieved. As a corollary, it has not yet been proven that toxins which are highly efficacious when delivered to lepidopterans by baculoviruses will be equally active in, for example, EPV-infected beetles. Nevertheless, the strategy of virally-vectored toxins is clearly applicable, and its implementation is unlikely to pose major difficulty.

A similar situation exists with iridoviruses with, as yet, no instance of an iridovirus being engineered to express a foreign protein. Over recent years, however, research effort has been applied to the molecular biology of insect-infecting members of this family (Fischer et al. 1988; Home et al. 1990; Tajbakhsh et al. 1990). Unlike baculoviruses and EPVs, the iridoviruses do not produce a polyhedral body but are still produced in very large amounts in infected insects (up to 25% of host dry weight; Williams and Smith 1957); proteins such as the major nucleocapsid protein are presumably produced in very large quantities. By analogy with other large dsDNA viruses such as the baculoviruses, the iridoviruses may have redundant areas in their genomes. In this case it might be possible to duplicate viral promoters, link them to foreign genes and insert the foreign gene construct into these redundant areas. The transgene should then be encapsidated and expressed following infection of an appropriate host.

As in the case of EPVs, development of iridovirus vectors would establish control potential for insect groups not otherwise susceptible to viral insecticides. For instance, a number of iridoviruses have been isolated from medically important dipteran species and from agriculturally important coleopteran species (Kelly 1985, for review).

3.1.2. RNA Viruses

The possibility of genetically engineering RNA viruses has largely been overlooked until recently, presumably because significant technical difficulties were anticipated. With the advent of efficient techniques for production of infectious RNA transcripts from cloned full-length cDNA copies of the viral genome, these viruses can now be manipulated relatively easily. Indeed, their (generally) small genomes may eventually prove to be more easily manipulated than the large ones, for example, the baculoviruses. Potential drawbacks which might hamper this approach, such as constraints on the amount of RNA which can be packaged into the virion, remain to be assessed.

The prospects for rapid development of engineering of insect RNA viruses are enhanced by the extensive experience already gained with mammalian and plant RNA viruses. To illustrate this, two of several demonstrably successful engineering strategies are described below, each of which prove that small RNA viruses can express heterologous genetic information. In these and most other reported cases genomic sequences not required for viral replication have been deleted and replaced with a "payload" gene. In several instances a reporter gene, such as the bacterial chloramphenicol acetyltransferase (CAT) gene, has been employed to study expression from the potential vector. For use in insect field control strategies, this gene would, of course, be replaced by another encoding an insecticidal factor.

Thus, the coat protein genes of two plant RNA viruses, tobacco mosaic virus (TMV, Takamatsu et al. 1987) and brome mosaic virus (BMV, French et al. 1986) have been replaced by the CAT gene. When RNA transcripts derived from the modified genomic templates were used to inoculate plants (in the case of TMV) or protoplasts (BMV), the virus was shown to be capable of replication, and the reporter gene was shown to be expressed.

In a similar strategy Xiong et al. (1989) replaced genes encoding the structural proteins of the mammalian Sindbis alphavirus with the CAT gene. When RNA transcripts derived from the modified template were transfected into cells they, too, showed expression of the CAT enzyme. In order to induce packaging of the engineered RNAs into virus particles, another complementary mutant RNA, carrying the structural protein genes but defective in others, was introduced into infected cells (Geigenmüller-Gnirke et al. 1991).

It is apparent from these examples that considerable knowledge about the structure, organization and replication of the RNA viral genome is required before strategies similar to those developed for DNA viruses can be implemented. Nevertheless, such information is becoming available and will facilitate further advances in this field.

As mentioned above, the potential of insect RNA viruses for use as vectors of insecticidal genes has not yet been exploited. In view of this and the firmly established position of baculoviruses, it seems likely that future development of RNA viruses will be seriously explored only for control of insect groups refractory to baculoviruses. We therefore concentrate our discussion here on small RNA viruses which infect grasshoppers and crickets (Orthoptera), aphids (Hemiptera), mites (Acari) and beetles (Coleoptera). This, of course, does not preclude the possibility that small RNA viruses, which infect other major groups such as lepidopterans, might be used when more is known of their biology. RNA virus families infecting lepidopterans include the *Nudaurelia*-β viruses and the so-called *Nudaurelia*-like viruses (reviewed by Moore et al. 1985; Moore 1991a).

The RNA viruses most likely to be useful in control of orthopterans are the members of the cricket paralysis virus (CrPV) complex (Scotti et al. 1981). The viruses of this complex are members of the Picornaviridae (Moore 1991b) and resemble mammalian enteroviruses in their biophysical characteristics. The potential of the CrPV complex for control of orthopteran pests is evidenced by the ability of some isolates to cause widespread mortality in laboratory cultures (Reinganum et al. 1970) and the high frequencies of some forms in field populations (Wigley and Scotti 1983). Given the wide host range of members of the complex, this group may also have potential to control pests from other orders. For example, one strain of CrPV originally isolated from the Australian field cricket, *Teleogryllus commodus* (Reinganum et al. 1970), will both infect and induce high mortality in the olive fruit fly, *Dacus oleae* (Manousis and Moore 1987; Manousis et al. 1988). These latter workers suggested that this isolate might be an effective component in a control strategy for *D. oleae*, which is a major pest of olives.

Another virus group, the nodaviruses, may prove useful in controlling some beetle pests. Members of this group, such as black beetle virus, are among the best characterized insect viruses at the molecular level (Kaesberg 1987; Kaesberg et al. 1990; Hendry 1991). The type member of this family (Nodamura virus) has, however, been shown to replicate in both insect and mammalian cells (Hendry 1991), and stringent safety testing would need to precede any use of this group of viruses in the field.

Some small RNA viruses have been shown to cause pathogenesis in aphids. These include two picorna-like viruses, *Rhopalosiphum padi* virus (RhPV) (Gildow and D'Arcy 1990) and aphid lethal paralysis virus (Hatfill et al. 1990), and an unclassified RNA virus, SAV, from *Sitobion avenae* (Allen and Ball 1990). RhPV has an unusual ability to utilize plants as passive reservoirs (Gildow and D'Arcy 1990), making it a plant-transmitted insect virus! Only one other insect virus, the leafhopper A reovirus from *Cicadulina bimaculata*, has been shown to utilize plants as passive reservoirs (Ofori and Francki 1985).

Few viruses have been described from mites. Reed and Desjardins (1978) found RNA in spherical virus-like particles obtained from the citrus red mite *Panonychus citri*, but these particles did not appear to be disease-causing agents. *P. citri* is, however, subject to significant control in the field by epizootics of a non-occluded, rod-shaped virus (Reed 1981), the classification of which is uncertain. A nonoccluded, rod-shaped virus was also visualized in fat cell nuclei of diseased European red mites, *Panonychus ulmi* (Bird 1967). In view of the importance of mites as agricultural pests, it is to be hoped that increased exploratory efforts will yield additional viral pathogens.

3.2. Viruses as Sources of Components for New Insecticidal Agents and Strategies

So far we have considered the use of insect viruses primarily as insecticides, either in wild-type (Section 2.2) or genetically engineered forms (Section 3.1). In each of these situations the virus must be in a replicative form in order to control the insect pest, either through direct pathogenesis associated with infection or through expression of an heterologous insecticidal protein. In the section that follows we consider examples of, and possibilities for, the use of isolated components of viruses in novel control agents and strategies.

The scheme shown in Figure 4.3 provides a paradigm of the manner in which most pest control programs function. While chemical insecticides provide good examples of the agent described therein, the scheme can be applied equally to a traditional bio-control agent. As described in Figure 4.3, the initial step in the control process involves production and

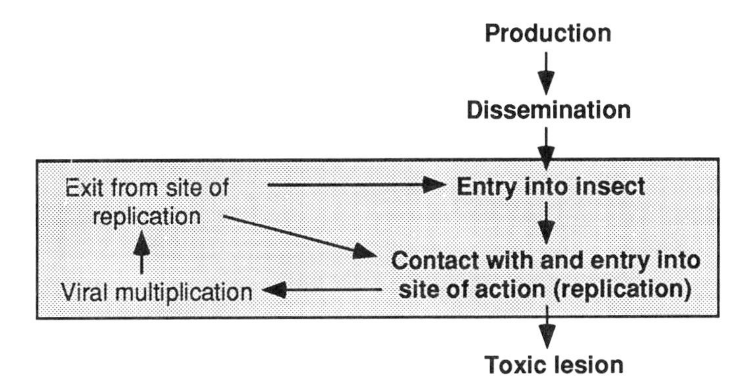

Figure 4.3. Generalised scheme for an insect control agent and its relationship to the replicative strategy of insect viruses.

distribution of the control agent into the environment; at this step the process is under human direction. Once released the agent must gain entry into the insect, reach its site of action, and finally, act to control the host. These latter processes closely parallel the replicative cycle of viruses (see Figure 4.1), and recognition of these similarities gives insight and access to many useful "tools" based on viral components. We now discuss how such viral components might be applied to the various stages of the control process outlined in Figure 4.3.

3.2.1. Entry into Insects

Unlike most chemical insecticides, proteinaceous control factors must enter the insect by ingestion, since their polar nature hampers efficient diffusion across the lipophilic insect epidermis. After ingestion such control factors must both survive the proteolytic digestive milieu of the gut and cross a series of membrane barriers before reaching cellularly-located target sites. This demanding requirement may explain why only one group of proteinaceous control factors is so far known to act effectively *via* the insect gut, that being the endotoxins of *Bacillus thuringiensis*. We anticipate that investigation of components of insect viruses will extend the number of orally active control agents, since viral capsids are composed largely of proteins yet are able to survive the gut environment and cross its membrane barriers. Determination of the manner in which they accomplish this might suggest new strategies and/or factors with which to deliver control agents to sites of action inside pest insects.

Our understanding of the process(es) governing viral entry into insects is still rather limited despite its importance and potential utility. In one well-studied example, the entry of AcMNPV into its insect host (a larval lepidopteran) has been shown to be through two distinct routes (Keddie et al. 1989). One of these routes involves viral entry from the gut lumen into midgut cells, where the virus replicates to produce new virions, which, in turn, bud into the haemocoel and infect other cells. The other, and less understood, route appears to be "between" gut cells. AcMNPV virions were observed to traverse the interstitial spaces between midgut cells and thus to enter the insect haemocoel directly from the gut lumen (Keddie et al. 1989).

The entry of viral pathogens into sucking insects, such as aphids or mirid bugs, can apparently be through either specific or nonspecific mechanisms. Some cases of nonspecific entry appear to result from the remarkable porosity of the gut to macromolecular particles (e.g., some dyes) (Gibb and Randles 1990). In other cases more specific mechanisms have been implicated. For example, plant luteoviruses can pass through the hindgut wall and into the haemolymph of aphids, but a specific interaction between the virus and the host salivary gland regulates uptake into that organ (Gildow 1987). In the case of potato yellow dwarf virus (a plant rhabdovirus), binding of the virus to insect cells has been shown to be mediated by the glycosylated G protein. A similar protein is known to be an acceptor molecule for vertebrate rhabdoviruses (Gaedigk-Nitschko et al. 1988).

Granados and co-workers have recently characterized a viral component which facilitates virion movement across one of the major membrane systems within the insect gut, the peritrophic membrane (Derksen and Granados 1990). This substance, viral enhancing factor (VEF), is a monomeric 101kDa protein component of the occlusion body of *Trichoplusia ni* granulosis virus. Proteins related to VEF also appear to be present in the occlusion bodies of other baculoviruses (Granados and Corsaro 1990). VEF has been shown to hydrolyse specific glycoprotein components of the peritrophic membrane, presumably leading to increased permeability of the structure and resulting in the observed enhancement of viral infectivity (Granados and Corsaro 1990).

The gene encoding this protein has recently been cloned (Granados and Hashimoto 1990), and several possibilities for its use are immediately apparent. A major use will probably be enhancement of the efficacy of other microbial pesticides, since VEF by itself is not toxic. It might be expected that the presence of VEF protein in transgenic plants, or in insecticide formulations, would reduce the necessary concentrations of insecticides, and thus, the cost of pest population control.

Granulosis viruses (GVs) have apparently evolved multiple protein components to assist their entry into insects. An infectivity enhancing factor from occlusion bodies of the *Pseudaletia unipunctata* GV has been identified and termed Synergistic Factor (SF; Tanada 1985). Although SF is less well characterized than VEF, it is clear that the two factors are different. In addition to certain distinct physical properties, SF has been shown to increase absorption of viruses to cells both *in vivo* and *in vitro*, where there is no peritrophic membrane barrier. The manner of action of SF is unknown, but it may find a use similar to that of VEF in genetically engineering microbial insecticides and plants resistant to insects.

3.2.2. Entry into Host Cells and Target Sites

Factors used by viruses to assist their attachment and entry into insect host cells are also likely to play important roles in the development of novel insect control measures. As mentioned previously, diffusion of proteinaceous control factors across cell membranes is limited, so that their effective entry into cells depends on specific mechanisms, perhaps similar to those used by viruses. By incorporating binding determinants of viral origin into the design of other control agents, it might be possible to enhance the ability of those agents to cross insect cell membrane barriers.

Furthermore, the use of determinants from insect viruses which display tropisms to specific cell types may allow control agents to be directed to sites critical to the agents' activity. Examples of such determinants include those possessed by the *Nudarelia*-β and cricket paralysis viruses, which display midgut cell and nerve cell tropisms, respectively (Hess et al. 1978; Moore 1991b).

As mentioned previously, little is known of the manner by which insect viruses enter host cells. However, extrapolation from investigations of vertebrate viruses suggests that they enter by a two-step process of attachment and membrane transit (Paulson 1985; Marsh and Helenius 1989). For vertebrate viruses it has been shown that attachment occurs when a structural determinant located on the surface of the virion, and sometimes referred to as a viral attachment protein, binds to a receptor on the cell membrane. Attachment of virus particles to host cell receptors brings them into intimate physical contact with the membrane, and transit can then occur by direct fusion with the plasma membrane (in the case of some enveloped viruses) or by adsorptive endocytosis.

Sivasubrahmanian et al. (1990) recently reported an example of the manner in which a viral element responsible for cell entry can be incorporated into the design of novel control agents. Those workers constructed a hybrid toxin gene whose amino-terminal portion was derived from an endotoxin gene of the bacterium *Bacillus thuringiensis* var. *tenebrionis* (*B.t.t.*), and whose carboxy-terminal was derived from the gp67 gene of the lepidopteran-specific AcMNPV baculovirus. This juxtaposition was shown to alter the specificity of the *B.t.t.* toxin from its normal coleopteran targets to the lepidopteran hosts of the virus. This was a result of great interest, since the product of the gp67 gene is normally found in the outer capsid structure of the non-occluded form of the virus, and has been implicated in the entry of virus into host cells.

The results of this hybrid-toxin experiment are best explained in the light of the known mode of action of *B.t.* toxins (Li et al. 1991). This and other work has suggested that the toxin moiety must first bind to an insect gut membrane receptor in a specific manner, and that this event leads to a conformational change in the toxin molecule, which in turn results in ion channel formation and subsequent death of the cell. The specificity displayed by *B.t.* toxins to different groups of insects is then thought to reflect specificity of the binding of toxins to receptors present in different insect groups. According to this model, *B.t.* is thus toxic only to coleopterans, because it can normally bind only to coleopteran gut membranes.

The implications of these ideas are far ranging for the *B.t.t.* industry. Finding new *B.t.* toxins with desirable species specificity has heretofore required expensive and time-consuming screening procedures. Conversely, viruses which infect specific insect pests, but which are unsuitable for control purposes, are relatively easy to find. The ability to alter the toxicity range of *B.t.* toxins through their combination with virally-derived gut binding determinants could greatly speed the discovery/development process and substantially reduce development costs.

Perhaps more importantly, however, the work of Sivasubrahmanian and colleagues may provide a paradigm for methods with which to counter resistance to *B.t.* insecticides, a phenomenon of increasing concern (Gibbons 1991). Most cases of insect resistance to *B.t.*-based insecticides have been correlated with decreased binding of the toxins to gut receptors. Alteration of the binding specificity so as to target different gut receptors in resistant species might allow new "models" of *B.t.* to be produced which can overcome this problem. In this manner the period over which *B.t.*-based insecticides remain efficacious might be greatly extended.

3.2.3. Direct Control by Viral Components

It is perhaps surprising to realize that many of the factors with which viruses exert control over host cells are relatively subtle in action, as compared to the acutely lethal toxins produced by entomopathogens such as some bacteria. However, such factors might still be of use for insect pest management if expressed by novel control agents.

An example of a "subtle" but potentially valuable factor is the product of the baculovirus AcMNPV *egt* gene which encodes an enzyme, ecdysteroid UDP-glucosyltransferase, that is expressed during the course of viral infection (O'Reilly and Miller 1989). This enzyme inactivates ecdysone, the molting hormone of insects, thereby enabling the virus to prevent further molting of the infected insect (Miller 1990). Perhaps the greatest potential for the use of the *egt* gene is in vectors derived from viruses which infect larvae of species that are pests as adults, particularly in cases where the virus is very slow to induce pathogenesis in the host. Insertion of the *egt* gene in these vectors might be expected to both limit host development (e.g., maintaining the larval stage of the mosquito life cycle) while allowing normal viral pathogenesis to continue.

Factors produced by polydnaviruses might also be of interest in design of insect control agents. Members of this group of viruses appear to be obligate symbionts of brachonid and ichneumonid parasitic wasps (Krell 1991). These viruses are transmitted when the female wasp oviposits, and they then affect the endocrine and immune systems of the wasps' hosts. Factors from a polydnavirus of *Apanteles kariyai* extend the temporal development of larvae of the wasp's host (Tanaka et al. 1987), while factors from a virus of a *Chenolus* species appear to shorten the time for larval development (Jones et al. 1986). Other factors from many of these viruses are able to suppress the host immune system and thereby prevent encapsulation of the developing parasitic wasp larva (see Guzo and Stolz 1987, who describe the activity of a virus from the brachonid wasp, *Cotesia melananoscela*). Precise knowledge of how the viruses exert these effects is still forthcoming. However, enough is already known to show that polydnaviruses possess a sophisticated repertoire of factors affecting the basic biological processes of insects. Further research on the biology of these viruses, with the aim of deriving factors of value to insect control, is certainly warranted.

3.3. Viruses as Tools for Developing Insect Control Strategies

Recent advances have led to the development and use of viruses as tools for manipulating and analyzing both prokaryote and eukaryote genomes. For instance, much of modern recombinant DNA technology has resulted from cloning and expression systems engineered from bacterial viruses. Viruses of eukaryotes have also found widespread use as expression systems and vectors for gene transfer, and a range of promoters from these viruses have been used to drive expression of foreign genes in mammalian and plant cells (e.g., the SV40 promoters, McKnight and Tjian 1986; Kaufman 1990; and the 35S promoter of cauliflower mosaic virus, Benfey and Chua 1989; Gordon 1990).

In this Section we address the potential for using insect viruses as tools in developing new pest control strategies. At present we envisage their use in two major areas. The first of these is the design of transformation vectors for generating transgenic insects, which will certainly aid our analyses of insect genomes and might also be of value in the implementation of genetic control strategies. The second lies in the use of viruses as cloning and screening systems. Once developed, such a system could be used to screen for insect-specific bioactive compounds.

3.3.1. Transformation of Insects

Stable transformation of cells by introduction of foreign genetic material has become an important strategy in studies of, for example, the control of gene expression, and in the development of processes for large scale *in vitro* production of proteins. There has, however, been limited success in the application of this approach to insects, and to date transformation has been reproducibly achieved only for drosophilids and mosquitoes (see O'Brochta and Handler, Chapter 12). In addition, the work so far reported has exploited only vectors derived from the P element of *Drosophila melanogaster* or other nonviral transposons.

If general transformation technologies could be developed for insects, one likely application would be in the efficient development of genetic control programs using, for example, sterile male release methodology.

While currently in use for some Diptera, extension of such programs to other orders would be of great value. This would, however, require development of alternative strategies for gene transfer, perhaps in the manner of those described below.

3.3.1.1. Retroviruses as Gene Transfer Vectors. Retroviruses and retrotransposons are members of the diverse family of mobile genetic elements which replicate and transpose DNA via an RNA intermediate (reviewed by Varmus 1988). The DNA products of the process are, in most cases, integrated into the host cell genome, from where they may then be further expressed *via* the normal transcription and/or translation processes.

Retroviruses have already found widespread application as vectors for the construction of stably transformed mammalian cell lines, and for the introduction of transgenes into germ-line or somatic tissues of organisms (reviewed by Palmiter and Brinster 1986). As discussed by those authors, our detailed knowledge of the retrovirus life cycle has allowed the design of vectors which integrate a foreign gene into a genome in a manner precluding further replication or spread; in such cases the transgene is then securely contained.

Strategies which lead to such genetic containment generally involve division of the viral genome in a manner which ensures that proteins required for viral replication are expressed from an otherwise replication-incompetent helper virus previously integrated into the cell line genome (the packaging line). Such division of the genome also eliminates the potential for regeneration of functional virus through homologous recombination between the helper virus and the gene transfer vector, so that the transgene can be considered contained within the initially infected target cell. Gene transfer efficiencies using these methods have been high (Miller and Rosman 1989; McLachlin et al. 1990). Identical strategies should have ready application to insect retroviruses for transformation of insect cells, as described below.

Recent work on the yeast *Ty* retrotransposon further illustrates the power of retrotransposon-based transformation strategies (Kingsman and Kingsman 1988). In particular, it has proven possible to introduce foreign proteins into predetermined regions of the retrovirus core protein, thus allowing synthesis of hybrid virus-like particles (Kingsman et al. 1991). It may prove possible to target such hybrid retroviral cores to specific cell types if appropriate combinations of the core virus protein (*gag*) and the virally-encoded membrane protein (*env*) can be devised.

3.3.1.2. Retroviruses and Retrotransposons in Insects. The only insect species from which retrotransposons have been extensively characterized is *Drosophila melanogaster*. Ten retrotransposon families have been described from that species (Bingham and Zachar 1990), and some of these families have been shown to contain elements which infect other drosophilid species. These retrotransposons could, in theory, be manipulated as gene transfer vectors, much as the yeast *Ty* elements have been.

Virus-like particles (VLPs) resembling retroviruses have been described from *Lucilia cuprina* and *Drosophila* species (Akai et al. 1967; Binnington et al. 1987), and those from the latter were found to contain RNA transcribed from the *copia* retrotransposon (Shiba and Saigo 1983), indicating that the retrotransposon was able to complete the intracellular viral replication cycle. It may then be possible to manipulate insect retrotransposons as vectors through use of strategies similar to those employed with the yeast *Ty* elements and described in Section 3.3.1.1 above.

VLPs found in *Drosophila* have been implicated in the transformation of host cells into tumorous lines (Gateff et al. 1984). Discovery of similar elements in other species might facilitate construction of immortalized cell lines for use in the mass production of virally-based insect control agents as described in Section 3.1.

3.3.2. Viral Cloning and Screening Systems—"Biostochastics"

One of the key requirements for an "ideal" insect control agent is ease of discovery and development (Table 4.1). Previous experience has shown that new control agents must constantly be developed to replace older ones which lose efficacy or acceptance. We envisage that insect viruses could play an important role in this development process, in particular as an integral part of a novel "biostochastic" cloning and screening system for the identification of proteinaceous control factors.

As described below, this biostochastic strategy contrasts with both major strategies previously applied to the search for insect control agents, namely, the random screening and biorational methods (summarised in Figure 4.4). In the random screening process, factors (usually chemicals) are bioassayed for insect toxicity without prior knowledge of their biological properties. Compounds which show toxicity are selected, further tests are used to determine their degree of insect selectivity, and other analogues of the molecule are synthesised and screened in turn. In this way an active compound is found and then "refined" to suit the criteria listed in Table 4.1. While the random screening process has proven successful, it is expensive to implement and its output in the recent past has been limited.

The alternate biorational approach was devised as a way to take advantage of scientific understanding of biological systems. This process aims to predict target sites at which lesions might produce selective mortality of particular pest species. Agents are then designed to produce those lesions, and once again are refined, perhaps through use of a bioassay more sophisticated than a toxicity test. As an example,

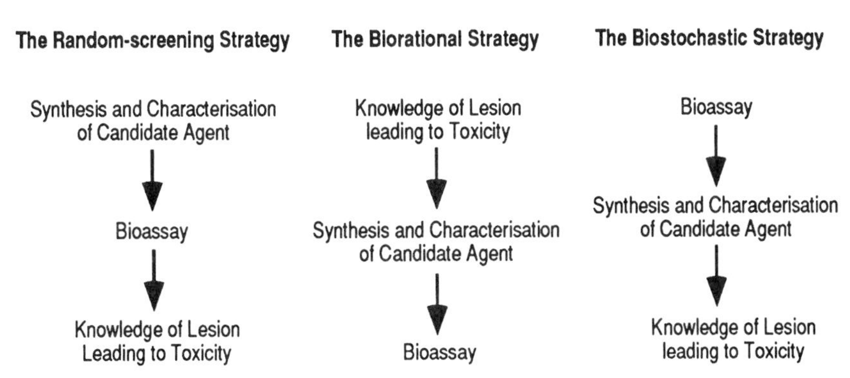

Figure 4.4. Order of steps in three strategies used for the discovery and development of insect control agents.

Hammock and colleagues identified an insect hormonal system as a target site, and then, in an attempt to produce the desired lesion, designed a recombinant baculovirus expressing a hormone-metabolizing enzyme (Hammock et al. 1990).

The biostochastic approach may be considered as a combination of the two previous strategies, since it would involve use of biological knowledge in conjuction with a stochastic screening process in order to isolate control factors from a large number of candidates. Implementation of this process will require a bioassay designed to simultaneously screen large numbers of candidate agents for bioactivity. This bioassay might take the form of a simple toxicity assay or could be designed to assess effects on a predetermined target site, e.g., blockage of a sodium channel. Once bioactivity is detected in a particular group of candidates, a reiterative process of group division and bioassay can be used to isolate the agent for characterization.

Integral to the biostochastic process is the availability of a system which allows rapid production and bioassay of large numbers of candidate agents. Insect viruses are ideally suited to this important role, since they have the capacity to enter the insect host efficiently, target particular cells and, following infection, express heterologous DNA inserted into their genomes. Development of efficient cloning and expression systems based around insect viruses will thus allow the biostochastic strategy to be both sensitive and efficient.

As an example of such a system, cDNA from a source such as the venom sac of a parasitic wasp could be cloned into a virus, and aliquots of this population of recombinant viruses used to infect a host of interest. The bioassay might then simply be the more rapid death of the insect, or might involve a more sensitive assay such as detection of a nervous system malfunction. Further subdivision of the progeny virus population would allow the "active" recombinant(s) to be isolated.

Section 3.3 described how a number of viruses are now used as the bases of efficient cloning and expression systems. To date, only a few of these have been insect viruses, and once again, most are baculoviruses. In principle, however, any of the viruses discussed in Sections 3.1.1 and 3.1.2 could be employed in a biostochastic screening system, and their use might then also facilitate discovery of bioactive compounds suitable for the control of different and important insect pests.

4. Challenges to the Future Use of Insect Viruses for Biological Control

In previous Sections we discussed ways in which insect viruses could be utilized to design novel pest control agents and strategies, and concentrated on the use of molecular techniques for enhancement of efficacy. However, Table 4.1 listed a series of other important criteria relating to the successful use of such agents, and it is those which we discuss in this Section.

4.1. Target Species Resistance

ffrench-Constant et al. (Chapter 1) present a detailed discussion of target-pest resistance to chemical insecticides. Until recently it was widely believed that entomopathogens, as components of co-evolved host-parasite systems, were less likely to engender resistance than were chemical insecticides. It is now, however, obvious that when entomopathogens are intensively utilized for pest control purposes, the processes which normally constrain selection for resistance are removed. For instance, rapid selection for resistance has been recorded in situations where *B. thuringiensis* has been used indiscriminately (Gibbons 1991).

Although less data are available detailing resistance to insect viruses, there are several well-documented examples. For instance, resistance to baculoviruses (Briese 1986), CPVs (Watanabe 1968), flacherie disease of silkworms (Uzigawa and Aruga 1966) and sigma and C viruses of *Drosophila* (Fleuriet 1976; Plus and Golubovsky 1980) have all been described.

For most strategies described in this chapter, it is then not a question of whether resistance will arise, but of when and to what extent it will occur. Given this expectation it is clear that future development of any new pest control agents or strategies must proceed hand-in-hand with development of efficient management plans for their use. In some cases, such as that of *Heliothis* species in the Australian cotton industry, strategies which attempt to limit the increasing problems of target-pest resistance to chemical insecticides have already been implemented and appear to be successful (Forrester 1991).

Implementation of management plans for insecticides does not present great problems since various widely acceptable options, such as rotation of insecticides within and between growing seasons, are available. Furthermore, as examples of efficient insecticide management receive wider publicity, their implementation is likely to become easier. More problematic, however, will be implementation of methods with which to limit resistance development when the control agents are constitutively expressed in plants or animals. Recent laboratory experience with the *B.t.* toxin suggests that insects can rapidly develop resistance if exposed to such constant forms of selection pressure (Stone et al. 1989).

4.2. User and Public Acceptance

A consequence of the use of chemical pesticides has been high user expectations regarding flexibility of control options and rapid knock-down effect on pest populations (Perkins 1988). It is therefore not surprising that users have traditionally distrusted biological insecticides, since their application was often restricted to narrow time "windows," and they were relatively slow to control the pest population. However, increasing public pressure and the obvious environmental damage caused by widespread chemical use have meant that many users now accept that sustainable forms of insect pest control must be implemented. At the same time, improvement in both efficacy and cost-effective production of biological control agents would encourage user acceptance. Taken together, these facts suggest that the challenges facing exploitation of biological control agents will be overcome.

As mentioned in Section 1, the safety of some viral insecticides has been questioned in the past, and this has limited their development. In one case the use of a *Nudaurelia*-β group virus to control oil-palm pests in Malaysia was discontinued following discovery of antibodies to the virus in sera of trial area residents (McCallum et al. 1979). The significance of the antibody response is not clear, as no adverse effects on health were reported, nevertheless, it was sufficient to justify cessation of the project. In other cases concerns expressed about safety of viral insecticides have proven unfounded. For example, the safety of baculoviruses was initially questioned, although there is now extensive data demonstrating their safety (see Summers et al. 1975, Vlak Chapter 3, and references therein).

Like chemical insecticides, all virally-based control agents and strategies must pass rigorous safety tests before their registration is approved, and well-defined safety tests for this purpose are already in place in many countries (Betz, 1986). Public concern about the safety of using insect viruses on a large scale is an important factor which should not be ignored; nevertheless, we believe that education about the safety requirements and testing procedures will satisfy most concerned parties.

Concern about the use of "new" engineered viruses, and genetically modified organisms (GMOs) in general, will be addressed in a similar manner. Many countries have now enacted a legislative framework to govern the release of GMOs, and in New York State (USA), a small-scale field trial using an engineered baculovirus has been completed without incident. GMOs have now been approved for release in nine field trials in Australia (Anon 1991). One genetically engineered bacterium is already commercially available in Australia (Wright 1989) and has been used for some years without incident.

4.3. Production and Implementation

Table 4.3 summarizes how the three general delivery methods for viral or virally-derived agents relate to the control strategies shown in Figure 4.1. In the remainder of this Section we discuss particular problems associated with production and implementation of viral insecticides and host-mediated virally-based control strategies. We do not discuss insect-mediated delivery further, since it is at present an experimental protocol only, and requirements for large-scale implementation are as yet unknown.

4.3.1. Production and Delivery of Viral Insecticides

Production of viral insecticides has, to date, been primarily *in vivo* in facilities built to raise and process very large numbers of infected hosts (Shapiro 1986). Current methodologies for these production systems are,

Table 4.3. Potential applications of new virally-derived agents and technologies to pest control strategies

Strategy	Delivery System	New Agents and Technologies[a]
INSECTICIDES	PATHOGEN	Viral vectors Viral genes Viral-derived tools
HOST RESISTANCE	PLANT	Viral genes Viral vectors Viral-derived tools
TRADITIONAL BIOCONTROL	PREDATOR/ PARASITE	Viral-derived tools Viral genes Viral vectors

[a] The relative potential of the agents and technologies discussed in Section 3 are shown by the size of the type-face. The greater the potential value, the greater the type-face.

however, labor intensive, and unless labor is relatively cheap, the products cannot be produced at a cost competitive with that of chemical insecticides.

Over the last five to 10 years, there has been greatly increased interest in large-scale fermentation of insect cells, primarily for use in production of pharmaceuticals by baculovirus expression systems (Maiorella et al. 1988; Vlak Chapter 3). Until recently the necessity of expensive sera supplements in insect cell-culture media made large-scale production of virus in such systems prohibitively expensive. Development of cheap, serum-free growth media is now being reported (Weiss and Vaughn 1986; Weiss et al. 1988), and such work has substantially increased the potential for use of large-scale *in vitro* production techniques. Furthermore, as interest in biological insecticides increases, there are likely to be other developments which improve formulation of biological agents. We believe that evolution of cheap production systems and improved formulation methodologies will result in the availability of new generations of genetically engineered, highly efficacious and competitively priced viral insecticides.

4.3.2. Host Mediated Delivery Systems

The most effective way to disseminate control agents to herbivorous insects of cropping systems is, of course, through the crop plants that they eat. This strategy has been used effectively in delivering *B.t.* toxins to insects under experimental conditions, achieving high levels of control (Vaeck et al. 1987; Barton et al. 1987). It is obvious that this would be an attractive delivery method, because many control agents described earlier (Section 3.2) are designed to interact either directly with gut cells, or to facilitate transport of other factors across the gut barrier. However, other types of agents might also benefit from this approach to delivery. The feasibility of putting whole genomes of insect pathogenic viruses into plants has recently been demonstrated (Selling et al. 1990). Those workers found that RNA from the Flock House nodavirus readily infected both cultured plant cells and plants themselves, and that the virions produced were infectious to *Drosophila* cells *in vitro*. If this phenomenon proves widely applicable to RNA viruses pathogenic to insects, it might form the basis of a new delivery system for those viruses. It is conceivable that the genomes of viruses engineered for increased efficacy against insect pests might themselves be engineered into crop plants.

There are currently two major limitations to plant-mediated delivery systems. First, it is still not possible to routinely and efficiently transform a large number of agriculturally important crops, most notably many of the cereals (Shimamoto, 1991). Second, even in species for which transgenic plants can be produced, the available technology is often not effective for commercial varieties. However, the rapid advances now being made in plant transformation technologies (Lycett and Grierson 1990) suggest that many crop varieties currently refractory to transformation will be amenable to manipulation in the near future.

5. Summary

As the human population of the globe continues to grow unchecked, the concomitant need for increased agricultural productivity has become more desperate. The associated necessity for effective and sustainable forms of insect control has thus become of paramount importance. In this Chapter we have attempted to describe potentially suitable control options based on the use of insect viruses. While we are optimistic that many of these possibilities will be realized over the next 10 to 15 years, it is also realistic to state that their successful implementation will require parallel achievements involving the biology and economics of production and formulation systems, and revised public perceptions of the safety of both wild type and genetically engineered forms of insect viruses.

Finally, we make no claim to have covered all possibilities for the utilization of viruses in design of new insect control strategies. We hope, however, to have portrayed insect viruses in a manner which encourages others to seriously consider the many options they offer for development of such methodologies.

References

Adams JRA, Bonami JR (1991) The Atlas of Invertebrate Viruses. CRC Press, Boca Raton, Florida, 684 pp

Akai H, Gateff E, Davis L, Schneiderman HA (1967) Virus-like particles in normal and tumourous tissues of *Drosophila*. Science 157:810–813

Allen MF, Ball BV (1990) Purification, characterization, and some properties of a virus from the aphid *Sitobion avenae*. J Invert Pathol 55:162–168

Anon (1991) Genetic Manipulation Advisory Committee Annual Report 1990–91. Dept of Administrative Services, Aust Gvt Publishing Services, Canberra

Ansardi DC, Porter DC, Morrow CD (1991) Coinfection with recombinant vaccinia viruses expressing poliovirus P1 and P3 proteins results in polyprotein processing and formation of empty capsid structures. J Virol 65:2088–2092

Arif BM, Kurstak E (1991) The entomopoxviruses. In Kurstak E (ed) Viruses of Invertebrates. Marcel Dekker, New York, pp 179–195

Barton KA, Whiteley HR, Yang N-S (1987) *Bacillus thuringiensis*-endotoxin expressed in *Nicotiana tabacum* provides resistance to lepidopteran insects. Plant Physiol 85:1103–1109

Belloncik S (1989) Cytoplasmic polyhedrosis viruses—Reoviridae. Adv Virus Res 37:173–209

Benfey PN, Chua NH (1989) Regulated genes in plants. Science 244:174–181

Betz FS (1986) Registration of baculoviruses as insecticides. In Granados RR, Federici BA (eds) The Biology of Baculoviruses Vol 2 Practical Application for Insect Control. CRC Press, Boca Raton, Florida, pp 204–215

Bingham PM, Zachar Z (1989) Retrotransposons and the FB transposon from *Drosophila melanogaster*. In Berg DE, Howe MM (eds) Mobile DNA. American Society for Microbiology, Washington DC, pp 485–502

Binnington KC, Lockie E, Hines E, van Gerwen ACM (1987) Fine structure and distribution of three types of virus-like particles in the sheep blow-fly, *Lucilia cuprina* and associated cytopathic effects. J Invert Pathol 49:175–187

Bird FT (1967) A virus disease of the European red mite *Panonychus ulmi* (Koch). Canad J Microbiol 13:1131–1134

Briese DT (1986) Insect resistance to baculoviruses. In Granados RR, Federici BA (eds) The Biology of Baculoviruses Vol 2 Practical Application for Insect Control. CRC Press, Boca Raton, Florida, pp 237–263

Brun G, Plus N (1980) The viruses of *Drosophila*. In Ashburner M, Wright TFR (eds) The Biology and Genetics of Drosophila Vol 3a. Academic Press, New York, pp 625–702

Büchen-Osmond C, Crabtree K, Gibbs A, McLean G (1988) Viruses of Plants in Australia. Australian National University Printing Service, Canberra

Carson R (1962) Silent Spring. Fawcett Publications Greenwich, Connecticut

Cockbain AJ, Jones P, Woods RD (1986) Transmission characteristics and some other properties of bean yellow vein-banding virus, and its association with pea enation mosaic virus. Ann Appl Biol 108:59–69

Cunningham JC (1988) Baculoviruses: their status compared to *Bacillus thuringiensis* as microbial insecticides. Outlook Agric 17:10–17

Derksen ACG, Granados RR (1988) Alteration of a lepidopteran peritrophic membrane by baculoviruses and enhancement of viral infectivity. Virology 167:242–250

Dobos P, Nagy E, Duncan R (1991) Birnaviridae. In Kurstak E (ed) Viruses of Invertebrates. Marcel Dekker, New York, pp 301–314

Doller G, Reimann R, Groner, A (1983) Seroreaction of human sera with baculovirus proteins without neutralising activity. Naturwissenschaften 70:370–371

Fazairy AAA, Hassan FA (1988) Infection of termites by *Spodoptera littoralis* nuclear polyhedrosis virus. Insect Sci Applic 9:37–39

Federici BA (1983) Enveloped double-stranded DNA insect virus with novel structure and cytopathology. Proc Natl Acad Sci USA 80:7664–7668

Fischer M, Schnitzler P, Scholz J, Rosen-Wolff A, Delius H, Darai G (1988) DNA nucleotide sequence analysis of the *Pvu* II DNA fragment L of the genome of insect iridescent virus type 6 reveals a complex cluster of multiple tandem, overlapping, and interdigitated repetitive DNA elements. Virology 167:497–506

Fleuriet A (1976) Presence of the hereditary rhabdovirus sigma and polymorphism for resistance to the virus in natural populations of *Drosophila melanogaster*. Evolution 30:735–739

Forgash AJ (1984) History, evolution and consequences of insecticide resistance. Pestic Biochem Physiol 22:178–186

Forrester NF (1991) Pyrethroid and endosulfan resistance in *Heliothis armigera* in Australia—1990/91. Resistant Pest Management 3:31–34.

French R, Janda M, Ahlquist P (1986) Bacterial gene inserted in an engineered RNA virus: efficient expression in monocotyledonous plant cells. Science 231:1294–1297

Gaedigk-Nitschko K, Adam G, Mundry KW (1988) Role of the spike protein from potato yellow dwarf virus during infection of vector cell mololayers. In Mitsohashi EJ (ed) Invertebrate Cell Systems in Application Vol 1. CRC Press, Boca Raton, Florida

Gateff E, Shristha R, Akai H (1984) Comparative ultrastructure of wild-type and tumorous cells of *Drosophila*. In King RC, Akai H (eds) Insect Ultrastructure Vol 2. Plenum, New York, pp 559–578

Geigenmüller-Gnirke U, Weiss B, Wright R, Schlesinger S (1991) Complementation between Sindbis viral RNAs produces infectious particles with a bipartite genome. Proc Natl Acad Sci USA 88:3253–3257

Georghiou GP, Lagunes A (1988) The Occurence of Resistance to Pesticides: Cases of Resistance Reported Worldwide through 1988. FAO, Rome, 325 pp

Gibb KS, Randles JW (1990) Distribution of velvet tobacco mottle virus in its mirid vector and its relationship to transmissibility. Ann Appl Biol 116: 513–521

Gibbons A (1991) Moths take the field against biopesticide. Science 254:646

Gildow FE (1987) Virus membrane interactions involved in circulation transmission of luteoviruses by aphids. In Harris KF (ed) Current Topics in Vector Research Vol 4. Springer-Verlag, New York, pp 93–120

Gildow FE, D'Arcy CJ (1990) Cytopathology and experimental host range of *Rhopalosiphum padi* virus, a small isometric RNA virus infecting cereal grain aphids. J Invert Pathol 55:245–257

Goodwin RH, Filshie BK (1975) Morphology and development of entomopoxviruses from two Australian scarab beetle larvae (Coleoptera: Scarabaeidae). J Invert Pathol 250:35–46

Gordon K (1990) Cauliflower mosaic virus: biology and applications. In Sangwan BA, Sangwan BS (eds) Impact of Biotechnology in Agriculture. Kluwer, Amsterdam, pp 381–390

Granados RR, Corsaro BG (1990) Baculovirus enhancing proteins and their implication for insect control. In Proc Fifth Int Colloq Invert Pathol Microbial Control. Society for Invertebrate Pathology, Adelaide, Australia, 446 pp

Granados RR, Federici BA (1986) The Biology of Baculoviruses. CRC Press, Boca Raton, Florida

Granados RR, Hashimoto Y (1990) Gene coded for polypeptide which enhances virus infection of host insects. European Patent Application No 90102939.7

Guzo D, Stoltz DB (1987) Observations on cellular immunity and parasitism in the tussock moth. J Insect Physiol 33:19–31

Hall RL, Moyer RW (1991) Identification, cloning, and sequencing of a fragment of *Amsacta moorei* entomopoxvirus DNA containing the spheroidin gene and three vaccinia virus-related open reading frames. J Virol 65:6516–6527.

Hammock BD, Bonning B, Possee RD, Hanzlik TN, Maeda S (1990) Expression and effects of the juvenile hormone esterase in a baculovirus vector. Nature 244:458–461

Harris KE (1981) Arthropod and nematode vectors of plant viruses. Ann Rev Phytopathol 19:391–426

Harrison BD (1987) Plant virus transmission by vectors: mechanisms and consequences. In Russell WC, Almond JW (eds) Molecular Basis of Virus Disease. Cambridge University Press, Cambridge, pp 319–344

Hatfill SJ, Williamson C, Kirby R, von Wechmar MB (1990) Identification and localization of aphid lethal paralysis virus particles in thin tissue sections of the *Rhopalosiphum padi* aphid by *in situ* nucleic acid hybridization. J Invert Pathol 55:265–271

Hendry DA (1991) Nodaviridae of invertebrates. In Kurstak E (ed) Viruses of Invertebrates. Marcel Dekker, New York, pp 227–276

Hendry DA, Hodgson V, Clark R, Newman J (1985) Small RNA viruses coinfecting the pine emperor moth *Nudaurelia cytherea capensis*. J Gen Virol 66: 627–663

Henry JE, Jutila JW (1966) The isolation of a polyhedrosis virus from a grasshopper. J Invert Pathol 8:417–418

Hess RT, Summers MD, Falcon LA (1978) A mixed virus infection in midgut cells of *Autographa californica* and *Trichoplusia ni* larvae. J Ultrastr Res 65:253–265

Hillman B, Morris TJ, Kellen WR, Hoffman D, Schlegel DE (1982) An invertebrate calici-like virus: evidence for partial virion disintegration in host excreta. J Gen Virol 60:115–123

Home WA, Tajbakhsh S, Seligy VL (1990) Molecular cloning and characterization of a late iridescent virus gene. Gene 94:243–248

Jackson RJ, Howell MT, Kaminski A (1990) The novel mechanism of initiation of picornavirus RNA translation. Trends Genet 15:477–483

Jones D, Gones G, Rudnicka M, Click A (1986) Precocious expression of the final larval instar developmental pattern in larvae of *Trichoplusia ni* pseudoparasitized by *Chelonus* spp. Comp Biochem Physiol 83B:339–346

Kaesberg P (1987) Organization of bipartite insect virus genomes: the genome of black beetle virus. In Rowlands DJ, Mayo MA, Mahy BWJ (eds) The Molecular Biology of the Positive Strand RNA Viruses. Academic Press, London, pp 207–218

Kaesberg P, Dasgupta R, Sgro J-Y, Wery J-P, Selling BH, Hosur MV, Johnson JE (1990) Structural homology among four nodaviruses as deduced by sequencing and X-ray crystallography. J Mol Biol 214:423–435

Kajigaya S, Fujii H, Field A, et al. (1991) Self-assembled B 19 parvovirus capsids, produced in a baculovirus system, are antigenically and immunogenically similar to native virions. Proc Natl Acad Sci USA 88:4646–4650

Kaufman RJ (1990) Vectors used for expression in mammalian cells. Methods Enzymol 185:487–511

Kawase S, Kurstak E (1991) Parvoviridae of invertebrates: densonucleosis viruses. In Kurstak E (ed) Viruses of Invertebrates. Marcel Dekker, New York, pp 315–343

Keddie BA, Aponte GW, Volkman LE (1989) The pathway of infection of *Autographa californica* nuclear polyhedrosis virus in an insect host. Science 243:1728–1730

Kelly DC (1985) Insect iridescent viruses. Current Topics Microbiol Immunol 116:24–35

Kingsman AJ, Adams SE, Burns NR, Kingsman SM (1991) Retroelement particles as purification, presentation and targeting vehicles. Trends Biotech 9:303–309

Kingsman AJ, Kingsman SM (1988) *Ty*: a retroelement moving forward. Cell 53:333–335

Krell PJ (1991) The polydnaviruses: multipartite DNA viruses from parasitic Hymenoptera. In Kurstak E (ed) Viruses of Inveretebrates. Marcel Dekker, New York, pp 141–177

Kurstak E (1991) Viruses of Invertebrates. Marcel Dekker, New York

Li J, Carroll J, Ellar DJ (1991) Crystal structure of insecticidal delta-endotoxin from *Bacillus thuringiensis* at 2.5 A resolution. Nature 253:815–821

Lycett GW, Grierson D (eds) (1990) Genetic Engineering of Crop Plants. Butterworths, London

Mackett M, Smith GL, Moss B (1982) Vaccinia virus: a selectable eukaryotic cloning and expression vector. Proc Natl Acad Sci USA, 79:7415–7419

Maeda S (1989) Increased insecticidal effect by a recombinant baculovirus carrying a synthetic diuretic hormone gene. Biochem Biophys Res Comm 165:1177–1183

Maeda S, Volrath SL, Hanzlik TN, Harper SA, Maddox DW, Hammock BD, Fowler E (1991) Insecticidal effects of an insect-specific neurotoxin expressed by a recombinant baculovirus. Virology 184:777–780

Maiorella B, Inlow D, Shauger A, Harano D (1988) Large-scale insect cell-culture for recombinant protein production. Bio/Technology 6:1406–1410

Manousis T, Arnold MK, Moore NF (1988) Electron microscopical examination of tissues and organs of *Dacus oleae* flies infected with cricket paralysis virus. J Invert Pathol 51:119–125

Manousis T, Moore NF (1987) Cricket paralysis virus, a potential control agent for the olive fruit fly, *Dacus oleae* Gmel. Appl Environ Microbiol 53:142–148

Marsh M, Helenius A (1989) Virus entry into animal cells. Adv Virus Res 36: 107–151

Martignoni ME, Iwai PJ (1986) A catalogue of viral diseases of insects, mites and ticks. US Department of Agriculture Forest Service, Gen Tech Report PNW 195

Martin RR, Keese PK, Young MJ, Waterhouse PM, Gerlach WL (1990) Evolution and molecular biology of luteoviruses. Ann Rev Phytopathol 28:341–363

Matthews REF (1982) Classification and nomenclature of viruses. Fourth Report of the International Committee on Taxonomy of Viruses. Karger, Basel, 199 pp

McCallum F, Brown G, Tinsley T (1979) Antibodies in human sera reacting with an insect pathogenic virus. Intervirology 11:234–237

McKnight S, Tjian R (1986) Transcriptional selectivity of viral genes in mammalian cells. Cell 46:795–805

McLachlin JR, Cornetta K, Eglitis MA, Anderson WF (1990) Retroviral-mediated gene transfer. Progress Nucl Acids Res Mol Biol 38:91–133

Miahle E, Croizier G, Veyrunes JC, Quiot JM, Rieb JP (1983) Etude d'un virus isole d'une population naturelle de Culicoides sp. (Diptera: Ceratopogonidae). Ann Virol (Inst Pasteur) 134:73–86

Miller AD, Rosman GJ (1989) Improved retroviral vectors for gene transfer and expression. Bio/Techniques 7:980–990

Miller LK (1990) Molecular baculovirology, regulation of ecdysis and improved viral pesticides. In Fifth Int Colloq Invert Pathol Microbial Control, Society for Invertebrate Pathology, Adelaide, Australia, p 446

Moore NF (1991a) The Nudaurelia β family of insect viruses. In Kurstak E (ed) Viruses of Invertebrates. Marcel Dekker, New York, pp 277–285

Moore NF (1991b) Identification, pathology, structure and replication of insect picornaviruses. In Kurstak E (ed) Viruses of Invertebrates. Marcel Dekker, New York, pp 287–299

Moore NF, Reavy B, King LA (1985) General characteristics, gene organization and expression of small RNA viruses of insects. J Gen Virol 66:647–659

Moscardi F (1988) Production and use of entomopathogens in Brazil. In Roberts DW and Granados RR (eds) Biotechnology, Biological Pesticides and Novel Plant-Pest Resistance for Insect Pest Management. Insect Pathology Resource Centre, Boyce Thompson Institute for Plant Research at Cornell University, Ithaca, New York, pp 53–60

Nault LR, Ammar ED (1989) Leafhopper and planthopper transmission of plant viruses. Ann Rev Entomol 34:503–529

Nuss DL, Dall DJ (1990) Structural and functional properties of plant reovirus genomes. Adv Virus Res 38:249–306

Ofori FA, Francki RIB (1985) Transmission of leafhopper A virus, vertically through eggs and horizontally through maize in which it does not multiply. Virology 144:152–157

O'Reilly DR, Miller LK (1990) Regulation of expression of a baculovirus ecdysteroid UDP-glucosyl-transferase gene. J Virol 64:1321–1328

Palmiter RD, Brinster RL (1986) Germ-line transformation of mice. Ann Rev Genet 20:465–499

Paulson JC (1985) Interactions of animal viruses with cell surface receptors. In Conn PM (ed) The Receptors Vol 2. Academic Press, Orlando, Florida, pp 131–219

Perkins JH (1988) The future history of biotechnology: a commentary. In Roberts DW and Granados RR (eds) Biotechnology, Biological Pesticides and Novel Plant-Pest Resistance for Insect Pest Management. Insect Pathology Resource Centre, Boyce Thompson Institute for Plant Research at Cornell University, Ithaca, New York, pp 168–175

Plus N, Golubovsky MD (1980) Resistance to *Drosophila* C virus of fifteen 1(2)gl/cy stocks carrying 1(2) lethals from different geographical origins. Genetika (Yugoslavia) 12:227–231

Purrini K, Rohde M (1988) Studies on a new disease in a natural population of migratory locusts, *Locusta migratoria*, caused by an entomopoxvirus. J Invert Pathol 51:284–286

Reed DK (1981) Control of mites by non-occluded viruses. In Burges HD (ed) Microbial Control of Pests and Plant Diseases 1970–1980. Academic Press, London, pp 427–432

Reed DK, Desjardins PR (1978) Isometric virus-like particles from citrus red mites, *Panonychus citri*. J Invert Pathol 31:188–193

Reinganum C, O'Loughlin GT, Hogan TW (1970) A non-occluded virus of the field crickets *Teleogryllus oceanicus* and *T. commodus* (Orthoptera: Gryllidae). J Invert Pathol 16:214–220

Sarver NP, Gruss M-F, Law G, Khoury G, Howley PM (1981) Bovine papilloma virus deoxyribonucleic acid: a novel eukaryotic cloning vector. Mol Cell Biol 1:486–496

Scotti PD, Longworth JF, Plus N, Croizier G, Reinganum C (1981) The biology and ecology of strains of an insect small RNA virus complex. Adv Virus Res 26:117–143

Selling BH, Allison RF, Kaesburg P (1990) Genomic RNA of an insect virus directs synthesis of infectious virions in plants. Proc Natl Acad Sci USA 87:434–438

Shapiro M (1986) In vivo production of Baculoviruses. In Granados RR, Federici BA (eds) The Biology of Baculoviruses Vol II. CRC Press, Boca Raton, pp 31–61

Shiba T, Saigo K (1983) Retrovirus-like particles containing RNA homologous to the transposable element *copia* in *Drosophila melanogaster*. Nature 302: 119–124

Shimamoto K (1991) Transgenic rice plants. In Dennis ES, Llewellen DJ (eds) Molecular Approaches to Crop Improvement. Springer-Verlag, New York, pp 1–15

Sivasubramanian N, Hice RH, Post CA, Chandrasekhar GN (1990) Use of baculovirus proteins for developing novel protein insecticides. In Proc Fifth Int Colloq Invert Pathol Microbial Control. Society for Invertebrate Pathology, Adelaide, Australia, p 161

Smith GE, Summers MD (1988) Method for producing a recombinant baculovirus expression vector. United States Patent No. 4,745,051

Smith GE, Summers MD, Fraser MJ (1983) Production of human beta interferon in insect cells infected with a baculovirus expression vector. Mol Cell Biol 3:2156–2165

Solnick D (1981) Construction of adenovirus SV40 recombinants producing SV40 T-antigen from and adenovirus late promoter. Cell 24:135–134

Steinhaus EA (1956) Microbial control—the emergence of an idea. A brief history of insect pathology through the nineteenth century. Hilgardia 26: 107–157

Stewart LMD, Hirst M, Lopez Ferber M, Merryweather AT, Cayley PJ, Possee RD (1991) Construction of an improved baculovirus insecticide containing an insect-specific toxin gene. Nature 352:85–88

Stone TB, Sims SR, Marrone PG (1989) Selection of tobacco budworm for resistance to a genetically engineered *Pseudomonas fluorescens* containing the δ-endotoxin of *Bacillus thuringiensis* subsp. *kurstaki*. J Invert Pathol 53:228–234

Sylvester ES (1985) Multiple acquisition of viruses and vector-dependent prokaryotes: consequences on transmission. Ann Rev Entomol 30:71–88

Tajbakhsh S, Lee PE, Watson DC, Seligy VL (1990) Molecular cloning, characterization, and expression of the *Tipula* iridescent virus capsid gene. J Virol 64:125–136

Takamatsu N, Ishikawa M, Meshi T, Okada Y (1987) Expression of bacterial chloramphenicol acetyltransferase gene in tobacco plants mediated by TMV-RNA. EMBO J 6:307–311

Tanada Y (1985) A synopsis of studies on the synergistic property of an insect baculovirus: a tribute to Edward A. Steinhaus. J Invert Pathol 45:125–138

Tanaka T, Agui N, Hiruma K (1987) The parasitoid *Apantales kariyai* inhibits pupation of its host, *Pseudaletia separata*, via disruption of prothoracicotropic hormone release. Gen Comp Endocrinol 67:364–371

Tomalski MD, Miller LK (1991) Insect paralysis by baculovirus-mediated expression of a mite neurotoxin gene. Nature 352:82–85

Uzigawa K, Aruga H (1966) On the selection of resistant strains to the infectous flacherie virus of the silkworm *Bombyx mori* L. J Sericult Sci Japan 35:23–26

Vaeck M, Reynaerts A, Hofte H, Jansens S, De Beuckeleer MD, Dean C, Zabeau M, Van Montagu MV, Leemans J (1987) Transgenic plants protected from insect attack. Nature 328:33–37

Varmus H (1988) Retroviruses. Science 240:1427–1435

Ward VK, Kalmakoff J (1991) Invertebrate Iridoviridae. In Kurstak E (ed) Viruses of Invertebrates. Marcel Dekker, New York, pp 197–225

Watanabe H (1968) Development of resistance in the silkworm *Bombyx mori* to peroral infection of a cytoplasmic polyhedrosis virus. J Invert Pathol 12: 310–320

Waterhouse PM, Murant AF (1983) Further evidence on the nature of the dependence of carrot mottle virus on carrot red leaf virus for transmission by aphids. Ann Appl Biol 103:455–464

Weiss SA, De Giovanni AM, Godwin CP, Kohler JP (1988) Large-scale cultivation of insect cells. In Roberts IW, Granados RR (eds) Biotechnology, Biological Pesticides and novel plant-pest resistance for insect pest management. Insect Pathology Resource Center, Boyce Thompson Institute for Plant Research at Cornell University, Ithaca, New York, pp 22–30

Weiss SA, Vaughn JL (1986) Cell culture methods for large-scale propagation of baculoviruses. In Granados RL, Federici BA (eds) The Biology of Baculoviruses Vol II. CRC Press, Boca Raton, pp 63–87

Wigley PJ, Scotti PD (1983) The seasonal incidence of cricket paralysis virus in a population of the New Zealand small field cricket, *Pteronemobius nigrovus* (Orthoptera: Gryllidae). J Invert Pathol 41:378–380

Williams RC, Smith KM (1957) A crystallizable insect virus. Nature 179:119–120.

Wright B (1989) Gene-spliced pesticide is uncorked in Australia. New Scientist 121(1654):23

Xiong C, Levis R, Shen P, Schlesinger S, Rice CM, Huang HV (1989) Sindbis virus: an efficient, broad host range vector for gene expression in animal cells. Science 243:1188–1191

Yearian WC, Young SY (1982) Control of insect pests of agricultural importance by viral insecticides. In Kurstak E (ed) Microbial and Viral Pesticides. Marcel Dekker, New York, pp 387–390

Yuen L, Binne J, Arif B, Richardson C (1990) Identification and sequencing of the spheroidin gene of *Choristoneura biennis* entomopoxvirus. Virology 175:427–433

Chapter 5

Molecular Methods for Insect Phylogenetics

Ross H. Crozier

1. Introduction

A well-founded phylogenetic scheme enlightens biology by enabling the drawing of inferences about the evolution of the organisms concerned (Pagel and Harvey 1988). The best characters for evolutionary inference would themselves evolve in ways explicable by simple, and hence, powerful models, and these data would be numerous, enabling rigorous statistical tests of phylogenetic hypotheses. Naturally, these ideal characters would be universally present in living things, so that the same empirical and analytical methods could be universally applied.

Alas, these ideal characters have yet to be found. But molecular data have powerful advantages over morphological information, especially in terms of sheer volume, universality (when nucleic acids are used), and a much better approximation of their evolution to that expected from simple models, even if this approximation is not perfect. These advantages do not mean the end of morphological studies. Not only does the explanation of morphological evolution remain a major rationale for phylogenetic studies, but the imperfections of phylogenetic inferences based on living forms mean that information from fossils remains crucial to understanding evolution, and most of this information remains strictly morphological.

Molecular approaches also have two potential disadvantages to systematists. The first of these is one of expertise. Systematists are not often endowed with molecular expertise, and it is a challenge for the field to both inject necessary molecular expertise in systematics while retaining the depth of general biological knowledge characteristic of most systematists. The second limitation is cost. Systematists used to materials costs of a few dollars for notebooks, slides, and coverslips, plus an allowance of computer time, will find alarming the notion of spending several thousand dollars on cloning and sequencing reagents!

The data needed for systematic analyses can be collected in two different forms: as individual character states and as distances between taxa. Character states are properties of the taxa from which they have been collected. A distance is not a property of any taxon, but only of the pair of taxa to which it pertains. Character state data can usually, perhaps always, be converted to distances, but the distances cannot be converted into character states. Most types of molecular data are character states, which represents another advantage, because there are two limitations to data collected initially as distances. One is that it is difficult to estimate the statistical confidence one can have in a tree as a whole, or in sections of it, if the distances were not originally based on characters. The second limitation is a logistical one: addition of a further taxon to a distance data set is best done by determining distances from it to each previously-studied taxon, whereas determining the character states of a new taxon automatically place it in relation to all previously studied taxa.

2. Overview of data types

2.1. Independent or Semi-independent Genes

Because of the single-gene phylogeny problem (section 2.3 below), it is desirable to use a number of gene loci for inferring phylogenies among closely-related taxa. Analysis is typically of changes in allele frequencies between the taxa and is limited ultimately by the proportion of loci at which there are alleles in common between taxa.

Ideally, the changes during evolution at one locus are independent of those occurring at other loci in order that the assumptions of the analytical methods be met. For closely-related forms, this makes it desirable that the loci be unlinked, or only loosely linked, and that there are no significant epistatic interactions for fitness between them.

The primary uses of data involving many independent loci are in studies on population structure (e.g., inbreeding, microgeographic variation) and of relationships between populations and closely related species. Relationships between genera are also often studied in this way, but in insects the proportion of loci with shared alleles will usually be too small to give confidence in the results.

Population structure studies use standard population genetic methods and are not really "phylogenetic." Analyses of the relationships between populations and species use either genetic distances (reviewed by Nei 1987, pp 208–253; and see also Hillis 1984) and various clustering methods or convert the data to a form suitable for parsimony analysis. These conversions vary from treating the alleles themselves as the characters, with either presence or absence of the alleles, or their frequencies, as the character states, or treating the loci as characters.

An entirely different form of analysis using many independent loci is in principle possible, by constructing an allele phylogeny for each locus. The overall phylogeny would be the majority vote between the various allele phylogenies. The labor required to derive multiple allele phylogenies make it unlikely that this method will become widely used soon.

2.2. Single-unit Systems

Single-unit systems are inherited in a unitary fashion, so that the alternatives behave as alleles. Such systems can be examined in great detail, in some cases by DNA sequencing, and include individual nuclear genes, the Y-chromosome and mitochondrial DNA. The last two cases each include a number of gene loci, but these loci are inherited without recombination and hence as a single unit.

Phylogenetic studies on nuclear genes bring out some problems with regard to simple interpretation of the organisms concerned, because they are not isolated entities. Thus, Stephens and Nei (1985) analyzed 11 *Adh* allele sequences from *Drosophila melanogaster* and found that the alleles were not related by a simple branching tree. The departures from dichotomy could be explained either by recombination or by gene conversion (see Section 7 below).

The remarks above concern autosomal genes. There are three other modes of inheritance possible in animals. In one of these, X-linked inheritance, biparental derivation of the genes in an individual (female) occurs, so that recombination is also a potential problem. For the other two, Y-linked genes and mitochondrial DNA (mtDNA), recombination is essentially absent. Gene conversion may still occur, especially in the case of Y-linked genes, but seems less likely than in the case of the other nuclear genes because of the lack of homologous chromosomes. Naturally, when a Y-chromosome has only recently been derived from an autosome, so that a high level of similarity remains between it and the neo-X, gene conversion may be more likely. That this is not an indefinitely effective process is shown by the well-known tendency of Y-chromosomes to become heterochromatic and reduced in size.

Y-linked and mitochondrial genes have unique and complementary attributes for population-level studies. Y-linked genes spread patrilineally, and mtDNA spreads matrilineally. These gene categories thus allow the tracing of hybrid ancestries when these are differentially affected by mating behavior. For example, mtDNA analysis identified which of the ancestral species is the male and which the female parent in the repeated origin of parthenogenetic species of the genus *Cnemidophorus* (Brown and Wright 1979). Probes for following Y-linked DNA are likely to be increasingly important in tracing the effects of differential mating success from the male side (e.g., Casanova et al. 1985; Vanlerberghe et al. 1986).

An issue arising out of consideration of Y-linked and mitochondrial genes is that it is wrong to label the former as "due to male behavior" and the latter as "due to female behavior." Differential diffusion of genes of these two types occurs when there are differences between the mating behavior of the two sexes: each sex cannot be considered apart from the other. A given case could be regarded as either due to greater reproductive success by females of taxon A, or by males of taxon B (as in Africanized honeybees: Crozier 1990).

2.3. The Single-Gene Phylogeny, or Time-Depth, Problem

The phylogeny of a segment of DNA is a gene-tree. Such a tree may depart from the phylogeny of the populations of individuals in which it occurs. Naturally, the phylogeny of the taxa is the sum of the trees of the constituent gene trees, but any individual tree may differ significantly from the whole picture. The problem has received broad attention elsewhere (e.g., Pamilo and Nei 1988; Avise 1989; Crozier 1990) but is sufficiently central to an understanding of the relationships between molecular data and phylogenetic inference to warrant attention here. It can be ascertained from Figure 5.1 that the chance of an incorrect conclusion about the branching order is determined by the ratio between generation time (t) and effective population size (2N).

From the treatment of Pamilo and Nei (1988), we can determine that the minimum value of t/2N consistent with a 95% probability of the gene tree giving the species tree is 2.59 (Crozier 1990). Mitochondrial and Y-linked genes have the advantage for such studies, because they generally have only one-quarter the effective population size of autosomal nuclear genes. (Avise et al. (1987) note that the probabilities of lineage loss depend also on stability of population size.) However, the problem is that, even if one knows that a given mitochondrial tree is much more likely to be correct (in terms of accurately reflecting the species tree) than an autosomal nuclear gene tree, considerable uncertainty can remain.

Naturally, the greater population subdivision expected with the smaller effective population sizes of mitochondrial and Y-linked genes makes these very sensitive indicators of such subdivision, as seen in *Drosophila* studies (e.g., Hale and Singh 1987). Given the need for verification of the findings of each gene tree, it seems desirable to match mtDNA trees with Y-chromosome trees for those organisms which have both.

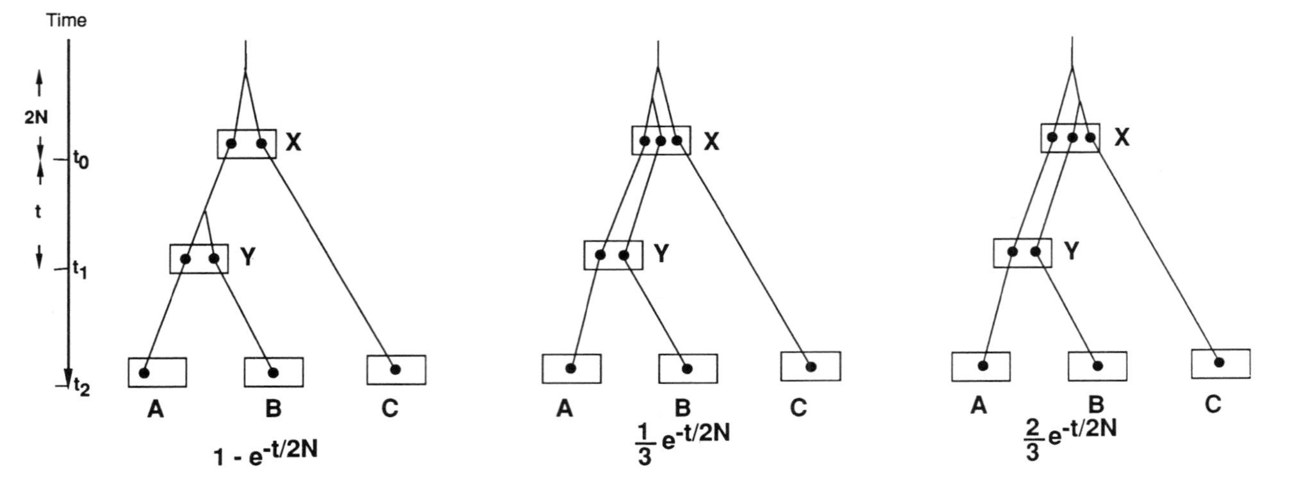

Figure 5.1. The possible relationships between gene tree and species tree for an autosomal gene among three species (after Pamilo and Nei 1988). The divergence of species C from the others occurs at time t_0, and that of species A and B from each other at time t_1, with t_2 being the present. The mean time since two copies of the gene in the ancestral population X diverged is 2N, with N being the effective population size. The two left-hand trees return the correct topology for the populations, so that the probability of a gene tree being topologically unlike the species tree is governed by the size of the ancestral population, N, and the time, t, between speciation events, and is $\frac{2}{3}e^{-t/2N}$

If one accepts that the "true tree" relating the taxa of interest is that followed by a majority of genes, and that there is a risk that not all genes will show the same tree, then phylogenies of five or more nuclear genes should be obtained. If all five phylogenies are identical, then from binomial theory, one can be 95% certain that this is the majority phylogeny.

When, as will often be the case, it is not possible to obtain phylogenies from many genes, additional confidence in the applicability of the one tree can be obtained by increasing the sample size. Takahata (1989) advocates using the most recent branch point involving gene copies from different populations as an estimate of the true branch point for the populations. He suggests that sampling 15 copies from each population would generally yield high confidence, although, of course, the true degree of confidence depends on the actual population tree itself, and this is unknown.

3. Autosomal Nuclear Genes

3.1. Allozymes

Allozymes are different forms of the same or similar enzymes resulting from allelic variation in amino acid sequence. The visualization of these gene products following separation on starch, acrylamide, cellulose acetate, or other supporting media is now well-understood and relatively easy to achieve for a relatively low cost by following one of the various excellent technique manuals now available (e.g., Harris and Hopkinson 1976; Richardson et al. 1986).

The advent of allozymes in evolutionary studies led to several decades of explosive growth of genetical studies on natural populations. In recent time, however, advances in molecular biology have called the utility of the method into question. Three objections can be raised to using allozymes in evolutionary studies. One is that it is not always possible to be certain of the mode of inheritance of the banding patterns observed. The second is that, even when modes of inheritance are known, the stochastic error associated with allozyme-derived phylogenies is higher than that associated with some other kinds of data now attainable. Last, specimens have to be treated more gently than when DNA sequence information is sought in order to preserve enzyme function. The lability of proteins means that all specimens should be treated in the same way because of the essentially analog nature of the data, electrophoretic mobilities, which are then digitized by the perceptions of the investigator as to whether or not two bands are the same. Against these objections must be set the cost-effectiveness of the method and its ready application to large numbers of specimens of a wide array of species.

Nei (1987, p 253) notes that "... electrophoresis is expected to survey about $1,200 \times 0.083 = 100$ nucleotides per locus. If we examine 60 loci by electrophoresis, it is equivalent to studying 6,000 nucleotides." Most investigators will not be able to stain for 60 loci, nor will the inheritance information always be sufficient for the task, but the message is clear: allozyme electrophoresis remains a powerful and cost-effective method for appropriate problems. But which problems are "appropriate"? For studies of the apportionment of genetic variation within populations, in which the sample sizes of both loci and individuals are important, it would be foolish not to consider using allozymes. For studying the relationships of very closely related populations or species, in which cases the time-depth problem (Section 2.3) might be important, the use of many independent loci afforded by allozymes is an important advantage. For distantly-related species and between genera, problems of confidence that the inheritance patterns are still the same between taxa reduce the number of usable loci and hence the power of the method.

3.2. RFLPs

3.2.1. Randomly Chosen Polymorphisms

Restriction fragment length polymorphisms (RFLPs) result when variations in nucleotide sequence lead to differences in the patterns of fragments observed after electrophoresis of DNA digested with a restriction enzyme. An observed difference may result either from a polymorphism for length in the target sequence (due to a deletion or insertion of material between restriction sites), or in gain or loss of one or more sites due to a nucleotide substitution.

Although in some cases involving highly-repeated DNA visualization by simple staining of the DNA will suffice to reveal the pattern, generally researchers use radioactive probes specific to particular regions to detect RFLPs. The method involves transferring the fragments of DNA after electrophoresis to a membrane. The probe is then allowed to anneal with homologous DNA bound to the membrane. Because probes can be washed off the membrane with little loss of the digest DNA, the same membrane can be probed for variation in a number of DNA sequences, greatly increasing the potential return in data.

Probes can be derived in a number of ways. For example, Bernatzky and Tanksley (1986) prepared clones of single-copy nuclear DNA from tomatoes in studies designed to increase knowledge of linkage maps in these plants. Because the DNA had previously been sheared to yield pieces no larger than 400 base pairs (bp), those clones typically showed only a small number of bands, and where polymorphisms occurred, they resulted from gain or loss of only one site, believed to occur in the spacer

DNA flanking the genes surveyed. Such variation is easily analyzed in terms very similar to those previously applied to monomeric allozymes.

Hall (1986) followed a slightly different course in searching for diagnostic differences in nuclear DNA between Africanized honeybees and other strains in the New World. Appropriate treatment of the DNA in this case yielded probes of honeybee DNA some 4 kilobase pairs (kb) long. These probes detected variation apparently diagnostic of Africanized bees. The price of increasing the amount of data, however, was to make it difficult to determine which polymorphic bands were allelic to others, i.e., an inheritance model becomes difficult when the pattern becomes complex.

Where a region of DNA presents RFLPs suitable for phylogenetic analysis, the choice is then between using the fragments directly or rather mapping the sites causing them. It is possible to derive distances in terms of estimates of the number of substitutions directly from fragment similarities (Nei and Li 1979). It is preferable, however, to use the presence or absence of sites as the phylogenetic data, not only because of better-developed models for phylogenetic analysis for sites (e.g., Smouse and Li 1987), but also because sites remain intact longer than do the fragments they cause.

3.2.2. Finger-Printing and VNTRs: Illusory Progress?

Probes need not be of single-copy nuclear DNA (scnDNA) but may involve high copy number nuclear DNA (hcnDNA). The copy number of cloned DNA can be assayed approximately by annealing total DNA radioactively labelled back to bacterial colonies carrying the clones. Clones showing high labelling contain DNA present many times in the total DNA, whereas light labeling indicates that the cloned DNA is present only once or a few times per genome.

The use of probes detecting hcnDNA has been called "genetic fingerprinting" (Jeffreys et al. 1985). The resulting gels typically display profuse variation. This variation arises spontaneously at a high rate, so that laboratory cultures of the moth *Ephestia kuhniella* showed bands in the progeny of crosses between various strains that were not present in either parental strain (Traut 1987). The level of variation is sufficiently great to have led to the method's now well-known importance in forensic and immigration cases. This importance arises because the chances of two humans being identical for a genetic fingerprint is remote, at odds of millions to one (unless they are twins), and because relatives also share bands at a much higher rate than nonrelatives.

Whence comes the variation for most "genetic fingerprinting"? Mostly it comes from variation in the number of repeated sequences that flank structural genes, so that these polymorphisms can also be described as being due to variable numbers of tandem repeats (VNTRs). VNTRs arise from errors in replication called "slippage replication" (see Section 7.1 below).

The probes for fingerprinting need not be derived from the organisms of interest. Human-derived probes have revealed variation in other vertebrates (Burke 1989), and the cloning vector M13, when used as a probe, also detects variation in a wide range of organisms (Westneat et al. 1988). For such wide-ranging efficacy, the sequences involved have to not only be conserved but also present, so that success is far from guaranteed for every organism!

Although valuable for detecting genetic identity between individuals and for tracing pedigrees involving close relatives, VNTRs are likely to have drawbacks for many evolutionary studies. First, the rapidity of change in VNTRs means that the required level of similarity between organisms is expected to be quickly lost, so that phylogenetic interpretations are unlikely to be forthcoming. Second, the difficulty in arriving at genetic interpretations for VNTR variation means that these cannot be subject to standard population genetic analyses. Thus Lynch (1988) noted that determination of relatedness levels using methods based on gene frequencies cannot be applied to standard genetic fingerprinting results (see also Pamilo 1989).

Something can be done about the second of these problems, in that the hcnDNA probe typically detects many linked scnDNA regions. Secondarily, cloning these and using the scnDNA probes yields genetically-interpretable information from single loci that are highly-variable because of the flanking-region polymorphism.

3.3. Sequencing

3.3.1. "Classical" Methods

The advent of sequencing has opened up the vast store of historical knowledge written in the genetic material of organisms to a much greater extent than any previous technical advance. Given that nucleic acid sequencing is now more easily accomplished than protein sequencing and yields more information, protein sequencing has lost its place as a leading source of evolutionary knowledge. This is to deny neither that protein sequencing is still important, such as for checking the operation of the translational machinery of divergent organisms, nor that converting DNA sequences into protein sequences can be extremely informative.

Historically, DNA sequencing follows cloning (Sambrook et al. 1989). Most such studies involve a known gene, thus enabling functional insights at the level of the gene as well as that of the organism, and also facilitating analysis through knowledge of the reading frame. However, a randomly-derived clone, identified as most likely to be scnDNA, could be used. The same DNA in different species could then be identified using the first clone as a probe. Once homologous DNA was cloned and recognized from different species, standard sequencing methods could be used for each such new clone. One risk to this procedure follows from the occurrence of duplicate genes, which it would be hard to control for if one could not identify the sequence being used. If there have been duplications during the history of the region studied, then the phylogenetic tree may reflect more of the history of these duplications than of the divergence of species. Large-scale errors in phylogeny estimation might then result.

Sequencing studies of nuclear genes are probably not practical for determining the phylogeny of intraspecific populations or closely-related species because of the time-depth problem. To avoid this difficulty many loci should be used, and it would seem more cost-effective to use allozymes or RFLPs. In addition, crossovers and gene conversions may disturb the resulting phylogeny at any phylogenetic level, but more seriously at lower levels where the smaller amount of difference between alleles would make them harder to detect. At higher levels, sharp changes in the level of similarity along the length of a gene, for some comparisons but not for others, would enable detection of gene conversions.

3.3.2. Short Cuts: Especially PCR

There are various drawbacks to the "classical" approach to sequencing involving cloning. One is the extra time and effort involved in cloning. Another is the need for substantial quantities of DNA to enable cloning to proceed. Both these difficulties are potentially obviated by two recent advances.

In both cases, good knowledge of the sequence of flanking DNA to the target region is needed. If sufficient DNA is available, then a primer made to the known flanking sequence can be used in direct, or genomic, sequencing. In this method, total DNA is denatured and the sequencing carried out directly using this mix (Sambrook et al. 1989). Although this appears to be a rather "temperamental" method, it has been used for phylogenetic studies in some cases.

The most exciting development for evolutionary sequencing studies has been the development of the polymerase chain reaction (PCR). Briefly, given sufficient knowledge of the flanking sequences, primers based on these can be used to amplify the region of choice from a very small starting amount of DNA (Saiki et al. 1985). This starting amount can be as small as the single gene copy contained in a sperm (Li et al. 1988).

Furthermore, this feature allows the use of preserved material from either museums or, in some cases, from nature (Paabo et al. 1989; Golenberg et al. 1990).

The DNA amplified by PCR can be investigated for RFLPs using restriction enzymes, which would enable population genetics of very small organisms. Li et al. (1988) were able to amplify more than one gene per sperm, so that amplifying several regions from a small insect should be possible. However, the most-used application of PCR in evolutionary studies involves sequencing. This can involve first cloning the amplified sequence, but more usually sequencing is carried out directly on the amplified DNA.

There are basically two ways in which such direct sequencing can be carried out. One is to perform an unbalanced synthesis, so that one strand of the amplified DNA is produced in much larger amounts than the other, thus facilitating sequencing in a manner similar to that involving M13 subcloning (Gyllensten 1989). The other sequencing method, so-called double-stranded sequencing, is performed after "balanced" synthesis in which approximately equal amounts of the two strands have been formed. This method is essentially the same as the direct sequencing mentioned above, in that a primer is introduced into a denatured preparation of the DNA.

Although it may seem that the advantage of direct sequencing from the amplified product is purely one of convenience, it is also more accurate than sequencing cloned PCR products. This greater accuracy stems from the fact that the *Taq* polymerase conventionally used in PCR has a relatively high error rate in synthesis; consequently DNA cloned from PCR products is liable to have a high rate of mutation relative to the original sample. Saiki et al. (1988) estimated this error to be 0.25% per nucleotide following a typical PCR procedure. Such an error rate would mean that the probability of finding an error-free piece of amplified DNA 1 kb long is only 0.08, and the probability that there are five or more errors is over 0.10. However, these errors are at different sites in different copies of the DNA; most copies are correct at any one position. Sequencing directly from the amplified product therefore leads to the swamping of these errors and an accurate result is expected (Paabo and Wilson 1988). Gyllensten (1989) discusses these considerations and the fact that PCR tends to produce "recombinant alleles" if the sample DNA is multiply heterozygous for the amplified region.

Some common sense is necessary in dealing with the cloning artifacts mentioned above. For inferring phylogeny such errors would be interpretable as unique derived character states (autapomorphies) and would therefore not bias the topology inferred, although branch lengths would be overestimated. These errors are potentially serious in studies of gene function, where it is important to get the sequence 100% right.

However, since the errors can be largely avoided while at the same time avoiding a tedious procedure, one might as well use direct sequencing if at all possible.

4. Single-unit Genes

4.1. Mitochondrial DNA

4.1.1. Special Characteristics

Animal mtDNA is a circular molecule usually containing 13 protein-encoding genes, two ribosomal RNA genes, 22 tRNA genes, and a control region containing the origin of replication (Avise et al. 1987; Moritz et al. 1987; Harrison 1989; Crozier 1990). In vertebrates the control region shows a displacement loop and hence is termed the D-loop region; in insects no such loop occurs, and the region is termed the A+T-rich region, reflecting its sequence composition. The minimum amount of DNA needed for these various regions is about 15 kb; typically 16.5 kb is the length of animal mtDNAs. MtDNAs with greater lengths generally have longer A+T-rich regions or duplications of some coding regions. Such duplications are probably evolutionarily ephemeral, with a possible exception in beetle species of the genus *Pissodes*, in which there are exceptionally large A+T-rich regions (Boyce et al. 1989). A further complication arises because of mutations yielding intra-individual size polymorphism in mtDNA molecules, as in some *Drosophila* (Hale and Singh 1987), cricket (Rand and Harrison 1986), and beetle (Boyce et al. 1989) species. Such heteroplasmy is probably ephemeral but can persist for many generations. For a succinct account of *Drosophila* mtDNA, the best-known among insects, see Clary and Wolstenholme (1985).

The characteristics which have attracted evolutionists to mtDNA are its presence in high copy number (up to 50% of the DNA in eggs), which facilitates its isolation and its maternal and clonal inheritance, meaning that conversion and similar processes affect it negligibly.

It is often assumed that maternal inheritance of mtDNA stems from a dilution effect: eggs have more mitochondria than sperm. An alternative explanation, that maternal inheritance results from specific exclusion of paternal mtDNA, is well-supported by studies on organelle inheritance in protists and plants (Crozier 1990). For example, inheritance of both chloroplast and mtDNA in *Chlamydomonas reinhardtii* (Boynton et al. 1987) and *Pinus taeda* (Neale and Sederoff 1989) is uniparental, but with one parent donating chloroplasts and the other mitochondria. Of course, where a specific exclusion mechanism is absent, then the dilution effect may still occur, making it difficult to distinguish between them.

Some extensive tests on animals have indicated no paternal transmission of mtDNA. For example, Lansman et al. (1983) carried out reciprocal backcrosses between two *Heliothis* species for many generations (45 generations in one case and 91 in the other) without detecting any "paternal leakage," concluding that any such leakage must be at a rate of no more than 4×10^{-5} per generation. More recently, Avise and Vrijenhoek (1987) studied a hybridogenic population of the fish *Poeciliopsis*, in which the paternal chromosomes are excluded every generation but in which sperm is required for development of the embryo. These fish showed no paternally-derived mtDNA despite constant potential for input every generation since the inception of the species.

From experiments such as those of Lansman et al. (1983) and Avise and Vrijenhoek (1987), it appeared clear that maternal inheritance is strictly observed in animals. However, such certainty was misplaced: Satta et al. (1988) found that two strains of *Drosophila simulans* were heteroplasmic for standard *D. simulans* mtDNA as well as that of the sympatric sibling species, *D. mauritiana*. Backcrossing 331 lines of *D. simulans* (initially homoplasmic for *D. simulans* mtDNA) to *D. mauritiana* for 10 generations led to three lines apparently containing only *D. mauritiana* mtDNA and one being heteroplasmic (Kondo et al. 1990). Kondo et al. (1990) estimate the probability of paternal transmission at 0.001 per cross.

Nevertheless, Kondo et al. (1990) note that this rate is not so high as to jeopardize the use of mtDNA as a good indicator of interpopulation differentiation. They also note the likelihood that intracellular selection is likely to be important in determining the rate of paternal transmission. This suggestion is strongly supported by findings of coadaptation between nuclear and mitochondrial genomes (Weide et al. 1982; Clark and Lyckegaard 1988; MacRae and Anderson 1988; Miller et al. 1986; Rand and Harrison 1986, 1989). The procedure followed by Kondo et al. (1990), in which *D. simulans* nuclear DNA would have been replaced by that of *D. mauritiana*, would thus facilitate the establishment of *D. mauritiana* mtDNA. Where hybridization is a rare event followed by backcrossing to the maternal parent, the conditions for establishment of paternally-derived mtDNA might be less favorable.

The numerous cases of heteroplasmy for mtDNA length have been plausibly explained as due to a high mutation rate for length variation, a conclusion reached by Harrison et al. (1987) for crickets. Buroker et al. (1990), working with sturgeon, came to the same conclusion and proposed a molecular model to explain the ready generation of length variants involving the D-loop of vertebrates. Certainly those cases in which novel length variants include sufficient deletions as to be safely assumed to be nonfunctional (e.g., in mice, Boursot et al. 1987), or occur in parthenogens (e.g., Densmore et al. 1985), are most easily explained as due to mutation rather than paternal leakage.

A further issue is whether or not mtDNAs remain unrecombined. In yeast and in plant tissue cultures recombination between mtDNAs from different parents do recombine (Birky 1978; Sederoff 1984). There is indirect evidence that recombination occurs between mtDNAs in mammalian somatic cell hybrids (Birky 1978), but if so this is an infrequent occurrence because the mtDNAs of the hybrids appear to retain their integrity, as judged by RFLP patterns (Hayashi et al. 1987). A similar result has also been observed in *Drosophila* (Satta et al. 1988). Further studies on *Drosophila* heteroplasmic systems, developing the methods established so far, should answer this question (Niki et al. 1989; Kondo et al. 1990). The observed integrity of mtDNA haplotypes in animal populations, frequently showing multiple base differences between sympatric variants, can therefore stem either from a general lack of paternal transmission, from failure to recombine when paternal transmission does occur, or from elimination through selection of recombined mitochondrial genomes.

Whatever the final conclusions regarding the occurrence of paternal transmission in nature, it is clear that absolute certainty of strict maternal inheritance is now lacking. It seems likely that the strictness of maternal inheritance will vary from taxon to taxon, but it is unlikely in most cases to be easy to estimate the rate of paternal leakage, leaving mtDNA studies on closely-related species with a slight trace of uncertainty concerning the applicability of the gene trees to those of the species, over and above the time depth problem.

MtDNA was also once thought to evolve more rapidly than nuclear DNA, but this does not appear to be true for insects (Powell et al. 1986). The attractive suggestion that the evolution of mtDNA is particularly "clock-like," and that it is the nuclear DNA which varies in evolutionary rate (e.g., Moritz et al. 1987), has also been disproved; mtDNA genes vary between lineages in overall substitution rates (Gillespie 1986; Crozier et al. 1989).

4.1.2. RFLPs

The earliest studies of mtDNA variation used RFLPs. The high copy number of mtDNA molecules per cell meant that it could be isolated without sequence information more readily than any kind of nuclear DNA.

The data provided from RFLP studies can be analyzed either in terms of the fragments produced or in terms of the sites which give rise to them. The latter approach requires a more lengthy procedure to map the sites but gives more precise information. This does not automatically impose an obligation to use sites rather than fragments, because the additional work in restriction site mapping may not be cost-effective for studies in which the results are sufficiently clear-cut using fragments alone.

Apart from the existence of more powerful methods of analysis for sites than for fragments, the question of time scale should also be borne in mind. Similarities between lineages persist longer for sites than for the fragments they produce—loss of one of the sites bounding a fragment leads to loss of the fragment even though one of the sites remains. As taxa approach the upper limit of analysis through RFLPs, therefore, use of sites becomes mandatory. Estimators of nucleotide substitutions are available for both sites and fragments (e.g., Nei and Li 1979; Kaplan 1983; Li 1986), although most effort has gone into the treatment of sites.

Most studies of RFLPs in insect mtDNAs (even the earliest—Shah and Langley 1979; Trick and Dover 1984; but see also Hale and Beckenbach 1985) have defined restriction sites as the character set. These studies have addressed questions of intraspecific population differentiation (e.g., DeSalle et al. 1987b; Hale and Singh 1987; Baba-Aissa et al. 1988), interspecific hybridization (e.g., Harrison et al. 1987; Powell 1983), and the phylogeny of related species (e.g., DeSalle et al. 1986a, 1986b; Latorre et al. 1988). The potential to use mtDNA for studies on microgeographic variation, separating female from male dispersal, has been exploited in studies on small mammals (Kessler and Avise 1985; Plante et al. 1989) but not, as yet, on insects. Harrison (1989) reviews the use of mtDNA markers in studies on hybrid zones.

One important question concerns the time scale over which mtDNA RFLP information is phylogenetically informative. This length of time depends on the substitution rate and the manner in which substitutions are distributed between possible base positions. Estimates of substitution rates for mtDNA in *Drosophila* vary from 0.5% per million years (my) (Latorre et al. 1988) to 1.7% per my (Caccone et al. 1988). A particularly detailed study, based on sequence information, was that of DeSalle et al. (1987a), who concluded that the rate for Hawaiian *Drosophila* is 1% per my. However, there is also good evidence that evolutionary rates for mtDNA vary not only among major insect groups (Crozier et al. 1989) but also among groups of *Drosophila* species, as discussed by Latorre et al. (1988).

If we take 1% per my as a representative substitution rate for mtDNA, then the probability that a 4-base site survives in each of two taxa for one million years is 0.9224, and for a six-base site this probability is 0.8858. If we take a probability of mutual site survival of 0.1 as setting a practical lower limit to the usefulness of site data, then for 4-base sites the maximum divergence time over which such data are useful would be 28 my and for 6-base sites, 15 my.

Substitutions are not, however, distributed randomly in time. DeSalle et al. (1987a) note that only 8% of base positions in the DNA sequence they studied changed at the maximum rate. They also noted that the initial rate of divergence between lineages is rapid, reflecting synonymous substitutions (largely transitions, i.e., purine to purine or pyrimidine to pyrimidine changes), and that after about 30 my divergence slows markedly. After about 50 my, divergence is mostly due to the accumulation of transversions (i.e., purine/pyrimidine changes or vice versa). Naturally, substitutions are not added randomly along the length of a DNA segment either, as is discussed further below.

The non-randomness of substitutions has a number of consequences that increase the utility of mtDNA sites for phylogenetic inference. The rapidity of divergence between closely-related taxa means that sites are generated and lost at a high rate, providing good information for such studies. The later deceleration in evolutionary divergence means that some sites are expected to be retained much longer than expected under our simple random model above. This expectation is also generated by the observation that stretches throughout the mitochondrial genome are highly conserved (Kocher et al. 1989). Sites falling in such stretches will be strongly conserved. Carr et al. (1987) found three sites to be invariant in "nearly all" vertebrate mtDNAs known.

There is thus no clear answer to the question of the upper limits of usefulness of mtDNA RFLP data, although it is likely that studies above the genus level would not be well-served by such data.

4.1.3. Sequencing: Old and New

The history of sequencing studies on insect mtDNA has followed that of the field of evolutionary genetics as a whole. Until very recently, sequencing required cloning, which required reasonably large quantities of tissue and a relatively labor-intensive procedure. Wilson et al. (1989) argued that using sites data is more cost-effective than sequences if the sequencing is by conventional methods. This conclusion seems correct for relatively closely-related forms where sites data are informative, but studies on higher levels would require sequence data, however acquired. For example, the relationships between insect subfamilies, families and orders can be studied using sequence information, whereas it is unlikely that there is sufficient site similarity left to provide useful phylogenetic information at these levels.

Wilson et al. (1989) argue that the development of PCR has made sequence studies more cost-effective than sites studies, reversing the conclusion made when only conventional sequencing methods were available. In particular, the advent of the universal primer approach (Kocher et al. 1989) has brought the bulk of animal species open to ready study by molecular methods.

4.2. Ribosomal Genes

4.2.1. Special Features

Ribosomal RNA makes up a large fraction of the RNA in cells because of its importance in the translational machinery. Consequently, efficiency in the production of ribosomal RNA (rRNA) is strongly favored by natural selection. This high rate of production has been achieved by having many copies of the genes for rRNA in the genome. These multiple copies are based on two types of repeating unit. In *D. melanogaster*, for example, one unit contains the 18S, 5.8S, 2S and 26S genes, plus a number of transcribed spacers and a nontranscribed spacer, whereas the other unit contains only the 5S gene plus one transcribed and one nontranscribed spacer (reviewed by Watson et al. 1987, p 653). The compositions of the two repeating units vary among eukaryote nuclei, and prokaryotes and mitochondria have different rRNA genes, although these are related to genes in the eukaryotic nucleus. Each unit occurs in tandem repetition hundreds to thousands of times, depending on the species.

When Brown and Sugimoto (1973) compared ribosomal regions within and between two *Xenopus* species, they found that different copies of an rDNA unit in the same individual showed high similarity, whereas these all differed from the copies in members of the other species. If the various copies were evolving independently, the different copies held by an individual should also have diverged. The finding for these and similar gene families of intrapopulation similarity, but interpopulation and interspecies divergence, has been described as being due to concerted evolution (see Section 7.1). Concerted evolution leads to the numerous copies of a gene family being similar enough to be regarded as replicates. Exceptions are expected when two diverged populations hybridize (Suzuki et al. 1987).

Ribosomal gene arrays may occur on more than one chromosome. In that case, it is possible for them to show different evolutionary dynamics and sequence divergence, as is the case in *Drosophila melanogaster*, where there are rDNA arrays on both the X and Y chromosomes (Williams et al. 1987). Such divergence would make analysis difficult. At an elementary level, the localization of rDNA arrays can be checked cytologically in that they tend to occur in discrete nucleolar organizer regions. Rarely, rDNA units are found separated from the rest of the family, and these lone elements have been termed orphons (Gerbi 1985).

The occurrence of rDNA (rRNA genes) in all forms of life means that they have universal applicability in phylogenetic reconstructions. Another advantage is that large amounts of the rRNA transcripts occur in cells, so that the RNA can be sequenced as an alternate strategy to direct sequencing of the DNA. This feature is explained further in Section 4.2.3

below. However, one problem is that the use of rDNA in phylogenetic studies is that changes in this system have a strong tendency to affect individual fitness. Such changes occur not only in the rRNA genes themselves but also in the spacer regions between them, probably reflecting their regulatory importance (De Winter and Moss 1986). They have been reported in *Drosophila* (Cluster et al. 1987) and barley (Saghai-Maroof et al. 1984), and mean that change in these regions is likely to be episodic rather than clock-like.

4.2.2. *RFLPs*

The spacer regions evolve much more rapidly than the rRNA genes themselves, as can be seen by comparing the positions of conserved and variable sites from *Rana* frogs (Figure 5.2; Hillis and Davis 1986). However, change in the spacer regions still occurs more slowly than in mtDNA: Hillis and Davis (1986) used spacer region data to construct a phylogenetic tree for *Rana* species whose divergence dates back 50 my. Change in the rRNA genes themselves might yield information about earlier divergences, but there are probably too few such changes for reliability (Hillis and Davis 1987).

It is uncertain how rapidly insect rDNA evolves, but the anuran case suggests that it could be useful over a wide time scale. Data on closely-related forms may be more useful for inferring the migrational interrelations of populations rather than their phylogenetic relationships, given the effects of gene flow on concerted evolution. Davis et al. (1990) found a polymorphic site in the wasp *Polistes exclamans* using a mouse rDNA probe and treated this as a one-locus two-allele system, augmenting an allozyme data set. These data were used to estimate population subdivision parameters, and a high degree of subdivision was found.

4.2.3. *Sequencing*

The development of a rapid sequencing method specific to rDNA has led to the production of a large volume of sequence information. Briefly, rDNA includes regions that are essentially invariant. Sequencing primers made homologous to these regions can then be used to sequence rRNA from bulk cellular preparations using reverse transcriptase (Lane et al. 1985; Pace et al. 1985). This method has a higher error rate than that expected from conventional cloning and sequencing (Lane et al. 1985) but still provides useful information for distantly-related forms (Lane et al. 1985; Pace et al. 1985; Yang et al. 1985; Pace et al. 1986; Hori and Osawa 1987; Field et al. 1988; Baroin et al. 1988).

Complete sequencing of rDNA, using either the reverse transcriptase method, cloning and sequencing, or (in future) PCR-based sequencing, is

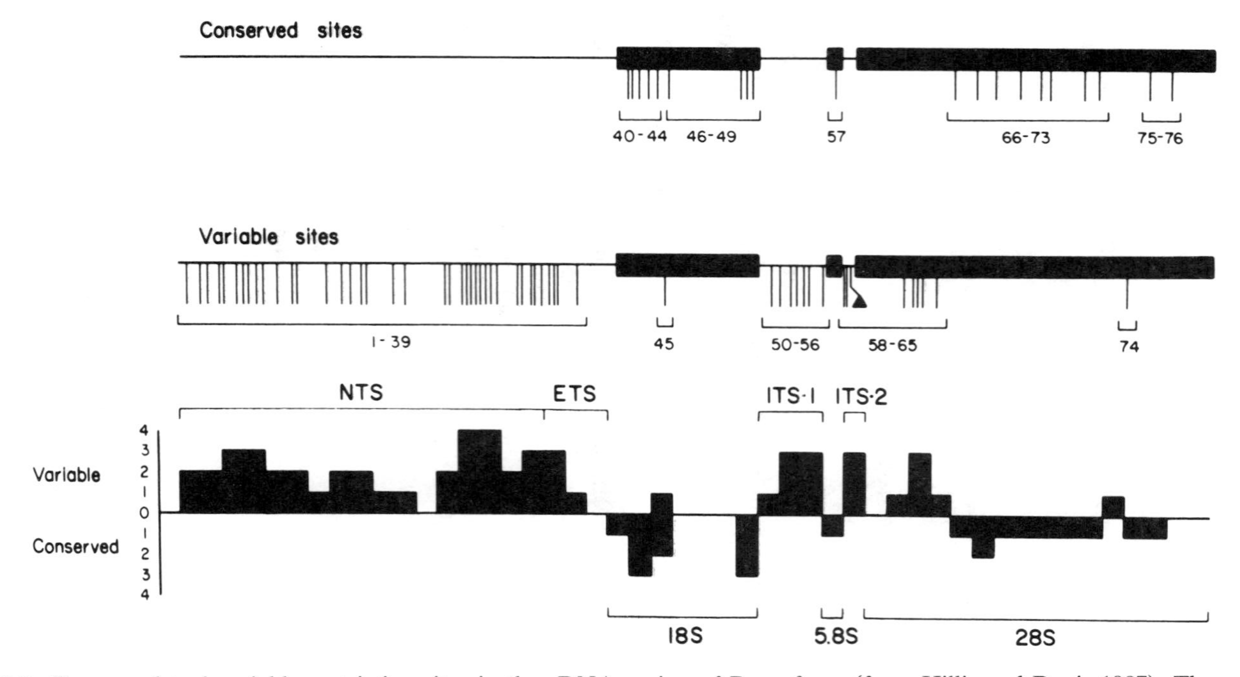

Figure 5.2. Conserved and variable restriction sites in the rDNA region of Rana frogs (from Hillis and Davis 1987). The regions coding for the 5.8S, 18S, and 28S subunits of rRNA are shown by bars; NTS = non-transcribed spacer, ETS = external transcribed spacer, ITS = internal transcribed spacer regions. The spacer regions are much more variable than those coding for rRNA subunits.

replacing the former use of oligonucleotide catalogs. The method of oligonucleotide catalogs involved digesting 16S RNA with a ribonuclease which cleaves at G positions, sequencing the resulting fragments and assessing similarity in terms of the fragments shared (Woese 1987). The difficulty with this method is that change of one base in a fragment destroys similarity between the taxa concerned, even though the remaining bases may be unchanged. Although rDNA sequences have been most used for comparisons of distantly related taxa, they can be used for more closely related forms. Larson and Wilson (1989) used large subunit sequences to infer relationships between seven families of salamanders. Closely-related forms (ca. 14 my divergence) lacked sequence differences, but these authors concluded that the region sequenced provides good information for divergence times in the 100–200 my range. This approach might therefore be useful for determining the relationships between insect orders.

4.3. Y-chromosomes: Paternal Gene Flow Markers

Y-linked genes provide a natural contrast to mtDNA for studies on population structure and differentiation. In both cases there is only one copy of the sequence when it occurs, and it is transmitted without recombination by only one of the two sexes. Given a 1:1 sex ratio, the effective population size of Y-linked and mtDNA genes will be the same, and only one quarter that of autosomal genes. There is also an obvious attraction of being able to trace gene dispersal through males as a counterpoint to tracing female-borne genes studied using mtDNA.

The chief hurdle to the use of Y-linked sequences is finding any. Even in species which have Y-chromosomes, it is well-known that there are few active genes on them, and it has been correspondingly hard to find sequences specific to them. However, screening of libraries to uncover clones that hybridize to male but not female-derived total DNA should succeed in yielding such markers. Use of a somatic cell line containing only the human Y-chromosome (plus a mouse background) produced six Y-linked sequences (Bishop et al. 1983; Casanova et al. 1985). These sequences were used to show that there has been extensive Y-chromosome introgression between mouse subspecies in the laboratory (Bishop et al. 1985) but not in nature (Vanlerberghe et al. 1986).

I do not know of any Y-specific sequences isolated from insects. However, Y-linked rDNA polymorphisms are known in *Drosophila melanogaster*, in which both X and Y chromosomes bear rDNA arrays. Williams and Strobeck (1986) used a tester stock with an X-chromosome with the rDNA largely deleted to assay field-caught females for mating frequency through progeny analysis.

5. Pure-distance Measures

5.1. Immunological Comparisons

"Pure-distance methods" mean those which can return only a distance between two species: no characteristics of a single species can be obtained.

Immunological comparisons can be made using extracts of the whole organism, but this approach potentially confounds changes in the relative dosage of various proteins with changes in their amino acid sequences. A more precise method is to monitor immunologically-detectable change in a single protein. Such a method is microcomplement fixation (MCF), which uses coupling to hemoglobin to enable a colorimetric assay (Champion et al. 1974). The method has been widely exploited in vertebrate phylogenetics using the abundant blood protein albumin (e.g., Roberts and Maxson 1985).

Beverley and Wilson (1982) were able to harness a major storage protein for MCF studies in *Drosophila*. The evolutionary rate of this protein was determined to be similar to that of mammalian hemoglobins (Beverley and Wilson 1984), and this calibration was used to infer an ancient origin of the Hawaiian *Drosophila* species prior to the existence of the current islands (Beverley and Wilson 1985).

While MCF allows greater precision in determining amino acid replacements, it faces a number of disadvantages, some of which have been alluded to already. These are: no characteristics of individual species are obtained; there are no means of determining the statistical sufficiency of the results; each new taxon added to the set should be tested against all previously-studied species, leading to an exponentially-increasing volume of work; and for closely-related forms, the gene tree-species tree problem applies. Given the advent of relatively easy DNA sequencing, MCF would seem to have had its day.

5.2. Single-Copy DNA Hybridization

The various methods discussed so far involve relatively small subsets of the genome, usually applicable to less than one hundred genes and sometimes just one gene. The notion of using a large fraction of the genome in systematic studies is therefore attractive, provided that such estimates can be made with appropriate precision and lack of bias. These advantages have been claimed for the method of DNA-DNA hybridization.

The DNA in the nuclei of higher organisms is made up of sequences differing in the number of copies present per genome. In addition to single-copy nuclear DNA (scnDNA), gene families can include many members, and high-copy number nuclear DNA (hcnDNA) is present hundreds of thousands of times or more. Comparisons based on repeated DNA are liable to involve changes in dosage as well as changes in sequence, and such dosage changes can occur rapidly and erratically; for example, the pea *Pisum sativum* has more than nine times as much total DNA as does the mung bean *Vigna radiata* (Sang 1984, p 46). The concept of scnDNA hybridization as a phylogenetic tool is thus to compare the sequence similarity of organisms using a class of DNA stable in copy number, so that sequence divergence alone, and not dosage, is the object of the comparison.

The use of the method has been clearly explained by Sibley and Ahlquist (1986; 1987a). Briefly, scnDNA is prepared by heating a DNA preparation to dissociate all DNA into the single-stranded form. As the solution cools, hcnDNA sequences associate more rapidly than scnDNA sequences and are removed by chromatography, leaving DNA enriched in scnDNA. The scnDNA of one species is labeled with radioactive iodine and allowed to hybridize with DNA from another species. The temperature of the hybrid DNA is raised in steps, and the proportion of dissociated DNA recorded at each step. Because the strength of DNA association depends in large part on the proportion of matches of the constituent nucleotides, hybrid DNA molecules made from homologous sequences from different species will dissociate at lower temperatures than such hybrids made from the same species. The more distantly two species are related, the more readily their DNAs will dissociate with heating. The difference between the stability of DNA hybrids from the same and from different species is therefore a measure of the evolutionary distance between them.

Sibley and Ahlquist (e.g., 1986; 1987) have been prolific in using DNA-DNA hybridization to infer the phylogenies of birds and primates. This body of work has recently come under strong attack (e.g., Cracraft 1987; Marks et al. 1988; Sarich et al. 1989). The critics raise problems that are mostly specific to the work of Sibley and Ahlquist, and which can be rectified in other studies. For example, the choice of divergence statistic used by Sibley and Ahlquist is widely held to be inappropriate, and more appropriate ones are available (Caccone et al. 1988; Sarich et al. 1989). Also, the buffers used by Sibley and Ahlquist lead to a dependence of the melting profiles on the GC content of the DNA, whereas this dependence can be effectively eliminated by use of the TEACL buffer (Hunt et al. 1981; Caccone et al. 1988).

Certainly, DNA hybridization has been used for a number of applications, such as evolutionary rate comparisons (e.g., Britten 1986; Vawter and Brown 1986; Powell et al. 1986) and for phylogenetic studies with insects (e.g., Hunt et al. 1981; Caccone and Powell 1987). Given that the above criticisms of Sibley's and Ahlquist's work can be rectified, it would seem that DNA hybridization remains a potentially powerful source of distance data for phylogenetic inference. Yet difficulties remain, especially with respect to insects.

One general difficulty often raised is that the data are indeed pure distances. Hence, it is difficult to determine the degree of confidence one can place in particular distances. To a considerable extent, however, when there is negligible intraspecific variation, the differences between replicates can be used to derive levels of confidence (Felsenstein 1987). In fact, Felsenstein (1987) argues that the discriminatory power of DNA hybridization is much greater than that of sequence data for realistic sequencing effort.

Even highly-replicable distances may not, however, accurately reflect evolutionary divergence between scnDNAs where there are differences between species in the proportion of DNA which is hcnDNA. Because the separation of scnDNA is not perfect, differences in hcnDNA dosage between species will lead to different degrees of contamination of the preparations and hence change the divergence estimates. On the other hand, Bledsoe and Sheldon (1989) suggest that these effects might be detected as non-reciprocity (different results depending on which species is used as the source for the labeled DNA).

Another limitation of scnDNA hybridization data concerns the upper limit to the degree of divergence which can be accurately estimated. Caccone and Powell (1987) note that DNA sequence evolution in vertebrates is unusually slow, so that the successes obtained for studies on vertebrates are probably not possible in other groups. Specifically, Caccone and Powell (1987) predict that DNA hybridization would not be useful for insect studies above the genus level.

Finally, there is the difficulty of intra-specific variation. In vertebrates, such divergence has been found to be minimal, whereas insects and other non-vertebrates show marked interpopulation and interindividual differences (Caccone and Powell 1987; 1990). Numerous comparisons among populations might therefore be necessary, greatly increasing the work load and concomitantly reducing the utility of the method.

6. Remarks on Analyses

6.1. Are Branching Trees Appropriate?

Most attention in considerations of phylogenetic inference is focussed on branching trees, but there are cases where the application of other models will be appropriate. The most common case where branching trees may be inappropriate is that of conspecific populations. In this case, a more accurate representation of the evolutionary relationship between the populations may be a net rather than a branching tree (Maynard Smith 1989). Another example would be when hybrids between two species give rise to a third, a common pattern in plant evolution.

How can one decide whether use of a branching tree method is justified? Eigen et al. (1988; 1989) provide rules of thumb for determining when the additional distances from a net should be regarded as significant. An alternative approach is that of Cavalli-Sforza and Piazza (1975; as discussed by Astolfi and Zonta-Sgaramella 1984), in which a "treeness" statistic is used to estimate how dependent two taxa are on their common ancestor rather than on OTUs (operational taxonomic units) elsewhere in the tree. Application of the treeness test depends on further assumptions, such as the evolutionary model underlying the tree under test. Consequently, decisions as to the appropriateness of a branching tree will usually rest on biological knowledge of what is reasonable in the data set.

6.2. Bias and Precision

Penny and Hendy (1986) took the view that ". . . it is unreasonable to publish an evolutionary tree derived from sequence data without giving an idea of the reliability of the tree." This standard should, in my view, be applied to phylogenetic inference generally and not just to molecular data. Numerous comparisons of different methods of phylogenetic inference have been made to assess them for both bias (tendency to produce the true tree) and precision (provision of statistical tests to compare trees). Two criteria have been used in these assessments: fit of the result to the true tree, and congruence (agreement between data sets) (Mickevich 1978; Rohlf et al. 1983).

The difficulty with the "true-tree" approach is that, for real organisms, this is seldom known in sufficient detail. The difficulty with the "congruence" approach is that the known biases of tree-building methods should apply to all kinds of data collected from a set of organisms. Congruence is therefore a risky guide to assessing bias. The use of simulated data sets based on known trees is therefore the test method of choice. These sets have mostly been produced by computers, but Fitch and Atchley (1987) provide a rare true-tree test using real data, namely gene frequency, metric and life-history data from mouse strains of known genealogy. This study actually proved a poor discriminator of analytical methods—all methods used returned the true tree from the genetic data, although the trees based on the other data sets were very widely at variance with the known tree.

The estimation of precision considers the organism as a bag of characters. The characters actually used are but a sample of the enormous number available if only they could be surveyed. The problem is then to estimate how likely different samples of characters are to yield different estimates of the phylogeny. Associated with this sampling problem is the fact that, even if characters are evolving at a constant rate, the stochastic nature of this change introduces a stochastic error into estimates of tree topology and branch lengths (Nei 1986), so that for a finite number of characters there will be a tendency for evolutionary rates to appear to be unequal. This spurious inequality results from the chance occurrence of more fixed mutations in some lineages than in others and is minimized as the sample size increases. Choice of an analysis method will be mediated first of all by the kind of data one has. Next, the criteria of bias and precision should come into play. Because the timing of evolutionary events will often, perhaps usually, be of interest, it is also desirable to select a method allowing the estimation of branch lengths in addition to branching order.

It is not always easy to find a method that will fulfil all three of the latter needs. Next, I will briefly discuss a few of the chief methods in the light of these criteria—Felsenstein (1988) provides a more detailed review of many of the issues. Computer packages are available for many of the operations. Two that include programs that not only estimate trees but also provide estimates of statistical sufficiency are PAUP (D Swofford) and PHYLIP (J Felsenstein).

6.3. An Eclectic Survey of Methods

6.3.1. Parsimony

Parsimony, or maximum parsimony, rests on the philosophical assumption that the most probable phylogeny is that which minimizes the number of postulated evolutionary changes. Parsimony methods can be applied to any data set with distinct characters. Although various authors still believe, following Hennigean tradition, that it is necessary to determine which character states are ancestral and which are derived before a parsimony analysis can be carried out, this is not the case (Colless 1985; Swofford and Berlocher 1987).

The most important bias of parsimony can be stated in the words of Hendy and Penny (1989) as "long edges attract." As first pointed out by Felsenstein (1978), under appropriate conditions of unequal evolutionary rates, the more rapidly-evolving taxa tend to be erroneously paired in the tree. Increasing the size of the data set simply increases the certainty of this outcome! According to Hendy and Penny (1989), a similar result can occur even when evolutionary rates are equal.

Simulation results confirm the unreliability of parsimony when evolutionary rates become markedly different among lineages. As expected, given enough rate differences and sufficient data, the rate of recovery of the true tree drops to zero with parsimony (Li et al. 1987). Otherwise, when evolutionary rates are the same for all lineages, parsimony is at least moderately efficient at recovering the true tree (e.g., Li et al. 1987; Rohlf and Wooten 1988; Kim and Burgman 1988; Saitou and

Imanishi 1989) but starts to fall behind other methods in efficiency when evolutionary rates are allowed to vary. But this fall-off in efficiency is in uniquely recovering the true tree. Sourdis and Nei (1988) found that, even when evolutionary rates vary, parsimony converges to an incorrect tree less than distance methods, often returning several equally-parsimonious trees including the true tree.

It is hard to know how serious for molecular studies is the tendency of parsimony algorithms to converge to an incorrect result when branch lengths differ sufficiently. When the true tree is not known, one cannot compare it with the result! However, given that evolutionary rates have to be very unequal before the problem becomes substantial (Felsenstein 1978; Li et al. 1987; Hendy and Penny 1989), rate differences are probably not a serious problem for most molecular studies. Exceptions may occur for distantly related organisms, such as those falling into different kingdoms evolving at markedly dissimilar rates (Lake 1988; but see also Gouy and Li 1989). If there is reason to suspect that evolutionary rates differ markedly among the lineages being examined, perhaps another method should be considered.

Strictly speaking, because the "best" parsimony tree is defined as that which involves the least character change of all the possible trees, one should examine all possible trees. For k species, the number of possible unrooted trees is (Cavalli-Sforza and Edwards 1967):

$$\prod_{x=3}^{k} (2x - 5)$$

For example, for 10 species the number of possible unrooted trees is 135,135. Clearly, for very many species at all, the number of trees to examine will become too large for an exhaustive examination to be practical.

Where an exhaustive examination of all possible trees is not possible, tree-construction algorithms are used in which species are successively added to the tree in junctions minimizing the tree length. In the branch-and-bound method (Hendy and Penny 1982), an approximate exhaustive search is performed in which construction of a tree begins but is abandoned when its length exceeds that of the shortest tree so far discovered. Of course, while the branch-and-bound approach greatly reduces the amount of computer time required, even it will become prohibitively expensive as the number of taxa increases. Otherwise, the only way to proceed is to sample the space of possible trees by using different input orders of the taxa, which starts the search in a different place each time.

Estimation of branch lengths is problematic for parsimony methods because character state changes inconsistent with the tree can usually be placed on several different alternative branches. However, many characters will be unambiguously assignable, allowing at least a rough estimate of branch lengths.

Felsenstein (1985b) proposed the use of the bootstrap statistical technique to infer confidence limits for phylogenetic trees. The method could be applied to any phylogenetic inference method using characters as against pure distances but has been chiefly seen as useful for parsimony analyses. Briefly, the method involves forming sets of the data by sampling with replacement from the original data. For each new set, a tree is determined. Groups of taxa which occur in 95% or more of these trees are taken to be statistically significant for *a priori* tests (i.e., for groups previously proposed). When the groups are recognized from the analysis itself (i.e., the test is an *a posteriori* one), Felsenstein notes that a more stringent criterion is needed, such as accepting groups as supported at the 95% level only if they occur in a proportion of at least $(1 - 5/(x - 2))$ of the replicates, where x is the total number of taxa.

Application of the bootstrap does not require that all characters evolve at the same rate, but it (and the parsimony method as a whole) requires that they are not correlated with each other (Felsenstein 1985b). The problem of correlation between characters is a general one for phylogenetic methods. It is possible to use methods such as principle components analysis to derive new, uncorrelated characters (the factor scores), as done for a genus of ants by Crozier et al. (1986), but then a new problem emerges. This problem is that characters can be correlated within a data set for two reasons: functional dependence of one character on another, and a common evolutionary history (Crozier 1983). It is desirable to retain the evolutionary correlation! Various molecular data probably have much lower intercharacter correlations than other kinds of data, but correlations are likely even for these (Section 7).

Tests of one tree as a whole against another can be carried out using simple probability models (Cann et al. 1987; Crozier et al. 1989; Williams and Goodman 1989). In such a test, the number of character changes supporting one tree are assessed against those supporting the other, with the null hypothesis that the two groups are equal. Following Templeton (1983), a more sophisticated test has been devised by Kishino and Hasegawa (1989), who derive the variance of the differences between character states in two trees to estimate the probability that one tree is a significantly worse fit to the data than another. This apparently powerful and computationally efficient (relative to bootstrapping) method is likely to be widely used but should be tested further with artificial data sets.

Use is often made of the "consistency index" of Kluge and Farris (1969):

$$\frac{\text{possible minimum number of steps}}{\text{actual number of steps}}$$

but this value depends on the number of taxa and the number and nature of the characters used. Worse yet, random data can yield consistency indices as high as many published for actual organisms (Archie 1989a). The situation can be rescued somewhat by randomizing the data to determine the distribution of values expected in the absence of true phylogenetic information and comparing the observed value with this (Archie 1989a). A significant result indicates that the data contain phylogenetic information but does not indicate that any one such tree is significantly better than another. Archie (1989b) proposes using the homoplasy observed in randomized-data trees to counter the above problems by using the *homoplasy excess* ratio.

Parsimony methods include a number differing in their assumptions about modes of change (reviewed by Felsenstein 1982). Most remarks above refer to the "Wagner network" approach, in which all changes are equally likely.

Swofford and Berlocher (1987) discuss applications of parsimony to electrophoretic data. Restriction fragments (e.g., Avise et al. 1979; Ovenden et al. 1987), restriction sites (e.g., DeSalle and Giddings 1986; Latorre et al. 1988), amino acid sequences (e.g., Wyss et al. 1987), and nucleic acid sequences (e.g., DeSalle et al. 1987a; Crozier et al. 1989) have all been used in parsimony analyses. Swofford and Olsen (1991) provide a comprehensive survey of phylogenetic methods, especially parsimony, as applied to molecular data.

6.3.2. *Invariants, or "Evolutionary Parsimony"*

The bias of parsimony when evolutionary rates are variable has prompted considerable search for alternative methods insensitive to such rate changes. One such method is that termed "evolutionary parsimony" by Lake (1987a, 1987b), and an "invariants" method by Felsenstein (e.g., Felsenstein 1988). Invariants are "functions calculated from the data that take one value for all trees of a given topology, irrespective of their branch length" (Felsenstein 1988).

Lake's method applies only to nucleic acid sequences and only analyzes four species at a time. For four species, A, B, C, and D, there are three unrooted trees. Following the notation of Jin and Nei (1990), we can call the tree ((A, B), (C, D)) Tx, denote by Ty the tree ((A, C), (B, D)), and by Tz the tree ((A, D), (B, C)). For any one position, the nucleotide present for species A is symbolized "1." For the other species, the same nucleotide as that in species A is denoted "1," one differing from this by a transition "2," and if by a transversion "3" or, if there is already a transversional difference, "4." Thus, the combination ATGC of nucleotides for the same position in the four species is denoted 1324; N_{1324} denotes the sum across all nucleotide positions that this and equivalent combinations (e.g., TACG) occur. The invariants X, Y, and Z, for trees Tx, Ty, and Tz are:

$$X = N_{1133} + N_{1234} - N_{1233} - N_{1134},$$
$$Y = N_{1313} + N_{1324} - N_{1323} - N_{1314}, \text{ and}$$
$$Z = N_{1331} + N_{1342} - N_{1332} - N_{1341}.$$

According to Lake, a tree may be accepted if the invariant pertaining to it is significantly greater than zero and the other two invariants are not significantly different from zero.

Simulation shows that this method is indeed the least perturbed of many by drastic differences in evolutionary rate (Li et al. 1987). However, it also appears that, when evolutionary rates are fairly constant, the maximum parsimony and neighbor-joining (a distance measure) methods are superior to the invariants approach (Gouy and Li 1989; Jin and Nei 1990).

This variation in relative efficiency of the invariants method depending on the degree of rate variation places the phylogeneticist in a difficult position. Perhaps the best approach, if extreme rate variation is suspected, is to use both invariants and at least one other method. If the invariants method returns a different answer to the others, simulate sets of data according to the trees returned by the invariants method and by the other method(s) and see if each method(s) can derive the true trees from the simulated data. Accept the result yielded by the method whose behavior is most robust with these simulated data sets. This was the approach taken by Gouy and Li (1989), who found that the maximum parsimony and neighbor-joining methods could more efficiently return a model tree from data simulated using either the invariants-generated tree or that returned by the other two than could the invariants method. Gouy and Li (1989) concluded, for the interkingdom data set, that the invariants method was the least efficient of the three used.

A reason for Lake's invariants method to fail would be violation of its assumption that the probabilities of the two transversions possible for any nucleotide are the same, e.g., an A can be replaced by T or G with equal probability. This assumption is clearly violated for many sequences, including insect mtDNA, and Jin and Nei (1990) note that sufficient deviation from this assumption could even lead to the wrong tree being strongly supported.

6.3.3. Maximum Likelihood

A weakness of both the invariants and the parsimony methods is that only some of the data are used in tree evaluation. For example, in parsimony analyses, character states unique to single taxa (autapomorphies in cladistic parlance) are discarded from the analysis. However, such occurrences do provide information on branching order, e.g., see Felsenstein (1988). The remaining two categories of methods take account of all information.

Maximum likelihood is a powerful statistical method in which the probability of obtaining the observed data is evaluated in terms of alternative hypotheses. In terms of phylogenetic inference, this probability, or likelihood, is assessed in the light of the particular tree and an evolutionary model of how characters change through time. The attractiveness of likelihood models stems not only from the complete use made of the data but also from the well-developed methodology for assessing the relative support of different hypotheses, especially the likelihood ratio test (e.g., Sokal and Rohlf 1981, p 695). The first application of likelihood methods to phylogenetic inference was made by Edwards and Cavalli-Sforza (1964), but most subsequent development has been by Felsenstein, as recently reviewed by Felsenstein (1988).

Despite its power, the maximum likelihood method has some difficulties in applications to phylogenetic inference. One difficulty is that a powerful underlying evolutionary model is necessary. While, as mentioned above, all phylogenetic inference methods require assumptions about the nature of evolutionary change, maximum likelihood requires complete specification of the factors involved. A second difficulty is that, strictly speaking, the method cannot be used to test the significance of one tree against another because of a lack of degrees of freedom for such a test (Felsenstein 1988). Felsenstein (1988) suggests that assuming that the degrees of freedom equal one may yield a conservative test, but notes this test needs evaluating, such as with simulated data. A practical problem with maximum likelihood methods is that they use large amounts of computer time, especially in studies of sequence evolution.

The problem of assessing the statistical significance of a tree could be approached using the bootstrap (M Hasegawa and H Kishino, pers. comm.) given the uncertainties of using the likelihood ratio test for this purpose, but the bootstrap method is also profligate of computer time. The likelihood ratio test can, however, be used to examine hypotheses about the underlying model of evolutionary change, such as the pattern of change between codon positions, the ratio of transitions to transversions, and the variation in evolutionary rate between different kinds of sequences (e.g., Ritland and Clegg 1987). Confidence limits for branch lengths can be estimated under the maximum likelihood methods, enabling evaluation of the reliability of different sections of the tree produced (e.g., Crozier et al. 1986).

The variance test mentioned above for parsimony methods can also be applied to maximum likelihood trees, and in fact was devised mainly with these in mind (Kishino and Hasegawa 1989). Because it needs very little computer time and apparently avoids the uncertainties of the likelihood ratio test, it is likely to be widely used. As noted above, testing using simulated data sets is desirable.

Maximum likelihood methods have been developed for gene frequencies (Edwards and Cavalli-Sforza 1964; Felsenstein 1973; 1981). The underlying model assumes that gene frequencies disperse through genetic drift and that loci evolve independently—the researcher does not have to supply any other information than the allele frequencies.

Smouse and Li (1987) developed a maximum likelihood method for restriction sites and applied it to the problem of the chimpanzee-human-gorilla relationship. The method requires information on the transition/transversion ratio, the frequencies of the nucleotides, and the probabilities of particular base changes. Felsenstein's restML algorithm (in the PHYLIP package) performs the method but assumes that the four bases are equally likely to occur, and that transitions and transversions are equally likely.

The application of maximum likelihood to DNA sequences requires the same complete specification of base changes as does the restriction site model. Felsenstein's DNAML algorithm (in the PHYLIP package) allows specification of the overall base frequencies, the transition/transversion ratio, and the partitioning of sites according to expected evolutionary rate. The method has been applied to problems in human (Bishop and Friday 1985; Hasegawa et al. 1987), bat (Bennett et al. 1988), plant (Ritland and Clegg 1987) and insect (Crozier et al. 1989) mtDNA evolution.

Ritland and Clegg (1990) report that sequences which have diverged between 25% and 50% are optimal for analysis with the maximum likelihood method. According to the model they used for base substitution, 75% divergence would be the maximum possible (representing random similarity), but higher values of divergence could occur when the taxa have diverged in base content. In fact, a difficulty with the present maximum likelihood methods is that it is hard to take account of changes in the proportions of bases.

Simulations showed that, when evolutionary rates vary, maximum likelihood inferences of phylogeny from gene frequencies were significantly closer to the true tree than those produced by maximum parsimony or UPGMA (a distance method, see below), and that the same may hold for constant evolutionary rates if the number of loci sampled is very large (Kim and Burgman 1988; Rohlf and Wooten 1988).

For DNA sequences, a simulation study (Saitou and Imanishi 1989) indicated that maximum likelihood was a little less efficient than two distance methods (neighbor-joining and minimum-evolution) but outperformed these and the others (Fitch-Margoliash and maximum parsimony) tested when evolutionary rates vary "drastically" among lineages. Hasegawa and Yano (1984) also found by simulation that maximum likelihood is more robust to variation in evolutionary rate than is maximum parsimony, the same result obtained by Saitou (1988) who also found, however, that various distance methods did even better than maximum likelihood under these circumstances.

6.3.4. Distance Methods

A drawback to using distance methods is the difficulty in determining confidence limits for the phylogenies. But this difficulty pertains only to "pure-distance" measures. Where distances have been derived from character state data, techniques such as the bootstrap can be used to derive confidence limits.

Use of distance methods entails two decisions: the method used to convert character data to a set of distances, and the method to analyze these to produce a tree. Distance methods can be used to their best advantage if the distances used reflect the passage of time rather than just the differences between the taxa, that is, if allowance is made for the saturation resulting from multiple mutations at the same site. There have been many distance metrics proposed for use with molecular data, and I will discuss only a few here.

The most used population genetic distance for gene frequencies is that of Nei (see Nei 1987, pp 219–222; Felsenstein 1984). Under appropriate assumptions and for a limited span of evolutionary time, Nei's distances are expected to have a linear relationship to time (Nei 1987, pp 230–253). Useful reviews of genetic distances are those of Felsenstein (1985a; 1985b) and Nei (1987, pp 208–222). Under the assumption that the loci sampled come from a homogeneous distribution, confidence limits for Nei's distances can be readily obtained (Nei 1987, pp 222–229; Tomiuk and Graur 1988).

For restriction fragments and sites, the preferred distance is usually d, the expected number of nucleotide differences per site, and not p, the observed proportion of differences between two sequences. Nei and Li (1979) present a formula for estimating d from fragment data and Nei (1987, pp 64–107) discusses the use of sites data. Nei (1987, pp 64–107) also discusses the use of sequence information to estimate d, but Sourdis and Krimbas (1987) and Sourdis and Nei (1988) note that, when sequence information is available, use of p actually yields better results than d. Of course, for evolutionary interpretations it may be preferable to convert the branch lengths of an estimated tree from p to d.

The algorithms to convert distance matrices into tree estimates fall into two broad categories: those that assume perfectly regular evolutionary rates and those that do not.

There are only two commonly used methods that assume constant evolution. The most used of these is UPGMA (unweighted pair-group method with arithmetic averaging) (Sneath and Sokal 1973, pp 228–234; Nei 1987, pp 293–298). Under this method, the two most similar taxa are joined first, and the distance of each of these to their common node is taken to be half that of the distance between them. The distances between any two groups is the mean of the distances between the members of one group and those of the other. In the closely-related WPGMA (weighted pair-group method with arithmetic averaging) (Sneath and Sokal 1973, pp 234–235), distances are calculated weighting the latest member of a group equally with all the previous members, thus avoiding distorting effects resulting from uneven sampling of species in different groups. Pamilo (1990) presents a modification of WPGMA in which account is taken of the independent evolutionary information carried by species joining a cluster before distances to other clusters are recalculated.

There are many distance methods that allow for variation in evolutionary rate. Prominent among these are the Fitch-Margoliash (FM: Fitch and Margoliash 1967), distance-Wagner (DW: Farris 1972), Faith (FA: Faith 1985), transformed-distance (TD: Farris 1977; Klotz and Blanken 1981; Li 1981), and neighbor-joining (NJ: Saitou and Nei 1987) methods. Several of these (FM, TD, and DW) are explained clearly by Nei (1987, pp 298–308). It is worth noting that the distance-Wagner method has nothing in common with the Wagner method used in deriving maximum parsimony trees. All these methods involve sequential addition of species during tree construction, and it is possible to obtain different trees, often according to the input order involved (as is also the case with UPGMA if there are tied distances).

Several papers have evaluated various distance methods using simulated data (Tateno et al. 1982; Hasegawa and Yano 1984; Faith 1985; Fiala and Sokal 1985; Saitou and Nei 1987; Sourdis and Krimbas 1987; Li et al. 1987; Kim and Burgman 1988; Rohlf and Wooten 1988; Saitou 1988; Sourdis and Nei 1988; Saitou and Imanishi 1989; Jin and Nei 1990; Pamilo 1990). Not surprisingly, none of these papers considered all of the possible methods. What emerges from these papers is that the relative efficiency of various methods varies with factors such as the variability of evolutionary rates and the length of evolutionary time (number of substitutions) involved.

The remaining methods vary in their ability to return the true tree, but the differences between them are not consistently noteworthy, allowing one to choose that method which minimizes computer time (such as the neighbor-joining method) or is readily available in a package such as PHYLIP (e.g., Fitch-Margoliash).

It is striking, although not surprising given the amount of computer time that would be called for, that tests of confidence intervals with the various methods remain to be performed for the above methods, with the exception of the constant-rate methods. All the simulations test bias, not precision, with the partial exception of Sourdis and Nei (1988). As noted above and by Pamilo (1990), the bootstrap method commonly applied to maximum parsimony analyses could be used with any of these methods.

6.3.5. A Unique Study in Which the True Tree was Known

Unless one uses computer simulations to generate an evolving group of organisms, the actual phylogeny of any group to be analyzed is by definition unknown. Such a situation leads to skepticism on the part of members of other disciplines as to how phylogeneticists can ever assess the likelihood that they have the right approaches to inferring past history. To a considerable extent computer simulation overcomes these difficulties, but does not entirely quell skepticism due to the uncertainty as to how well the characteristics of the evolutionary model used in the simulation mirrors that actually occurring in nature.

There have been two studies attempting to use a true tree, one of these (Fitch and Atchley 1987) used a known genealogy of mouse strains to compare the efficiencies of various tree-building methods analyzing allozyme and morphometric data. All the tree-building methods returned the true tree for the allozyme data, but did less well with the morphometric data. This study, while extremely valuable, has been criticized for its choice of morphometric data, and also lacked information on the character states of the internal nodes (the ancestral strains).

Hillis et al. (1992) used a purely experimental group of organisms: cultures of bacteriophage T7 serially propagated in the presence of the mutagen nitrosoguanidine. The study began with two strains. One of these was then subdivided twice in later generations to generate a tree of nine final cultures with all internal branches of the same expected length (except that going to the first culture, the outgroup). The seven ancestors occupying the internal nodes of the tree were thus completely known (unlike the mouse or any other study) and available for tests of the ability of analytic methods to reconstruct ancestral states.

The cultures were examined using restriction enzymes, and the presences or absences of restriction sites formed the data set.

The methods used were maximum parsimony and the distance methods Fitch-Margoliash, Cavalli-Sforza, Neighbor-Joining, and UPGMA. The methods all performed well, in that each returned as its estimate that tree known to be the true one. It is unfortunate, and a little puzzling, that Maximum Likelihood was not included among the methods used.

If all phylogenetic analysis methods produced the true topology, how well do they return branch lengths? Recall that the branch lengths, too, are known exactly, in the sense that the restriction sites of all descendant and ancestral (nodal) populations are known and hence the lengths can be determined by simple counting. Hillis et al. (1992) found that correlations between the actual and predicted branch lengths ranged from 0.91 (maximum parsimony) down to 0.82 (UPGMA). It would seem that all methods tested thus did well, although a randomization test indicated that UPGMA did significantly less well than the remaining methods (which did not differ significantly from each other).

A striking result is that Maximum Parsimony correctly inferred 97% of the variable sites for the internal nodes (ancestors). I infer from this remarkable finding that, although Hillis et al. (1992) did not discuss this inference, extraordinarily few site changes could not be assigned unequivocally to a branch (an unusual situation in most studies).

C. Cunningham (personal communication) has embarked on the logical extension to this study, sequencing a section of the T7 genome. The study is intended to be extended to cases of unequal branch lengths. So far, the original set of strains from the Hillis et al. (1992) study has been examined. This time UPGMA did not even return the correct tree topology, but the other methods all performed about equally well.

One is drawn towards the conclusion that UPGMA is not desirable as a method, and indeed in the previous section I recommended that Maximum Parsimony be used to estimate the topology and a distance method such as Fitch-Margoliash or Neighbor-Joining be used to estimate branch lengths. But it is worthwhile examining this unique and interesting study a little more carefully.

Firstly, a distinction should be made between the return of a good fit to a data set, and obtaining a good estimate of the phylogeny of the group. In this case, we know that the expectation of the model is that all branches should be of the same length (except that going to the outgroup). Two kinds of error conspire to produce distributions of characters producing unequal branch lengths. Obviously, the set of characters (section of sequence) one chooses will be a subset of the whole array possible (genome), and hence there will be sampling error. More subtle is the fact that character change is stochastic, so that even though the rates of change may be equal the number of changes per unit time may be unequal, producing what has been called "stochastic error" (Nei 1986).

Thus, if it is known that evolution has been proceeding in a clocklike manner for the organisms and characters under study, then it is appropriate to use a method returning equal evolutionary rates and not one which will mirror or perhaps even accentuate unequal branch lengths produced by sampling and stochastic errors. Thus, for the restriction site data, it might be that UPGMA had the best performance, and not the worst, although no test was reported for the fit of the results of the various methods to the expectation of equal branch lengths.

Secondly, the numbers of character state changes (fixed mutations) detected in the sequencing study were much less than in the restriction site study, even down to one mutation per branch (C. Cunningham, personal communications). The failure of UPGMA to return the correct topology for these data suggests that UPGMA can lead to incorrect topologies when the variance in branch lengths becomes great enough.

Thirdly, it is possible that for some reason the experimental procedure did not actually produce an underlying equality of evolutionary rates. Although the restriction sites and sequence data sets each separately met Poisson expectations, combining these led to a significant deviation from expectations under the Poisson distribution. The evolutionary model necessary for UPGMA to be completely appropriate may thus have been violated, although probably not by much.

6.4 Managing Uncertainty

What should one do to choose a method of phylogenetic analysis, given that the bias of the method depends on factors such as evolutionary rate variation which it may be the purpose of the study to determine? Sourdis and Nei (1988) note that there is another criterion of success than return of the true tree, and that is exclusion of incorrect trees. Maximum parsimony, which did poorly in returning the true tree, did well under their simulations in excluding incorrect trees. Consequently, maximum parsimony and maximum likelihood, which both do better than distance methods when evolutionary rates vary, should be included in any suite of analytical programs. Among distance methods, FM is less efficient than many others (Saitou and Imanishi 1989) but has the advantage of allowing a test of evolutionary rate variation (in the PHYLIP package). Otherwise, the NJ method is described as being very economical of computer time and also one of the more efficient in finding the true tree.

Given the inability of maximum parsimony methods to unequivocally estimate branch lengths, an approach to be considered would be to use FM or NJ to estimate the branch lengths of a tree whose topology has been determined by parsimony.

As recommended by Felsenstein (1988), consider using simulations based on your data structure to test the ability of different methods to evaluate it. This is the approach Gouy and Li (1989) took to test the robustness of their conclusions on the relationships of the eukaryotes.

What should you do if, after all your efforts collecting data for, say, 10 taxa, you still have 10 trees which cannot be separated statistically? Put up with it. To reduce the number of trees from the possible 135,135 to 10, eliminating over 99% of the possibilities, is no mean feat. If you must present just one tree, use a consensus tree that agrees with all of those in your confidence interval (Hillis 1987), and not one that disagrees with any. Similarly, if your bootstrap confidence intervals for the branch points generally fall below the 95% confidence limit, put up with that too, or else state that you are using, say, an 80% confidence level. Sadly, even large data sets generally have poor confidence levels for allozymes, although the situation improves somewhat for restriction sites data, and is yet further improved for sequence data.

7. Looking at the Data More Carefully

7.1. Genome-Level Processes: Effects on Phylogeny Inference

Most nuclear DNA is noncoding. For example, the typical mammalian genome is sufficient to encode 2,250,000 proteins, yet indicators such as the number of genes capable of mutating to lethal alleles yield estimates closer to 30,000 (see Crozier 1987). Dawkins (1976, p 47) suggested that much of the noncoding DNA could be "selfish," i.e., increased in copy number by selection processes within the genome, and molecular biologists agreed with this view some time later (e.g., Doolittle and Sapienza 1980; Orgel and Crick 1980).

Dover (1986; 1989) has argued that genome-level forces have probably played an important role in changing the adaptive nature of organisms. Put more generally, selection can be viewed as operating at a number of levels, ranging from selection on sequences to selection on ecosystems (e.g., Arnold and Fristrup 1982). The observed functioning of organisms can then be regarded as due to selection on the "community of genes" being stronger than that on individual sequences (Crozier 1987).

It is therefore entirely possible that opposing selection between levels (e.g., genome vs individual) could lead to outcomes optimal under neither level. The evidence so far, however, is that individual-level selection mediates evolution at the level of the genome, but not the reverse. For example, ecological factors control the frequency of a modified 28s rRNA gene in *Drosophila mercatorum* (Templeton and Johnston 1982; Templeton et al. 1990), the frequency of the *Responder* array of satellite repeats of *D. melanogaster* is probably due to a balance between different selective forces (Wu et al. 1989), and selection on developmental rate in this organism led to a change in rDNA spacer length (Cluster et al. 1987). Dover's (1989) suggestion that changes in *Drosophila* courtship behavior result from genome-level processes affecting the *period* locus is convincing but there is no evidence that these changes have been opposed by selection at the individual level. It has been conjectured (e.g., by Clayton and Robertson 1955) that some phenotypic variation in *Drosophila* populations is probably selectively neutral, and the courtship changes just might be a parallel case. Certainly, models incorporating known selection intensities (e.g., Lamb and Helmi 1982; Walsh 1986) indicate that individual-level selection is expected *usually* to prevail over biased gene conversion.

Even if genome-level processes (called "molecular drive" by Dover) do not in general affect the outward functioning of organisms, they certainly have implications for molecular evolutionists. The importance of "concerted evolution" in homogenising rRNA genes was discussed earlier, as was the role of slippage-replication in leading to repeat length variation detected by "genetic fingerprinting." The evolution of DNA sequences therefore results from the following mutational processes, potentially mediated by selection:

1. Transition and transversion mutations of single nucleotides (the main concern of most models of evolutionary divergence). Included here could be solitary duplications or deletions of one or a small group of nucleotides.
2. Gene conversion, especially biased gene conversion in which one variant has a higher chance of being used as the template than the other. Gene conversion can take place both within and between chromosomes (e.g., Baltimore 1981).
3. Mutation pressure, or more generally, change in the likelihood of occurrence or fixation of mutations (e.g., Sueoka 1988).
4. Transposition and reverse transcriptional insertion of DNA copied from mRNA (e.g., Temin 1985) in which additional copies of genes are inserted into novel chromosomal locations.
5. Unequal crossing-over, in which mispairing, especially involving genes already duplicated, leads to gain or loss of genes. It is rightly regarded as a major evolutionary process (e.g., Jeffreys and Harris 1982).
6. Slippage replication, in which repeated sequences are linearly amplified along the length of the chromosome due to mispairing between tandem repeats (Levinson and Gutman 1987).
7. Horizontal gene exchange, in which a sequence from one genome is incorporated into another, possibly quite distant, genome (e.g., Bannister and Parker 1985). Although Syvanen (1987) invokes inter-specific gene transfer as a major force in shaping metazoan genomes, the instances involved are both rare and generally specialized (viruses or transposable elements), so that this seems most unlikely to be an undetectable source of error.
8. Random deletion of sequences.

The standard view of evolution being due to a balance between mutation, selection, and random genetic drift can be applied to these phenomena (e.g., Van Valen 1983), but the necessary intrusion of higher levels of selection makes for difficult modeling. Even the total amount of DNA can be seen as due to a balance of selective forces (Loomis and Gilpin 1986), a view consistent with observations that DNA amounts vary between groups in ways suggestive of lifestyle correlates (e.g., Grime and Mowforth 1982; Novacek and Norell 1989). The implications for molecular studies of evolution of genome-level processes can be dichotomized as involving errors in estimating evolutionary rates, and in leading to reticulate evolution of gene trees.

Apparent evolutionary rates can be affected by mutation pressure, specifically by differences in the rates at which particular bases are substituted for others. For example, whether an A is replaced more often than expected by a T than a C or G. Evidence for such mediation comes from the occurrence of different stationary nucleotide frequencies in homologous sequences from different species (e.g., Sueoka 1962; Jukes and Bhushan 1986; Osawa and Jukes 1988). If there is a change between lineages in the stationary nucleotide frequencies, then the apparent number of substitutions will be increased because there will be a reduction in the number of back-mutations eliminating the effects of substitutions. In other words, even though the instantaneous fixation of mutations may be the same in two lineages, if the mutation repair enzymes in one have altered so as to lead to a different stationary array of nucleotide frequencies, divergence between the two taxa will be more rapid than if they had the same mutation dynamics.

Bias in mutation will be reflected in differences in nucleotide composition between different kinds of DNA, so that non-coding regions more closely reflect the actual rate at which different mutations occur than do coding regions, and the same will be true of the third positions of codons (at which base changes are less likely to cause amino acid replacements) compared with the first and second positions (e.g., Bernardi and Bernardi 1986; Sueoka 1988). Between genes, commonly transcribed genes are more subject to a further internal pressure, selection by tRNA abundance, than are rarely transcribed genes (Bulmer 1987).

Apparent evolutionary rates can also be affected by such factors as changes in non-coding scnDNA due to slippage replication, transposition, and gene conversion. Dover (1987) suggests that variation in the rates of these processes between lineages would greatly reduce the reflection by DNA hybridization data of true divergences, but this concern is not borne out by the fairly good agreement between sequence and DNA-hybridization divergence reported by Maeda et al. (1988).

The inference of phylogenetic trees from DNA sequences can be made difficult by gene conversion, in which part, perhaps all, of a gene can be replaced by a sequence copied from a related sequence elsewhere. The phylogeny of one gene then becomes meshed with that of another. How seriously do these effects interfere with attempts at phylogenetic inference? The question of dealing with rate variation has been discussed above; more serious is the possibility of illegitimacy of sequence comparisons. Conversion events are expected to occur most often between regions with high similarity, such as duplicate genes. Thus, Slightom et al. (1985) adduce evidence for conversions between fetal globin genes in primates, and Dod et al. (1989) report the same for mouse hcnDNA arrays, whereas Maeda et al. (1988) found little evidence of conversions occurring in an intergenic space in the primate globin cluster. It therefore seems that choice of the appropriate region and attention to such details as variation in similarity between taxa along the length of a sequence should enable detection and correction for conversion in many instances. Noteworthy also is the apparent lack of conversion events in other instances involving duplicate genes (Yokoyama and Yokoyama 1989). Mitochondrial genomes, lacking gene families, would seem to be particularly immune to conversions.

7.2. Violation of the Assumption of Independence

A basic condition of quantitative studies of phylogeny is that the characters be functionally uncorrelated or, failing that, that the correlations be well understood (see Section 6.2 above). Molecular data are more likely to fulfill this condition than most other data, but there are still cases of functional correlation. While it is possible that the full range of correlation occurs, it is convenient to describe cases of strong versus weak correlation.

Strong correlation occurs between sequences coding for paired stretches of rRNA or tRNA. When a mutation occurs that disturbs a stem or hairpin loop structure, any further mutation that restores the integrity of the structure will be positively selected. This further mutation may be a reversion (which would not be phylogenetically detectable) or one restoring the Watson-Crick pair (Figure 5.3). Mutations therefore effectively come in pairs. A knowledge of the structure of the gene product is necessary so that one member of the pair is used in phylogenetic analyses, otherwise determination of the level of confidence in the phylogeny estimated would be spurious (Wheeler and Honeycutt 1988).

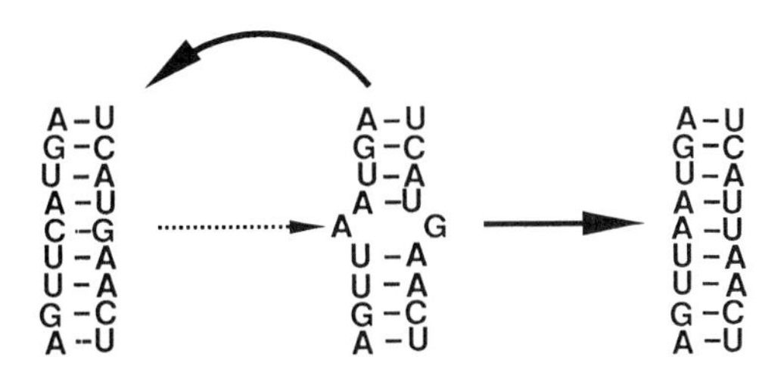

Figure 5.3. A mutation leading to a reduction in pairing of a paired RNA structure, such as in rRNA or tRNA, will select for either a reversion or a counterbalancing mutation that restores the lost Watson-Crick pair. Mutations at these paired sites are therefore correlated in evolutionary time.

Weak correlation between nucleotide positions is widespread. Paralleling Almagor (1985), we can describe the factors acting on coding genes as protein-based and DNA-based. The DNA-based factors include both the relative rates at which mutations occur and selective factors. These DNA-based factors lead to compositional differences between DNA classes and to regularities in sequence structure in addition to those specifying amino acid sequence. The nature of the selective factors is quite unclear, and these probably include not only selection components at the level of the individual organism (or higher) but also components due to genome-level processes.

The influence of DNA-based factors on sequence make-up is mediated by the strength of protein-based factors. Shields et al. (1988) report that GC content is reduced in regions of *Drosophila* nuclear DNA that are associated with reduced protein-based selection. Within protein-coding genes, this effect is clearly seen in different base compositions between the three codon positions, reflecting the relative probabilities of substitutions in leading to amino acid replacements. For example, according to the genetic code for *Drosophila* mtDNA, the proportion of base substitutions (other than those involving termination codons) that lead to an amino acid replacement is 95% for the first codon position, 100% for the second, but only 27% for the third. These constraints presumably underlie differences in the nucleotide compositions of the three positions in insect mitochondrial genes (Table 5.1) and, in turn, the observed rates of substitution at the different codon positions (Table 5.2).

The strength of protein-level processes in mediating DNA-level ones should not be overplayed. Sueoka (1961; discussed by Osawa et al. 1992) noted that the amino acid makeup of proteins is correlated with the nucleotide content of the corresponding genes, and this observation has been abundantly born out by subsequent studies (e.g., D'Onofrio et al. 1991; Osawa et al. 1992). A particularly extreme case has emerged in honeybee mtDNA, whose protein-encoding genes are 83.3% AT, with the inferred associated proteins showing numerous differences in amino acid composition to those of the less biassed mtDNA of *Drosophila*

Table 5.1. Composition of the gene (CO-I) for cytochrome c oxidase subunit 1 in honeybee (*Apis mellifera*) mitochondrial DNA (data from Crozier et al. 1989)

Position	A	T	G	C
1	0.33	0.35	0.18	0.15
2	0.20	0.45	0.14	0.21
3	0.51	0.44	0.01	0.04
Overall	0.35	0.41	0.11	0.13

Table 5.2. Similarities between the CO-I genes of *Drosophila yakuba* and *Apis mellifera* (data from Clary and Wolstenholme 1985; Crozier et al. 1989)*

Position	All substitutions		Transversions	
	Obs	Exp	Obs	Exp
1	0.74	0.27	0.83	0.50
2	0.88	0.30	0.92	0.55
3	0.59	0.43	0.69	0.50
Overall	0.74	0.30	0.82	0.50

* The similarities, counting only transversions, are shown because of the lower rate at which transversions occur relative to transitions; using transversions alone is therefore believed to reduce evolutionary noise due to back mutations (e.g., Hasegawa and Yano 1984; Lake 1987a). Random expectations are presented based on the nucleotide proportions at each position in each species.

in directions predictable from base composition (RH Crozier and YC Crozier, in prep.). There are, of course, examples of the apparent effects of adaptive evolution leading to difficulties in obtaining the phylogeny widely regarded as correct, but these appear to be extremely rare. One example concerns the digestive lysozyme of leaf-eating monkeys, whose amino acid sequence is sufficiently similar to that of 'true' ruminants that a cladistic analysis would place it with the cow and not with other monkeys (Stewart et al. 1987)! However, even in this case, the composition of the gene itself make the false tree ((cow, leaf-eater), other monkeys) no more likely than the expected one (cow, (leaf-eater, other monkeys) (Swanson et al. 1991).

A related phenomenon is the correlation between the identities of adjacent bases, which has been observed in both coding and noncoding DNA of vertebrate nuclei (Lipman and Wilbur 1983; Blaisdell 1985; Alff-Steinberger 1987). These correlations include a dependence of the third base of codons on the second base and on the first base of the next codon. In addition, there is in nuclear DNA a substantial deficiency of CG and TA doublets (Alff-Steinberger 1987), and when these do occur, they are removed at a high rate by subsequent substitutions (Maeda et al. 1988). Lipman and Wilbur (1983) state that these correlations are very much weaker for mitochondrial genes than for eukaryotic nuclear or prokaryotic genes but such correlations do indeed occur, at least in our studies of honeybee mtDNA (Crozier and Crozier, in prep.). It therefore appears that more independent information is given by the different codon positions of mitochondrial genes than in nuclear ones but that the third mtDNA codon position of animals as remotely related as bees and flies still carries phylogenetic information (Figure 5.4), although this information may be derived from the other positions (the alternative possibility, of a failure of saturation, seems much less likely).

In the case of nuclear genes, the correlation between adjacent nucleotides is a potential source of error in estimating the confidence intervals of the unknown true tree. Yokoyama and Yokoyama (1989) used each codon position separately in inferring the phylogeny of primate globin genes and obtained the same tree from each analysis. This procedure would give one more confidence in the estimation of precision for any one tree, but the trees would tend to be the same because of the dependence of adjacent nucleotides on each other. It remains to determine how serious the problem of internucleotide correlation is for nuclear genes; in all likelihood the correlation, while statistically detectable, is not sufficient to cause serious error.

7.3. Fine-Tuning the Region Chosen for Sequencing

Genes vary in evolutionary rate. These differences have long been studied in terms of amino acid replacement rates and understood in terms of variation between proteins in functional constraints (e.g., Nei 1987, pp 50–51). For example, fibrinopeptides are the most rapidly evolving protein-coding sequences known, with an average of nine replacements per billion (10^9) years, whereas histone H4 is the slowest at 0.01 replacements per billion years, with intermediate rates shown by hemoglobin (1.2) and cytochrome b_5 (0.45) (Nei 1987, p 50, after Dayhoff 1978). At the amino acid level, it would be preferable to use fibrinopeptides to study interspecific relationships in mammals rather than histone H4!

Genes also vary in evolutionary rate at the DNA level, as expected. Although noncoding regions show the highest overall rates, there are advantages to using protein-encoding genes even for closely-related taxa (Crozier 1990). These advantages include the ability to use internal checks of the accuracy of the sequence due to its coding function, and the variation in rate between the three codon positions (Figure 5.4). The rRNA genes are suitable for studying highly divergent groups, but their slow rates of change mean that they may not be as suitable for studying differences among insects (at any systematic level) as protein-encoding genes. Alcohol dehydrogenase gene sequences have been used to study *Drosophila* phylogeny (Cohn et al. 1984; Starmer and Sullivan 1989), as have mtDNA sequences (DeSalle et al. 1987a).

As discussed above, the relative independence of the codon positions of mtDNA genes make them particularly suitable for phylogenetic analysis. A table of similarities between *Drosophila* and mammalian mitochondrial genes is given in Table 5.3.

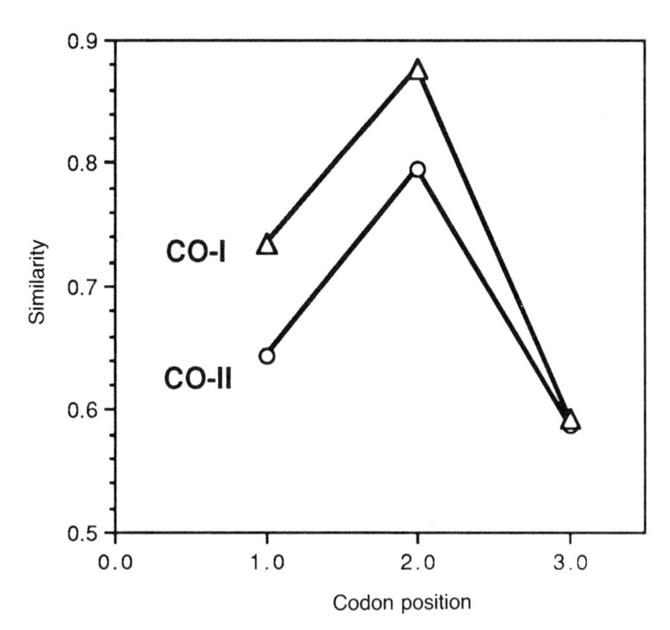

Figure 5.4. Similarities between *Apis mellifera* and *Drosophila yakuba* for the mitochondrial genes CO-I and CO-II, by codon position (data from Clary and Wolstenholme 1985; Crozier et al. 1989), showing the greater similarity in evolutionary divergence at the third position than at the other two.

Table 5.3. Similarities between protein-encoding mitochondrial genes of *Drosophila yakuba* and mouse (after Clary and Wolstenholme 1985)*

Gene	% Nucleotide Similarity	% Amino Acid Similarity
CO I	72	75
CO III	67	64
Cyt b	65	67
CO II	64	56
ATPase 8	59	26
ND 1	52	46
ND 3	52	42
ND 4	51	42
ATPase 6	49	36
ND 2	47	35
ND 5	42	33
ND 6	38	17
ND 4L	34	40

* Cyt b denotes the cytochrome b gene; CO I, CO II, and CO III denote the genes for subunits 1, 2, and 3 of cytochrome c oxidase; and ND 1–6 denote genes for enzymes in the NADH dehydrogenase system. The genes are listed in order of nucleotide similarity.

7.4. Mix and Match

A phylogeny derived from a single gene will be a good estimate of the phylogeny of the taxa concerned if separation times are sufficient to avoid the time depth problem. Special problems at the population level, such as tests of gene flow and differential male and female dispersal, may still be approached using single gene phylogenies (Crozier 1990), but phylogenetic studies on conspecific populations and closely-related species may produce misleading results if the taxa are very closely related. In such a case, studies on a number of genes is desirable. From the studies of Pamilo and Nei (1988), it would be ideal to use the sequences from, say, 10 independently assorting genes (Crozier 1990) to be sure all genes follow the same tree, but the technical difficulties of this approach will usually be prohibitive. Gene frequency data provide a good guide to intraspecific relationships, although the number of loci needed is depressingly large (Nei and Livshits (1990) analyzed 186 loci). Nevertheless, allozymes and simple RFLPs remain methods of choice for studying conspecific populations.

8. Phylogenetics to Systematics?

Phylogenies do not intrinsically lead to systematics. For the recognition of species, criteria such as gene flow or evolutionary independence are applied, which cannot be read into the phylogeny. Species can differ by exceedingly few gene substitutions (Val 1977), and conspecific populations may show large genetic distances (Johnson et al. 1984). Tests for gene flow remain the only certain test for species status in bisexual species, although when a systematic decision has to be made and gene flow information is lacking, noone could be blamed for applying rules of thumb about the genetic distances characteristic of species-status in a particular group.

The relationship of phylogenies to the systematics of higher categories depends on the criteria of the systematist. A systematics founded on a definitive cladogram will enable the retrieval of that cladogram, but such a scheme ignores branch-length information. Ideally, systematic schemes should enable retrieval of total similarity and not just the branching pattern of history, and such similarity should be at the phenotypic level and not at the level of gene sequences. Of course, it remains possible that morphological evolution will be sufficiently clock-like in some instances that cladistic criteria will lead to overall predictivity, but the observed discrepancies between morphological and genetic distance (e.g., Cherry et al. 1982) make this seem unlikely as a universal rule.

Acknowledgments. I thank Scott Davis, Paul Gadek, Rees Griffiths, Tony Mackinlay, Jenny Ovenden, Chris Quinn, and Mike Westerman for comments on the manuscript; Naoyuki Takahata for discussions; and Pekka Pamilo and Naoyuki Takahata for sending me papers before publication. My work on the evolutionary genetics of social insects is supported by grants from the Australian Research Council, the Honey Research Council, and the Ian Potter Foundation, to all of whom I am grateful.

References

Alff-Steinberger C (1987) Codon usage in *Homo sapiens*: Evidence for a coding pattern on the non-coding strand and evolutionary implications of dinucleotide discrimination. J Theor Biol 124:89–95

Almagor H (1985) Nucleotide distribution and the recognition of coding regions in DNA sequences: an information theory approach. J Theor Biol 117: 127–136

Archie JW (1989a) Phylogenies of plant families: A demonstration of phylogenetic randomness in DNA sequence data derived from proteins. Evolution 43: 1796–1800

Archie JW (1989b) Homoplasy excess ratios: new indices for measuring levels of homoplasy in phylogenetic systematics and a critique of the consistency index. Syst Zool 38:253–269

Arnold AJ, Fristrup K (1982) The theory of evolution by natural selection: An hierarchical expansion. Paleobiology 8:113–129

Astolfi P, Zonta-Sgaramella L (1984) Phylogenetic trees: An analysis of the treeness test. Syst Zool 33:159–166

Avise JC (1989) Gene trees and organismal histories: A phylogenetic approach to population biology. Evolution 43:1192–1208

Avise JC, Arnold J, Ball RM, Bermingham E, Lamb T, Neigel JE, Reeb CA, Saunders NC (1987) Intraspecific phylogeography: The mitochondrial DNA bridge between population genetics and systematics. Ann Rev Ecol Syst 18:489–522

Avise JC, Lansman RA, Shade RO (1979) The use of restriction endonucleases to measure mitochondrial DNA sequence relatedness in natural populations. I. Population structure and evolution in the genus *Peromyscus*. Genetics 92: 279–295

Avise JC, Vrijenhoek RC (1987) Mode of inheritance and variation of mitochondrial DNA in hybridogenetic fishes of the genus *Poeciliopsis*. Mol Biol Evol 4:514–525

Baba-Aissa F, Solignac M, Dennebouy N, David JR (1988) Mitochondrial DNA variability in *Drosophila simulans*: quasi absence of polymorphism within each of the three cytoplasmic races. Heredity 61:419–426

Baltimore D (1981) Gene conversion: Some implications for immunoglobulin genes. Cell 24:592–594

Bannister JV, Parker MW (1985) The presence of a copper/zinc superoxide dismutase in the bacterium *Photobacterium leiognathi*: A likely case of gene transfer from eukaryotes to prokaryotes. Proc Natl Acad Sci USA 82: 149–152

Baroin A, Perasso R, Qu L-H, Brugerolle G, Bachellerie J-P, Adoutte A (1988) Partial phylogeny of the unicellular eukaryotes based on rapid sequencing of a portion of 28S ribosomal RNA. Proc Natl Acad Sci USa 85:3474–3478

Bennett S, Alexander LJ, Crozier RH, Mackinlay AG (1988) Are megabats flying primates? Contrary evidence from a mitochondrial DNA sequence. Aust J Biol Sci 41:327–332

Bernardi G, Bernardi G (1986) Compositional constraints and genome evolution. J Mol Evol 24:1–11

Bernatzky R, Tanksley SD (1986) Toward a saturated linkage map in tomato based on isozymes and random cDNA sequences. Genetics 112:887–898

Beverley SM, Wilson AC (1985) Ancient origin for Hawaiian Drosophilinae inferred from protein comparisons. Proc Natl Acad Sci USA 82:4725–4757

Beverley SM, Wilson AC (1984) Molecular evolution in *Drosophila* and the higher Diptera. II. A time scale for fly evolution. J Mol Evol 21:1–13

Beverley SM, Wilson AC (1982) Molecular evolution in *Drosophila* and higher Diptera. I. Micro-complement fixation studies of a larval hemolymph protein. J Mol Evol 18:251–264

Birky CW (1978) Transmission genetics of mitochondria and chloroplasts. Ann Rev Genet 12:471–512

Bishop CE, Boursot P, Baron B, Bonhomme F, Hatat D (1985) Most classical *Mus musculus domesticus* laboratory mouse strains carry a *Mus musculus domesticus Y* chromosome. Nature 315:70–72

Bishop CE, Guellaen G, Geldwerth D, Voss R, Fellous M, Weissenbach J (1983) Single-copy DNA sequences specific for the human Y chromosome. Nature 303:831–832

Bishop MJ, Friday AE (1985) Evolutionary trees from nucleic acid and protein sequences. Proc Roy Soc Lond Ser B 226:271–302

Blaisdell BE (1985) Markov chain analysis finds a significant influence of neighboring bases on the occurrence of a base in eucaryotic nuclear DNA sequences, both protein-coding and noncoding. J Mol Evol 21:278–288

Bledsoe AH, Sheldon FH (1989) The metric properties of DNA-DNA hybridization dissimilarity measures. Syst Zool 38:93–105

Boursot P, Yonekawa H, Bonhomme F (1987) Heteroplasmy in mice with deletion of a large coding region of mitochondrial DNA. Mol Biol Evol 4:46–55

Boyce TM, Zwick ME, Aquadro CF (1989) Mitochondrial DNA in the bark weevils: size, structure and heteroplasmy. Genetics 123:825–836

Boynton JE, Harris EH, Burkhart BD, Lamerson PM, Gillham NW (1987) Transmission of mitochondrial and chloroplast genomes of *Chlamydomonas*. Proc Natl Acad Sci USA 84:2391–2395

Britten RJ (1986) Rates of DNA sequence evolution differ between taxonomic groups. Science 231:1393–1398

Brown DD, Sugimoto K (1973) The structure and evolution of ribosomal and 5S DNAs in *Xenopus laevis* and *Xenopus mulleri*. Cold Spring Harb Symp Quant Biol 38:501–505

Brown WM, Wright JW (1979) Mitochondrial DNA analyses and the origin and relative age of parthenogenetic lizards (Genus *Cnemidophorus*). Science 203:1247–1249

Bulmer M (1987) Coevolution of codon usage and transfer RNA abundance. Nature 325:728–730

Burke T (1989) DNA finger printing and other methods for the study of mating success. TREE 4:139–144

Buroker NE, Brown JR, Gilbert TA, O'Hara PJ, Beckenbach AT, Thomas WK, Smith MJ (1990) Length heteroplasmy of sturgeon mitochondrial DNA; an illegitimate elongation model. Genetics 124:157–163

Caccone A, Amato GD, Powell JR (1988) Rates and patterns of scnDNA and mtDNA divergence within the *Drosophila melanogaster* subgroup. Genetics 118:671–683

Caccone A, Powell JR (1990) Extreme rates and heterogeneity in insect DNA evolution. J Mol Evol 30:273–280

Caccone A, Powell JR (1987) Molecular evolutionary divergence among North American cave crickets. II. DNA-DNA hybridization. Evolution 41: 1215–1238

Cann RL, Stoneking M, Wilson AC (1987) Mitochondrial DNA and human evolution. Nature 325:31–36

Carr SM, Brothers AJ, Wilson AC (1987) Evolutionary inferences from restriction maps of mitochondrial DNA from nine taxa of *Xenopus* frogs. Evolution 41:176–188

Casanova M, Leroy P, Boucekkine C, Weissenbach J, Bishop CE, Fellous M, Purello M, Fiori G, Siniscalco M (1985) A human Y-linked DNA polymorphism and its potential for estimating genetic and evolutionary distance. Science 230:1403–1406

Cavalli-Sforza LL, Edwards AWF (1967) Phylogenetic analysis. Amer J Human Genet 19:233–257

Cavalli-Sforza LL, Piazza A (1975) Analysis of evolution: Evolutionary rates, independence, and treeness. Theor Pop Biol 8:127–165

Champion AB, Prager EM, Wachter D, Wilson AC (1974) Microcomplement fixation. In Wright CA (ed) Biochemical and Immunological Taxonomy of Animals. Academic Press, London, pp 397–416

Cherry LM, Case SM, Kunkel JG, Wyles JS, Wilson AC (1982) Body shape metrics and organismal evolution. Evolution 36:914–933

Clark AG, Lyckegaard EMS (1988) Natural selection with nuclear and cytoplasmic transmission. III. Joint analysis of segregation and mtDNA in *Drosophila melanogaster*. Genetics 118:471–481

Clary DO, Wolstenholme DR (1985) The mitochondrial DNA molecule of *Drosophila yakuba*: nucleotide sequence, gene organization, and genetic code. J Mol Evol 22:252–271

Clayton G, Robertson A (1955) Mutation and quantitative variation. Amer Nat 59:151–158

Cluster PD, Marinkovic D, Allard RW, Ayala FJ (1987) Correlations between developmental rates, enzyme activities, ribosomal DNA spacer-length phenotypes, and adaptation in *Drosophila melanogaster*. Proc Natl Acad Sci USA 84:610–614

Cohn VH, Thompson MA, Moore GP (1984) Nucleotide sequence comparison of the *Adh* gene in three drosophilids. J Mol Evol 20:31–37

Colless DH (1985) On the status of outgroups in phylogenetics. Syst Zool 34: 364–366

Cracraft J (1987) DNA hybridization and avian phylogenetics. In Hecht MK, Wallace B, Prance GT (eds) Evolutionary Biology, Vol 21, Plenum, New York, pp 47–96

Crozier RH (1990) From population genetics to phylogeny: Uses and limits of mitochondrial DNA. Aust Syst Bot 3:111–124

Crozier RH (1987) Selection, adaptation, and evolution. Proc Roy Soc NSW 120:21–37

Crozier RH (1983) Genetics and insect systematics: retrospect and prospect. In Highley E, Taylor RW (eds) Australian Systematic Entomology: a Bicentenary Perspective. CSIRO, Melbourne, pp 80–92

Crozier RH, Crozier YC, Mackinlay AG (1989) The CO-I and CO-II region of honeybee mitochondrial DNA: evidence for variation in insect mitochondrial rates. Mol Biol Evol 6:399–411

Crozier RH, Pamilo P, Taylor RW, Crozier YC (1986) Evolutionary patterns in some putative Australian species in the ant genus *Rhytidoponera*. Aust J Zool 34:535–560

Davis SK, Strassmann JE, Hughes C, Pletscher LS, Templeton AR (1990) Population structure and kinship in *Polistes* (Hymenoptera, Vespidae): An analysis using ribosomal DNA and protein electrophoresis. Evolution 44: 1242–1253

Dawkins R (1976) The Selfish Gene. Oxford University Press, Oxford

Dayhoff MO (1978) Survey of new data and computer methods of analysis. In Dayhoff MO (ed) Atlas of Protein Sequence and Structure. Vol 5, Supp 3. National Biomedical Research Foundation, Silver Spring, Maryland pp 41–45

Densmore LD, Wright JW, Brown WM (1985) Length variation and heteroplasmy are frequent in mitochondrial DNA from parthenogenetic and bisexual lizards (Genus *Cnemidophorus*). Genetics 110:689–707

DeSalle R, Freedman T, Prager EM, Wilson AC (1987a) Tempo and mode of sequence evolution in mitochondrial DNA of Hawaiian *Drosophila*. J Mol Evol 26:157–164

DeSalle R, Templeton AR, Mori I, Pletscher S, Johnston JS (1987b) Temporal and spatial heterogeneity of mtDNA polymorphisms in natural populations of *Drosophila mercatorum*. Genetics 116:215–223

DeSalle R, Giddings LV (1986) Discordance of nuclear and mitochondrial DNA phylogenies in Hawaiian *Drosophila*. Proc Natl Acad Sci USA 83:6902–6906

DeSalle R, Giddings LV, Templeton AR (1986a) Mitochondrial DNA variability in natural populations of Hawaiian *Drosophila*. I. Methods and levels of variability in *D. silvestris* and *D. heteroneura* populations. Heredity 56: 75–85

DeSalle R, Giddings LV, Kaneshiro KY (1986b) Mitochondrial DNA variability in natural populations of Hawaiian *Drosophila*. II. Genetic and phylogenetic relationships of natural populations of *D. silvestris* and *D. heteroneura*. Heredity 56:87–96

De Winter RFJ, Moss T (1986) Spacer promoters are essential for efficient enhancement of *X. laevis* ribosomal transcription. Cell 44:313–318

Dod B, Mottez E, Desmarais E, Bonhomme F, Roises G (1989) Concerted evolution of light satellite DNA in genus *Mus* implies amplification and homogenization of large blocks of repeats. Mol Biol Evol 6:478–491

D'Onofrio G, Mouchiroud D, Aïssani B, Gautier C, Bernardi G (1991) Correlations between the compositional properties of human genes, codon usage, and amino acid composition of proteins. J Mol Evol. 32:504–510

Doolittle WF, Sapienza C (1980) Selfish genes, the phenotype paradigm and genome evolution. Nature 284:601–603

Dover GA (1989) Slips, strings and species. Trends Genet 5:100–102

Dover GA (1987) DNA turnover and the molecular clock. J Mol Evol 26:47–58

Dover GA (1986) Molecular drive in multigene families: How biological novelties arise, spread and are assimilated. Trends Genet 2:159–165

Edwards AWF, Cavalli-Sforza LL (1964) Reconstruction of evolutionary trees. In Heywood VH, and McBeill J (eds) Phenetic and Phylogenetic Classification. Syst Assoc Publ No. 6, London pp 67–76

Eigen M, Lindemann BF, Tietz M, Winckler-Oswatitisch R, Dress A, von Haeseler A (1989) How old is the genetic code? Statistical geometry of tRNA provides an answer. Science 244:673–679

Eigen M, Winckler-Oswatitisch R, Dress A (1988) Statistical geometry in sequence space: A method of quantitative comparative sequence analysis. Proc Natl Acad Sci USA 85:5913–5917

Faith DP (1985) Distance methods and the approximation of most parsimonious trees. Syst Zool 34:312–325

Farris JS (1977) On the phenetic approach to vertebrate classification. In Hecht MK, Goody PC, and Hecht BM (eds) Major Patterns in Vertebrate Evolution. Plenum, New York, pp 823–850

Farris JS (1972) Estimating phylogenetic trees from distance matrices. Amer Nat 106:645–668

Felsenstein J (1988) Phylogenies from molecular sequences: inference and reliability. Ann Rev Genet 22:521–565

Felsenstein J (1987) Estimation of hominoid phylogeny from a DNA hybridization data set. J Mol Evol 26:123–131

Felsenstein J (1985a) Phylogenies from gene frequencies: a statistical problem. Syst Zool 34:300–311

Felsenstein J (1985b) Confidence limits on phylogenies: an approach using the bootstrap. Evolution 39:783–791

Felsenstein J (1984) Distance methods for inferring phylogenies: a justification. Evolution 38:16–24

Felsenstein J (1982) Numerical methods for inferring evolutionary trees. Quart Rev Biol 57:379–404

Felsenstein J (1981) Evolutionary trees from gene frequencies and quantitative characters: finding maximum likelihood estimates. Evolution 35:1229–1242

Felsenstein J (1978) Cases in which parsimony or compatibility methods will be positively misleading. Syst Zool 27:401–410

Felsenstein J (1973) Maximum-likelihood estimation of evolutionary trees from continuous characters. J Human Genet 25:471–492

Fiala KL, Sokal RR (1985) Factors determining the accuracy of cladogram estimation: evaluation using computer simulation. Evolution 39:609–622

Field KG, Olsen GJ, Lane DJ, Giovannoni SJ, Ghiselin MT, Raff EC, Pace NR, Raff RA (1988) Molecular phylogeny of the animal kingdom. Science 239: 748–753

Fitch WM, Atchley WR (1987) Divergence in inbred strains of mice: a comparison of three different types of data. In Patterson C (ed) Molecules and Morphology in Evolution: Conflict or Compromise? Cambridge University Press, Cambridge pp 203–216

Fitch WM, Margoliash E (1967) Construction of phylogenetic trees. Science 155:279–284

Gerbi SA (1985) Evolution of ribosomal DNA. In MacIntyre RJ (ed) Molecular Evolutionary Genetics. Plenum, New York, pp 419–517

Gillespie JH (1986) Variability of evolutionary rates of DNA. Genetics 112: 1077–1091

Golenberg EM, Giannasi DE, Clegg MT, Smiley CJ, Durbin M, Henderson D, Zurawski G (1990) Chloroplast DNA sequence from a Miocene Magnolia species. Nature 344:656–658

Gouy M, Li W-H (1989) Phylogenetic analysis based on rRNA sequences supports the archaebacterial rather than the eocyte tree. Nature 339:145–147

Grime JP, Mowforth MA (1982) Variation in genome size—an ecological interpretation. Nature 299:151–153

Gyllensten U (1989) Direct sequencing of in vitro amplified DNA. In Erlich HA (ed) PCR Technology. Stockton, New York, pp 45–60

Hale LR, Beckenbach AT (1985) Mitochondrial DNA variation in *Drosophila pseudoobscura* and related species in Pacific Northwest populations. Canad J Genet Cytol 27:357–364

Hale LS, Singh RS (1987) Mitochondrial DNA variation and genetic structure in populations of *Drosophila melanogaster*. Mol Biol Evol 4:622–637

Hall HG (1986) DNA differences found between Africanized and European honeybees. Proc Natl Acad Sci USA 83:4874–4877

Harris H, Hopkinson DA (1976) Handbook of Enzyme Electrophoresis in Human Genetics. North Holland, Oxford, UK

Harrison RG (1989) Animal mitochondrial DNA as a genetic marker in population and evolutionary biology. TREE 4:6–11

Harrison RG, Rand DM, Wheeler WC (1987) Mitochondrial DNA variation in field crickets across a narrow hybrid zone. Mol Biol Evol 4:144–158

Hasegawa M, Kishino H, Yano T (1987) Man's place in Hominoidea as inferred by molecular clocks of DNA. J Mol Biol 26:132–147

Hasegawa M, Yano T-A (1984) Maximum likelihood method of phylogenetic tree inference from DNA sequence data. Bull Biometric Soc Japan 5:1–7

Hayashi J-I, Yonekawa H, Murakami J, Tagashira Y, Pereira-Smith OM, Shay JW (1987) Mitochondrial genomes in intraspecies mammalian cell hybrids display codominant or dominant/recessive behavior. Exper Cell Res 172: 218–227

Hendy MD, Penny D (1989) A framework for the quantitative study of evolutionary trees. Syst Zool 38:297–309

Hendy MD, Penny D (1982) Branch and bound algorithms to determine minimal evolutionary trees. Math Biosci 59:277–290

Hillis DM (1987) Molecular versus morphological approaches to systematics. Ann Rev Ecol Syst 18:23–42

Hillis DM (1984) Misuse and modification of Nei's genetic distance. Syst Zool 33:238–240

Hillis DM, Bull JJ, White ME, Badgett MR, Molineux IJ (1992) Experimental phylogenetics: generation of a known phylogeny. Science 255:589–592

Hillis DM, Davis SK (1987) Evolution of the 28S ribosomal RNA gene in anurans: regions of variability and their phylogenetic implications. Mol Biol Evol 4:117–125

Hillis DM, Davis SK (1986) Evolution of ribosomal DNA: fifty million years of recorded history in the frog genus *Rana*. Evolution 40:1275–1288

Hori H, Osawa S (1987) Origin and evolution of organisms as deduced from 5S ribosomal RNA sequences. Mol Biol Evol 4:445–472

Hunt JA, Hall TJ, Britten RJ (1981) Evolutionary distances in Hawaiian *Drosophila* measured by DNA reassociation. J Mol Evol 17:361–367

Jeffreys AJ, Harris S (1982) Processes of gene duplication. Nature 296:9–10

Jeffreys AJ, Wilson V, Thein SL (1985) Individual-specific "fingerprints" of human DNA. Nature 316:76–79

Jin L, Nei M (1990) Limitations of the evolutionary parsimony method of phylogenetic analysis. Mol Biol Evol 7:82–102

Johnson MS, Stine OC, Murray J (1984) Reproductive compatibility despite large-scale genetic divergence in *Cepaea nemoralis*. Heredity 53:655–665

Jukes TH, Bhushan V (1986) Silent nucleotide substitutions and G + C content of some mitochondrial and bacterial genes. J Mol Evol 24:39–44

Kaplan N (1983) Statistical analysis of restriction enzyme map data and nucleotide sequence data. In Weir BS (ed) Statistical Analysis of DNA Sequence Data. Dekker, New York, pp 75–106

Kessler LG, Avise JC (1985) Microgeographic lineage analysis by mitochondrial genotype: variation in the cotton rat (*Sigmodon hispidus*). Evolution 39: 831–837

Kim J, Burgman MA (1988) Accuracy of phylogenetic estimation methods under unequal evolutionary rates. Evolution 42:596–602

Kishino H, Hasegawa M (1989) Evaluation of the maximum likelihood estimate of the evolutionary tree topologies from DNA sequence data, and the branching order in Hominoidea. J Mol Evol 29:170–179

Klotz LC, Blanken RL (1981) A practical method for calculating evolutionary trees from sequence data. J Theor Biol 91:261–272

Kluge AG, Farris JS (1969) Quantitative phyletics and the evolution of anurans. Syst Zool 18:1–32

Kocher TD, Thomas WK, Meyer A, Edwards SV, Paabo S, Villeblanca FX, Wilson AC (1989) Dynamics of mitochondrial DNA evolution in animals: Amplification and sequencing with conserved primers. Proc Natl Acad Sci USA 86:6196–6200

Kondo R, Satta Y, Matsuura ET, Ishiwa H, Takahata N, Chigusa SI (1990) Incomplete maternal transmission of mitochondrial DNA in *Drosophila*. manuscript.

Lake JA (1988) Origin of the eukaryotic nucleus determined by rate-invariant analysis of rRNA sequences. Nature 331:184–186

Lake JA (1987a) Determining evolutionary distances from highly diverged nucleic acid sequences: operator metrics. J Mol Evol 26:59–73

Lake JA (1987b) A rate-independent technique for analysis of nucleic acid sequences: evolutionary parsimony. Mol Biol Evol 4:167–191

Lamb BC, Helmi S (1982) The extent to which gene conversion can change allele frequencies in populations. Genet Res 39:199–217

Lane DJ, Pace B, Olsen GJ, Stahl DA, Sogin ML, Pace NR (1985) Rapid determination of 16S ribosomal RNA sequences for phylogenetic analyses. Proc Natl Acad Sci USA 82:6955–6959

Lansman RA, Avise JC, Huettel MD (1983) Critical experimental test of the possibility of "paternal leakage" of mitochondrial DNA. Proc Natl Acad Sci USA 80:1969–1971

Larson A, Wilson AC (1989) Patterns of ribosomal RNA evolution in salamanders. Mol Biol Evol 6:131–154

Latorre A, Barrio E, Moya A, Ayala FJ (1988) Mitochondrial DNA evolution in the *Drosophila obscura* group. Mol Biol Evol 5:717–728

Levinson G, Gutman GA (1987) Slipped-strand mispairing: A major mechanism for DNA sequence evolution. Mol Biol Evol 4:203–221

Li H, Gyllensten UB, Cui X, Saiki RK, Erlich HA, Arnheim N (1988) Amplification and analysis of DNA sequences in single human sperm and diploid eggs. Nature 335:414–417

Li W-H (1986) Evolutionary change of restriction cleavage sites and phylogenetic inference. Genetics 113:187–213

Li W-H (1981) Simple method for constructing phylogenetic trees from distance matrices. Proc Natl Acad Sci USA 78:1085–1089

Li W-H, Wolfe KH, Sourdis J, Sharp PM (1987) Reconstruction of phylogenetic trees and estimation of divergence times under nonconstant rates of evolution. Cold Spring Harb Symp Quant Biol 52:847–856

Lipman DJ, Wilbur WJ (1983) Contextual constraints on synonymous codon choice. J Mol Biol 163:363–376

Loomis WF, Gilpin ME (1986) Multigene families and vestigial sequences. Proc Natl Acad Sci USA 83:2143–2147

Lynch M (1988) Estimation of relatedness by DNA fingerprinting. Mol Biol Evol 5:584–599

MacRae AF, Anderson WW (1988) Evidence for non-neutrality of mitochondrial DNA haplotypes in *Drosophila pseudoobscura*. Genetics 120:485–494

Maeda N, Wu C-I, Bliska J, Reneke J (1988) Molecular evolution of intergenic DNA in higher primates: Pattern of DNA changes, molecular clock, and evolution of repetitive sequences. Mol Biol Evol 5:1–20

Marks J, Schmid CW, Sarich VM (1988) DNA hybridization as a guide to phylogeny: Relations of the Hominoidea. J Human Evol 17:769–786

Maynard Smith J (1989) Trees, bundles or nets? TREE 4:302–304

Mickevich MF (1978) Taxonomic congruence. Syst Zool 27:143–158

Miller SG, Huettel MD, Davis M-TB, Weber EH, Weber LA (1986) Male sterility in *Heliothis virescens x H. subflexa* backcross hybrids. Mol Gen Genet 203:451–461

Moritz C, Dowling TE, Brown WM (1987) Evolution of animal mitochondrial DNA: relevance for population biology and systematics. Ann Rev Ecol Syst 18:269–292

Neale DB, Sederoff RR (1989) Paternal inheritance of chloroplast DNA and maternal inheritance of mitochondrial DNA in loblolly pine. Theor Appl Gent 77:212–216

Nei M (1987) Molecular Evolutionary Genetics. Columbia University Press. New York

Nei M (1986) Stochastic errors in DNA evolution and molecular phylogeny. In Gershowitz H, Rucknagel RL, Tashain RE (eds) Evolutionary Perspectives and the New Genetics. AR Liss, New York, pp 133–147

Nei M, Li W-H (1979) Mathematical model for studying genetic variation in terms of restriction endonucleases. Proc Natl Acad Sci USA 76:5269–5273

Nei M, Livshits G (1990) Evolutionary relationships of Europeans, Asians, and Africans at the molecular level. In Takahata N, Crow JF (eds) Population Genetics of Genes and Molecules. Baifuhan, Tokyo, pp 251–265

Niki Y, Chigusa SI, Matsuura ET (1989) Complete replacement of mitochondrial DNA in *Drosophila*. Nature 341:551–552

Novacek MJ, Norell MA (1989) Nuclear DNA content in bats and other organisms: implications and unanswered questions. TREE 4:285–286

Orgel LE, Crick FHC (1980) Selfish DNA: the ultimate parasite. Nature 284: 604–607

Osawa S, Jukes TH (1988) Evolution of the genetic code as affected by anticodon content. Trends Genet 4:191–198

Osawa S, Jukes TH, Watanabe K, Muto A (1992) Recent evidence for evolution of the genetic code. Microbiol Rev 56:229–264

Ovenden JR, Mackinlay AG, Crozier RH (1987) Systematics and mitochondrial genome evolution of Australian rosellas (Aves: Platycercidae). Mol Biol Evol 4:526–543

Paabo S, Higuchi RG, Wilson AC (1989) Ancient DNA and the polymerase chain reaction. J Biol Chem 264:9709–9712

Paabo S, Wilson AC (1988) Polymerase chain reaction yields cloning artefacts. Nature 334:387–388

Pace NR, Olsen GJ, Woese CR (1986) Ribosomal RNA phylogeny and the primary lines of evolutionary descent. Cell 45:325–326

Pace NR, Stahl DA, Lane DJ, Olsen GJ (1985) Analyzing natural microbial populations by rRNA sequences. ASM News 51:4–12

Pagel MD, Harvey PH (1988) Recent developments in the analysis of comparative data. Quart Rev Biol 63:413–440

Pamilo P (1990) Statistical tests of phenograms based on genetic distances. Evolution 44:689–697

Pamilo P (1989) Estimating relatedness in social groups. TREE 4:353–355

Pamilo P, Nei M (1988) Relationships between gene trees and species trees. Mol Biol Evol 5:568–583

Penny D, Hendy MD (1986) Estimating the reliability of evolutionary trees. Mol Biol Evol 3:403–417

Plante Y, Boag PT, White BN (1989) Microgeographic variation in mitochondrial DNA of meadow voles (*Microtus pennsylvanicus*) in relation to population density. Evolution 43:1522–1537

Powell JR (1983) Interspecific cytoplasmic gene flow in the absence of nuclear gene flow: evidence from *Drosophila*. Proc Natl Acad Sci USA 80: 492–495

Powell JR, Caccone A, Amato GD, Yoon C (1986) Rates of nucleotide substitution in *Drosophila* mitochondrial DNA and nuclear DNA are similar. Proc Natl Acad Sci USA 83:9090–9093

Rand DM, Harrison RG (1989) Molecular population genetics of mtDNA size variation in crickets. Genetics 121:551–569

Rand DM, Harrison RG (1986) Mitochondrial DNA transmission genetics in crickets. Genetics 114:955–970

Richardson BJ, Baverstock PR, Adams M (1986) Allozyme Electrophoresis: a Handbook for Animal Systematics and Population Studies. Academic Press, Sydney

Ritland K, Clegg MT (1990) Optimal DNA sequence divergence for testing phylogenetic hypotheses. In Clegg MT, O'Brien SJ (eds) Molecular Evolution. UCLA Symposium on Molecular and Cellular Biology, New Series, Vol 122. AR Liss, New York, in press

Ritland K, Clegg MT (1987) Evolutionary analysis of plant DNA sequences. Amer Nat 130:S74–S100

Roberts JD, Maxson LR (1985) Tertiary speciation models in Australian anurans: Molecular data challenge Pleistocene scenario. Evolution 39:325–334

Rohlf FJ, Colless DH, Hart G (1983) Taxonomic congruence reexamined. Syst Zool 32:144–158

Rohlf FJ, Wooten MC (1988) Evaluation of the restricted maximum likelihood method for estimating phylogenetic trees using simulated gene frequency data. Evolution 42:581–595

Saghai-Maroof MA, Soliman KM, Jorgensen RA, Allard RW (1984) Ribosomal DNA spacer-length polymorphisms in barley: Mendelian inheritance, chromosomal location, and population dynamics. Proc Natl Acad Sci USA 81:8014–8018

Saiki RK, Gelfand DH, Stoffel S, Scharf SJ, Higuchi R, Horn GT, Mullis KB, Erlich HA (1988) Primer-directed enzymatic amplification of DNA with a thermostable DNA polymerase. Science 239:487–491

Saiki RK, Scharf S, Faloona F, Mullis KB, Horn GT, Erlich HA, Arnheim N (1985) Enzymatic amplification of β-globin genomic sequences and restriction site analysis for diagnosis of sickle cell anemia. Science 230:1350–1354

Saitou N (1988) Property and efficiency of the maximum likelihood method for molecular phylogeny. J Mol Evol 27:261–273

Saitou N, Imanishi T (1989) Relative efficiencies of the Fitch-Margoliash, maximum-parsimony, maximum-likelihood, minimum-evolution, and neighbor-joining methods of phylogenetic tree construction in obtaining the correct tree. Mol Biol Evol 6:514–525

Saitou N, Nei M (1987) The neighbor-joining method: a new method for reconstructing phylogenies. Mol Biol Evol 4:406–425

Sambrook J, Fritsch EF, Maniatis T (1989) Molecular Cloning, second edition. Cold Spring Harbor Lab Press, Cold Spring Harbor

Sang JH (1984) Genetics and Development. Longman, New York

Sarich VM, Schmid CW, Marks J (1989) DNA hybridization as a guide to phylogenetics: a critical analysis. Cladistics 5:3–32

Satta Y, Toyohara N, Ohtaka C, Tatsuno Y, Watanabe TK, Matsura ET, Chigusa SI, Takahata N (1988) Dubious maternal inheritance of mitochondrial DNA in *D. simulans* and evolution of *D. mauritiana*. Genet Res 52:1–6

Sederoff RR (1984) Structural variation in mitochondrial DNA. Adv Genet 22:1–108

Shah DM, Langley CH (1979) Inter- and intraspecific variation in restriction maps of *Drosophila* mitochondrial DNAs. Nature 281:696–699

Shields DC, Sharp PM, Higgins DG, Wright F (1988) "Silent" sites in *Drosophila* genes are not neutral: Evidence of selection among synonymous codons. Mol Biol Evol 5:704–716

Sibley CG, Ahlquist JE (1989) Avian phylogeny reconstructed from comparisons of the genetic material, DNA. In Patterson C (ed) Molecules and Morphology in Evolution: Conflict or Compromise? Cambridge University Press, Cambridge, pp 95–121

Sibley CG, Ahlquist JE (1986) Reconstructing bird phylogeny by comparing DNAs. Sci Amer 254(2):82–92

Sibley CG, Ahlquist JE (1987) DNA hybridization evidence of hominoid phylogeny: results from an expanded data set. J Mol Evol 26:99–121

Slightom JL, Change L-Y, Koop BF, Goodman M (1985) Chimpanzee fetal $^{G}\gamma$ and $^{A}\gamma$ globin gene nucleotide sequences provide further evidence of gene conversions in hominine evolution. Mol Biol Evol 2:370–389

Smouse PE, Li W-H (1987) Likelihood analysis of mitochondrial restriction-cleavage patterns for the human-chimpanzee-gorilla trichotomy. Evolution 41:1162–1176

Sneath PHA, Sokal RR (1973) Numerical Taxonomy. The Principles and Practice of Numerical Classification. Freeman, San Francisco

Sokal RR, Rohlf FJ (1981) Biometry, second edition. Freeman, San Francisco

Sourdis J, Krimbas C (1987) Accuracy of phylogenetic trees estimated from DNA sequence data. Mol Biol Evol 4:159–166

Sourdis J, Nei M (1988) Relative efficiencies of the maximum parsimony and distance-matrix methods in obtaining the correct phylogenetic tree. Mol Biol Evol 5:298–311

Starmer WT, Sullivan DT (1989) A shift in the third-codon-position nucleotide frequency in alcohol dehydrogenase genes in the genus *Drosophila*. Mol Biol Evol 6:546–552

Stephens JC, Nei M (1985) Phylogenetic analysis of polymorphic DNA sequences at the *Adh* locus in *Drosophila melanogaster* and its sibling species. J Mol Evol 22:289–300

Stewart C-D, Schilling JW, Wilson AC (1987) Adaptive evolution in the stomache lysozymes of foregut fermenters. Nature 330:401–404

Sueoka N (1988) Directional mutation pressure and neutral molecular evolution. Proc Natl Acad Sci USA 85:2653–2657

Sueoka N (1962) On the genetic basis of variation and heterogeneity of DNA base composition. Proc Natl Acad Sci USA 48:582–592

Sueoka N (1961) Correlation between base composition of deoxyribonucleic acid and amino acid composition of proteins. Proc Natl Acad Sci USA 47: 1141–1149

Suzuki H, Moriwaki K, Nevo E (1987) Ribosomal DNA (rDNA) spacer polymorphism in mole rats. Mol Biol Evol 4:602–610

Swanson KW, Irwin DM, Wilson AC (1991) Stomache lysozyme gene of the langur monkey: tests for convergence and positive selection. J Mol Evol 33:418–425

Swofford DL, Olsen GJ (1990) Phylogeny reconstruction. pp 411–501 in: "Molecular Systematics" eds. Hillis DM, Moritz C Sinauer, Sunderland

Swofford DL, Berlocher SH (1987) Inferring evolutionary trees from gene frequency data under the principle of maximum parsimony. Syst Zool 36: 293–325

Syvanen M (1987) Molecular clocks and evolutionary relationships: Possible distortions due to horizontal gene flow. J Mol Evol 26:16–23

Takahata N (1989) Gene genealogy in three related populations: Consistency probability between gene and population trees. Genetics 122:957–966

Tateno Y, Nei M, Tajima F (1982) Accuracy of estimated phylogenetic trees from molecular data. I. Distantly related species. J Mol Evol 18:387–404

Temin HM (1985) Reverse transcription in the eukaryotic genome: retroviruses, pararetroviruses, retrotransposons, and retrotranscripts. Mol Biol Evol 2:455–468

Templeton AR (1983) Phylogenetic inference from restriction endonuclease cleavage site maps with particular reference to the evolution of humans and the apes. Evolution 37:221–244

Templeton AR, Hollocher H, Lawler S, Johnston JS (1990) The ecological genetics of abnormal abdomen in *Drosophila mercatorum*. In Barker JSF, Starmer WT, MacIntyre RJ (eds) Ecological and Evolutionary Genetics of *Drosophila*. Plenum, New York, pp 17–35

Templeton AR, Johnston JS (1982) Life history evolution under pleiotropy and K-selection in a natural population of *Drosophila mercatorum*. In Barker JSF, and Starmer WT (eds) Ecological Genetics and Evolution: The Cactus-Yeast *Drosophila* Model System. Academic Press, Sydney, pp 225–239

Tomiuk J, Graur D (1988) Nei's modified genetic identity and distance measures and their sampling variances. Syst Zool 37:156–162

Traut W (1987) Hypervariable Bkm DNA loci in a moth, *Ephestia kuehniella*: does transposition cause restriction fragment length polymorphism? Genetics 115:493–498

Trick M, Dover GA (1984) Genetic relationships between subspecies of the tsetse fly *Glossina morsitans* inferred from variation in mitochondrial DNA sequences. Canad J Genet Cytol 26:692–697

Val FC (1977) Genetic analysis of the morphological differences between two interfertile species of Hawaiian *Drosophila*. Evolution 31:611–629

Vanlerberghe F, Dod B, Boursot P, Bellis M, Bonhomme F (1986) Absence of Y-chromosome introgression across the hybrid zone between *Mus musculus domesticus* and *Mus musculus musculus*. Genet Res 48:191–197

Van Valen LL (1983) Molecular selection. Evol Theory 6:297–298

Vawter L, Brown WM (1986) Nuclear and mitochondrial DNA comparisons reveal extreme rate variation in the molecular clock. Science 234:194–196

Walsh JB (1986) Selection and biased gene conversion in a multigene family: consequences of interallelic bias and threshold selection. Genetics 112: 699–716

Watson JD, Hopkins NH, Roberts JW, Steitz JA, Weiner AM (1987) Molecular Biology of the Gene, fourth edition. Benjamin Cummings, Menlo Park

Weide LG, Clark MA, Rupert CS, Shay JW (1982) Detrimental effect of mitochondria on hybrid cell survival. Somatic Cell Genet 8:15–21

Westneat DF, Noon WA, Reeve HK, Aquadro CF (1988) Improved hybridization conditions for DNA "fingerprints" probed with M13. Nucl Acids Res 16:41–61

Wheeler WC, Honeycutt RL (1988) Paired sequence difference in ribosomal RNAs: evolutionary and phylogenetic implications. Mol Biol Evol 5:90–96

Williams SA, Goodman M (1989) A statistical test that supports a human/chimpanzee clade based on noncoding DNA sequence data. Mol Biol Evol 6:325–330

Williams SM, Furnier GR, Fuog E, Strobeck C (1987) Evolution of the ribosomal DNA spacers of *Drosophila melanogaster*: different patterns of variation on X and Y chromosomes. Genetics 116:225–232

Williams SM, Strobeck C (1986) Measuring the multiple insemination frequency of *Drosophila* in nature: use of a Y-linked molecular marker. Evolution 40:440–442

Wilson AC, Zimmer EA, Prager EM, Kocher TD (1989) Restriction mapping in the molecular systematics of mammals: a retrospective salute. In Fernholm B, Bremer K, Jornvall H (eds) The Hierarchy of Life. Elsevier, Amsterdam, pp 407–419

Woese CR (1987) Bacterial evolution. Microbiol Rev 51:221–271

Wu C-I, True JR, Johnson N (1989) Fitness reduction associated with the deletion of a satellite DNA array. Nature 341:248–251

Wyss AR, Novacek MJ, McKenna MC (1987) Amino acid sequence versus morphological data and the interordinal relationships of mammals. Mol Biol Evol 4:99–116

Yang D, Oyaizu Y, Oyaizu H, Olsen GJ, Woese CR (1985) Mitochondrial origins. Proc Natl Acad Sci USA 82:4443–4447

Yokoyama S, Yokoyama R (1989) Molecular evolution of human visual pigment genes. Mol Biol Evol 6:186–197

Note added in proof

A recently popular method of studying nuclear markers (sections 3.2 and 3.3.2) is to use a single primer per PCR run which amplifies many regions of the genome and reveals large amounts of polymorphism (a mixture of substitutions and length variants) (Welsh and McClelland 1990; Williams et al. 1990). This attractive approach does have some problems, such as the generation of spurious bands (Riedy et al. 1992) and in some cases the dependence of the pattern produced on the concentration of the initial DNA sample.

Riedy MF, Hamilton WJ, Aquadro CF (1992) Excess of non-parental bands in offspring from known primate pedigrees assayed using RAPD PCR. Nucleic Acids Res 20:918.

Welsh J, McClelland M (1990) Fingerprinting genomes using PCR with arbitrary primers. Nucleic Acids Res 18:7213–7218.

Williams JGK, Kubelik AR, Livak KJ, Rafalski JA, Tingey SV (1990) DNA polymorphisms amplified by arbitrary primers are useful as genetic markers. Nucleic Acids Res 18:6531–6535.

Chapter 6

Molecular Population Genetics of *Drosophila*

Charles F. Aquadro

1. Introduction

Knowledge of the nature, level and distribution of nuclear DNA sequence polymorphism present within natural populations is a prerequisite to a complete understanding of organic evolution. To date, studies of molecular variation within populations have been carried out primarily in one dipteran species, *Drosophila melanogaster*. The reason for this focus has been the practical one that elegant genetic tools are available for *D. melanogaster*, such as chromosome balancers that allow chromosomes to be manipulated and stocks made homozygous for genes or whole chromosomes. Also, an increasing number of molecular clones of genes with interesting and diverse effect have become available to use as molecular probes for restriction map variation and to isolate additional copies of the genes for direct sequencing. Furthermore, the ability to transform the germline of *D. melanogaster* with altered or foreign genes has increased the utility of this species for studies addressing the functional significance of naturally occurring molecular variants (e.g., Laurie-Alhberg and Stam 1987).

In this chapter, I review some of what we have learned about the level and nature of naturally occurring DNA variation in the single copy, nuclear genome of *D. melanogaster*. (The literature covered in this review included that available as of January 1991, although some more recent papers have been included at the proof stage of this manuscript.) I contrast these results with recent findings from the closely related species *D. simulans* and several more distantly related species of *Drosophila*.

These studies suggest that we must be cautious about making generalizations from *D. melanogaster* to other *Drosophila*, much less to insects in general. I also discuss how the contrasts in molecular variation between the two sibling species, *D. melanogaster* and *D. simulans*, provide a particularly good opportunity to test current debates about the evolutionary forces acting on DNA sequence, protein sequence, transposable elements and chromosome arrangement. Finally, I discuss prospects for future studies: what are the questions and how can they best be approached? What lessons can we learn from studies carried out to date in *Drosophila*?

2. Methods for Assaying Nuclear DNA Sequence Variation

2.1. Full Sequence Data

Ideally, DNA sequence variation is assayed by direct determination of nucleotide sequences representing independent copies of a gene region or other homologous segment of the genome sampled from a natural population. I will refer to such multiple copies of single regions as "alleles." The effort required to accurately determine the sequence of a several kilobase (kb) region for one allele is substantial in its own right. An estimate of the level and distribution of sequence polymorphism requires multiple alleles to be sampled and sequenced. The sequences must also be determined with great accuracy, since a 1% error rate would generate an apparent level of nucleotide polymorphism higher than that actually seen for many regions of the *D. melanogaster* genome. The large sampling variances of most of the descriptive statistics of variation require samples of perhaps 10 or more alleles in order to obtain statistically reasonable estimates. This laborious task has not yet been accomplished for a truly random or representative, large sample of alleles sampled from a single locality.

Only two samples of nucleotide sequences for nuclear genes from natural populations in *Drosophila* had been published as of December 1990: one of 11 alleles of a 2.7 kb region including the structural gene for alcohol dehydrogenase (*Adh*; Kreitman 1983), and a second analysis of 13 isolates representing 10 different protein allozymes produced by the esterase-6 locus (*Est*-6; Cooke and Oakeshott 1989). Both of these studies involved making recombinant DNA libraries for each gene isolate, cloning the region from each library, and subsequently sequencing each region. Improvements in sequencing strategies and the introduction of the polymerase chain reaction (PCR) to selectively amplify regions of interest for direct sequencing promise to increase the number of sequence level population samples in the near future (e.g., Berry et al. 1991; McDonald and Kreitman 1991).

2.2. Six-cutter Restriction Endonuclease Analyses

An alternative to full sequence analysis is to use clones of gene regions of interest as molecular probes of restriction endonuclease digested nuclear DNA. These studies usually involve obtaining samples of individuals from a single location and establishing a set of homozygous lines by using chromosome balancers or by inbreeding (sib mating). Genomic DNA is isolated from each of these lines, cut with restriction enzymes, size fractionated on gels, transferred (Southern blotted) to a filter, and probed with clones from the genomic region of interest. Restriction maps are constructed using single and double digests for each of the lines. Comparisons of maps for a sample of gene isolates provide estimates of the type, level and distribution of DNA sequence polymorphism among the sample of alleles. The data available on nuclear DNA variation in *Drosophila* have largely been obtained using such restriction map surveys. A complete description of the methods used in our laboratory can be found in Aquadro et al. (1992).

Most of these studies have used restriction endonucleases with six-base recognition sequences (so-called "six-cutters"). These enzymes cut on average once every 4096 base pairs, yielding DNA fragments of a size easily resolved by standard agarose gel electrophoresis. In addition, bacteriophage clones of genomic regions that are generally used as molecular probes contain 10–20 kb of sequence and often completely span the gene of interest. The number of cleavage sites for any one enzyme in the probed region is usually in the range of four to eight, and these sites can readily be mapped relative to one another by comparisons to DNA cut with different enzymes singly and in combination with other enzymes. The use of four to 10 endonucleases with different recognition sequences allows the "sampling" of sequence variation across the region probed.

Two general classes of DNA variation can be assayed by comparisons of these restriction site maps: sequence length variation and nucleotide substitutions. All sequence length variation more than approximately 50–100 bp in size can be scored by the average six-cutter restriction map survey. In fact, these surveys are perhaps the most efficient way to screen large regions of the genome for the frequency and distribution of large insertions and deletions, particularly those due to transposable elements. More refined techniques, discussed below, and direct sequencing often show extensive size variation as small as one base pair in length.

There are two approaches to summarizing nucleotide substitutions. The first is a direct estimate of the average sequence divergence, or nucleotide diversity (see Nei 1987). This can be estimated from the proportion of cleavage sites shared (S_{ij}) between the restriction site maps of lines compared two at a time (e.g., Nei and Li 1979). This estimate can be viewed as the average probability that two randomly chosen sequences will differ at any one nucleotide site. For complete sequences, the statistic π_{ij} is calculated as the proportion of nucleotide differences between the ith and jth alleles. From the restriction maps of a sample of alleles,

$$\pi_{ij} = \frac{-\log_e S_{ij}}{r}$$

where $S_{ij} = 2m_{ij}/(m_i + m_j)$, m_i and m_j are the number of cleavage sites in the ith and jth allele, m_{ij} is the number of sites shared between these alleles, and r is the number of nucleotides in the endonuclease recognition sequence. Average nucleotide diversity (π) is calculated as,

$$\pi = \frac{n}{n-1}\Sigma_{ij}x_i x_j \pi_{ij}$$

where x_i and x_j are the sample frequencies of the ith and jth type of sequence. Both the number of variable sites and the frequencies of variants at each site contribute to the estimate of π. For the levels of variation typically found within populations, π is equal to the heterozygosity per nucleotide. Engels (1981) provides an alternative approach to calculating this measure. In a population at equilibrium with respect to mutation and drift, heterozygosity per nucleotide is approximately equal to the quantity $4N\mu$ (called theta, θ), where N is the effective population size and μ is the neutral mutation rate per nucleotide site. An alternative statistic is the number of segregating sites (S_n) in a sample of n alleles. This is the number of polymorphic nucleotide sites for a sample of complete sequences, and is related to θ by

$$\theta = S_n\left\{n_s\left[1 + \frac{1}{2} + \frac{1}{3} + \cdots + \frac{1}{n-1}\right]\right\}^{-1}$$

for a random mating population of constant size at a genetic equilibrium and where n_s is the size in base pairs of the region surveyed. From restriction maps, one can estimate S_n, following Hudson (1982), as

$$S_n = p(n_s) \text{ and } p = \frac{k}{(2m - k)r}$$

where p is the proportion of nucleotide sites polymorphic in a sample, m is the number of cleavage sites that cut at least once in the sample, k is the number of those sites that were not cut in all lines (the number of polymorphic restriction sites), and r is the number of nucleotides in the endonuclease recognition sequence. Ewens et al. (1981) and Hudson (1982) have proposed estimating θ by

$$\theta = \frac{p}{\log_e(n)} = \frac{S_n}{[n_s \log_e(n)]}$$

The quantity $\log_e(n)$ is used as an approximation to $[1 + \frac{1}{2} + \frac{1}{3} + \cdots + 1/(n-1)]$ for large n. Most studies have used $\log_e(n)$; however, the latter would actually be preferred (R Hudson pers comm). Under neutrality and at equilibrium for mutation and drift, an estimate of θ based on the number of segregating sites should be identical to one based on average pairwise nucleotide divergence. Tajima (1989a) has developed this into a test of the neutral theory. When the two estimates are not equal to each other, then either the sequences are not evolving neutrally or the population size has recently changed and the population is not at equilibrium. We will return to an application of this test below.

As discussed by numerous authors (e.g., Nei and Jin 1989; Weir 1990; Kreitman 1991), the variances of the above estimates are large, due to both the sampling process and to variance in the evolutionary process. Thus it is important to examine as many nucleotides as possible for a given region.

In addition to the nature and level of variation segregating in natural populations, we are interested in the organization of that variation; that is, are variants at adjacent sites or regions correlated in a statistical sense? Variants showing nonrandom associations are considered to be in "linkage disequilibrium." The ability to establish lines of flies homozygous for individual genes or chromosomes allows the unambiguous determination of each multilocus genotype, or haplotype, and its frequency. Linkage disequilibrium is estimated as $D = x_{11} - p_1q_1$, where x_{11} is the observed frequency of the A_1B_1 haplotype, and p_1 and q_1 are the frequencies of the A_1 and B_1 alleles at sites A and B. The significance of the nonrandom association is usually assessed by a 2×2 contingency table of expected and observed frequencies of each two-locus haplotype using Fisher's exact test or the Chi-Square test of homogeneity (see Weir 1990).

The "six-cutter" approach has the advantage of requiring little in the way of sophisticated molecular tools or techniques in order to rapidly screen a large number of lines or chromosomes for DNA sequence variation (Aquadro et al. 1992). Filters can be reprobed with clones representing adjacent or unlinked regions of interest. All sequence length variation over approximately 50–100 bp in length is detectable, and the presence/absence of sites scored across lines allows the assessment of the extent of linkage disequilibrium among variants along the chromosome. However, only a small proportion of the individual nucleotide sites in a region is actually assayed. Many surveys have used four to six different restriction endonucleases, giving a density of roughly one to two cleavage sites per kilobase studied. Taking into account that one screens for changes at the six nucleotides at each recognition site cleaved (scored as the loss of the site), as well as approximately six additional potential changes from sites one nucleotide off from being a restriction site (scored as the gain of the site), then it can be estimated that many of the six-cutter surveys are assaying only 12 to 24 nucleotides per kilobase. Variances on estimates derived from this small proportion of nucleotides are expected to be very large. Nonetheless, more detailed studies of these regions have revealed that the original six-cutter estimates are remarkably accurate as to the *average* level of nucleotide substitution variation across a region. Studies of variation in the level of sequence polymorphism on a fine scale, such as between different regions of a gene, however, require more sophisticated approaches or direct sequencing.

2.3. Four-cutter Restriction Endonuclease Analyses

One solution to the low number of sites screened in a typical six-cutter survey has been to use endonucleases that cleave more often. For example, "four-cutter" enzymes have four base recognition sites, and are expected to cleave on average once every 256 nucleotides. However, four-cutter sites cannot readily be mapped with the traditional restriction mapping strategy of single and double digests because of the large number of cleavage sites, and because many of the fragments generated are below the size range well resolved on agarose gels. As an alternative strategy that overcomes this problem, Kreitman and Aguadé (1986a) used knowledge of the complete DNA sequence of the *Adh* gene region to predict the location of fragments separated and sized on high resolution gels typically used for DNA sequencing. Coupling this approach with electroblotting and UV crosslinking the DNA to nylon filters has led to a very rapid and general method for surveying sequence variation (see Kreitman 1987; Aquadro et al. 1992). The use of molecular probes two to four kilobases long leads to the detection of a manageable number of fragments. Larger sized regions are best studied by successive probings of the filters with adjacent or slightly overlapping probes. Thus, the virtually impossible task of mapping the many four-cutter sites by the traditional single and double digest method is avoided. The location of site gains and losses can also generally be determined based on the sequence.

Kreitman and Aguadé (1986a) estimated that the use of 10 different "four-cutter" restriction endonucleases allows the screening of roughly 19% of the nucleotides in the *Adh* region of *D. melanogaster*. Given that some 50–60 chromosomes can be screened per gel for each enzyme and that the filters can be probed repeatedly with different probes, the large population samples desired are obtainable. In addition, the high density of sites increases the power to detect local variation in levels of polymorphism and linkage disequilibrium.

As powerful as the "four-cutter" approach is, however, only direct sequencing allows the determination of the exact base substitution in all cases. This information is of course essential for determining whether a change in coding sequence would be amino acid changing (nonsynonymous) or "silent" (synonymous).

Turning briefly to consider technical issues relating to interspecific comparisons, the four-cutter approach by itself is generally not suitable if more than a few percent divergence exists, since fragment homologies become too difficult to determine. However, six-cutter maps can be used to this end, at least up to roughly 5–10% divergence, because restriction maps can still be determined directly for each species by single and double digestion. Even so, small insertions and deletions between species can make such comparisons difficult and large numbers of differences between maps make assignment of homologous sites difficult. Direct sequence comparisons are clearly the method of choice.

A word as to the choice and availability of probes for both six- and four-cutter surveys is in order here. Population geneticists working within *D. melanogaster* have often had the advantage of drawing on the clones generated by their molecular geneticist colleagues. The 10–20 kb bacteriophage clones generated in chromosomal "walks" or initial genomic library screenings have been ideal for six-cutter surveys. Smaller plasmid subclones of 2–4 kb are ideal as probes for four-cutter blots. However, these *D. melanogaster* clones have limited use as direct probes outside of the *D. melanogaster* species group because of reduced sequence similarity, particularly in noncoding regions. Fortunately, *D. melanogaster* clones work well for *D. simulans*, for example, where average sequence divergence from *D. melanogaster* averages only 4–5%. Gene organization is also very similar between these species, allowing us very rapidly to extend our analyses to homologous regions in *D. simulans*.

The analysis of more distantly related species usually necessitates the isolation of species-specific clones for use as probes or to sequence directly. While this is a relatively straightforward process now, particularly with the widespread commercial availability of the necessary reagents and biologicals, differences in gene and genome organization can mean that substantial characterization of the new species' clones is necessary before one can proceed with the population genetic work. Two examples are instructive in this regard. The first is the α-amylase locus in *D. pseudoobscura*. While this locus is duplicated in *D. melanogaster*, previous allozyme work suggested it was a simple, single gene system in *D. pseudoobscura*. However, cloning and characterization of this locus (Brown et al. 1990; Aquadro et al. 1991) revealed a multigene family with one highly expressed gene, one lowly expressed gene and a pseudogene. These three genes are located adjacent to each other and have extremely similar DNA sequences, making the interpretation of genomic Southern blots difficult until the organization of the gene family was resolved. A second example of potential problems comes from the study of molecular variation in *D. athabasca*. CK Yoon and CF Aquadro (unpublished) have cloned the homologous sequences in this species for the *period*, *Adh* and *rosy* (xanthine dehydrogenase) genes. However, every genomic sequence cloned appears to be interspersed with repetitive sequences that are also present in multiple copies elsewhere in the genome. Thus we have had to initiate a very careful characterization and subcloning of the regions in order to produce useful molecular probes to study variation in this species.

Studies of other genes have not encountered these problems (e.g., *Adh* and *Xdh* in *D. pseudoobscura*; Schaeffer et al. 1987; Riley et al. 1989), but I raise the preceding two examples as cautionary illustrations of the types of problems that may occur. The extent to which similar problems arise with other species is unknown. Of course, for a clear understanding of the reasons for the distribution of sequence variation seen in any species, a careful characterization of the molecular and gene organization of the region of interest in that species is essential anyway.

3. Patterns of Variation in *D. Melanogaster*

3.1. Historical Perspective

Initial studies in *Drosophila* focused on the alchohol dehydrogenase structural gene (*Adh*), located on the left arm of the second chromosome. The rich literature on the temporal and geographic variation in the level and distribution of ADH protein polymorphism assayed by protein electrophoresis prompted much of the interest in this gene. Particularly intriguing was the suggestion that the two allozymes commonly segregating in natural populations of *D. melanogaster* were maintained by balancing selection (reviewed in Van Delden 1982). The cloning of the *Adh* structural gene from this species (Goldberg 1980) paved the way for the analysis of naturally occurring variation underlying the allozyme polymorphism.

Langley et al. (1982) were the first to estimate levels of nuclear DNA sequence variation in *Drosophila*. They examined restriction map variation in a 12 kb region centered on the *Adh* gene among 18 isochromsomal lines of *D. melanogaster* representing four populations in eastern Unites States. Two findings were notable. First, they estimated that any two randomly chosen sequences differed by 0.6% of their nucleotides in this region ($\pi = 0.006$). With the exception of the third exon of *Adh*, this level of nucleotide heterozygosity was also observed in a sample of 11 alleles from around the world for which the complete sequence was determined for 2.7 kb by Kreitman (1983). Second, in addition to what were presumed to be nucleotide substitutions underlying the restriction site changes, they observed extensive variation due to large insertions and

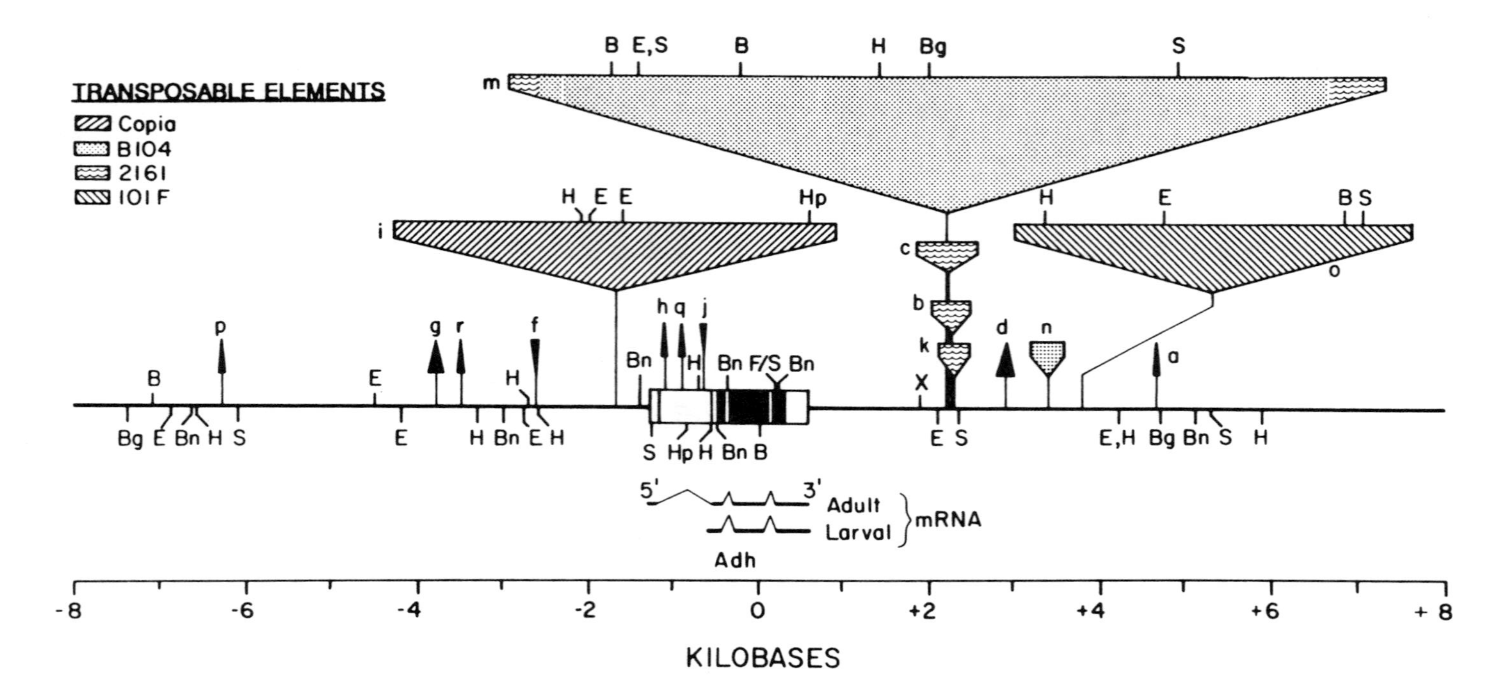

Figure 6.1. Summary restriction site map of naturally occurring DNA sequence variation in the *Adh* region of *D. melanogaster*. The *Adh* gene is located approximately in the center of the region examined, as indicated by the boxed region (filled boxes indicate the three coding exons). Insertions and deletions, relative to the most common restriction map, are indicated approximately to scale by triangles pointing toward or away from the map, respectively. Each size insertion/deletion is denoted by a lower case letter. Insertion/deletions of unique sequence as identified by Southern blot analysis are indicated in solid black. The identity of transposable element insertions is indicated in the upper left of the figure. Variant and invariant restriction sites are indicated above and below the line, respectively, as follows: B (*Bam*HI), Bg (*Bgl*II), Bn (*Ban*II), E (*Eco*RI), H (*Hin*dIII), Hp (*Hpa*I), S (*Sal*I), and X (*Xho*I). (From Aquadro et al. 1986.)

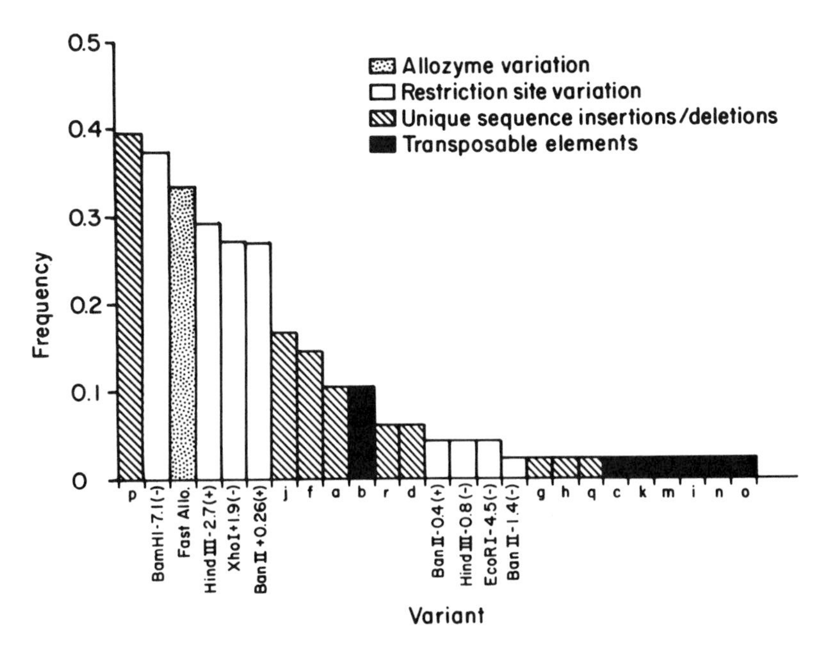

Figure 6.2. Frequency spectrum for naturally occurring DNA sequence variants in the 13 kb *Adh* region of *D. melanogaster*. Shown is the less frequent character at each variable site. (From Aquadro et al. 1986.)

deletions, all of which flanked the *Adh* transcriptional unit. Subsequent, more extensive studies of additional natural populations using four- and six-base restriction enzymes have supported these results (e.g., Aquadro et al. 1986; Cross and Birley 1986; Kreitman and Aguadé 1986a; 1986b).

Cloning and characterization of all sequence length variants detected in 48 lines from the eastern United States (Aquadro et al. 1986) revealed that all sequence length variants over roughly 250 bp were transposable elements (Figure 6.1). Furthermore, the frequency in the sample of any one transposable element insertion was very low or unique (Figure 6.2). Small insertions (<250 bp) and all deletions (defined relative to the most common restriction map length) appeared to represent insertions or deletions of unique sequence not homologous to transposable elements and are found at both low and intermediate frequency in populations.

Studies of the level, distribution and significance of molecular variation at *Adh* within and between species of *Drosophila* continue to play a leading role in the development of our understanding of the roles of different evolutionary forces in shaping polymorphism within species and rates and patterns of molecular evolution. The interested reader is referred to Kreitman (1987; 1991), Sullivan et al. (1989) and Laurie et al. (1990) for an entree to this literature.

The cloning of additional genes and their subsequent use to survey population variation has led to a growing data set on naturally occurring DNA sequence variation in *D. melanogaster*. Table 6.1 summarizes data from gene regions on the two major autosomes and the X chromosome in this species available as of January 1991. These regions span from 12 to 120 kb of contiguous sequence, the total summarized here being 675 kb, or roughly 0.5% of the single copy nuclear genome.

3.2 Estimates of Nucleotide Variability

Estimates of average heterozygosity per nucleotide vary from less than 0.001 for the *yellow-achaete-scute* region of some populations to 0.011 for the *white* region. The average estimate of heterozygosity per nucleotide is 0.4% (average weighted by the size of the region surveyed). This means that two randomly chosen chromosomes will differ, on average, by four nucleotides for every kilobase compared, and that there are perhaps 400,000 differences between the two haploid genome complements in every fly (considering only the single-copy portion of the genome, approximately 103,000 kb in size, and that the individual is a female with two X chromosomes). Every fly has a unique genome.

That the six-cutter estimates are representative of average levels of sequence variation in the assayed regions is supported by their comparison to estimates from the same regions using four-cutter surveys. Using a battery of 10 four-cutter restriction enzymes, WA Noon and CF Aquadro (unpublished) have screened an 8.2 kb region including *rosy* in 52 lines of *D. melanogaster* sampled from a single location in Raleigh, North Carolina. Comparison of the fragments probed with those predicted from the DNA sequence of one allele available for *D. melanogaster* (Lee et al. 1987; Keith et al. 1987) allowed the scoring of approximately 1,770 nucleotide sites, corresponding to 21% of the base pairs in the region. Variation among lines included 48 restriction sites and four small sequence length variants occurring in intron and flanking regions. The level of nucleotide heterozygosity estimated from the four-cutter analysis is 0.005 (both π and θ), which compares to $\pi = 0.003$ and $\theta = 0.004$ estimated for the same region by our six-cutter work (Aquadro et al. 1988). This level is also virtually identical to the estimates of π of 0.004, both for 6.9 kb of the *Adh* region (Kreitman and Aguadé 1986b) and for 9.7 kb covering the *white* locus (Miyashita and Langley 1988). Taken together, the results from these four-cutter surveys on three unlinked regions (*white* on the X chromosome, *Adh* on the second and *rosy* on the third) support the conclusion reached from the six-cutter surveys that the per-nucleotide heterozygosity for *D. melanogaster* averages about 0.004.

Many of these estimates of nucleotide heterozygosity and insertion frequency come from surveys of only a few populations in the United

Table 6.1. Summary of DNA sequence and large insertion variation in *D. melanogaster*[*]

Chromosome/genes	Examined		Het./nt.		Insertions (>250 bp)[†]				
	Chrom.	kb	θ	π	Total	Diff.	Density	Freq.	Ref.
X Chromosome									
yellow, ac-sc[††]	64	106	0.001	0.0003	41	14	0.006	0.046	1
	49	120	0.002	0.002	15	9	0.003	0.034	2
	109	31	0.003	0.002	10	10	0.003	0.009	3
period	66	52	0.002	0.001	8	8	0.002	0.015	4
zeste-tko	64	20	0.004	0.004	2	2	0.002	0.016	5
white	38	45	0.013	0.011	16	12	0.009	0.035	6
	64	45	0.008	0.009	29	11	0.010	0.038	7
	64	10	0.004	0.004	0	0	0	0	8
Notch	37	60	0.007	0.005	4	3	0.002	0.036	9
forked	64	25	0.004	0.002	na	na	0.002	0.020	10
vermilion	64	24	0.006	0.003	na	na	0.002	0.020	10
G6pd[††]	126	13	0.004	0.001	25	8	0.015	0.025	11
Su(f)[††]	64	24	0.000	0.000	na	na	0.004	0.020	10
Second Chromosome									
Adh	48	13	0.007	0.006	11	8	0.018	0.029	12
	58	12	0.005	na	6	2	0.009	0.052	13
	81	3	na	0.004	na	na	na	na	14
amylase	85	15	0.006	0.008	11	10	0.009	0.012	15
Ddc	46	65	0.004	0.005	8	8	0.003	0.022	16

(*continued*)

Table 6.1. *Continued*

Chromosome/genes	Examined		Het./nt.		Insertions (>250 bp)[†]				
	Chrom.	kb	θ	π	Total	Diff.	Density	Freq.	Ref.
Third Chromosome									
87A heat shock	29	25	0.002	0.002	4	4	0.006	0.034	17
rosy (Xdh)	60	100	0.005	0.005	16	9	0.003	0.030	18
Est-6	42	22	0.010	na	6	4	0.006	0.036	19
Weighted average	675 kb		0.004	0.004			0.004	0.028	

* Only studies of natural populations are included. References are: 1) Aguadé et al. 1989; 2) Beech and Leigh Brown 1989 and Macpherson et al. 1990; 3) Eanes et al. 1989b; 4) D Stern, WA Noon, E Kindahl and CFA, unpublished; 5) Aguadé et al. 1989b; 6) Langley and Aquadro 1987; 7) Miyashita and Langley 1988 with six-cutters, 8) Miyashita and Langley 1988 with four-cutters; 9) Schaeffer et al. 1988; 10) cited in Langley 1990; 11) Eanes et al. 1989a; 12) Langley et al. 1982; Aquadro et al. 1986; 13) Jiang et al. 1988; 14) Kreitman and Aguadé 1986b with four-cutters; 15) Langley et al. 1988; 16) Aquadro et al. 1992; 17) Leigh Brown 1983; 18) Aquadro et al. 1988 and CFA, K Lado, V Bansal and WA Noon unpublished; 19) Game and Oakeshott 1990.

† For insertions: Total is the total number of insertions observed. Diff indicates the number of detectably different insertions (different size or location of insertion). Density is calculated as the number of insertions observed per chromosome per kilobase surveyed. Frequency refers to the proportion of chromosomes surveyed having individual insertions.

†† Refers to region of reduced rate of recombination.

na, not available.

States. Several studies have included limited worldwide collections or populations sampled from Europe, Africa, Asia and Australia, and differences have been observed in the frequency of variants and in the level of variability among samples from different geographic regions (e.g., Kreitman 1987; Jiang et al. 1988; Aguadé et al. 1989a; 1989b; Beech and Leigh Brown 1989; Eanes et al. 1989a; 1989b; Simmons et al. 1989; Macpherson et al. 1990). Studies by Stephan and Langley (1989) and Stephan (1989) indicate that this is particularly true for *D. ananassae* and perhaps other *Drosophila* species. At present the data are insufficient to assess geographic patterns of level of DNA variation, in part because of the nonoverlapping sets of populations studied, and also because the estimates are so low that the presence or absence of a single polymorphic restriction site has a dramatic influence on the estimate of variation. The examination of patterns of DNA sequence variation underlying protein electrophoretic variation will be of particular interest in understanding the causes of the clines in allozyme frequency observed at several loci (e.g., *Adh*, Simmons et al. 1989). Nonetheless, the levels of variation summarized here appear at this time to be approximately representative of *average* levels for the species as a whole.

Less neutral variation would be expected at equilibrium for genes on the X chromosome compared to the autosomes because the effective population size for genes on the X is approximately three quarters that of the autosomes. Selection is also greater against recessive deleterious variants on the X chromosome. However, no clear evidence of such a reduction is evident from Table 6.1 (see particularly Langley and Aquadro 1987; Miyashita and Langley 1988; Schaeffer et al. 1988; Aguadé et al. 1989b; Beech and Leigh Brown 1989). The comparison of different genes with different levels of constraint or evolutionary histories on the different chromosomes complicates this comparison, so that a firm conclusion should not be drawn at this time.

While estimates from a diverse array of genes on all three major chromosomes appear to converge on an *average* heterozygosity per nucleotide, the variation for base pair substitutions, like that for large insertions, is not distributed uniformly across regions. There is some evidence for reduced nucleotide variation in some populations, particularly a reduction of intermediate frequency variants, in regions of reduced recombination near the telomere and centromere of the X chromosome (Table 6.1). For example, Eanes et al. (1989a) surveyed the 13 kb region surrounding the glucose 6 phosphate dehydrogenase gene (*G6pd*) located near the base of the X chromosome in 126 copies of the gene isolated from *D. melanogaster* from North America, Europe and Africa, and found nucleotide heterozygosity (π) to be only 0.00065. An estimate of θ for the same data is higher, at 0.0035, which, compared to the estimate of π, suggests a deficiency of intermediate frequency variants relative to that expected for selectively neutral variation in an equilibrium population. Variation in genes near the tip of the X chromosome are also very low in at least some populations. Aguadé et al. (1989b), Eanes et al. (1989b), Beech and Leigh Brown (1989) and Macpherson et al. (1990) have each surveyed six-cutter map variation across the *yellow-achaete-scute* region which maps genetically to position 0.0 near the telomere of the X chromosome. Aguadé et al. (1989b) estimated π to be only 0.0003, while θ was estimated to be 0.0011 averaged across collections from North Carolina, Texas and Japan. Similar results have been reported by Begun and Aquadro (1991). The other studies have not revealed such an extreme reduction in level of variation in the populations examined (see particularly Macpherson et al. 1990). These studies do, nonetheless, suggest a trend towards a deficiency of intermediate frequency site variants in regions of reduced recombination.

Significant variation in levels of nucleotide polymorphism has also been seen on a much finer scale. For example, Kreitman and Aguadé (1986b) observed that the third exon of *Adh* was significantly more variable than the surrounding sequence, and Miyashita and Langley (1988) have found a significant nonrandom distribution of variable four-cutter restriction sites across the *white* region. Before we explore the reasons for such variation in the levels of polymorphism, discussion of sequence length variation and of the organization of sequence variation, or linkage disequilibrium, is warranted.

3.3. Sequence Length Variation

Extensive variation for sequence length is observed in many regions of the *D. melanogaster* genome. Six-cutter surveys have revealed variation ranging in size from 21 bp (the lower limit of resolution for a detailed survey) to more than 20 kb. Four-cutter and sequence analyses demonstrate the presence of a large number of small insertions and deletions ranging down to single nucleotide differences. As one might expect from their effect on reading frame and the encoded protein, these are found largely in noncoding regions (introns and flanking sequences). However, some genes show variation in the number of repeats of codons found within the structural gene. A particularly striking example is the extensive variation for the number of Threonine-Glycine pairs found within a large repeat region in the period locus among lines of *D. melanogaster* (and related species) from natural populations (Yu et al. 1987; Kyriacou this volume; Wheeler et al. 1991; Costa et al. 1991; E Bermingham and CF Aquadro unpublished).

Cloning and characterization by Southern blot analysis or by direct sequencing has indicated that the sequence length variation could be separated into two classes: that due to the presence of transposable elements (generally insertions >250 bp in size), and that due to gains or losses of nonmiddle repetitive sequences (generally insertions and deletions <250 bp in size; Leigh Brown 1983; Aquadro et al. 1986; Beech and Leigh Brown 1989; Eanes et al. 1989a). This latter class ranges in size from single nucleotides to roughly 200 bp in length, and in the cases studied did not involve sequences found elsewhere in the genome. This variation is often associated with adjacent repeats which presumably contribute to variation in sequence length via unequal crossing over or template slippage during replication (e.g., Jeffreys et al. 1985). More complex mechanisms, including those associated with secondary structure in the DNA, are clearly also involved in generating variation (e.g., Golding and Glickman 1986). In addition to the fundamental distinction between transposable elements and unique-sequence length variation, these two classes differ in the frequency with which individual variants of each class are found segregating in natural populations. The smaller unique-sequence insertions and deletions are not only found in low frequency but occasionally occur in moderate to intermediate frequency (Figure 6.2). This is particularly true for variants less than 50 bp in length. Deletions relative to the most common genotype of more than a few hundred base pairs in length are seen only rarely.

Since much of the available data reviewed here comes from six-cutter map surveys where small length variations are not scorable, I will restrict discussion to insertions greater than 250 bp in length which are most likely to be transposable elements. My discussion will be brief as Charlesworth and Langley (1989) have recently published an extensive review of the population genetics of transposable elements in *Drosophila* (see also Langley 1990).

Large insertions, defined here as a class >250 bp in length, are occasionally quite common in a region, but are usually present individually in only low frequency, such that any one insertion is usually represented only once or twice in a sample. For example, consider the 13 kb *Adh* region studied by Aquadro et al. (1986). While 22% of the 49 chromosomes sampled had a transposable element within the 13 kb region (Figure 6.1), no individual insertion was found in more than 6% of the chromosomes (Figure 6.2). The density of large insertion variation is quantified in Table 6.1 as the number of insertions per chromosome per kilobase surveyed. Densities in *D. melanogaster* range from 0.002 for the *Notch*, *zeste-tko*, *forked* and *vermilion* regions to 0.018 for the *Adh* region, with an average density of 0.004 large insertions per chromosome per kilobase. These densities suggest that there is one transposable element every 460 kb on average in the euchromatic portion of the *D. melanogaster* genome. These results parallel those obtained using clones of known transposable elements as probes for the location and distribution of these elements on the polytene chromosomes of *D. melanogaster* (Montgomery and Langley

1983; Leigh Brown and Moss 1987; Montgomery et al. 1987; Ronsseray and Anxolabéhère 1987; Ajioka and Eanes 1989; Charlesworth and Lapid 1989).

Virtually every study cited in Table 6.1 has observed spatial heterogeneity in the distribution of insertion variation across regions of the genome. This is in part due to the fact that large insertions have not been observed in known coding sequences, consistent with the reduced fitness expected from such interruptions. There are, however, several instances of insertions in introns that do not appear to significantly disrupt expression (e.g., *white*, Miyashita and Langley 1988; *G6pd*, Eanes et al. 1989a; dopa decarboxylase (*Ddc*), Aquadro et al. 1992). Several studies have observed very strong clustering of insertions in small segments of the genome (e.g., near *Adh*, Aquadro et al. 1986; in *G6pd*, Eanes et al. 1989a). Whether the accumulation of insertions at these sites is due to relaxed functional constraint in these regions or to increased accessability, due perhaps to alterations in chromatin structure, remains untested. Surprisingly, a significant *average* decrease in large insertion variation is not apparent in regions for which a high density of lethal complementation groups, transcripts and genes have been detected (e.g., Aquadro et al. 1988; Aquadro et al. 1992).

Many transposable element insertions cause loss-of-function mutations by disrupting normal gene structure or expression (Finnegan and Fawcett 1986). It is thus reasonable to expect these mutations to be at least partially recessive and deleterious. Increased selection against these mutations on the X chromosome would predict a lower density and frequency of transposable elements on the X chromosome versus the autosomes (Langley and Aquadro 1987). The data summarized in Table 6.1 do not show the predicted reduction, although the comparison is complicated by substantial heterogeneity in density and frequency across both X chromosomal and autosomal gene regions. The data support the hypothesis that the apparently deleterious nature of transposable elements in *Drosophila* (Golding et al. 1986) is due not entirely to alteration in expression of adjacent genes, but may be related to selection limiting overall copy number (Langley et al. 1988a; Eanes et al. 1988).

Langley et al. (1988a) proposed that much of the selection limiting the copy number of transposable elements is due to unequal crossing over between members of the same transposable element family in nonhomologous sites on the chromosome. Such crossover events between transposable elements have been observed in mutants at the *white* locus of *D. melanogaster* (Goldberg et al. 1983; Davis et al. 1987) and in the 6F region of its X chromosome (Lim, 1988). These can result in duplications, deletions and translocations, all of which will likely be deleterious and eliminated by natural selection. This mechanism is proposed to limit copy number of transposable elements and thus keep individual insertion frequencies low in regions of normal recombination.

This mechanism would operate at a reduced level in regions of reduced recombination, such as near the centromeres and telomeres of chromosomes in *Drosophila*. Individual elements in those regions could therefore increase in frequency by genetic drift. In accordance with this prediction, in situ experiments show that euchromatic regions adjacent to the centromeres of the X have accumulated transposable elements (Montgomery et al. 1987; Eanes et al. 1988; Langley et al. 1988a; Charlesworth and Lapid 1989). A similar increase in frequency of individual large insertions is seen for some variants in restriction map surveys of genes located near the centromeric base of the X chromosome (Table 6.1; see particularly Aguadé et al. 1989a; Beech and Leigh Brown 1989; Eanes et al. 1989a).

Results from both restriction map and in situ studies of the distribution of transposable elements in natural populations indicate that transposable elements are deleterious parasites of the *Drosophila* genome (e.g., Golding et al. 1986; Charlesworth and Langley 1989). In no case are individual transposable element insertions seen in the high frequency expected if their insertion leads to adaptive change in gene expression. Similar conclusions come from mutation-accumulation experiments at *Amy* and *Adh* in *D. melanogaster* (Tachida et al. 1989; Aquadro et al. 1990). It remains to be demonstrate the extent to which the insertion/excision process contributes to evolutionary significant base pair variation.

3.4. Linkage Disequilibrium and the Organization of Sequence Variation

In the absence of linkage disequilibrium, we might expect a tremendous number of multilocus genotypes, or haplotypes, to be sampled for a given region. For example, for n polymorphic nucleotide sites and with four possible nucleotides at each site, there are $(4)^n$ possible haplotypes. For the 13 kb *Adh* region surveyed by Aquadro et al. (1986) and given two alternative states (presence versus absence) of the 24 restriction site and length variants scored, we would have expected $(2)^{24}$, or approximately 1.7×10^7 different restriction map haplotypes to be possible. We would have expected all 48 of the alleles sampled to be unique. However, only 29 distinct haplotypes were observed. Recombination had not randomized variants at adjacent sites. Thus, we were able to attempt reconstructing the evolutionary relatedness of the haplotypes observed (Figure 6.3). While this phylogeny must be viewed as only a summary of possible haplotype derivations and recombination events, it illustrates that sufficient linkage disequilibrium exists in certain regions of the *D. melanogaster* genome so that vestiges of the evolutionary history of the alleles segregating in the population have been retained.

The strong linkage disequilibrium observed in the *Adh* region may be due, in part, to strong selection acting at that locus. However, significant nonrandom associations are also seen throughout the 65 kb *Ddc* region on

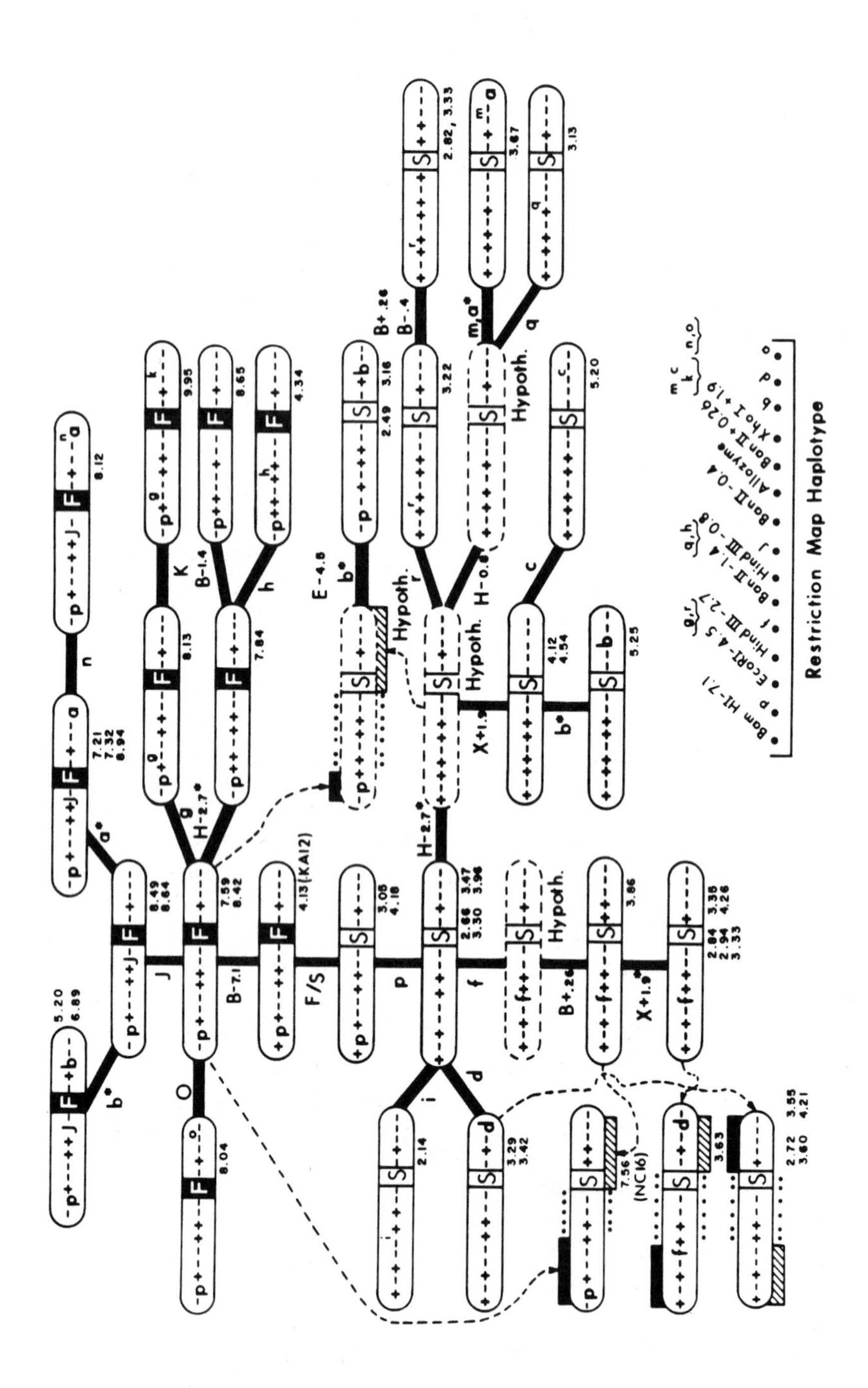
Restriction Map Haplotype

the second chromosome (38% of pairwise comparisons involving sites 48 kb apart; Aquadro et al. 1992). We have also observed significant linkage disequilibrium between variants several hundred kilobases apart in the *rosy* region on the third chromosome (CF Aquadro and WA Noon unpublished data). Less linkage disequilibrium is apparent around the *Amy* and *Est-6* regions on the second and third chromosomes, respectively (Langley et al. 1988b; Game and Oakeshott 1990), however, variation in the frequencies of variants in different regions means that the power to detect differences in linkage disequilibrium is not uniform.

In contrast, analysis of restriction map data for several regions with average rates of recombination on the X chromosome has indicated virtually no linkage disequilibrium between variants more than a few kilobases apart. The regions assayed included the *white* locus (map position 1.5; Langley and Aquadro 1987; Miyashita and Langley 1988), the 65 kb *Notch* region (map position 3.0; Schaeffer et al. 1988), 52 kb around the *per* locus (located just distal to *white* and *Notch*; Begun and Aquadro 1991; D Stern, WA Noon, E Kindahl and CF Aquadro unpublished) and an adjacent 20 kb including *zeste* and *tko* at map position 1.0 (Aguadé et al. 1989b). All of these regions show approximately average levels of recombination.

Significant linkage disequilibrium is, however, observed near both the tip and the base of the X chromosome. For example, substantial linkage disequilibrium, including some between variants more than 80 kb apart, has been reported in the *yellow-achaete-scute* region at map position 0.0 near the tip of the X (Aguadé et al. 1989a; Beech and Leigh Brown 1989;

←

Figure 6.3. Phylogenetic tree for the *Adh* region restriction map haplotypes in *D. melanogaster*. This particular tree represents one of several most parsimonious networks constructed to minimize the total number of mutational and recombination events required to relate all 29 restriction map haplotypes observed among the 49 lines. Each haplotype is composed of restriction map variants shown 5′ to 3′ (see key in lower right of figure). Further explanation of these variants can be found in Figure 6.1. Solid haplotype "balloons" are observed haplotypes, and dashed "balloons" represent hypothetical intermediate haplotypes. Haplotypes are connected by solid lines to designate mutation or dotted lines to indicate hypothetical recombination/gene conversion events. For mutation events, the variant changing along each branch of the tree is indicated, with those apparently showing parallel or convergent (multiple) changes indicated by an asterisk. Putative recombinant haplotypes have solid and hatched bars above and below the haplotype to indicate the two segments that have been joined by recombination; dotted lines indicate the approximate region where the recombination event occurred. Dashed lines and arrows indicate the possible ancestral haplotype for each segment of the recombinant haplotype. (From Aquadro et al. 1986.)

Eanes et al. 1989b; Macpherson et al. 1990). The *G6pd* region near the base of the X chromosome also appears to show substantial nonrandom association across the 13 kb region examined, although a very low frequency of variation substantially weakened statistical tests (Eanes et al. 1989a). Both of these regions have in common a significantly reduced rate of recombination per generation. Consequently, any linkage disequilibrium generated, whether by selection, drift or population admixture, will take a long time to decay. Determining the forces generating the initial linkage disequilibrium remains an important but difficult problem. However, correlations among linked nucleotides implies that they share and reflect a common evolutionary history. As we will discuss below, this correlated evolutionary history provides significant insight into the forces shaping variation in the genome.

3.5. Why Are Some Regions So Variable and Others So Monomorphic?

The neutral theory of molecular evolution holds that the vast majority of molecular variation and evolution is determined not by natural selection but by the chance drift of selectively equivalent (neutral) mutations (Kimura 1983). Regions of high and low variability are often interspersed within a single gene or genomic region and might be interpreted as simply reflecting differences in selective constraint—conserved or invariant regions being critically important for gene (or some unknown) function. While this may often be true, a recent fixation of a favored, newly-arisen mutation will also result in a local reduction (or even elimination) of adjacent sequence variability (e.g., Maynard Smith and Haigh 1974; Thomson 1977; Kaplan et al. 1989). This results from "hitchhiking" of adjacent sequences to fixation along with the favored mutation undergoing directional selection. The consequence is a reduction of variation, even for linked selectively neutral mutations, compared to that which accumulates as a balance of mutation and genetic drift. Thus, a region that is found to be invariant within a population sample from a particular organism may not necessarily be under strong constraint; it could be virtually free of constraint, but simply have contained a mutation strongly favored and recently driven to fixation in the population by directional selection.

By analogy, balancing selection, where two or more alternative nucleotides are maintained in the population by some form of selection for longer than expected under a simple neutral model, can lead to an "excess" of intraspecific variability at closely linked nucleotide sites (Sved 1968; Strobeck 1983; Kaplan et al. 1988; Hudson and Kaplan 1988). While these processes may seem clear conceptually, in practice it continues to be extremely difficult to discriminate between the different forms of selection (directional selection, balancing selection and what is referred to as purifying selection, reflecting selective constraint and the purging of the less fit mutations that periodically arise). The ability to now obtain reasonably large samples of DNA sequences from within populations and from closely related species, ..owever, has opened up several elegant approach towards addressing this classic problem in populations genetics.

3.6. Tests for Departure from Selective Neutrality

3.6.1. The HKA Test

One of the most powerful conceptual frameworks to arise from population genetics in the last several years is the comparison of within-species DNA sequence variation to between-species DNA sequence differentiation. Kreitman and Aguadé (1986b) noted a strong disparity between the pattern of nucleotide polymorphism across the *Adh* gene of *D. melanogaster* compared to sequence divergence between *Adh* in *D. melanogaster* and a single *Adh* sequence of *D. simulans*. They pointed out that under the hypothesis of selective neutrality, interspecific divergence is simply and positively correlated with intraspecific polymorphism. Highly polymorphic regions are expected to be rapidly evolving due to a higher mutation rate to selectively neutral alternatives (Kimura 1983). Hudson et al. (1987) formalized this concept into a test of departure from the expectation of selective neutrality, given intraspecific and interspecific estimates of DNA sequence variation (referred to hereafter as the HKA test). This approach and extensions of it will clearly revolutionize molecular population genetics.

The basis of the HKA test is as follows. Under the infinite-site, neutral model of molecular evolution, the expected number of segregating sites in a sample of DNA sequences randomly chosen from a single natural population is

$$E(S_n) = 4N\mu n_s\left[1 + \frac{1}{2} + \frac{1}{3} + \cdots + \frac{1}{n-1}\right]$$

The expected number of nucleotide differences (D) between two alleles randomly chosen from two species is

$$E(D) = 4N\mu n_s\left(T + \frac{(1+f)}{2}\right)$$

Where $T = t/2N$, with t being the time in generations since the two species shared a common ancestor, and f is the factor by which the effective population sizes differ between the two species. Variation within and between species both are direct functions of the mutation rate, μ. Genes or regions with high mutation rates or low functional constraint will be both highly variable within species and highly divergent between species. Estimates of 4Nμ and T are obtained that best fit the observed values of S_n and D. The expected variances for S_n and D are obtainable and allow a Chi Square goodness-of-fit test to neutrality (for details, see Hudson et al. 1987; Hudson 1990; Kreitman 1991).

To date, the only DNA sequence data appropriate for this test have been for the *Adh* gene of *Drosophila*, and here it was concluded that at least two sites were under some form of balancing selection (see reviews in Kreitman 1987; 1991; Hudson 1990). However, comparisons of restriction site data within and between *D. melanogaster* and *D. simulans* suggest that *Adh* is not unique, and that evidence for natural selection may be found in many gene regions.

Our evidence for this includes analyses of six-cutter restriction map variability within and between *D. melanogaster* and *D. simulans* across a 100 kb segment on the second chromosome containing the *rosy*, *l(3)S12*, *snake* and *Heat shock cognate-2* (*Hsc2*) genes (Aquadro et al. 1988; CF Aquadro, V Bansal and WA Noon unpublished). There is striking variation across these regions of the genome in the levels of nucleotide heterozygosity in each species. For example, heterozygosity varies as much as 12-fold across the *rosy* region (Figure 6.4; Aquadro et al. 1988). The "density," or abundance, of transcriptional units across regions does not appear to be strongly associated with either divergence or polymorphism, suggesting that the patterns are not simply due to a comparison of coding versus non-coding sequences. In fact, the most variable regions sometimes correspond with known transcripts.

Divergence between *D. simulans* and *D. melanogaster* averages 4% across the regions surveyed, consistent with previous direct sequence information (Cohn et al. 1984; Bodmer and Ashburner 1984) and DNA-DNA hybridization between single copy DNA (Zwiebel et al. 1982). Divergence is not uniformly distributed, however, varying as much as 10-fold between adjacent regions. No sites are polymorphic in both species,

Figure 6.4. Summary of intra- and interspecific DNA sequence variation estimated for the *rosy* locus region in *D. melanogaster* and *D. simulans*. Nucleotide heterozygosity and divergence are graphed for a 10 kb "window" slid along the 40 kb region 5 kb at a time: (**a**) heterozygosity per nucleotide (π) among 60 lines of *D. melanogaster*; (**b**) heterozygosity per nucleotide among 30 lines of *D. simulans*; (**c**) average nucleotide divergence (Nei and Li 1979) between *D. melanogaster* and *D. simulans* (uncorrected and corrected for intraspecific variation indicated by dark and light gray, respectively); (**d**) genes and transcripts detectable from the 40 kb region. Specific transcripts for the indicated segments are also shown. Error bars indicate the 95% confidence interval on the estimates. (From Aquadro et al. 1988.)

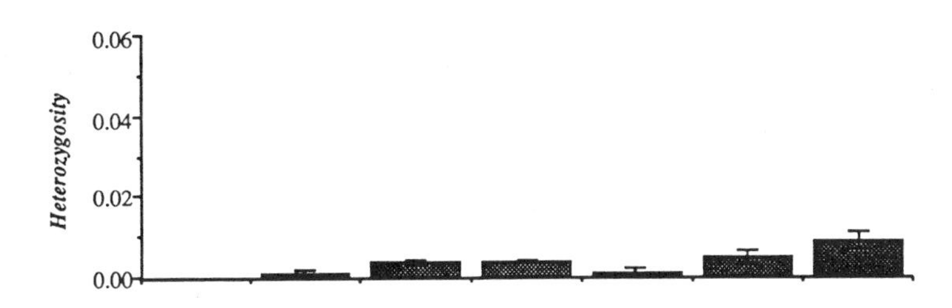
Heterozygosity
0.06
0.04
0.02
0.00

b

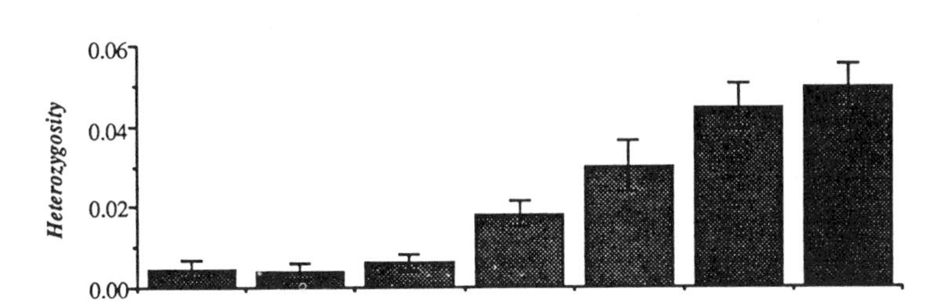
Heterozygosity
0.06
0.04
0.02
0.00

c

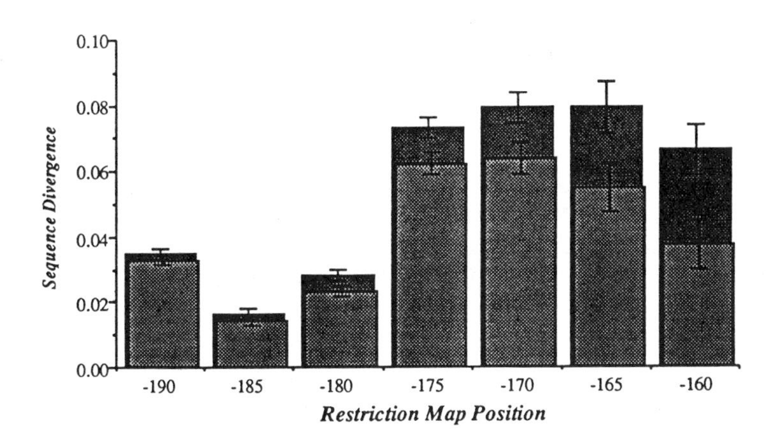
Sequence Divergence
0.10
0.08
0.06
0.04
0.02
0.00
-190
-185
-180
-175
-170
-165
-160
Restriction Map Position

d

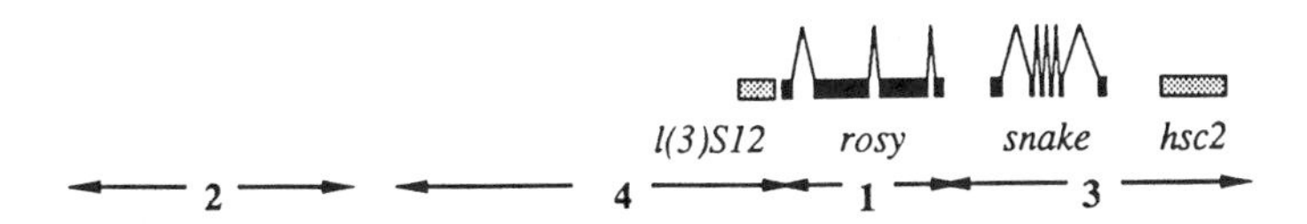
l(3)S12
rosy
snake
hsc2
2
4
1
3

indicating the complete fixation or loss of variants polymorphic in their common ancestral species. (The same result was obtained for the *Adh* locus when comparing *D. melanogaster* to another sibling species, *D. sechellia*; Coyne and Kreitman 1986.)

Application of the HKA test to these data suggests departures from the expectation based on neutrality that variability within the two species will be proportional to divergence between them. These departures can be visualized from the ratio of heterozygosity to divergence plotted for a "sliding window" along the sequence as in Figure 6.5. These plots visualize the concept of the HKA test. Strict neutrality predicts a constant ratio across regions, since variation in constraint is equivalent to varying the mutation rate and will affect heterozygosity and divergence proportionally. This can be shown as follows. Following Hudson et al. (1987), heterozygosity under neutrality is equal to $4N\mu$ while divergence is equal to $2\mu t + 4N\mu$ for autosomal genes in diploid species, where t is the time since divergence of the two species. Thus, in the ratio $4N\mu/(2\mu t + 4N\mu)$, μ cancels out, and the ratio is a function only of time since divergence and population size, both of which are identical across a region for a given comparison under neutrality. Peaks and valleys in Figure 6.5 thus represent regions of "too much" and "too little" intraspecific polymorphism, consistent with the effects of balancing and recent directional selection, respectively. Selective constraint has been factored out and cannot account for the differences observed. The development of suitable statistical tests remain a challenge for this approach, but several very interesting regions are apparent as candidates for four-cutter analysis and direct sequencing.

As a further example of the use of the HKA test, Begun and Aquadro (1991) have found that, despite the tendency towards reduced variation in the *yellow-achaete* gene region within the populations examined of both *D. melanogaster* and *D. simulans*, sequence divergence is not reduced between *D. melanogaster* and *D. simulans*. This falsifies two hypotheses to explain reduced variation in this region of reduced recombination, these being reduced mutation rate and strong functional constraint (Langley 1990). The virtual lack of intraspecific variation detected in the region in *D. simulans*, which typically has several fold more variation than *D. melanogaster* (detailed below), further supports the generality of a trend towards reduced variation in regions of reduced recombination, possibly due to the widespread effects of hitchhiking in these regions associated with directional selection. Such a process would be expected to lead to high levels of linkage disequilibrium across these regions, precisely as is observed (Aguadé et al. 1989a; Beech and Leigh Brown 1989; Eanes et al. 1989b; Macpherson et al. 1990; Begun and Aquadro 1991). Similar results have been reported for the *cubitus interruptus Dominant* region on the nonrecombining fourth chromosome of *D.*

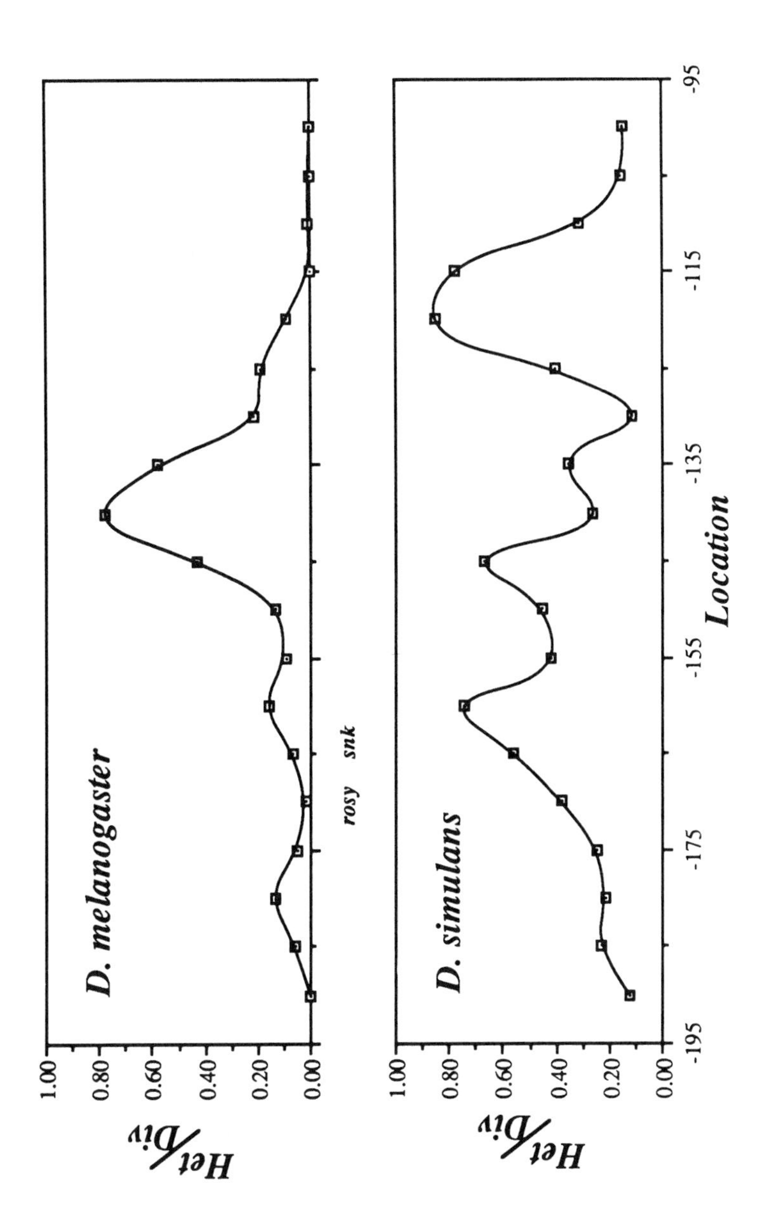

Figure 6.5. Ratio of variation within species to divergence between *D. melanogaster* and *D. simulans* for 100 kb around the *rosy* gene. Plotted are the ratios for a 10 kb "window" slid along the 100 kb region, 5 kb at a time. Data from Aquadro et al. (1988) and CF Aquadro, V Bansal and WA Noon (unpublished).

melanogaster and *D. simulans* (Berry et al. 1991). The extent to which such events affect levels of variation across that genome remains to be demonstrated. However, occasional instances of directional selection have the potential to impose an upper limit on the level of sequence polymorphism that can accumulate in a region, even if the species' effective population size is very large (Maynard-Smith and Haigh 1974; Kaplan et al. 1989). Begun and Aquadro (1992) have in fact shown that levels of nucleotide polymorphism correlate with regional rates of recombination for 20 gene regions across the *D. melanogaster* genome. They also show that the correlation cannot be due to differences in mutation rate or functional constraint, but appears to result from genetic hitchhiking associated with the fixation of advantageous mutants. Aquadro and Begun (1992) estimate that as few as one out of every 4,000 fixations need be selectively driven to give the pattern observed. Such a rate is generally consistent with estimates of selective fixations at the *Adh* locus between *Drosophila* species (McDonald and Kreitman 1991).

3.6.2. Tajima's Test—Frequency Distributions of DNA Polymorphisms

It is important to note that the HKA test only detects departures from selective neutrality. It remains problematic why the departure has occurred (for example, is there a deficiency of polymorphism suggesting a recent directional selection event, or too much polymorphism due perhaps to balancing selection). One approach, which Hudson et al. (1987) applied to a departure in the third exon of *Adh*, was to reason that the departure was due to an excess of intraspecific polymorphism because the surrounding regions had levels of polymorphisms similar to the average for most other regions surveyed.

An alternative, formulated into a test of selective neutrality by Tajima (1989a), is to compare the estimate of θ calculated from the number of segregating sites with that estimated from the number of pairwise nucleotide differences among a sample of sequences. Under the hypothesis of selective neutrality, and in a population at equilibrium for mutation and drift, these two estimates should be identical. Balancing selection leads to an apparent excess of intermediate frequency variants and thus $\pi > \theta$. The recent fixation of a favorable mutation (directional selection) causes the elimination of intermediate frequency variants expected at equilibrium in the region surrounding the favored mutation, and thus leads to an apparent excess of rare variants in the region. This leads π to be less than estimates of θ derived from the number of segregating sites. This is precisely the pattern seen in regions of reduced recombination where recent directional selection events have been hypothesized to have swept away significant portions of the standing variation (e.g., *G6pd* and *yellow-achaete-scute* in *D. melanogaster*, *yellow-achaete* in *D. simulans*, *vermilion* in *D. ananassae*). A similar effect is apparent in the *Adh-Dup* region, adjacent to *Adh* in *D. melanogaster* (Kreitman and Hudson 1991).

In summary, *Adh* in *D. melanogaster* does not appear to represent a unique instance of the action of strong natural selection. Sequencing of a selection of 13 alleles representing 10 electrophoretically different allozymes of *Est-6* (Cooke and Oakeshott 1989) demonstrates that, like *Adh*, strong purifying selection associated with functional constraint is acting—synonymous (nonamino-acid-changing) polymorphisms significantly outnumber nonsynonymous (amino-acid-changing) polymorphisms. Unlike ADH, EST-6 shows extensive amino acid polymorphism within and between allozymes. Like at *Adh*, latitudinal clines in allozyme frequency are seen at *Est-6* on a worldwide scale, and thus it is likely that at least some of the extensive amino acid polymorphism revealed by sequencing is the target of natural selection (Cooke and Oakeshott 1989). The application of formal tests of neutrality such as the HKA or Tajima test, however, must await the sequencing of a random or representative sample of alleles from what can be considered a single population.

Six-cutter analyses would be unlikely to pick up narrow zones of excess polymorphism such as that detected at *Adh* by a combination of direct DNA sequencing and detailed four-cutter restriction site analysis (Kreitman and Aguadé 1986b). The small number of sites scored per region in six-cutter surveys substantially weakens the formal application of the HKA and Tajima tests. The use of a large battery of four-base restriction endonucleases and direct DNA sequence data are needed to test whether the contrasting patterns of variation and differentiation observed in the six-cutter surveys depart significantly from neutrality expectations.

Despite the few applications of the HKA and Tajima tests to nucleotide variation, it already appears likely from allozyme data that much of the genome has been influenced by periods of balancing or directional selection. Mean heterozygosity at allozyme loci within *D. melanogaster* and *D. simulans* is also reported not to be positively correlated with Nei's genetic distances estimated from allozyme data between these two species suggesting balancing or other forms of selection acting on at least some proteins (Singh 1989). The number of nucleotide variants directly under the influence of natural selection could be proportionately quite small, though their dynamics influence a large number of adjacent neutral sites through linkage and hitchhiking effects.

4. Contrasting Patterns of Variation Among *Drosophila* Species

Interest in the generality of the results from *D. melanogaster* has led several groups to analyze nuclear DNA restriction map variation in related species of *Drosophila*. The results of these studies indicate that there are significant differences in the level, distribution and source of DNA sequence variation between these species. I will review only data for which a comparison to the homologous region in *D. melanogaster* is available and will focus mainly on the comparison of the sibling species *D. melanogaster* and *D. simulans*.

4.1. Species Differences in Large Insertions

Data for large insertions (>250 bp) for several different gene regions surveyed in *D. melanogaster* and *D. simulans*, *D. ananassae*, or *D. pseudoobscura* are presented in Table 6.2. In *D. melanogaster*, insertions have been observed at densities of 0.002 to 0.018 per chromosome examined per kilobase surveyed, with an average density of 0.004. In contrast, only three large sequence length variants were detected (each only once) over the 165 kb summarized here among 30 lines of *D. simulans*, leading to an average density of only 0.0005 per line per kilobase (Table 6.2). This density is eight fold lower than that for the homologous regions in *D. melanogaster*. These results support the hypothesis of fewer copies per genome of dispersed repetitive, nomadic DNA in *D. simulans* compared to *D. melanogaster* (Dowsett and Young 1982). Large insertion variation likely to be due to transposable elements is also rare (in fact undetected at this point) in *D. pseudoobscura* (Schaeffer et al. 1987; Riley et al. 1989; Aquadro et al. 1991).

The pattern for *D. ananassae* is not clear. A high frequency of one 900 bp insertion at *vermilion* leads to a very high density and frequency of insertions. However, the density and frequency of large insertions at *forked* is approximately comparable to the average for *D. melanogaster*. The examination of additional regions should help provide an accurate picture for *D. ananassae*.

These results are of particular interest given that many of the phenotypic mutations of *D. melanogaster* have been shown due to the insertion of transposable elements in or near structural genes (see Finnegan and Fawcett 1986; Green 1988). Transposable elements clearly can cause mutations in non-*melanogaster* species. For example, spontaneous phenotypic mutants at both the *Notch* and *white* loci in *D. simulans* (Kidd and Young 1986; Inoue and Yamamoto 1987) and at the *Om(1D)* locus in *D. ananassae* (Tanda et al. 1989) have been found to be due to the insertion

Table 6.2. Large insertion variation in homologous regions of *Drosophila* species*

Gene Region/Species	Examined		Total	No.	Density	Freq.	Ref.
	No. Chrom.	kb	No. Ins.	Diff. Ins.			
period							
D. melanogaster	66	52	8	8	0.002	0.015	1
D. simulans	38	52	0	0	0	0	1
alcohol dehydrogenase							
D. melanogaster	48	13	11	8	0.018	0.029	2
D. simulans	38	13	0	0	0	0	3
D. pseudoobscura	19	32	0	0	0	0	4
amylase†							
D. melanogaster	85	15	11	10	0.009	0.012	5
D. pseudoobscura	28	26	0	0	0	0	6
rosy (Xdh)							
D. melanogaster	60	100	16	12	0.003	0.030	7
D. simulans	30	100	3	3	0.001	0.033	7
D. pseudoobscura	58	5	0	0	0	0	8
forked							
D. melanogaster	64	25	na	na	0.002	0.020	9
D. ananassae	39	14	2	2	0.004	0.026	10
vermilion							
D. melanogaster	64	24	na	na	0.002	0.020	9
D. ananassae	39	18	18	10	0.026††	0.046††	10
AVERAGES		Ave. Insertion Density				Ave. Insertion Freq.	
D. melanogaster (229 kb)		0.004				0.023	
D. simulans (165 kb)		0.0005				0.033	
D. ananassae (32 kb)		0.016††				0.037††	
D. pseudoobscura (63 kb)		0				0	

* References are: 1) D Stern, WA Noon, E Kindahl and CF Aquadro unpublished; 2) Aquadro et al. 1986; 3) CF Aquadro, K Sykes and P Nelson unpublished; 4) Schaeffer et al. 1987; 5) Langley et al. 1988b; 6) Aquadro et al. 1991; 7) Aquadro et al. 1988 and CF Aquadro, K Lado, V Bansal and WA Noon unpublished; 8) Riley et al. 1989; 9) Langley 1990; 10) Stephan and Langley 1989.

† Does not include variation due to differences in the number of amylase structural genes in *D. pseudoobscura*.

†† *vermilion* in *D. ananassae* is high due to one 0.9 kb insertion being present in high frequency. Note also for *D. ananassae* that only the two Asian populations are included since Brazil has only recently been colonized (Stephan and Langley 1989).

of transposable elements. The results summarized here do indicate that transposable elements are a much less prevalent source of sequence variation in natural populations of *D. simulans* and *D. pseudoobscura* than *D. melanogaster*.

Table 6.3. Nucleotide heterozygosity* in species of *Drosophila*†

	D. melanogaster	*D. simulans*	*D. ananassae*	*D. pseudoobscura*
X chromosome				
per	0.001	0.007	—	—
forked	0.004	—	0.010	—
Om(lD)	—	—	0.009	—
vermilion	0.006	—	0.003††	—
Autosome				
Adh	0.006	0.015	—	0.026
Amy	0.008	—	—	0.019
rosy	0.005	0.018	—	0.013
Average (normal recomb. regions)	0.004	0.014	0.010	0.022

* Estimates are π if available.
† References are in Tables 6.1 and 6.2 with the exception of *Om(1D)*, which is from Stephan (1990).
†† denotes region with reduced rate of recombination per generation, not included in average for the species.

4.2. Levels of Nucleotide Heterozygosity

Our analyses of restriction site variation in the *rosy*, *Adh* and *per* regions of *D. simulans* lead to an average estimate of heterozygosity per nucleotide of 0.014, three and a half times higher than what has been observed in *D. melanogaster* (Table 6.3). A recent study of the *y-ac* and *Pgd* gene regions in both *D. melanogaster* and *D. simulans* do not show this pattern, but appear to reflect differences in the action of natural selection between the two species (Begun and Aquadro 1991). Heterozygosity in the *Adh* and *Amy* gene regions of *D. pseudoobscura* averages 0.023, six times higher than *D. melanogaster*. Results of detailed four-cutter surveys of the *rosy* region in both *D. melanogaster* (WA Noon and CF Aquadro unpublished) and *D. pseudoobscura* (Riley et al. 1989) indicate that the six-cutter estimates accurately reflect differences in sequence variation. Estimates for *D. ananassae* from Southeast Asia and Brazil show differences among populations; on average the Asian populations are intermediate between *D. melanogaster* and *D. simulans* in terms of variability (Stephan and Langley 1989; Stephan 1989).

Evidence that *D. melanogaster* is perhaps among the least variable *Drosophila* species also comes from DNA-DNA hybridization studies of single-copy, nuclear DNA in *D. mercatorum*, where approximately 1.3% base-pair mismatches were found among three homozygous strains (Caccone et al. 1987). This is roughly equivalent to the nucleotide diversity observed by the restriction site surveys in *D. simulans* and *D. pseudoobscura*.

Thus the average level of nucleotide heterozygosity in regions of normal recombination is three- to sixfold higher in *D. simulans* and *D. pseudoobscura* than in *D. melanogaster*. This is in contrast to the eight-fold higher density of large insertion variation in *D. melanogaster* compared to *D. simulans* and *D. pseudoobscura*. The variances on estimates of sequence variation are quite large, with the largest component due to the variance in the evolutionary process. It is thus significant that the contrasts in level of variation are consistent among regions on three and two chromosomes of *D. simulans* and *D. pseudoobscura*, respectively. Hence, while we have admittedly little data relative to the total size of the genome, the patterns suggest that there exist genome-wide differences in the average level and nature of DNA sequence variation among these species of *Drosophila*.

4.3. Hypotheses to Account for Species Differences

Several hypotheses can be proposed to account for the differences in levels of nucleotide heterozygosity and large insertion (transposable element) variation among *Drosophila* species. It is unclear whether the contrasting levels of nucleotide and large insertion variation have a common underlying cause. Below, I will focus on the contrasts in levels of nucleotide heterozygosity, and in particular, examine hypotheses to account for the contrasting levels of nuclear DNA and allozyme variation in *D. melanogaster* and *D. simulans*.

A strictly neutral, infinite sites model predicts the average heterozygosity at equilibrium to be approximately equal to $4N\mu$. Thus, under the simplest neutral model, difference in the mutation rate and/or the effective population size among the species could account for the different levels of nucleotide heterozygosity. The highest mutation rate and/or effective population size would be found in *D. pseudoobscura*, followed in decreasing order by *D. simulans*, *D. ananassae* and *D. melanogaster*.

A simple difference in species effective population size has been considered as a likely explanation for higher levels of allozyme variation observed in *D. pseudoobscura* compared to *D. melanogaster* (Choudhary and Singh, 1987a) and is consistent with the higher nucleotide variation observed at *Adh* and *Xdh* in *D. pseudoobscura* (Schaeffer et al. 1987; Riley et al. 1989). A reduction in population size associated with a historically recent founding of South America has been invoked to explain the reduced nucleotide variation in *D. ananassae* from Brazil compared to those sampled from southeast Asia (Stephan and Langley 1989; Stephan 1989).

This simple hypothesis is difficult to reconcile, however, with the substantial body of data (reviewed in Choudhary and Singh 1987b) on protein variation, that indicates that *D. simulans* is significantly *less* polymorphic and geographically differentiated for allozymes than *D. melanogaster*. Allozyme polymorphism at both *Xdh* and *Adh* fit this pattern, with fewer alleles observed in *D. simulans* compared to *D. melanogaster* (reviewed in Choudhary and Singh 1987b and Aquadro et al. 1988). In fact, prior to the DNA data it was proposed that the reduced allozyme variation and differentiation in *D. simulans* was a result of a recent reduction in population size and/or a recent worldwide expansion of the species (Choudhary and Singh 1987b; Singh et al. 1987). Clearly a simple neutral model of different effective population sizes cannot explain both the DNA and protein data.

A hypothesis of differences in mutation rate faces similar difficulties since a positive, linear relationship would be expected, but is not observed, between nucleotide heterozygosity and protein polymorphism across the species. Data on empirically determined mutation rates are very limited, although no difference in lethal mutation rates was detected between *D. simulans* and *D. melanogaster* (Eeken et al. 1987). Additional studies are needed, but await the construction of chromosome balancers in *D. simulans*.

Thus the contrast between *D. melanogaster* and *D. simulans* is difficult to reconcile with a strictly neutral model of molecular evolution. Several forms of selection could, however, predict the observed patterns. Singh and his colleagues (Choudhary and Singh 1987b; Singh 1989) have favored a strong role for natural selection in proposing that the two species have different "strategies of adaptation." This hypothesis poses various kinds and magnitudes of balancing and directional selection to account for the observed patterns and levels of allozyme variation. The application of the HKA and Tajima tests to DNA variation at several of the allozyme loci should allow this hypothesis to be directly tested. There are presently insufficient sequence data to make such a comparison.

An alternative, proposed by Aquadro et al. (1988), hypothesizes that differences in population size, coupled with mild selection against many of the segregating but low frequency allozyme variants, could lead to the observed patterns. This model assumes that a significant portion of the allozyme alleles are not strictly neutral but slightly deleterious (Ohta 1976; Ohta and Tachida 1990), whereas the DNA site polymorphisms detected by the restriction map surveys are largely neutral or nearly neutral. In fact, all that is required is that the DNA polymorphisms scored are simply less subject to selection, a reasonable assumption compared to amino acid differences that change protein charge and/or confirmation and are thus detectable by protein electrophoresis. Under this model, even a dramatic increase in effective population size could lead to very little increase in allozyme heterozygosity (Ohta 1976; Kimura 1983), while restriction site polymorphism could show a sizable increase. This contrast results from the increased efficiency of selection against slightly deleterious allozyme variants in a larger population where the magnitude of genetic drift is decreased.

Ohta and Tachida (1990) have recently examined in detail the slightly deleterious mutation model in light of the data from *D. melanogaster* and *D. simulans* and have shown that a reasonable fit to the data can be obtained. Under this scenario, slightly deleterious electrophoretic variants are more likely to be eliminated or dramatically reduced in frequency in *D. simulans* because of its larger population size and reduced genetic drift. Consistent with this proposition, the twofold higher proportion of allozyme loci polymorphic in *D. melanogaster* relative to *D. simulans* is in fact due largely to the presence of a larger number of low frequency alleles in *D. melanogaster* compared to *D. simulans* (Choudhary and Singh 1987b).

The lower copy number of transposable elements may also be related to a difference in effective population size and in the associated efficacy of natural selection, since transposable elements appear to behave as though slightly deleterious on average in natural populations of *D. melanogaster* (Golding et al. 1986). However, Charlesworth and Lapid (1989) suggest that the differences in population size between these two species would be insufficient to cause such an effect, based on current models of transposable element population biology. Whether the contrast in transposable element density and copy number between closely related species such as *D. melanogaster* and *D. simulans* has a molecular, genetic or population biological basis is clearly an important area for future study.

Clearly population size is a key parameter in several of the above models to account for the different patterns of nucleotide versus allozyme polymorphism in *D. melanogaster* and *D. simulans*. Studies of the geographic distribution of allozyme frequencies have led to estimates of Nm (the product of effective population size, N, and migration rate, m) that are two to four times higher for *D. simulans* than *D. melanogaster* (Choudhary and Singh 1987b). These authors and Singh and Rhomberg (1987) concluded that the differences lay in migration rate (m), although they conceded one would have thought *D. melanogaster* to have a higher rate of movement due to its close association with the highly mobile human species. On the other hand, Aquadro et al. (1988) noted that a larger N of the direction and magnitude proposed from the DNA data for *D. simulans* would be consistent with these estimates of Nm. Moreover, if the differences in variation are due to effective population size, we might expect more mitochondrial DNA variation in *D. simulans* than *D. melanogaster*. While there is often less variation within any one population of *D. simulans* than is seen for *D. melanogaster*, there are, however, several significantly more divergent lineages within *D. simulans* than within *D. melanogaster* (e.g., Baba-Aissa et al. 1988; Satta and Takahata

1990). This pattern is not inconsistent with the hypothesis of a larger long-term species effective population size for *D. simulans* compared to *D. melanogaster*.

The hypothesis of Aquadro et al. (1988) that *D. simulans* has a larger effective population size and that many low frequency allozyme variants are slightly deleterious thus appears to be consistent with much of the available data. Nevertheless, support for its validity awaits comparisons of levels of synonymous and nonsynonymous polymorphism in random samples of alleles from several populations of both species. In addition, the proposition of slightly deleterious allozyme variants has its detractors (e.g., Gillespie 1987), and we must consider how other models, particularly spatially and temporally varying selection, might fit the observed data. Comparison of DNA sequence variation, within and between species, and the comparison of synonymous and nonsynonymous sequence polymorphism (e.g., Hughes and Nei 1988) offer hope of testing some of these hypotheses. Consideration must also be given to the predictions of nonequilibrium population genetic models (e.g., Tajima 1989b).

The sharp contrasts in nucleotide heterozygosity and transposable element variation between the sibling species *D. simulans* and *D. melanogaster* provide a particularly good opportunity to examine the roles of various evolutionary forces acting to create, maintain and shape the patterns and levels of genotypic and phenotypic variation in *Drosophila*. The results obtained to date clearly cast doubt on the generality of an extensive contribution by transposable elements to naturally occurring sequence variation and caution against attempts to generalize levels of variation between species.

4.4. Linkage Disequilibrium in *D. simulans* and *D. pseudoobscura*

In contrast to the significant linkage disequilibrium observed for several regions on the second and third chromosomes of *D. melanogaster*, virtually no significant linkage disequilibrium is observed between sites more than 10–20 kb apart on the homologous autosomes of *D. simulans* (e.g., 13 kb *Adh* region; CF Aquadro, K Sykes and P Nelson unpublished; 100 kb *rosy* region; CF Aquadro, K Lado, V Bansal and WA Noon unpublished). Neither species showed appreciable linkage disequilibrium over more than a few kilobases in the *per* locus region on the X chromosome (e.g., the *per* locus region of the X chromosome; Begun and Aquadro 1991; D Stern, WA Noon, E Kindahl and CF Aquadro unpublished).

Extensive linkage disequilibrium exists among restriction map variants within the 35 kb amylase gene cluster on the third chromosome of *D. pseudoobscura* (Aquadro et al. 1991). In contrast, virtually all variants in the 32 kb *Adh* region on the fourth chromosome of *D. pseudoobscura* appear to be in linkage equilibrium (Schaeffer et al. 1987), as was the case for the 5.2 kb *rosy* region on the second chromosome examined by Riley et al. (1989).

An examination of chromosomes and species with and without polymorphism inversions indicates a positive correlation between the presence of polymorphic chromosome inversions and the extent of linkage disequilibrium. Only the second and third chromosomes of *D. melanogaster* and the third of *D. pseudoobscura* show extensive polymorphism for paracentric inversions. The X of *D. melanogaster* and fourth of *D. pseudoobscura* are free of inversions, as are all of the chromosomes in *D. simulans* populations. Additional data are obviously needed from other sites on these chromosomes in these species.

Several instances of linkage disequilibrium have been observed between allozyme alleles and polymorphic chromosomal inversions in a number of species of *Drosophila*. More recently, numerous instances of associations between inversions and DNA sequence or restriction map variants or haplotypes have been found (e.g., Aquadro et al. 1986; Langley et al. 1988b; Aguadé 1988; Aquadro et al. 1991). Many of the associations between inversions and allozyme variants have been interpreted as reflecting selection for favored combinations of alleles (reviewed recently by Sperlich and Pfriem 1986). However, Nei and Li (1975; 1980) and Ishii and Charlesworth (1977) have argued that it was not possible with data then at hand to rule out a recent origin of the inversions. This implies that the linkage disequilibrium observed was simply due to chance and a lack of sufficient time for associations generated by mutation to have decayed by recombination and gene conversion. Information on the age of inversions could help resolve this issue. The ability to obtain empirical estimates of genetic exchange between inversion heterozygotes and to obtain DNA sequence and restriction map data from regions within different chromosome arrangements may give us our first real opportunity to test these ideas. Such data would allow us to estimate the relative ages of different inversions and provide clues as to the evolutionary forces acting on them in natural populations (Aquadro et al. 1986; Aguadé 1988; Rozas and Aguadé 1990; Aquadro et al. 1991). In the case of the third chromosome inversions polymorphic in *D. pseudoobscura*, comparison of the *Amy* region located in the middle of the inverted region suggests an age of the inversions too great for drift to be consistent with the extent of linkage disequilibrium observed (Aquadro et al. 1991).

Clearly, an understanding of these fascinating yet complex patterns of linkage disequilibrium will require not only extensive scoring of linkage disequilibrium among genetically mapped genes in large samples of chromosomes but also fine structure genetic analyses of the regions under study in both inversion and standard arrangement chromosomes.

5. Prospects

One puzzle which has emerged from the past quarter century of population genetics research has been the narrow range of genic heterozygosities observed with protein electrophoresis for an incredibly diverse array of organisms. These different organisms certainly appear to vary tremendously in their abundance and hence population sizes, differences that should manifest themselves as similarly dramatic differences in genic heterozygosity if much of the variation is selectively neutral. It has been argued (e.g., Ohta 1976; Kimura 1983) that the disparity between expectation and observation is due to the nature of the distribution of selective effects of new mutations (most being slightly deleterious, and the effect being dependent upon effective population size). An alternative to this model is that virtually all species undergo periodic reductions in population size (population bottlenecks) which lead to the loss of genetic variability (e.g., Nei and Graur 1984). Others view these data as damning to any neutral model and argue for natural selection to account for the genic heterozygosities observed (e.g., Gillespie 1987). This controversy remains unresolved, but is of central importance to our understanding of molecular evolution and the significance of molecular variation. The contrasts seen between *D. melanogaster* and *D. simulans* open the way to the examination of several competing hypotheses concerning the distribution of selective effects of new mutants and standing variation. The positive correlation between nucleotide heterozygosity and regional rates of recombination for 20 gene regions from across the genome of *D. melanogaster* indicate hitchhiking must play some role in limiting variability in a significant portion of the genome (Begun and Aquadro 1992; Aquadro and Begun 1992).

The power of any test of neutrality is dependent on the level of variation detected. The finding of three to six times more variation in *D. simulans* and *D. pseudoobscura* than in homologous regions of *D. melanogaster* means that tests of the neutrality of the variation in these species will have more statistical power. The close relatedness and similar genome structure of *D. simulans* and *D. melanogaster*, in which most of the molecular biology has been done, makes *D. simulans* perhaps the species of choice for further studies of the extent of natural selection across the genome. Interest in a particular gene or situation in a particular species, of course, will lead to studies in other species.

Population genetic theory predicts that when the product of the rate of recombination between sites and the effective population size is very small, then significant linkage disequilibrium is expected by chance due to genetic drift of chromosome segments. Certainly much of the linkage disequilibrium observed between DNA variants several base pairs to a few kilobases apart is largely due to genetic drift. However, the explanation for linkage disequilibrium over the much larger distances we have sometimes observed remains unclear. Could it be due simply to variation in rates of recombination in different parts of the genome? Could inversion polymorphism and the suppression of recombination in inversion heterozygotes be sufficient to account for the patterns? To what extent have selection and genetic hitchhiking contributed to the nonrandom associations observed?

The data gathered to date have generally been from scattered, relatively small segments of the genome, and it has been very difficult interpreting the different patterns and levels of linkage disequilibrium observed. Analyses of large numbers of lines over large regions for which many intermediate frequency sites can be scored (e.g., with four-cutter surveys, to maximize the statistical power of tests for linkage disequilibrium) and for which empirically determined rates of recombination are available should allow us to answer some of these and related questions.

By combining a population genetic approach to molecular biology and a molecular approach to population genetics, we will continue to advance our understanding both of the structure/function relationships in the genome and of the processes by which the genome evolves. One key goal is to gain a clearer understanding of the reasons for variation in levels of DNA sequence polymorphism and linkage disequilibrium in different gene and genome regions. We are now equipped, both technically and analytically, to assess the relative contributions of balancing, purifying and directional selection and of stochastic processes to the often remarkable variation in sequence polymorphism observed in different gene regions and between adjacent genes. By coupling estimates of linkage disequilibrium with knowledge of recombination rates, inversion polymorphism and the ages of inversions (derived from the levels and patterns of DNA sequence variation in genes contained within the inversion breakpoints), we will also be able to assess the roles of various evolutionary forces acting to create, maintain, and/or slow the decay of the linkage disequilibrium observed.

Acknowledgments. I thank the members of my laboratory, particularly David Begun, Eric Kindahl, Bill Noon, and Carol Yoon, for insightful discussion and constructive criticism of this chapter, and for their important contributions to the science presented herein. I also appreciate the criticism on this manuscript provided by Tim Prout, Thom Boyce, Monty Slatkin and the editors. The work in my laboratory on *Drosophila* molecular population genetics has been supported by N.I.H. grant number GM 36431.

References

Aguadé M (1988) Restriction map variation of the *Adh* locus of *Drosophila melanogaster* in inverted and noninverted chromosomes. Genetics 119: 135–140

Aguadé M, Miyashita N, Langley CH (1989a) Reduced variation in the *yellow-achaete-scute* region in natural populations of *Drosophila melanogaster*. Genetics 122:607–615

Aguadé M, Miyashita N, Langley CH (1989b) Restriction-map variation at the *Zeste-tko* region in natural populations of *Drosophila melanogaster*. Mol Biol Evol 6:123–130

Ajioka J, Eanes WF (1989) The accumulation of P-elements on the tip of the X chromosome in populations of *Drosophila melanogaster*. Genet Res 53:1–6

Aquadro CF, Begun DJ (1992) Evidence for and implications of genetic hitchhiking in the *Drosophila* genome. In Takahata N, Clark A (eds) Molecular Paleo-population Biology. Springer-Verlag, Berlin (in press)

Aquadro CF, Deese SF, Bland MM, Langley CH, Laurie-Ahlberg CC (1986) Molecular population genetics of the alcohol dehydrogenase gene region of *Drosophila melanogaster*. Genetics 114:1165–1190

Aquadro CF, Jennings RM, Bland MM, Laurie CC, Langley CH (1992) Patterns of naturally occurring restriction map variation, DDC activity variation and linkage disequilibrium in the dopa decarboxylase gene region of *Drosophila melanogaster*. Genetics (in press)

Aquadro CF, Lado KM, Noon WA (1988) The *rosy* region of *Drosophila melanogaster* and *D. simulans*. I. Contrasting levels of naturally occurring DNA restriction map variation and divergence. Genetics 119:875–888

Aquadro CF, Noon WA, Begun DJ (1992) RFLP analysis using heterologous probes. In Hoelzel AR (ed) Molecular Genetic Analysis of Populations—A Practical Approach. IRL Press, Oxford, pp 115–157

Aquadro CF, Tachida H, Langley CH, Harada K, Mukai T (1990) Increased variation in ADH enzyme activity in *Drosophila* mutation–accumulation experiment is not due to transposable elements at the *Adh* structural gene. Genetics 126:915–919

Aquadro CF, Weaver AL, Schaeffer SW, Anderson WW (1991) Molecular evolution of inversions in *Drosophila pseudoobscura*: the amylase gene region. Proc Natl Acad Sci USA 88:305–309

Baba-Aissa F, Solignac M, Dennebouy N, David JR (1988) Mitochondrial DNA variability in *Drosophila simulans*: quasi absence of polymorphism within each of the three cytoplasmic races. Heredity 61:419–426

Beech RN, Leigh Brown AJ (1989) Insertion-deletion variation at the *yellow-achaete-scute* region in two natural populations of *Drosophila melanogaster*. Genet Res 53:7–15

Begun DJ, Aquadro CF (1991) Molecular population genetics of the distal portion of the X chromosome in *Drosophila*: evidence for genetic hitchhiking of the *yellow-achaete* region. Genetics 129:1147–1158

Begun DJ, Aquadro CF (1992) Levels of naturally occurring DNA polymorphism correlate with recombination rates in *D. melanogaster*. Nature 356:519–520

Berry AJ, Ajioka JW, Kreitman M (1991) Lack of polymorphism on the *Drosophila* fourth chromosome resulting from selection. Genetics 129: 1111–1117

Brown CB, Aquadro CF, Anderson WW (1990) DNA sequence evolution in the amylase multigene family in *Drosophila pseudoobscura*. Genetics 126: 131–138

Bodmer M, Ashburner M (1984) Conservation and change in the DNA sequence coding for alcohol dehydrogenase in sibling species of *Drosophila*. Nature 309:425–430

Caccone A, Amato GD, Powell JR (1987) Intraspecific DNA divergence in *Drosophila*: A study on parthenogenetic *D. mercatorum*. Mol Biol Evol 4:343–350

Charlesworth B, Langley CH (1989) The population genetics of *Drosophila* transposable elements. Ann Rev Genet 23:251–287

Charlesworth B, Lapid A (1989) A study of ten families of transposable elements on X chromosomes from a population of *Drosophila melanogaster*. Genet Res 54:113–125

Choudhary M, Singh RS (1987a) Historical effective size and the level of genetic diversity in *Drosophila melanogaster* and *Drosophila pseudoobscura*. Biochem Genet 25:41–51

Choudhary M, Singh RS (1987b) A comprehensive study of genic variation in natural populations of *Drosophila melanogaster*. III. Variations in genetic structure and their causes between *Drosophila melanogaster* and its sibling species *Drosophila simulans*. Genetics 117:697–710

Cohn VM, Thompson MA, Moore GP (1984) Nucleotide sequence comparisons of the *Adh* gene in three Drosophilids. J Mol Evol 20:31–37

Cooke PH, Oakeshott JG (1989) Amino acid polymorphisms for esterase-6 in *Drosophila melanogaster*. Proc Natl Acad Sci USA 86:1426–1430

Costa R, Peixoto AA, Thackeray JR, Dalgleish R, Kyriacou CP (1991) Length polymorphism in the Threonine-Glycine-encoding repeat region of the period gene in *Drosophila*. J Mol Evol 32:238–246

Coyne JA, Kreitman M (1986) Evolutionary genetics of two sibling species, *Drosophila simulans* and *D. sechellia*. Evolution 40:673–691

Cross SRH, Birley AJ (1986) Restriction endonuclease map variation in the *Adh* region in populations of *Drosophila melanogaster*. Biochem Genet 24: 415–433

Davis PS, Shen MW, Judd BH (1987) Assymmetrical pairings of transposons in and proximal to the white locus of *Drosophila* account for four classes of regularly occurring exchange products. Proc Natl Acad Sci USA 84:174–178

Dowsett AP, Young MW (1982) Differing levels of dispersed repetitive DNA among closely related species of *Drosophila*. Proc Natl Acad Sci USA 79: 4570–4574

Eanes WF, Wesley C, Hey J, Houle D, Ajioka JW (1988) The fitness consequences of P element insertion in *Drosophila melanogaster*. Genet Res 42:17–26

Eanes WF, Ajioka JW, Hey J, Wesley C (1989a) Restriction-map variation associated with the G6PD polymorphism in natural populations of *Drosophila melanogaster*. Mol Biol Evol 6:384–397

Eanes WF, Labate J, Ajioka JW (1989b) Restriction-map variation with the *yellow-achaete-scute* region in five populations of *Drosophila melanogaster*. Mol Biol Evol 6:492–502

Eeken JCJ, de Jong AWM, Green MM (1987) The spontaneous mutation rate in *Drosophila simulans*. Mutation Res 192:259–262

Engels WR (1981) Estimating genetic divergence and genetic variability with restriction endonucleases. Proc Natl Acad Sci USA 78:6329–6333

Ewens WJ, Spielman RS, Harris H (1981) Estimation of genetic variation at the DNA level from restriction endonuclease data. Proc Natl Acad Sci USA 78:3748–3750

Finnegan DJ, Fawcett DH (1986) Transposable elements in *Drosophila melanogaster*. Oxford Surv Eukaryotic Genes 3:1–62

Game AY, Oakeshott JG (1990) The association between restriction site polymorphism and enzyme activity variation for Esterase 6 in *Drosophila melanogaster*. Genetics 126:1021–1031

Gillespie JH (1987) Molecular evolution and the neutral allele theory. Oxford Surv Evol Biol 4:10–37

Goldberg DA (1980) Isolation and partial characterization of the *Drosophila* alcohol dehydrogenase gene. Proc Natl Acad Sci USA 77:5794–5798

Goldberg ML, Sheen J-Y, Gehring WJ, Green MM (1983) Unequal crossing over associated with asymmetrical synapsis between nomadic elements in the *Drosophila melanogaster* genome. Proc Natl Acad Sci USA 80:5017–5021

Golding GB, Glickman BW (1986) Evidence for local DNA influences on patterns of substitutions in the human α-interferon gene family. Canad J Genet Cytol 28:483–496

Golding GB, Aquadro CF, Langley CH (1986) Sequence evolution within populations under multiple types of mutation. Proc Natl Acad Sci USA 83:427–431

Green MM (1988) Mobile DNA elements and spontaneous gene mutation. Banbury Report 30: Eukaryotic Transposable Elements as Mutagenic Agents, pp 41–50

Hudson RR (1982) Estimating genetic variability with restriction endonucleases. Genetics 100:711–719

Hudson RR (1990) Gene genealogies and the coalescent process. Oxford Surv Evol Biol 7:1–44

Hudson RR, Kaplan NL (1988) The coalescent process in models with selection and recombination. Genetics 120:831–840

Hudson RR, Kreitman M and Aguadé M (1987) A test of neutral molecular evolution based on nucleotide data. Genetics 116:153–159

Hughes AL, Nei M (1988) Pattern of nucleotide substitution at major histocompatibility complex class I loci reveals overdominant selection. Nature 355:167–170

Inoue YH, Yamamoto M-T (1987) Insertional DNA and spontaneous mutation at the white locus in *Drosophila simulans*. Mol Gen Genet 209:94–100

Ishii K, Charlesworth B (1977) Associations between allozyme loci and gene arrangements due to hitch-hiking effects of new inversions. Genet Res 30: 93–106

Jeffreys AJ, Wilson V, Thein SL (1985) Hypervariable "minisatellite" regions in human DNA. Nature 314:67–73

Jiang C, Gibson JB, Wilks AV, Freeth AL (1988) Restriction endonuclease variation in the region of the alcohol dehydrogenase gene: a comparison of null and normal alleles from natural populations of *Drosophila melanogaster*. Heredity 60:101–107

Kaplan NL, Darden T, Hudson RR (1988) The coalescent process in models with selection. Genetics 120:819–829

Kaplan NL, Hudson RR, Langley CH (1989) The "hitchhiking effect" revisited. Genetics 123:887–899

Keith TP, Riley MA, Kreitman M, Lewontin RC, Curtis D, Chambers G (1987) Sequence of the structural gene for xanthine dehydrogenase (rosy locus) in *Drosophila melanogaster*. Genetics 116:67–73

Kidd S, Young MW (1986) Transposon-dependent mutant phenotypes at the Notch locus of *Drosophila*. Nature 323:89–91

Kimura M (1983) The Neutral Theory of Molecular Evolution. Cambridge University Press, Cambridge, 367 pp

Kreitman M (1983) Nucleotide polymorphism at the alcohol dehydrogenase locus of *Drosophila melanogaster*. Nature 304:412–417

Kreitman M (1987) Molecular population genetics. Oxf Surv Evol Biol 4:38–60

Kreitman M (1991) Detecting selection at the level of DNA. In Selander R, Clark A, Whittam T (eds) Evolution at the Molecular Level. Sinauer, Sunderland, Massachusetts, pp 204–221

Kreitman M, Aguadé M (1986) Genetic uniformity in two populations of *Drosophila melanogaster* as revealed by filter hybridization of four-nucleotide-recognizing restriction enzyme digests. Proc Natl Acad Sci USA 83:3562–3566

Kreitman M, Hudson RR (1991) Inferring the evolutionary histories of the *Adh* and *Adh-dup* loci in *Drosophila melanogaster* from patterns of polymorphism and divergence. Genetics 127:565–582

Kreitman ME, Aguadé M (1986) Excess polymorphism at the *Adh* locus in *Drosophila melanogaster*. Genetics 114:93–110

Langley CH (1990) The molecular population genetics of *Drosophila*. In Takahata N, Crow JF (eds) Population Biology of Genes and Molecules. Baifukan, Tokyo, pp. 75–91

Langley CH, Aquadro CF (1987) Restriction map variation in natural populations of *Drosophila melanogaster*: white locus region. Mol Biol Evol 4:651–663

Langley CH, Montgomery E, Quattlebaum WF (1982) Restriction map variation in the *Adh* region of *Drosophila*. Proc Natl Acad Sci USA 79:5631–5635

Langley CH, Montgomery E, Hudson R, Kaplan N, Charlesworth B (1988a) On the role of unequal exchange in the containment of transposable element copy number. Genet Res 52:223–235

Langley CH, Shrimpton AE, Yamazaki T, Miyashita N, Matsuo Y, Aquadro CF (1988b) Naturally occurring variation in the restriction map of the *Amy* region of *Drosophila melanogaster*. Genetics 119:619–629

Laurie-Ahlberg CC, Stam LF (1987) Use of P-element-mediated transformation to identify the molecular basis of naturally occurring variants affecting *Adh* expression in *Drosophila melanogaster*. Genetics 115:129–140

Laurie CC, Heath EM, Jacobson JW, Thomson MS (1990) Genetic basis of the difference in alcohol dehydrogenase expression between *Drosophila melanogaster* and *D. simulans*. Proc Natl Acad Sci USA 87:9674–9678

Lee CS, Curtis D, McCarron M, Love C, Gray M, Bender W, Chovnick A (1987) Mutations affecting expression of the rosy locus in *Drosophila melanogaster*. Genetics 116:55–66

Leigh Brown AJ (1983) Variation at the 87A heat shock locus in *Drosophila melanogaster*. Proc Natl Acad Sci USA 80:5350–5354

Leigh Brown AJ, Moss JE (1987) Transposition of the I element and *copia* in a natural population of *Drosophila melanogaster*. Genet Res 49:121–128

Lim JK (1988) Intrachromosomal rearrangements mediated by *hobo* transposons in *Drosophila melanogaster*. Proc Natl Acad Sci USA 85:9153–9157

Macpherson JN, Weir BS, Leigh Brown AJ (1990) Extensive linkage disequilibrium in the *achaete-scute* complex of *Drosophila melanogaster*. Genetics 126:121–129

Maynard Smith J, Haigh H (1974) The hitch-hiking effect of a favorable gene. Genet Res 23:23–35

McDonald JH, Kreitman M (1991) Adaptive protein evolution at the *Adh* locus in *Drosophila*. Nature 351:652–654

Miyashita N, Langley CH (1988) Molecular and phenotypic variation of the white locus region in *Drosophila melanogaster*. Genetics 120:199–212

Montgomery E, Charlesworth B, Langley CH (1987) A test for the role of natural selection in the stabilization of transposable element copy number in a population of *Drosophila melanogaster*. Genet Res 49:31–41

Montgomery EA, Langley CH (1983) Transposable elements in Mendelian populations. II. Distribution of three copia-like elements in a natural population of *Drosophila melanogaster*. Genetics 104:473–483

Nei M (1987) Molecular Evolutionary Genetics. Columbia University Press, New York, 512 pp

Nei M, Chakraborty R, Fuerst PA (1976) Infinite allele model with varying mutation rate. Proc Natl Acad Sci USA 73:4164–4168

Nei M, Graur D (1984) Extent of protein polymorphism and the neutral mutation theory. Evol Biol 17:73–118

Nei M, Jin L (1989) Variances of the average number of nucleotide substitutions within and between populations. Mol Biol Evol 6:290–300

Nei M, Li W-H (1975) Probability of identical monomorphism in related species. Genet Res 25:31–43

Nei M, Li W-H (1979) Mathematical model for studying genetic variation in terms of restriction endonucleases. Proc Natl Acad Sci USA 76:5269–5273

Nei M, Li W-H (1980) Non-random association between electromorphs and inversion chromosomes in finite populations. Genet Res 35:65–83

Nei M, Tajima F (1981) DNA polymorphism detectable by restriction endonucleases. Genetics 97:145–163

Ohta T (1976) Role of very slightly deleterious mutations in molecular evolution and polymorphism. Theor Pop Biol 10:254–275

Ohta T, Tachida H (1990) Theoretical study of near neutrality. I. Heterozygosity and rate of mutant substitution. Genetics 126:219–229

Riley MA, Hallas ME, Lewontin RC (1989) Distinguishing the forces controlling genetic variation at the *Xdh* locus in *Drosophila pseudoobscura*. Genetics 123:359–369

Ronsseray S, Anxolabéhère D (1987) Chromosomal distribution of P and I transposable elements in a natural population of *Drosophila melanogaster*. Chromosoma 94:433–440

Rozas J, Aguadé M (1990) Evidence of extensive genetic exchange in the *rp49* region among polymorphic chromosome inversions in *Drosophila subobscura*. Genetics 126:417–426

Satta Y, Takahata N (1990) Evolution of *Drosophila* mitochondrial DNA and the history of the *melanogaster* subgroup. Proc Natl Acad Sci USA 87:9558–9562

Schaeffer SW, Aquadro CF, Anderson WW (1987) Restriction map variation in the alcohol dehydrogenase gene region of *Drosophila pseudoobscura*. Mol Biol Evol 4:254–265

Schaeffer SW, Aquadro CF, Langley CH (1988) Restriction map variation in the Notch locus of *Drosophila melanogaster*. Mol Biol Evol 5:30–40

Simmons GM, Kreitman ME, Quattlebaum WF, Miyashita N (1989) Molecular analysis of the alleles of alcohol dehydrogenase along a cline in *Drosophila melanogaster*. I. Maine, North Carolina, and Florida. Evolution 43:393–409

Singh RS (1989) Population genetics and evolution of species related to *Drosophila melanogaster*. Ann Rev Genet 23:425–453

Singh RS, Choudhary M, David JR (1987) Contrasting patterns of geographic variation in the cosmopolitan sibling species *D. melanogaster* and *D. simulans*. Biochem Genet 25:27–40

Singh RS, Rhomberg LR (1987) A comprehensive study of genic variation in natural populations of *Drosophila melanogaster*. II. Estimates of heterozygosity and patterns of geographic differentiation. Genetics 117: 255–271

Sperlich D, Pfriem P (1986) Chromosomal polymorphisms in natural and experimental populations. In Ashburner M, Carson HL, Thompson JN Jr (eds) The Genetics and Biology of *Drosophila*, Vol 3e. Academic Press, London, pp. 257–309

Stephan W (1989) Molecular genetic variation in the centromeric region of the X chromosome in three *Drosophila ananassae* populations. II. The *Om(1D)* locus. Mol Biol Evol 6:624–635

Stephan W, Langley CH (1989) Molecular genetic variation in the centromeric region of the X chromosome in three *Drosophila ananassae* population. I. Contrasts between the *vermilion* and *forked* loci. Genetics 121:89–99

Strobeck C (1983) Expected linkage disequilibrium for a neutral locus linked to a chromosomal rearrangement. Genetics 103:545–555

Sullivan DT, Atkinson PW, Starmer WT (1989) Molecular evolution of the alcohol dehydrogenase genes in the genus *Drosophila*. Evol Biol 24:107–147

Sved JA (1968) The stability of linked systems of loci with a small population size. Genetics 59:543–563

Tachida H, Harada K, Langley CH, Aquadro CF, Yamazaki T, Cockerham CC, Mukai T (1989) Restriction map and α-amylase activity variation among *Drosophila* mutation accumulation lines. Genet Res 54:197–203

Tajima F (1989a) Statistical method for testing the neutral mutation hypothesis by DNA polymorphism. Genetics 123:585–595

Tajima F (1989b) The effect of change in population size on DNA polymorphism. Genetics 123:597–601

Tanda S, Shrimpton AE, Hinton CW, Langley CH (1989) Analysis of the *Om(1D)* locus in *Drosophila ananassae*. Genetics 123:495–502

Thomson G (1977) The effect of a selected locus on linked neutral loci. Genetics 85:753–788

van Delden W (1982) The alcohol dehydrogenase polymorphism in *Drosophila melanogaster*. Selection at an enzyme locus. Evol Biol 15:187–222

Weir BS (1990) Genetic Data Analysis. Sinauer, Sunderland, Massachusetts, 377 pp

Wheeler DA, Kyriacou CP, Greenacre ML, Yu Q, Rutila JE, Rosbash M, Hall JC (1991) Molecular transfer of a species-specific behavior from *Drosophila simulans* to *Drosophila melanogaster*. Science 251:1082–1085

Yu Q, Colot HV, Kyriacou CP, Hall JC, Rosbash M (1987) Behaviour modification by in vitro mutagenesis of a variable region within the *period* gene of *Drosophila*. Nature 326:765–769

Zwiebel LJ, Cohn VH, Wright DR, Moore GP (1982) Evolution of single-copy DNA and the *Adh* gene in seven Drosophilids. J Mol Evol 19:62–71

Chapter 7

The Regulation of Cellular Pattern Formation in the Compound Eye of *Drosophila melanogaster*

Rick G. Tearle, Trevor J. Lockett, Wayne R. Knibb, Jeremy Garwood, and Robert B. Saint

1. Introduction

The development of complex, multicellular organisms from a single-cell zygote is a fundamental biological process in all metazoans. Regulation of insect development is a very efficient process if the evolutionary success of this group of animals is any guide. Fortunately for entomologists, one particular insect species, the dipteran *Drosophila melanogaster*, has provided an outstanding model system for the study of the molecular biology of developmental processes. As a result, developmental control strategies are being elucidated first in this insect, providing paradigms for the study of development in all animals.

Unravelling the nature of development will have a wider significance than to simply provide a solution to an intrinsically fascinating problem. For example, limitations in the flexibility of developmental processes may result in restrictions to the extent of morphological change that can be generated by genetic variation. Thus it is interesting to note that while insects display a vast array of forms, the primary structures such as segments are strikingly conserved. While this is clearly due in part to the remarkable success of the mature insect body plan, it may also reflect functional constraints that have limited the extent of change permitted in developmental programs.

While the problem of development seemed indecipherable two decades ago, major advances in molecular and cellular biology, and in immunochemistry and histochemistry, have combined with the established strengths of genetics and cytology to revolutionize our understanding of the process in *D. melanogaster*. The purpose of this Chapter is to highlight this progress using, as an example, the development of the compound eye.

The unit of development is the cell. Development can be viewed as a series of regulatory events in which cells undergo one of a range of alternative "behaviors," including proliferation, migration, biochemical differentiation or programmed cell death. Such cellular "behaviors" are under molecular regulation, and a vast body of cell biological and genetic research has revealed the types of molecules that are involved in these processes. These molecules can be grouped into functional classes and include:

1. diffusible regulatory factors (e.g., hormones/growth factors) which play a role in intercellular communication,
2. membrane-bound or soluble receptors for hormones/growth factors,
3. cell surface molecules (e.g., cell adhesion molecules, substrate adhesion molecules and molecules comprising the cell matrix) that are involved in intercellular or substrate adhesion,
4. molecules involved in signal transduction pathways (e.g., G proteins, protein kinases) that transfer information to the nucleus, and
5. regulators of gene expression (e.g., transcriptional regulators and regulators of RNA processing and translation) that generate the intercellular signals or alter the biochemical state of the cell upon receipt of the signal.

The molecular approach to developmental regulation seeks to determine how these regulators of cell behavior are coordinated during the complex process of development.

Perhaps the best characterized example of developmental regulation occurs during early embryogenesis in *D. melanogaster*. Following fertilization and during the final cleavage divisions, localized maternal products induce the expression of a set of regulatory genes which interact in a hierarchical fashion. The result of this hierarchy is the cellular blastoderm embryo (comprised of an epithelium of about 5,000 morphologically identical cells) which is a mosaic of expression of different regulatory genes (reviewed by Nüsslein-Volhard et al. 1987, and Ingham 1988). These regulatory genes are responsible for the induction of the major anterior-posterior and dorsal-ventral structures found in the mature larva. While this is a particularly exciting example of a developmental regulatory hierarchy, it is also a specialized one, since much of the hierarchy occurs while the embryo is a syncytium, i.e., while there are no cell membranes present to complicate the process with a requirement for intercellular communication. The regulatory events of gastrulation, neurogenesis and organogenesis, which follow the cellular blastoderm stage, occur in cellular tissues and are responsible for generating the characteristic pattern of differentiated cells of the mature larva. These post-blastoderm regulatory events are fascinating in their own right, as rules governing cellular pattern formation may prove to be somewhat different to those governing syncytial regulatory gene interaction.

The process whereby an epithelial sheet of undifferentiated cells is converted into a discrete pattern of differentiated cells is the subject of this Chapter. Neurogenesis and the formation of the epidermis in the *Drosophila* embryo is proving to be one good model system for the analysis of the events regulating cellular pattern formation (Campos-Ortega and Knust 1990). The other outstanding model system is that provided by development of the compound eye. In this chapter we describe basic features of eye development, approaches to the identification and study of genes which regulate eye development, and properties of some of the regulatory genes. We intend this chapter to be an introduction to the developing eye as a model system rather than a comprehensive description of the state of research in this field. Excellent detailed reviews have been published recently by Tomlinson (1988) and Ready (1989).

2. Cell Biology of the Compound Eye

The compound eye is a neurepithelium comprised of approximately 800 light-sensing units, termed ommatidia. These are arranged in a precise hexagonal array which gives the eye a crystalline appearance. The compound eye is composed of a few well defined cell types arranged in an invariant pattern within the ommatidium, permitting the formation of the regular arrangement of ommatidia. The cell types (whose arrangement within the ommatidium is shown diagrammatically in Figure 7.1) include photoreceptor cells, cone cells from which the lens is secreted, pigment cells which optically insulate one ommatidium from the next, and cells which produce a mechanosensory bristle. Photoreceptor cells are neural cells which extend axons into the brain. The photoreceptor cells contain a light sensing organelle termed the rhabdomere. These organelles, seen in cross-section in the electron micrograph shown in Figure 7.2, are packed with microvilli containing the light-receiving rhodopsin molecules. The pattern of photoreceptor cells, and therefore of rhabdomeres, is identical in all ommatidia (although a mirror image of the pattern occurs about a dorsal-ventral plane through the eye). This permits specific identities to be assigned to each photoreceptor cell, as indicated in Figure 7.1. The cells are termed Rl to R8. Rl to R6 have the outer rhabdomeres which extend the full length of the epithelium. The rhabdomeres of R7 and R8 lie centrally within the ommatidium and one on top of the other, the R7 rhabdomere being apical and the R8 rhabdomere basal.

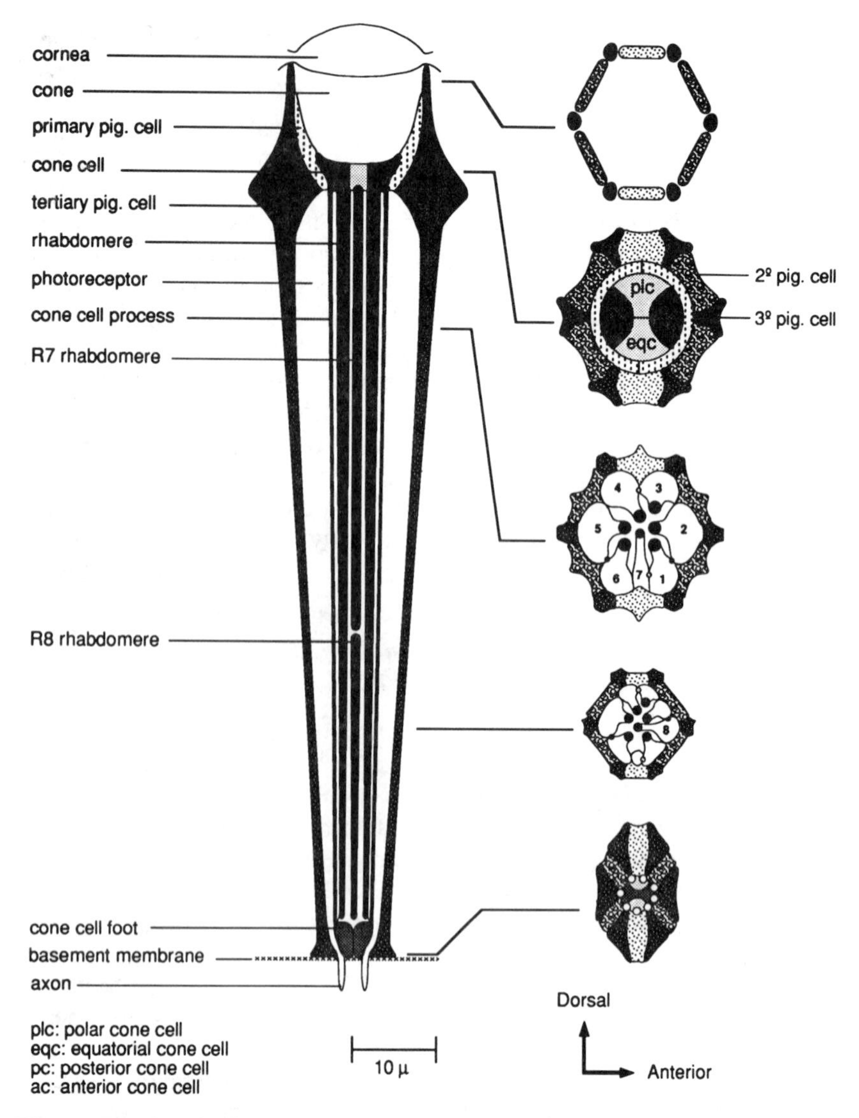

Figure 7.1. A schematic diagram of the *Drosophila* ommatidium showing the invariant arrangement of cell types within the ommatidium (from Ready 1989).

The compound eye arises from the progeny of a small group of cells which are set aside early in embryonic development. These cells (termed the eye imaginal disc) form a single-cell layer epithelium. The cells of the imaginal disc proliferate exponentially during larval life while remaining in an apparently undifferentiated state. By the time morphological dif-

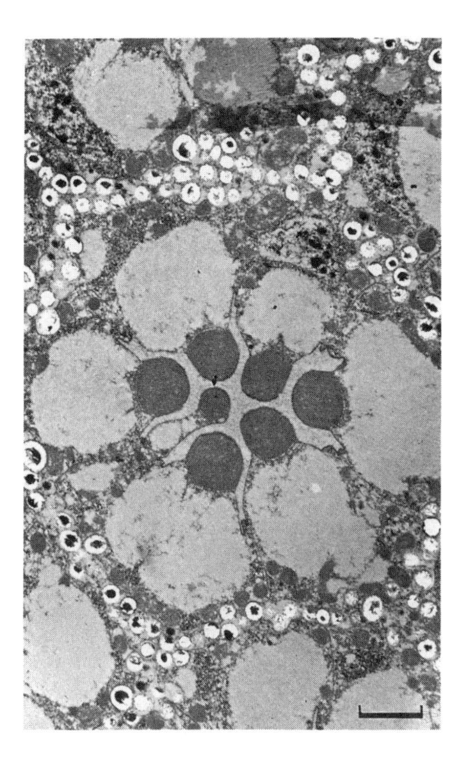

Figure 7.2. Electron micrograph of a transverse section of a *Drosophila* ommatidium magnified 10,000 times. The seven dark regions are the seven rhabdomeres of photoreceptor cells R1 to R7. The rhabdomeres are dense stacks of microvilli containing the light-receiving rhodopsin molecules. The granules surrounding the photoreceptor cells are pigment granules. The section corresponds to the middle transverse section shown in Figure 7.1. The bar corresponds to 2 microns.

ferentiation occurs, there are approximately 30,000 cells in the eye disc. The first visible sign of cellular differentiation occurs towards the end of larval life. A furrow (termed the morphogenetic furrow) begins to move across the disc from the posterior to the anterior edge. During this process the cells in the vicinity of, and immediately behind, the furrow begin to differentiate in a specific order, as indicated in Figure 7.3.

The study of the cellular basis of eye development was initiated by Benzer and his colleagues with the description of the preommatidial clusters and with the generation and use of a series of monoclonal antibodies which recognized cells at different stages during eye development (Ready et al. 1976; Zipursky et al. 1984). Examples of staining with the monoclonal antibody 22C10 are shown in Figure 7.3. The detailed cellular

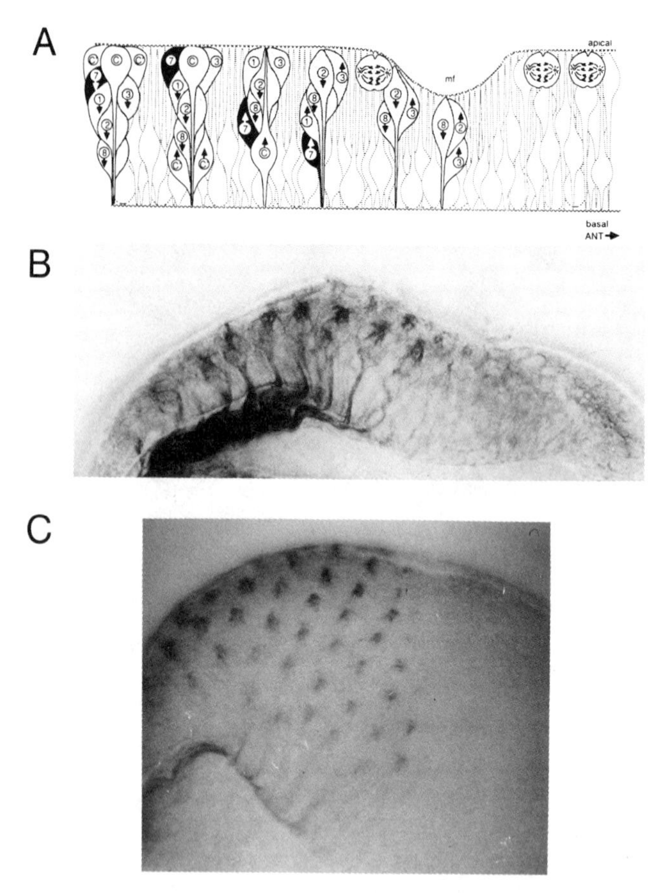

Figure 7.3. **A**. Schematic diagram of a longitudinal section through a developing eye disc, indicating stages in ommatidial development. Ommatidial assembly begins at the morphogenetic furrow (mf), which moves anteriorly. Cells are incorporated into the growing ommatidia in a defined order and at specific times. Movements of the cell nuclei during this process are indicated by arrows. A band of dividing cells between stages B and C is indicated by a mitotic nucleus (from Tomlinson and Ready 1987a). **B**. A developing eye disc stained with the monoclonal antibody MAb22C10 (Zipursky et al. 1984). This antibody detects an antigen specifically expressed on the cell membranes of newly differentiated neural cells, in this case the differentiating photoreceptor cells. The plane of focus reveals a longitudinal section through the disc. The growing clusters of photoreceptor cells and their trailing axons can be seen. **C**. A view of the apical epithelial surface of a developing eye disc stained as in panel B. The regular pattern of preommatidial clusters forming behind the morphogenetic furrow can be seen. Antibody binding in panels B and C is visualized by virtue of a horseradish peroxidase conjugated secondary antibody and incubation with Diaminobenzidene.

events were exquisitely described by Tomlinson (1985), Tomlinson and Ready (1987a), and Cagan and Ready (1989a) using a number of electron microscopic and immunohistochemical techniques. The following is, in the main, a summary of their description of the events occurring during cellular differentiation of the eye.

The first cells to identifiably undergo differentiation are the presumptive R8 cells which are spaced along a row of cells immediately behind the morphogenetic furrow as it sweeps across the eye. These cells appear to be the seeds for the formation of the ommatidia. The process by which the presumptive R8 cell is induced to differentiate is not understood. Soon after R8 is committed to its developmental fate, two more photoreceptor cells, R2 and R5, begin synchronously to differentiate. These are followed by the R3 and R4 pair of cells. A short, tight burst of cell division then occurs, followed by the addition of the Rl and R6 cells and lastly the R7 cell. The addition of the photoreceptor cells follows the same order in all ommatidia and is not clonally restricted (Ready et al. 1976; Lawrence and Green 1979), i.e., the differentiation of any particular cell is not a result of its ancestry. Rather it appears to be a consequence of the position of the cell within the tissue. Commitment of cells to the growing pre-ommatidium is associated with specific movements of the cell nucleus. The nucleus of an undifferentiated cell is not located at any particular point in the apical-basal plane, but prior to differentiation, the nucleus moves basally. As the cell initiates the process of differentiation, the nucleus moves apically. Basal movement usually follows when another cell is about to be added to the ommatidium. After commitment of the photoreceptor cells, the cone cells are added to the growing ommatidium, again in a specific order.

Passage of the zone of primary morphological differentiation has been followed by Basler and Hafen (1989a; 1989b), who used heat shock induced expression of the *sevenless* gene in a *sevenless* mutant background (see below) to produce an R7 cell at defined times following pupariation. The region of R7 cell recruitment advances with increasing speed, from an approximate initial rate of one row every two hours to a final rate of one row every hour. At 16 hours after pupariation, the morphogenetic furrow has moved right across the eye disc, and differentiation of all of the photoreceptor cells has been initiated.

Two more cells, the primary pigment cells (PPCs), are added to each ommatidium. Although the movements associated with the genesis of the PPCs have not been described, the PPCs themselves become apparent when they wrap around the cone cells near the apical surface and round up, losing their contacts with the basal surface of the epithelial monolayer. They are the only cell types to lose contact with the basal surface. Although addition of the PPCs occurs in a posterior to anterior series like that of the photoreceptor and cone cells, the timing is much more relaxed, with some anteriorly located cells differentiating earlier than more posteriorly located cells.

With the addition of the PPCs, the number of cells which form each individual ommatidium is complete. However, there are still a number of cell types which are shared between ommatidia and must be generated to form the complete eye. These cells are formed between ommatidia and serve to lock each ommatidium in with its neighbors. The four cells which form the ommatidial bristle appear to be derived from a single mother cell which is located in the interommatidial space (each ommatidium has four neighboring ommatidia at this point). The cell divisions that generate these cells follow a radial rather than posterior-anterior distribution, with bristle cells appearing first in the centre of the eye and progressing out to the periphery. The two remaining cell types, the secondary (SPC) and tertiary (TPC) pigment cells, also form from the pool of uncommitted interommatidial cells. A number of cells are not incorporated into the developing ommatidium and die. On average, about 35 cells are available to form each ommatidium but only 21 do so. The choice of which cells form pigment cells and which undergo programmed cell death appears to depend on the contacts they make. Only those cells which make contacts with PPCs of neighboring ommatidia will become SPCs and TPCs, and where there are more cells making contact than are required, only the requisite number will persist. The nuclei of SPCs and TPCs do not move apically during their formation. The differentiation of these pigment cells, which involves an alteration of shape, pulls the ommatidia together into a hexagonal array.

3. Genes Involved in Cellular Pattern Formation During Eye Development

Three approaches have been used to identify genes involved in regulating the process of cellular patterning described above. They are: (a) mutant analysis which reveals genes required for correct pattern formation by the disruption of the normal pattern in mutants, (b) identification of genes based on sequence homology to known regulatory genes, and (c) random insertion of a transposable element engineered to detect regulatory sequences which direct specific patterns of expression during eye development. It should also be noted that genes required for cellular morphogenesis following primary differentiation have also been identified and cloned using a fourth method, the generation of monoclonal antibodies directed against epitopes found in the developing eye disc (e.g., the *chaoptic* gene, Van Vactor et al. 1988), but the discussion here is restricted to examples of genes required for the generation of the correct pattern of cell types.

The following are examples of the isolation of genes using each approach. The properties of each gene will be discussed later in the Chapter.

3.1. Mutant Analysis

The *sevenless* (*sev*) gene was cloned on the basis of the effect of the mutant on eye development. The mutant was identified by Benzer and his colleagues in a screen for flies which could not see at a particular wavelength of ultraviolet light. The cellular basis of this mutant phenotype was the absence of the R7 cell, a cell which expresses a unique opsin required for the fly's normal response to ultraviolet light. Cloning of the gene was achieved in two laboratories based on this mutant phenotype. Banerjee et al. (1987) generated a mutant by the insertion of the transposable P element and then used this transposon "tag" to isolate the gene. Hafen et al. (1987) used microdissection of the region of a polytene chromosome containing the *sev* gene. The cloned DNA was shown to encode the *sev* gene by genetic transformation (Spradling and Rubin 1982; O'Brochta and Handler this volume) and subsequent rescue of the mutant phenotype.

3.2. Isolation of a Gene Based on Homology to known Regulatory Proteins

As described above, many developmental regulatory genes fall into a small number of functional classes. The functional relationships are often associated with conservation of sequences which encode the regions of the protein involved in the conserved function. Such sequence homologies can be used to isolate new regulatory genes. An example of this is our own isolation of the *rough* gene, a member of the homeobox class of genes (Saint et al. 1988).

The homeobox class of genes share a conserved sequence which encodes a DNA binding domain (for a review see Scott et al. 1989). Many members of this gene family play vital roles in the development of *D. melanogaster*. The best known of these are the homeotic genes of the Bithorax and Antennapedia Complexes which are responsible for directing the development of segmental structures (for a review see Akam 1987). The DNA sequence of the homeobox in many genes of this class is sufficiently well conserved to permit the formation of hybrids between DNA encoding different members of the family. This is how the gene family was originally identified (McGinnis et al. 1984a; 1984b). We took advantage of this homology by chemically synthesizing an oligonucleotide corresponding to the most conserved region of the homeobox and using it as a probe for *Drosophila* genomic DNA sequences which were capable

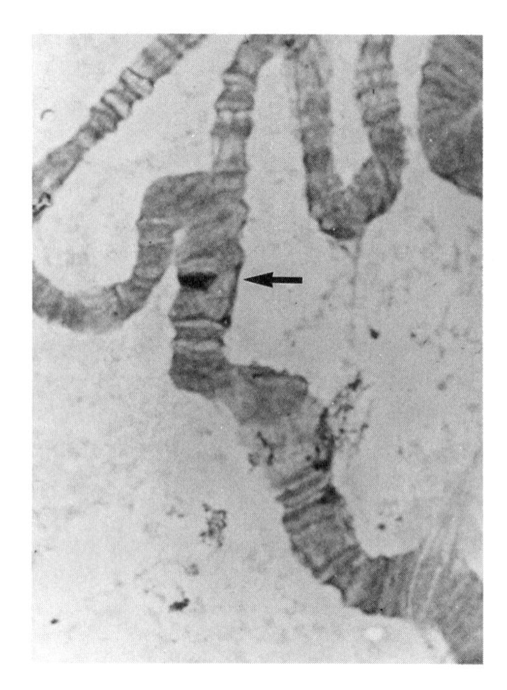

Figure 7.4. In situ hybridization of a biotinylated DNA fragment of the *rough* gene to polytene chromosomes from a strain heterozygous for a wild-type chromosome and a Df(3R)ro^{XB3} chromosome deficient for DNA in the region of *rough*. The arrow indicates a band of hybridization to the wild-type chromosome which is absent in the deficiency chromosome.

of hybridizing to this oligonucleotide. The advantages of working with an experimentally manipulable organism can be clearly seen from the relative ease with which we characterized the homologous DNA fragment isolated in this way.

First, the cloned DNA was localized to region 97D2-5 of the third chromosome using in situ hybridization. As a result of the wealth of mutant stocks available, the cloned DNA was shown to fall within *Df(3R)ro*^{XB3}, a small deficiency in region 97D (Figure 4). The deficiency was then used to identify a variety of lethal and visible mutants, generated on an isochromosomal line, which were located within the deficiency. Molecular analysis of one of these mutants (*ro*^{W}) and of a pre-existing mutant (*ro*^{1}) revealed the presence of mutant lesions within this gene (Saint et al. 1988), confirming the identity of this gene as the *rough* gene. The *rough* gene was isolated independently by Tomlinson et al. (1988) using the transposon tagging method described in the preceding section.

More recently, the Polymerase Chain Reaction (PCR) technique (Saiki et al. 1988) has greatly increased the ease with which such related genes can be isolated. A variety of genes which encode new members of the functional classes described above and which are expressed during eye development are currently being isolated.

3.3. Isolation Based on the Enhancer Activity of the Regulatory Region of the Gene

Bellen et al. (1989), Wilson et al. (1989) and Bier et al. (1989) recently described a method for the in vivo detection of transcription regulatory sequences termed enhancers using an artificial "reporter" gene, in this case the β-galactosidase gene derived from *Escherichia coli*. The method involves the random insertion into the *Drosophila* germline of a DNA construct which will only be expressed if activated by a regulatory sequence adjacent to the point of insertion. Mlodzik et al. (1990) used this method to search for enhancers which drove expression of the β-galactosidase gene in subsets of cells in the developing eye. One such insert was found to be expressed only in the photoreceptor cells Rl, R3, R4 and R6 cells, suggesting that the gene, which was normally regulated by this enhancer, would be expressed in the same precise pattern in the developing eye. The inserted DNA was used to isolate the adjacent DNA, and the gene regulated by the enhancer in question identified. Insertion of the DNA also created a mutation within the gene concerned, allowing ready genetic analysis of the locus. The gene identified and cloned in this way was given the name *seven-up*. All genes whose isolation is described here are discussed in more detail in the following section.

4. Molecular Characterization of Genes that Regulate Pattern Formation in the Developing Eye

The isolation of genes of interest described in the preceding section does not depend on a knowledge of the biochemical role of the gene products. A number techniques are widely used to analyze the genes and determine the biochemical role of their products. Many of techniques involve standard molecular biological analyses such as the analysis of the RNA products of the gene, which is achieved using such techniques as hybridization of labeled probes to mRNA species separated by gel electrophoresis (Northern blot analysis), and cDNA cloning and characterization by nucleotide sequence analysis. Sequence analysis of the cDNA reveals the presence of open translational reading frames which in turn can be conceptually translated to reveal the amino acid sequence of the encoded protein. Frequently, this sequence provides clues to the function of the protein.

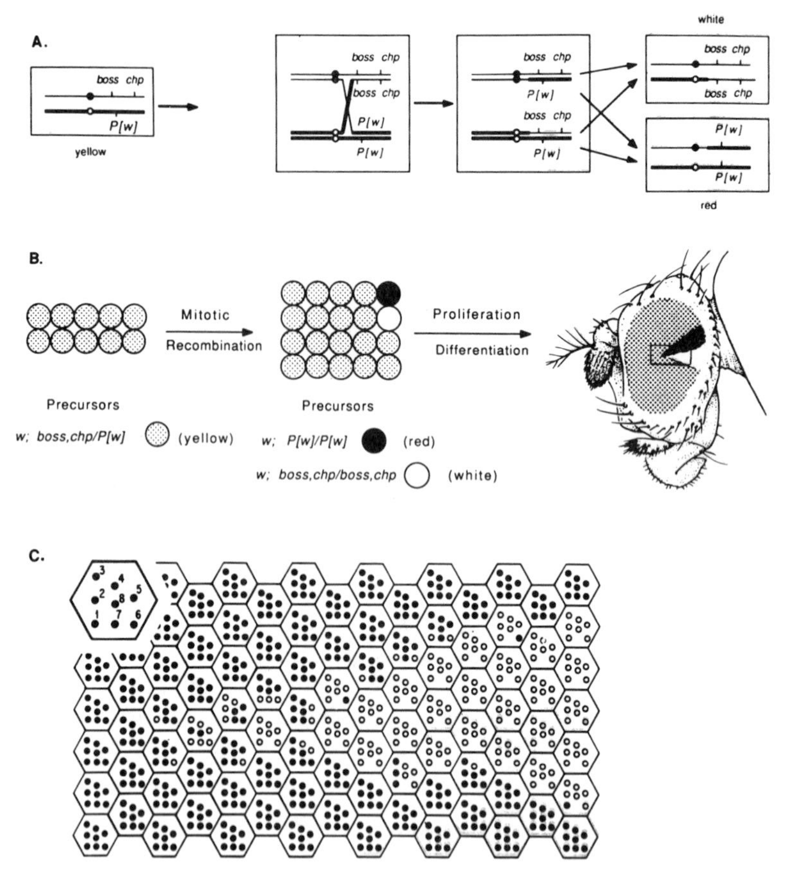

Figure 7.5. A schematic diagram showing the induction of mosaic clones in the developing eye and their final appearance in the differentiated adult eye. **A.** A mitotic recombination event is induced by irradiation in a cell that is heterozygous for: (l) a chromosome carrying two mutations, one in the gene of interest, in this case the *bride of sevenless* (*boss*) gene, and a second in a cell-autonomous mutant, *chaoptic* (*chp*), which alters cell morphology; and (2) a chromosome carrying a partially active copy of the *white* gene (*P[w]*) inserted in the region of the *boss* and *chp* genes. The *white* genes in the normal X chromosome location are non-functional. After mitotic recombination, one daughter cell now carries two copies of both *boss* and *chp* and no copies of *P[w]*, while the other daughter cell carries two copies of *P[w]*. **B.** During eye development the cells proliferate, generating sectors of the eye derived from the mitotic recombinant cells. The region derived from the cell lacking *white* shows up as a white patch in an otherwise yellow eye. The region derived from the cell carrying two copies of *P[w]* shows up as a red patch. The ommatidia in the white patch are of interest because they are mutant for the *boss* and *chp* gene. **C.** Because there is no strict lineage relationships between cells in the eye, ommatidia at the border between white and yellow patches can be made up of a mixture of *boss*$^+$ and *boss*$^-$ cells. The patch of

Analysis of spatial and temporal expression of the gene is achieved using in situ hybridization of labeled DNA directly to the RNA products of the genes in the developing eye discs (for example, see Figure 7.7). Analysis of the spatial and temporal patterns of protein production can be achieved by engineering the expression of the protein coding sequences in a bacterial system, purifying the protein product and raising antibodies against it by vaccinating rabbits. The antibodies are then incubated with histochemically fixed embryos where they bind to the gene product if it is present in the embryo. Binding is visualized by virtue of a fluorescent or enzyme conjugate bound to a secondary antibody (for example, see Figure 7.3C).

Genetic analysis provides powerful tools for the dissection of the role of the gene under study. In the case of eye development, by far the most powerful of these is clonal analysis, illustrated diagrammatically in Figure 7.5. Here X-ray induced mitotic recombination generates a single cell which is homozygous for a mutant in the gene of interest. This cell gives rise to a clone of cells which is mutant for the gene in a background of normal cells. The clone is generated such that it is also mutant for cell autonomous markers like the *white* mutant or *chaoptic* mutant, so that the individual cells can be scored for presence or absence of the marker gene and thus the gene of interest. The relationship between mutant cells and the developmental effect can therefore be studied. For example, in the *bride of sevenless* (*boss*) mutant, the R7 photoreceptor cell does not form (see below). Clonal analysis reveals that normal ommatidia (i.e., ommatidia with the R7 cell present) can form even if certain of the photoreceptor cells are mutant. A normal ommatidium is never seen, however, in which the R8 photoreceptor is mutant for the *boss* gene. The conclusion that can be drawn is that the R8 cell requires a normal *boss* gene for the R7 photoreceptor cell to form. Conversely, no other cell type in the developing ommatidium requires the *boss* gene for formation of the R7 cell. Thus the requirements for particular genes in specific cells of the developing ommatidium can be determined using clonal analysis.

The following is a brief discussion of the molecular genetic characterization, using the approaches described above, of a number of genes

←

mosaic ommatidia shown here corresponds to the area of the eye outlined in panel B. At this level of analysis the $boss^-$ cells can be determined because they are also chp^- w^- and have both altered cellular morphology and no pigmentation. A comparison of which cell types are chp^- w^- within each ommatidium and whether they are affected by the *boss* mutation facilitates the assignment of a requirement for the $boss^+$ gene to a particular cell type. (●) indicates a cell carrying $boss^+$ *P[w]* chp^+ and (○) represents a cell carrying $boss^-$ *P[w]* chp^- (From Reinke and Zipursky 1988.)

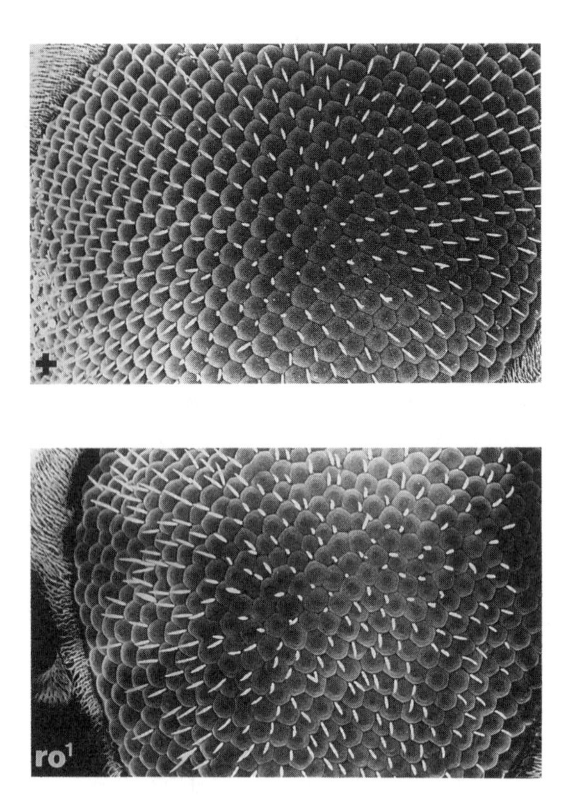

Figure 7.6. Scanning electron micrograph of a normal (+) and *rough* (ro) mutant eye. The normal eye shows the crystalline appearance of the pattern of ommatidia, while the pattern in the *rough* mutant is disrupted.

which have play a role in cellular pattern formation in the developing compound eye.

4.1. Genes Encoding DNA Binding Proteins: the *rough* Gene

The *rough* gene is so named because of the abnormal roughened appearance of the ommatidia in *rough* mutants (Figure 7.6). The role of the *rough* gene is restricted to eye development, since a null allele of *rough* is viable and exhibits an eye phenotype only (Saint et al. 1988). Tomlinson et al. (1988) carried out a detailed analysis of the cellular phenotype of the developing eye in a *rough* mutant. Addition of R3 and R4 to the precluster of R8/R2/R5 requires the expression of the wild-type *rough* gene. In the eye discs of *rough* mutant flies, R3 and R4 are not added to the precluster at the appropriate time. Instead, there is a delay, followed by the addition of either none, one, two or three photoreceptor cells of

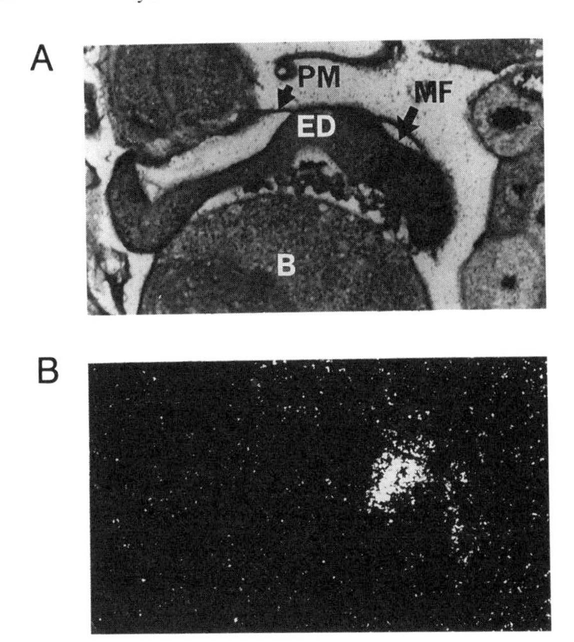

Figure 7.7. In situ hybridization of a radioactively labeled complementary RNA fragment of the *rough* gene to a tissue section through a wandering third instar larva. **A.** A bright field photograph of the tissue section. The brain (B), eye disc (ED), peripodial membrane (PM) and morphogenetic furrow (MF) are indicated. **B.** A dark field photograph of the same section revealing hybridization of the *rough* probe to the region of the morphogenetic furrow indicated in panel A.

undetermined identity. The number of photoreceptor cells added differs from ommatidium to ommatidium, disrupting the regular architecture of the ommatidia and hence the lattice structure of the resulting eye.

In situ hybridization using a radioactively labeled *rough* probe revealed that the *rough* gene is expressed in the developing eye at a time consistent with a role in the commitment of R3 and R4 to their fates, i.e., just after the passage of the morphogenetic furrow (Figure 7.7). Interestingly, in light of the apical movement of the cell nucleus soon after incorporation into the preommatidium, *rough* gene transcripts are localized at the apical region of cells expressing the gene (Saint et al. 1988). Immunohistochemical staining with antibodies directed against the *rough* protein product (Kimmel et al. 1990) showed that the *rough* protein product is found in photoreceptors R2/R5 and R3/R4. Mosaic analysis (Tomlinson et al. 1988) has shown that although the gene is expressed in these four photoreceptor cells, it is only required in R2/R5 to ensure that R3/R4 are committed properly. Why it is also expressed in R3 and R4 is not clear, but we postulate that *rough* plays a role in these cells independent of its developmental role, e.g., in the regulation of genes not involved in morphogenesis, or that it reflects some more ancient role in the regulation of eye development that has since been subsumed by another gene.

As described above, the *rough* gene encodes a protein which contains a homeodomain. The predicted activity of such proteins is an ability to bind in a sequence specific manner to DNA. Figure 7.8 shows an example of such sequence-specific DNA binding activity. In this example, a chimeric protein containing the *rough* homeodomain linked to the β-galactosidase protein of *E. coli* (J Garwood, N Dixon and RB Saint unpublished observations) has been incubated at various concentrations with radioactively labeled DNA containing sequences previously characterized as consensus binding sites for the *Ultrabithorax* homeodomain protein (Beachy et al. 1988). The bound DNA-protein complex (tracks B,C) is retarded during electrophoresis compared to the free DNA (track A). While we do not yet know the significance of the binding of the *rough* protein product to consensus sequences defined for other homeodomain proteins (a common feature of many homeodomain proteins), we can postulate that the *rough* protein in R2/R5 interacts directly with the regulatory regions of other genes to regulate their expression. The ultimate result of this regulation appears to be mediation of the communication between R2/R5 and the adjacent presumptive Rl, R6 and possibly R7 cells.

4.2. Membrane Bound Tyrosine Kinases: the *sevenless* Gene

The *sevenless* (*sev*) gene controls a later event in the developmental cascade of eye formation: the addition of photoreceptor cell R7 to the growing preommatidium. When the *sev* gene is mutant, R7 fails to form. The cell that would have become R7 moves into the correct position and its nucleus moves basally and then apically, but it does not adopt the R7 fate. Instead, the cell becomes a cone cell, and the cell that would have become that cone cell disappears (Tomlinson and Ready 1986; 1987b). Mosaic analysis has demonstrated that *sev* is cell-autonomous, i.e., the *sev* gene product is required by the presumptive R7 cells to enable them to adopt their appropriate developmental fate (Harris et al. 1976; Campos-Ortega et al. 1979).

The *sev* gene encodes a membrane-spanning protein that contains a cytoplasmic tyrosine kinase domain. The protein is expressed in a complex sequence of transient patterns during photoreceptor formation. This expression is not restricted to R7. Indeed, every photoreceptor cell expresses the *sev* protein at some stage. The irrelevance of much of this pattern of expression for eye development was conclusively demonstrated by the indiscriminate expression of a transformed *sev* gene under heat shock promoter control (Bowtell et al. 1989; Basler and Hafen 1989).

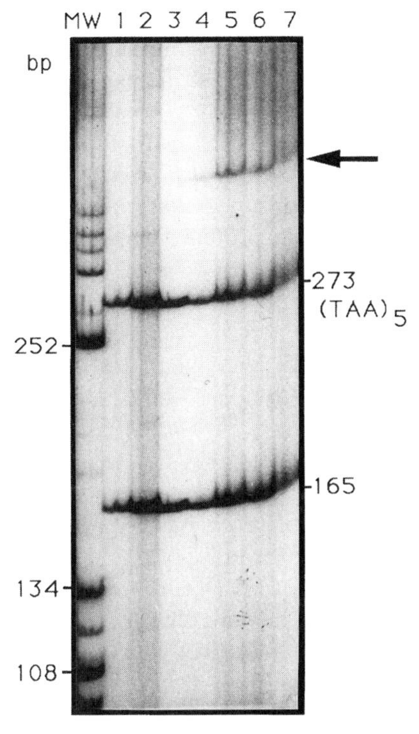

Figure 7.8. The interaction of an *Ultrabithorax* consensus DNA binding site (TAATAATAATAATAA) and the *rough* homeodomain. ^{32}P-labeled DNA containing the consensus sequence (the 273 base pair fragment) was incubated with increasing concentrations of a fusion protein comprised of the *rough* homeodomain linked to the β-galactosidase protein of *E. coli.* The homeodomain-β-galactosidase fusion protein was purified from *E coli* engineered to express the protein. The appearance of a new band, indicated by the arrow, in samples containing both the labeled DNA and the *rough* homeodomain is caused by the binding of the homeodomain to the labeled DNA fragment, altering its migration properties during polyacrylamide gel electrophoresis. The 165 base pair fragment is a second fragment in the plasmid from which the 273 base pair fragment is derived. Tracks are: (l) and (2) ^{32}P- labeled fragments alone; (3) and (4) labeled fragments plus *E. coli* β-galactosidase protein without the fused *rough* homeodomain; (5), (6) and (7) labeled fragments incubated with the homeodomain β-galactosidase fusion protein showing the presence of a retarded band (arrow). The MW lane shows radioactively labeled size markers.

When this construct is placed in a *sev* mutant fly, ubiquitous expression of the gene results in normal formation of R7 cells and has no discernible effect on the formation of other cell types. The restricted domain of *sev* gene function in contrast with the widespread expression of the gene cautions against interpretation of patterns of gene expression in the absence of genetic analysis of the requirements for a particular gene. The presence of extracellular, transmembrane and cytoplasmic tyrosine kinase domains suggests that the *sev* protein plays a role in the transduction of a signal from another cell or cells to instruct the presumptive R7 cell to adopt an "R7" developmental pathway. According to this model, a gene or genes should exist that produce the ligand which interacts with the extracellular domain of the *sev* protein, while a separate class of genes are activated by the tyrosine kinase domain to regulate the development of the R7 identity.

One gene which has properties consistent with a gene encoding a product required to produce the ligand for the *sev* protein has been discovered. It is the *bride of sevenless* gene.

4.3. The *bride of sevenless* Gene

As for *sev*, when the *bride of sevenless* (*boss*) gene is non-functional the R7 cell fails to develop. However, mosaic analysis has shown that *boss* must be expressed by R8, not R7 (Reinke and Zipursky 1988). The *boss* gene product must therefore be necessary for generation of the signal that R8 sends to the presumptive R7 cell to induce differentiation into the R7 cell. The signaling process could involve secretion of a diffusible factor or direct interaction between cell surface molecules such as that observed for the products of the *Notch* and *Delta* genes (Fehon et al. 1990). The presumptive R7–specific requirement for the *sev* gene product, in contrast to the general distribution of this product, would appear to argue against a diffusible signal model. Mediation via direct cell-cell contact is supported by the observation that the presumptive R7 cell contacts the R8 cell as its nucleus undergoes the movements which accompany commitment of the cell to its R7, or in the *sev* mutant, cone cell, fate (Tomlinson 1985; Tomlinson and Ready 1987a; 1987b).

4.4. The EGF Receptor Family of Genes: the *Ellipse* Gene

One of the first decisions that must be made in setting up the pattern of photoreceptor cells is the spacing of R8 cells in a line behind the morphogenetic furrow. One gene, *Ellipse* (*Elp*), has been implicated in determining this spacing on the basis of its mutant phenotype. Unlike the genes identified above, *Elp* was recovered as a dominant gain-of-function mutation and has been analyzed by Baker and Rubin (1989). In adults carrying two different dominant mutations (Elp^{1}/Elp^{B1}) the eyes are smaller than usual and there are areas without ommatidia, but the number and arrangement of photoreceptor cells within the remaining ommatidia are normal. The spaces devoid of ommatidia are filled with pigment and bristle cells. This mutant phenotype could be overcome by placing the *Elp* mutation over a chromosome carrying a deletion for the gene (e.g., Elp^{1}/Elp^{-}), suggesting that the derangement of eye formation is caused by overexpression of the *Elp* gene.

Baker and Rubin disrupted *Elp* gene function by mutagenizing adults carrying the gene, thus reverting the dominant small eye phenotype. Such revertants were embryonic lethal, suggesting that the *Elp* gene function was required for normal embryonic development. The *Elp* gene was subsequently shown to be allelic to a previously identified embryonic pattern forming gene, *faint little ball* (Price et al. 1989; Schejter and Shilo 1989). This gene encodes a member of the epidermal growth factor (EGF) receptor family of proteins which was cloned by virtue of its homology to the mammalian EGF receptor (Schejter and Shilo 1989). One possible explanation for the effect of *Elp* on eye development suggested by Baker and Rubin (1989) is that the cells which will give rise to R8 do so in regions where the concentration of the ligand is low. Overexpression of the *Elp* encoded receptor could lead to an increase in the total number of ligand-bound receptors, with a consequent enhancement of the signaling responsible for suppressing R8, and thus ommatidial, formation. Although this model is attractive, it is undoubtedly oversimplified. The EGF receptor must do more than just determine R8 spacing, as genetic mosaics, in which cells completely lacking the *Elp* gene were generated in an otherwise normal developing eye disc, show that the mutant cells do not contribute to the mature eye.

4.5. The Steroid Receptor Family of Genes: the *seven-up* Gene

Although the six photoreceptor cells (Rl–R6), which form an outer ring within the ommatidium, appear to be physiologically identical, they are incorporated at different times and play different roles in ommatidium formation. The *seven-up* (*svp*) gene is necessary for four of these six cells to adopt the correct cell fate (Mlodzik et al. 1990). As described above, the gene was identified by virtue of the insertion of a β-galactosidase gene into the regulatory region of the gene, causing expression of the reporter gene in R3/R4 and Rl/R6 cells. This suggested a specific role for this gene in determining photoreceptor cell identity. The gene was cloned and shown to correspond to a previously identified locus that mutated to embryonic lethality. To determine what effect absence of gene function would have on eye development, genetic mosaics were generated in which small *svp* mutant mosaic patches were induced in the eyes of otherwise normal flies. Two types of aberrant events were found in these patches. First, cells destined to become R3/R4 and Rl/R6 became R7 cells instead, and second, two "mystery" cells which, in the normal developing ommatidium, transiently associate with the ommatidial precluster and disappear soon after, appeared sometimes to remain attached and adopt an outer cell fate (i.e., one of Rl–R6). The *svp* gene product thus helps to specify the correct fate of the four cells in which it is expressed (R3/R4 and Rl/R6), and in its absence, these cells default to an R7 fate, with concomitant effects on the fate of the "mystery" cells.

Analysis of *sev*; *svp* double mutant patches has further revealed that, although up to five R7 cells may be found in a mutant ommatidium, these R7 states are not equivalent. When *sev* function alone is missing (see above), the R7 cell becomes a cone cell. However, when both *sev* and *svp* functions are missing, only the R7 cells, which are derived from the normal R7 and R3/R4, become cone cells. The R7 cells which derived from the normal Rl/R6 remain as R7 cells. Thus there appear to be at least two different pathways a cell can follow to adopt an R7 fate, one of which depends on expression of *sev* and the other of which does not. This in turn raises the question of whether these R7 cells are really the same at a biochemical and physiological level, a question that cannot be answered until the whole regulatory cascade of genes controlling eye development is unraveled.

Of particular significance to the study of pattern formation in this self-organizing epithelium, the *svp* gene appears to be a member of the steroid receptor family of genes, such as those which bind steroid hormones and retinoic acid-related molecules. The ligands for members of this class of receptor are membrane-soluble and can therefore diffuse through cellular tissues. The characteristics of the receptors are that they are intracellular molecules which contain a ligand binding domain and a DNA binding domain (reviewed by Beato 1989). Upon binding the ligand, the receptors locate to the nucleus where they regulate the expression of other genes. The particular significance of the *svp* protein, therefore, is that it is strong evidence for the involvement of diffusible steroid-like factors in the process of eye development, where previously it had appeared that direct cell-cell interactions may have been sufficient to account for the assembly of the cells of the ommatidia. The ligand for the *svp* gene has yet to be identified.

4.6. Other Genes

Many other genes are implicated in eye development by virtue of their mutant phenotypes. Dominant eye mutations can be the result of ectopic expression of genes which, during normal development, are never expressed in the eye but which disrupt the normal regulatory events of eye formation. The *Irregular facets* mutant, for example, is the result of a mutation in the embryonic segmentation gene *Kruppel* which leads to ectopic expression of this gene in the developing eye disc. The resultant eye is reduced in size and contains roughened facets (RG Tearle and RB Saint unpublished; NE Baker pers comm).

Another class of genes which have been implicated are the neurogenic genes (reviewed by Campos-Ortega and Knust 1990). These genes are required for the developmental decision between alternative epidermal and neural fates of embryonic ectoderm cells. In the absence of any member of this gene class, all the cells adopt a neural fate. The genetic evidence suggests that the gene products are required for an intercellular signaling event in which the primary neuroblast inhibits the adjacent ectodermal cells from following a similar fate, resulting in the adoption of the alternative epidermal fate. Although mutations in these genes cause embryonic lethality, their effect on eye development can be gauged by analyzing temperature-sensitive alleles (Cagan and Ready 1989b) or by generating small patches of homozygous mutant cells in otherwise wild-type adults (G. Jürgens pers comm), as described previously in this Chapter. In both cases eye development is disrupted, demonstrating that these genes also play a role in the formation of neural tissues other than the embryonic nervous system.

Two of the neurogenic genes, *Notch* and *Delta*, encode protein products that contain extracellular EGF-like repeats, a transmembrane domain and an intracellular domain (Wharton et al. 1985; Kidd et al. 1986; Vassin et al. 1987). These features are consistent with their proposed role in intercellular signaling. Recently Fehon et al. (1990) have shown that the products of these two genes can induce clumping in cultured cells, suggesting that the genes may act to mediate cell-cell adhesion and that the intercellular signaling event in which they are involved occurs between cells in contact. Cagan and Ready (1989b) showed that all eye cell types require normal *Notch* gene activity when establishing their fate. Using a temperature-sensitive allele of *Notch*, they showed that different cells adopted inappropriate developmental pathways at different developmental stages in response to the inactivation of the *Notch* gene.

5. Conclusions

From the above discussion of compound eye formation in *Drosophila melanogaster*, it should be evident that the cellular basis of the development of this organ is well understood, thanks to a number of elegant cell biological studies, particularly those of Tomlinson and Ready (1987a; 1987b). This detailed characterization of the cellular events occurring during eye development is the foundation on which studies of the molecular regulation of pattern formation in this tissue are based. A second principal feature of compound eye formation is that intercellular signaling events, rather than programmed differentiation of particular cell lineages, are responsible for the induction of differentiation in uncommitted cells. Thus, although the process by which the first of these cells, the R8 photoreceptor cell, is induced to differentiate is not understood, subsequent differentiation proceeds by virtue of the nature of cell-cell contacts made between the various cells and by virtue of the activity of a variety of molecules involved in intercellular signaling, signal transduction and gene regulation.

In relation to this, it should be noted that the cascade of events which bring about eye development appears very different in character to the cascade of events regulating two other well characterized systems in *Drosophila*. The early events in embryonic segment formation consist of a regulatory hierarchy of transcriptional regulation, with intercellular signaling becoming important only in the latter stages of the hierarchy (Ingham 1988 for a review). The regulation of sex determination (Belote this volume) is initially controlled by a mechanism operating at the transcriptional and RNA processing (splicing) level. In the case of segment formation, the primary events occur in a syncytium that allows gene products, which are normally exclusively intracellular, to diffuse between nuclei. In the case of sex determination, the initial developmental decision is an either/or choice between alternative pathways that later affects a range of developmental and physiological processes. Eye development differs from these examples, then, in the nature of the starting tissue, the eye imaginal disc being a single cell layer epithelium, and in the complexity of the developmental decisions made by individual cells within a short period of time.

Eye development, therefore, represents a particularly good system for studying the generation of cellular pattern formation in a single-cell layer epithelium. Although only a handful of the many regulatory genes involved have been identified as yet, the knowledge already generated has highlighted the role of cell communication in this tissue and has led to the identification of genes which play roles in the differentiation of particular cell types. The prospect, then, is that the continued molecular genetic analysis of eye development will result in the integration of the regulatory events that generate the different cell types during the development of this organ and in so doing, provide a model for the analysis of differentiation in a wide variety of epithelial tissues.

Acknowledgments. The authors wish to thank Hilary Mende and Michael Calder for their technical assistance. We also thank Brandt Clifford for assistance with photography.

References

Akam ME (1987) The molecular basis for metameric pattern in the *Drosophila* embryo. Development 101:1–22

Baker NE, Rubin GM (1989) Effect on eye development of dominant mutations in *Drosophila* homologue of the EGF receptor. Nature 340:150–153

Banerjee U, Renfranz PJ, Pollack JA, Benzer S (1987) Molecular characterization and expression of *sevenless*, a gene involved in neuronal pattern formation in the *Drosophila* eye. Cell 49:281–291

Basler K, Hafen E (1989) Dynamics of *Drosophila* eye development and temporal requirements of *sevenless* expression. Development 107:723–731

Basler K, Hafen E (1989) Ubiquitous expression of *sevenless*: position-dependent specification of cell fate. Science 243:931–934

Beachy PA, Krasnow MA, Gavis ER, Hogness DS (1988) An Ultrabithorax protein binds sequences near its own and the Antennapedia Pl promoters. Cell 55:1069–1081

Beato M (1989) Gene regulation by steroid hormones. Cell 56:335–344

Bellen HJ, O'Kane CJ, Wilson C, Grossnicklaus U, Kurth-Pearson R, Gehring WJ (1989) P-element mediated enhancer detection: a versatile method to study development. Genes Devel 3:1288–1300

Bier E, Vässin H, Shephard S, Lee K, McCall K, Barbel S, Ackerman L, Carretto R, Uemura T, Grell EH, Jan LY, Jan YN (1989) Searching for pattern and mutations in the *Drosophila* genome with a P-lacZ vector. Genes Devel 3:1273–1287

Bowtell DDL, Simon MA, Rubin GM (1989) Ommatidia in the developing *Drosophila* eye require and can respond to *sevenless* for only a restricted period. Cell 56:931–936

Cagan RL, Ready DF (1989a) The emergence of order in the *Drosophila* pupal retina. Devel Biol 136:346–362

Cagan RL, Ready DF (1989b) Notch is required for successive cell decisions in the developing *Drosophila* retina. Genes Devel 3:1099–1112

Campos-Ortega JA, Jürgens G, Hofbauer A (1979) Cell clones and pattern formation: Studies on *sevenless*, a mutant of *Drosophila melanogaster*. Wilhelm Roux Arch Devel Biol 181:227–245

Campos-Ortega JA, Knust E (1990) Molecular analysis of a cellular decision during embryonic development of *Drosophila melanogaster*: epidermogenesis or neurogenesis. Eur J Biochem 190:1–10

Desplan C, Theis J, O'Farrell PH (1988) The sequence specificity of homeodomain-DNA interaction. Cell 54:1081–1090

Fehon RG, Kooh PJ, Rebay I, Regan CL, Xu T, Muskavitch MAT, Artavanis-Tsakonas S (1990) Molecular interactions between the protein products of the neurogenic loci Notch and Delta, two EGF-homologous genes in *Drosophila*. Cell 61:523–534

Hafen E, Basler K, Edström J-E, Rubin G (1987) *Sevenless*, a cell-specific homeotic gene of *Drosophila*, encodes a putative transmembrane receptor with a tyrosine kinase domain. Science 236:55–63

Harris WA, Stark WS, Walker JA (1976) Genetic dissection of the photoreceptor system in the compound eye of *Drosophila melanogaster*. J Physiol 256: 415–439

Ingham PW (1988) The molecular genetics of embryonic pattern formation in *Drosophila*. Nature 335:25–34.

Kidd S, Kelley MR, Young MW (1986) Sequence of the *Notch* locus of *Drosophila*: relationship of the encoded protein to mammalian clotting and growth factors. Mol Cell Biol 6:3094–3108

Kimmel BE, Heberlein U, Rubin GM (1990) The homeo domain protein rough is expressed in a subset of cells in the developing *Drosophila* eye where it can specify photoreceptor cell subtype. Genes Devel 4:712–727

Lawrence PA, Green SM (1979) Cell lineage in the developing retina of *Drosophila*. Devel Biol 71:142–152

McGinnis W, Garber RL, Wirz J, Kuroiwa A, Gehring WJ (1984a) A homologous protein coding sequence in *Drosophila* homeotic genes and its conservation in other metazoans. Cell 37:403–408

McGinnis W, Levine M, Hafen E, Kuroiwa A, Gehring WJ (1984b) A conserved DNA sequence found in homeotic genes of the *Drosophila Antennapedia* and *Bithorax* complexes. Nature 308:428–433

Mlodzik M, Hiromi Y, Weber U, Goodman CS, Rubin GM (1990) The *Drosophila seven-up* gene, a member of the steroid receptor gene superfamily, controls photoreceptor cell fates. Cell 60:211–224

Nüsslein-Volhard C, Frohnhöfer HG, Lehmann R (1987) Determination of anteroposterior polarity in *Drosophila*. Science 238:1675–1681

Price JV, Clifford RJ, Schüpbach T (1989) The maternal ventralizing locus *torpedo* is allelic to *faint little ball*, an embryonic lethal, and encodes the *Drosophila* EGF receptor homolog. Cell 56:1085–1092

Ready DF (1989) A multifaceted approach to neural development. Trends Neuroscience 12:102–110

Ready DF, Hanson TE, Benzer S (1976) Development of the *Drosophila* retina, a neurocrystalline lattice. Devel Biol 53:217–240

Reinke R, Zipursky S (1988) Cell-cell interaction in the *Drosophila* retina: *the bride of sevenless* gene is required in photoreceptor cell R8 for R7 cell development. Cell 55:321–330

Saiki RK, Gelfand DH, Stoffel S, Scharf SJ, Higuchi R, Horn GT, Mullis KB, Erlich HA (1988) Primer-directed enzymatic amplification of DNA with a thermostable DNA polymerase. Science 239:487–491

Saint R, Kalionis B, Lockett TJ, Elizur A (1988) Pattern formation in the developing eye of *Drosophila melanogaster* is regulated by the homeobox gene, *rough*. Nature 334:151–154

Schejter ED, Shilo B-Z (1989) The *Drosophila* EGF receptor homologue (DER) gene is allelic to *faint little ball*, a locus essential for embryonic development. Cell 56:1093–1104

Scott MP, Tamkun JW, Hartzell GW (1989) The structure and function of the homeodomain. Biochem Biophys Acta 989:25–48

Spradling AC, Rubin GM (1982) Transposition of cloned P elements into *Drosophila* germline chromosomes. Science 218:341–347

Tomlinson A (1985) The cellular dynamics of pattern formation in the eye of *Drosophila*. J Embryol Exper Morphol 89:313–331

Tomlinson A (1988) Cellular interactions in the developing *Drosophila* eye. Development 104:183–189

Tomlinson A, Kimmel BE, Rubin GM (1988) *rough*, a *Drosophila* homeobox gene required in photoreceptors R2 and R5 for inductive interactions in the developing eye. Cell 55:771–784

Tomlinson A, Ready DF (1986) *Sevenless*: a cell specific homeotic mutation of the *Drosophila* eye. Science 231:400–402

Tomlinson A, Ready DF (1987a) Neuronal differentiation in the *Drosophila* ommatidium. Devel Biol 120:366–376

Tomlinson A, Ready DF (1987b) Cell fate in the *Drosophila* ommatidium. Devel Biol 123:264–275

Van Vactor Jr D, Krantz DE, Reinke R, Zipursky SL (1988) Analysis of mutants in chaoptin, a photoreceptor cell-specific glycoprotein in *Drosophila*, reveals its role in cellular morphogenesis. Cell 52:281–290

Vässin H, Bremer KA, Knust E, Campos-Ortega JA (1987) The neurogenic gene Delta of *Drosophila melanogaster* is expressed in neurogenic territories and encodes a putative tranomembrane protein with EGF-like repeats. EMBO J 6:3431–3440

Wharton KA, Johansen KM, Xu T, Artavanis-Tsakonas S (1985) Nucleotide sequence from the neurogenic locus Notch implies a gene product that shares homology with proteins containing EGF-like repeats Cell 43: 567–581

Wilson C, Kurth-Pearson R, Bellen HJ, O'Kane CJ, Grossnicklaus U, Gehring WJ (1989) P-element mediated enhancer detection: an efficient method for isolating and characterizing developmentally regulated genes in *Drosophila*. Genes Devel 3:1301–1313

Zipursky SL, Venkatesh TR, Teplow DB, Benzer S (1984) Neuronal development in the *Drosophila* retina: monoclonal antibodies as molecular probes. Cell 36:15–26

Chapter 8

Molecular Genetics of Sex Determination in *Drosophila melanogaster*

John M. Belote

1. Introduction

Many schemes for the biological control of insect pests rely on generating large, unisexual populations of sterile individuals. If one has the ability to manipulate the sexual development and fertility of the progeny from controlled crosses, then generating such populations should be facilitated. Thus, there is potential practical application of studies aimed at elucidating the genetic mechanisms controlling insect sexual development. However, practical considerations aside, the study of sex differentiation is also a very attractive system from the standpoint of basic molecular genetics research. In particular, sex differentiation of the fruitfly, *Drosophila melanogaster*, has proven to be one of the most accessible systems for addressing the fundamental question, "How does the genome of an organism orchestrate developmental events that occur during its life cycle?" The chromosomal signal that determines whether a zygote follows the male or the female pathway of development has been described (Bridges 1921; 1925). In addition, a set of regulatory genes that assesses this initial signal and implements the appropriate program of sexual development has been identified and characterized by genetic analyses (for reviews, see Baker and Belote 1983; Cline 1985; Nöthiger and Steinmann-Zwicky 1987). Several of these genes have been cloned and are now being analyzed at the molecular level (for reviews, see Hodgkin 1989; Baker 1989). Finally, some of the structural genes whose expressions are sexually dimorphic have been isolated, and the nature of their control by the pathway regulating sex differentiation is now amenable to study (see Wolfner 1988). Thus, this is a model system in which one can potentially trace and define at the molecular level all of the steps from initial commitment (i.e., sex determination, per se) through to terminal differentiation of sexually dimorphic tissues (i.e., sex differentiation). It is this property that makes sex differentiation of the fruitfly one of the most advantageous systems for examining in detail how genes control development in a higher eukaryote.

Drosophila melanogaster adults exhibit sexual dimorphism at many levels of organization. The most obvious sexual differences are external, seen in the abdominal pigmentation patterns, the structures of the genitalia and analia, and the presence or absence of sex combs on the basitarsus of the foreleg (Figure 8.1). Internally, the gonads and reproductive duct systems of the two sexes are likewise easily discernible. In addition to these overt differences, males and females differ with respect to several biochemical criteria. For example, there are enzymes, such as glucose

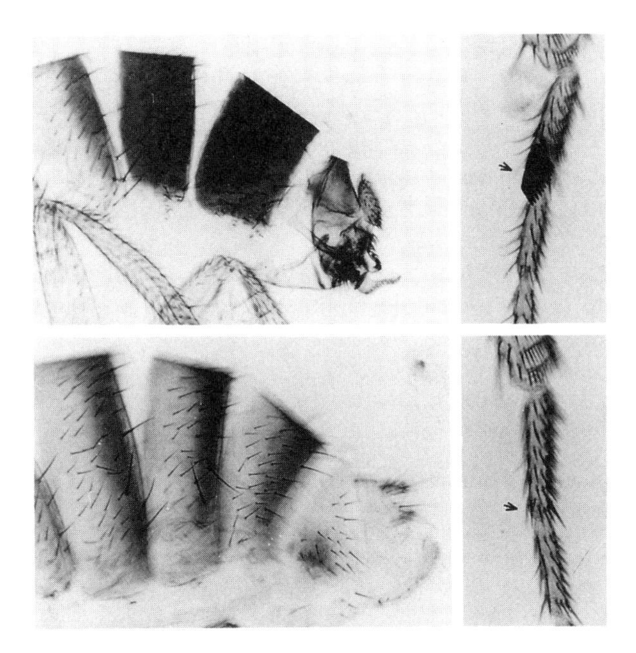

Figure 8.1. Sexual dimorphism in *Drosophila melanogaster*. At the left are shown the posterior abdominal segments and genitalia of an adult male (top) and adult female (bottom). In the male the fifth and sixth abdominal tergites are completely pigmented, while in the female these segments are pigmented only along the posterior border. The genital structures are also obviously dimorphic. At the right are shown the forelegs of a male (top) and a female (bottom). Males have a row of specialized bristles called a sex comb on the basitarsus of each foreleg that is missing in females.

dehydrogenase (Cavener et al. 1986) and esterase-6 (Oakeshott et al. 1987), that are present in adult males, but not detected in adult females. Conversely, the three yolk polypeptides, YPl, YP2, and YP3, are very abundant in mature females, but totally absent in males (Jowett and Postlethwait 1981; Bownes and Nöthiger 1981). Another cryptic difference between males and females concerns their innate behavioral patterns. For example, adult males exhibit an elaborate, male-specific courtship ritual that is inborn. Thus, there must be aspects of the nervous system's structure and/or function that are sexually dimorphic. Ultrastructural studies of the adult brain (Technau 1984) and peripheral sensory neurons (Taylor 1989) have, in fact, documented such sexual dimorphisms, although their relationships to the observed behavioral differences are not known.

Given this extensive sexual dimorphism, it is clear that there are distinct, though overlapping, pathways of development in male versus female *Drosophila*. Implicit in the above examples is the idea that the expression of individual genes must be regulated at some level in a sex-differential or sex-specific manner. In this chapter, I will focus on what has been learned about the genetic factors that are responsible for generating this sexual dimorphism. First, I will present an overview of the regulatory pathway controlling sexual differentiation. Much of our understanding of this has come from classical genetic studies, primarily using mutational approaches. I will then go through the pathway in more detail, focusing on more recent molecular analyses that have given us a better understanding of how these genes and their products interact with one another to control sexual development.

2. Genetic Studies of Sexual Differentiation

2.1. The Regulatory Hierarchy Controlling Sex

The genetic functions that control whether an individual develops as a male or a female can be arranged in hierarchical fashion (Figure 8.2). At the top of the regulatory hierarchy is the primary determinant of sex. In *Drosophila* this is provided by the chromosomal constitution of the fly, or more specifically, the number of X chromosomes relative to the number of autosomal sets (i.e., the X/A ratio). At some point fairly early in development, this signal is assessed and it acts to influence the sex-specific expression of an important regulatory gene known as *Sex lethal* (*Sxl*), which plays a pivotal role in several aspects of the fly's sexual development. The effect of *Sxl* on somatic sexual differentiation is mediated through its effect on the expression of a relatively small set of down-stream regulatory genes that make up the third level in this

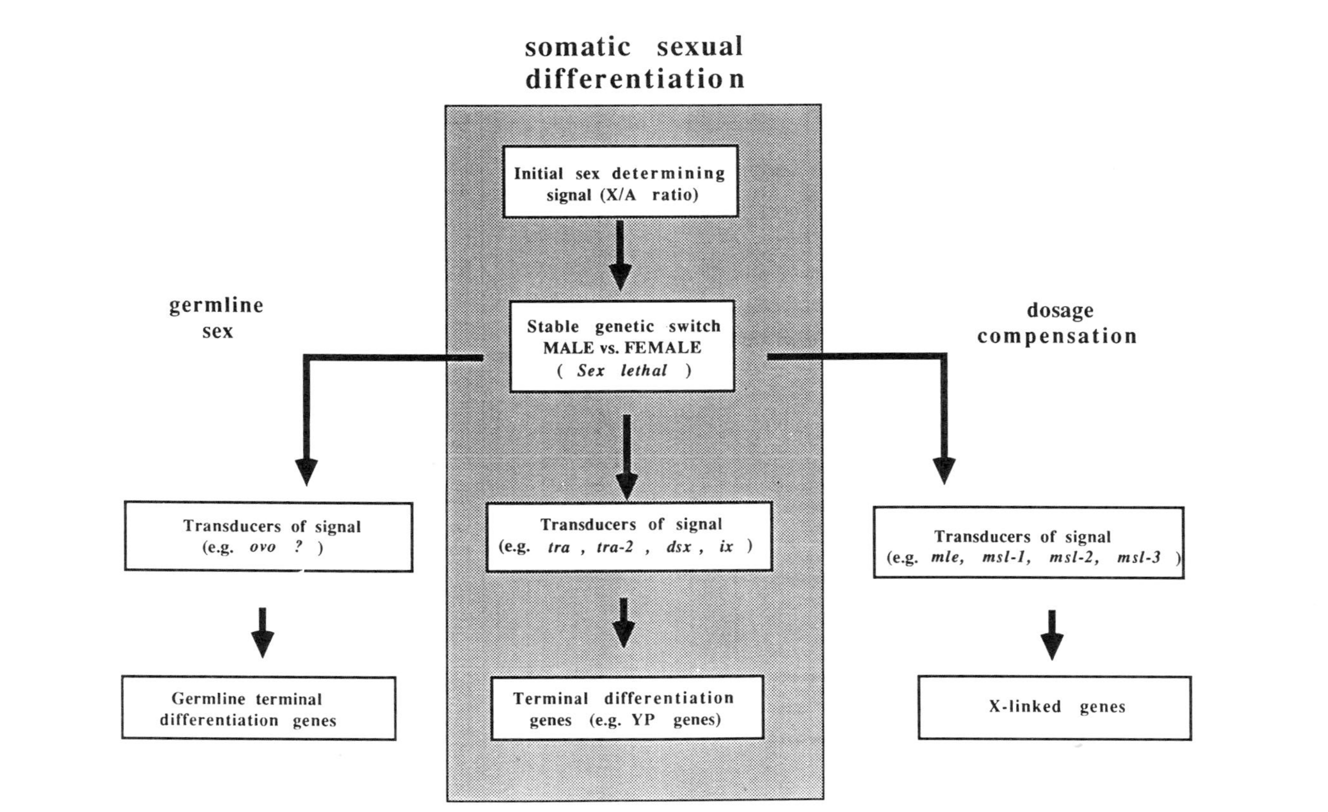

Figure 8.2. Regulatory hierarchy controlling sexual development in *Drosophila melanogaster*. There are three processes that are controlled by the X/A ratio and the key regulatory gene, *Sex lethal*: germ line sex differentiation, somatic sexual differentiation, and dosage compensation. This chapter focuses on the genetic pathway governing somatic sex (shaded area). For details, see the text.

hierarchy. These sex differentiation regulatory genes include *transformer* (*tra*), *transformer-2* (*tra-2*), *double sex* (*dsx*), and *intersex* (*ix*). At the bottom of the regulatory hierarchy controlling sexual differentiation are the many different structural genes that are expressed in a sex-differential manner and whose products are directly responsible for the structure and function of sexually dimorphic tissues. It has been shown that in at least some cases, e.g., the female-specific yolk protein gene expression, the target genes are regulated by the sex differentiation regulatory hierarchy at the level of their transcription (Kraus et al. 1988).

2.2. Dosage Compensation

As in many species with heteromorphic sex chromosomes, *Drosophila melanogaster* has a mechanism by which X-linked gene expression is equalized in females, with two X chromosomes per diploid cell, and in males, with one X per diploid cell. This regulatory process, termed dosage compensation, involves a twofold hyperactivation of X-linked genes in males, with the increased gene expression seen as an increase in the levels of steady-state RNA from X-linked genes (Mukherjee and Beermann 1965; Birchler et al. 1982; Ganguly et al. 1985; Breen and Lucchesi 1986). Sexual differentiation and dosage compensation are both controlled by the same primary determinant, the X/A ratio, and they share some of the early sex determination regulatory signals such as the key regulatory gene *Sex lethal* (Cline 1978; 1983a; 1985; Lucchesi and Skripsky 1981), but the pathways governing these two processes are genetically separable. For example, mutations in genes such as the *transfomer* locus (see below) that affect somatic sexual differentiation have no effect on the state of X chromosome activity (Smith and Lucchesi 1969), and mutations in genes such as *male-specific lethal-2* (see below) that affect the dosage compensation process have no effect on sexual differentiation (Belote 1983). Thus, the two pathways controlling sexual differentiation and dosage compensation must diverge at some point after the primary sex determination event. For a thorough discussion of dosage compensation in *Drosophila*, see Lucchesi and Manning (1987).

2.3. Sex Determination of the Germ Line

Another aspect of sexual development that is under the control of the X/A ratio is the sex-specific differentiation of the germline. In *Drosophila* the germline precursor cells, the pole cells, form very early in embryogenesis at the posterior pole of the preblastoderm embryo. Pole cell transplantation studies, in which the developmental fates of XX and XY pole cells in male and female hosts have been followed, have shown that the subsequent differentiation of pole cells into oocytes or spermatocytes is dependent on two kinds of signals, one cell autonomous and the other inductive (Steinmann-Zwicky et al. 1989). The cell-autonomous signal is provided by the X/A ratio. In ovaries, XX cells undergo oogenesis and XY cells undergo spermatogenesis, suggesting that these cells follow the sexual pathway determined by their own X/A ratio. In testes, however, the germ cells do not develop in a cell autonomous manner. In this case, both XX and XY cells undergo spermatogenesis, suggesting that there is an inductive signal provided by the male soma that causes the female germ cells to enter the spermatogenic pathway in spite of their XX:AA genotype. That the regulatory pathway controlling sexual differentiation of somatic cells is different from that controlling germline sex is inferred from genetic studies in which the effects of mutations affecting somatic sexual differentiation have been examined in the germline (Marsh and Wieschaus 1978; Schupbach 1982). These analyses have revealed that the *tra*, *tra-2*, *dsx*, and *ix* genes, which are essential for normal sex differentiation of somatic tissues, do not participate in the sexual differentiation of the germline. There is at least one gene, however, *Sxl*, that does play a critical role in the sexual development of both germline and soma (Cline 1983b; Schupbach 1985; Steinmann- Zwicky et al. 1989).

In light of the above, one can view the regulatory hierarchy governing the three sex-specific processes, somatic sexual differentiation, germline sex and dosage compensation, as being a branched pathway, with all three sharing the same primary determinant (i.e., the X/A ratio) and all relying on the same key gene, *Sex lethal*, but with the three regulatory pathways diverging at some point after the functioning of *Sxl* (see Figure 8.2). In this chapter I will be focusing on the regulatory pathway governing somatic sexual differentiation and will cover germline sex and dosage compensation only in passing.

2.4. The Primary Determinant of Sex

2.4.1. The X/A Ratio

In *Drosophila melanogaster*, an individual with two X chromosomes and two sets of autosomes (i.e., XX; AA) develops as a female, while one with a single X chromosome and two sets of autosomes (XY; AA) develops as a male. Although a Y chromosome is normally carried by a male, it is required only for the completion of spermatogenesis and plays no role in sex determination. Thus, XO and XXY flies, which can occasionally arise as the result of X chromosome nondisjunction in the mother, develop as sterile males and fertile females, respectively (Bridges 1916). Examination of the sexual phenotypes of other aneuploid types has helped define the nature of the chromosomal determinant of sex. For instance, individuals with two Xs and three sets of autosomes (XX;

AAA) develop as sexually intermediate flies with intermingled patches of male and female tissue (Stern 1966). The intermediate sexual phenotype of such triploid intersexes, taken together with observations on the phenotypes of other X chromosome aneuploid types, led Bridges to propose that it is not merely the number of X chromosomes that is important for determining sex, but rather, it is the balance between the number of Xs and the number of haploid sets of autosomes, i.e., the X/A ratio (Bridges 1921, 1925), that is the relevant parameter.

Another informative genotype concerns the situation arising when an X chromosome is lost from a cell during the early cleavage divisions of the embryo. When this happens, the resulting fly is a mosaic that is comprised of large patches of XO male tissue in a background of XX female tissue, i.e., it is a gynandromorph (Morgan 1916). The existence of distinct boundaries separating male and female cells in these mosaics indicates that the sexual phenotype in *Drosophila* is not determined by hormones that influence sexual differentiation of the fly as a whole, but rather each cell, or small group of cells, determines its sex irrespective of the genotype of neighboring cells. Some exceptions to the cell-autonomous nature of sex determination have been found, however, such as in the case of the development of a male-specific abdominal muscle that requires an inductive signal provided by male neurons (Lawrence and Johnston 1986). As mentioned above, it has also been shown that sex differentiation of the germ line is not strictly cell-autonomous (Steinmann-Zwicky et al. 1989).

2.4.2. *When Is the X/A Ratio Counted in Order to Determine Sex?*

One can imagine two very different ways in which the X/A ratio signal might control sexual differentiation. It could be the case that, as the many different sexually dimorphic tissues differentiate, they must continually monitor their cells' X/A ratio in order to follow the appropriate program of sexual differentiation. An alternative model is that there is an early, irreversible step in which the X/A ratio is assessed and at which time the sexual state of each cell is stably set in either the male or female differentiation mode. One observation favoring the latter comes from the phenotype of XX:AAA, triploid intersexes. As mentioned above, these individuals develop as sexual mosaics, with coarse-grained intermingling of male and female tissues. Within these patches the cells differentiate as either male or female and do not exhibit an intermediate sexual morphology at the level of individual cells. This phenotype can be easily understood if, at an early stage in development, the cells monitor their X/A ratio and make a heritable commitment to either the male or female pathway and then differentiate accordingly. In this case, since the X/A ratio is intermediate between the male and female X/A ratios, it may be sufficiently close to the border between male and female so that different cells within the same individual might perceive the sex determination signal differently. Once the signal is read by a cell, however, that cell and its descendants follow either the male or the female program of differentiation, thereby leading to clones of male and female tissue. The relatively large size of the clones in these triploid intersexes suggests that the sex determination decision occurs fairly early in development, before there are a large number of cuticle progenitor cells.

The hypothesis that sex determination involves an early, irreversible step is also supported by experiments in which the X/A ratio was altered at various times during development by inducing X chromosome loss in XX:AA females, and then the resulting XO:AA clones that encompassed the sexually dimorphic regions scored for sexual phenotype in the adult (Sanchez and Nöthiger 1983; Baker and Belote 1983). In these experiments, XO clones that arose early in embryogenesis were phenotypically male and of the large size expected for clones produced early in development. However, XO:AA clones produced later during development (e.g., during the pupal period) were very small, indicating that they grew poorly, and they appeared to be female in phenotype (e.g., they exhibited the female pattern of abdominal pigmentation). This result is expected if the X/A ratio is read by each cell at some point relatively early in development, and as a result of that assessment, the pathway of sexual differentiation irreversibly set for that cell and its descendants. In this case, the counting of the X/A ratio would occur when there were 2 Xs per cell, and the sex of each cell would thus be irreversibly set as female. The slow growth of the XO:AA "female" cells can be interpreted as meaning that the level of X chromosome activity (i.e., dosage compensation) is also irreversibly set at the time of sex determination. In this case, the XO:AA "female" cells that arose after the X/A ratio was monitored would be expressing their X-linked genes at an inappropriately low level characteristic of female cells and would therefore have reduced levels of X-linked gene products.

A third line of evidence supporting an early, heritable sexual pathway commitment comes from experiments reported by Cline (1985). In these experiments, sexually mosaic intersexes were generated using a combination of mutant alleles that affects the choice of sexual state that the cells make, and then the technique of mitotic recombination was used to tag cells at some point after embryogenesis was completed but before sexual differentiation of the adult structures had occurred. The adults generated were sexually mosaic individuals carrying clones of genetically marked cells that had been generated at a known time in development. The observation that none of the marked clones encompassing the sexually dimorphic sex comb region (51 clones examined) ever included both male and female tissue suggests that at the time the sexual state was chosen in these flies, it was irreversibly set in either the male or female mode, and that there was no switching of sexual state later in development.

Taken together, all of the above observations support the idea that sex determination in *Drosophila* involves an early, irreversible step that is made at the cellular level and is then "remembered" thereafter throughout the remainder of development.

2.4.3. What Is the Nature of the X/A Signal?

Early genetic studies on the nature of the chromosomal signal determining sex involved searching for the regions of the X chromosome and the autosomes, whose dosage mattered for determining the sexual phenotype of the fly (reviewed in Baker and Belote 1983). These studies identified several regions of the X chromosome that appeared to contain loci that were important for determining sex, however, searches for autosomal loci that were involved in specifying the X/A ratio were inconclusive. More recent studies have focused on specific X-linked loci that appear to be involved in the setting up or monitoring of the X/A ratio (Cline 1986; 1987; 1988; Steinmann-Zwicky and Nöthiger 1985). For example, Cline has identified two discrete X-linked loci, named *sisterless-a* (*sis-a*) and *sisterless-b* (*sis-b*), that have the properties of X/A "numerator" elements—sites on the X whose dosage determines what the X/A ratio is perceived as. Mutations in either of these genes are lethal to females due to a failure to activate the *Sxl* gene, which, as discussed below, is essential for female development (Cline 1986; 1988). The exact role of these genes in setting up or assessing the X/A ratio is unclear. One possibility is that these loci are genes that produce *trans*acting products that interact with the *Sxl* gene, with their dosage determining how much of the *trans*acting factors are produced. Another possibility is that the *sis* loci might not be genes in the usual sense, but rather, they might be sites on the DNA that bind factors that are present in the early embryo in limiting amounts. In this case, the number of copies of *sis-a* and *sis-b* determines how much of the factor remains unbound and it is this, and not the products of the *sis* genes themselves, that interacts with *Sxl* (Cline 1988; 1989).

2.5. The *Sex Lethal* Gene

One of the earliest responses to the signal provided by the X/A ratio is the setting of the activity state of *Sex lethal*, a key gene in the regulatory pathways controlling both sex differentiation and dosage compensation (Cline 1978; 1979; 1983a; 1983b; 1984; 1985; Lucchesi and Skripsky 1981; Sanchez and Nöthiger 1982; Gergen 1987). This gene was originally identified by the discovery of a semidominant female-lethal mutant called *Female lethal* (Muller and Zimmering 1960). It wasn't until the extensive genetic characterization and mutational analysis of this gene by Cline, however, that its pivotal role in sexual development was recognized and appreciated (1978; 1983a; 1983b; 1984; 1985).

The most striking property of the *Sxl* gene is the sex-specific lethality exhibited by many of its mutant alleles. Complete or partial loss-of-function *Sxl* mutations (such as Sxl^{F1}) are recessive and cause chromosomally female individuals (i.e., XX; AA) to inappropriately follow the male pathways of sexual differentiation and dosage compensation (Cline 1979; 1983a; Lucchesi and Skripsky 1981; Sanchez and Nöthiger 1982; Maine et al. 1985a; Gergen 1987). Normally, the female > male sex transformation that occurs as the result of the Sxl^{F1} mutation cannot be seen because the mutant females die before reaching the adult stage. However, genetic manipulations, such as the production by X-ray induced mitotic recombination of Sxl^{F1}/Sxl^{F1} mutant clones in an otherwise normal $Sxl^{F1}/+$ female, can be used to circumvent this problem (Cline 1979; Sanchez and Nöthiger 1982). The female-specific lethality exhibited by many of these alleles is thought to be due to the abnormally high levels of X-linked gene products resulting from abnormal dosage compensation (Lucchesi and Skripsky 1981; Cline 1883a; Gergen 1987). *Sxl* null mutations (e.g., deletions of the *Sxl* gene) do not affect XY; AA individuals (Salz et al. 1987), suggesting that the *Sxl* gene plays no major role in male development, and that it might be in a functionally "off" state in males.

In contrast to what is seen with loss-of-function mutations, alleles that behave as gain-of-function, constitutive mutants (such as Sxl^{M1}) have the opposite phenotype: they have no effect on XX; AA flies but cause chromosomally male individuals (i.e., XY; AA) to follow the inappropriate female pathways of sex differentiation and dosage compensation (Cline 1979; 1983a). Such alleles are dominant and are lethal to males. As with Sxl^{F1} mutants, the sex transformation phenotype cannot be observed directly because of the lethal phenotype. However, the male > female sex transformation associated with Sxl^{M1} can be seen in XX//XO mosaics in which the XO cells are mutant for Sxl^{M1}. In this case, the patch of mutant cells develops as female in spite of its male chromosome constitution (Cline 1979). The male-specific lethal phenotype associated with Sxl^{M1} is interpreted as being the result of a failure of the X chromosome hyperactivation that normally occurs in males, although direct evidence on this is lacking.

The extensive genetic analysis of *Sxl* by Cline has led him to propose the following model. If the X/A ratio is 0.5 (i.e., one X, two sets of autosomes), as in a male, then the *Sxl* gene is functionally "off," and as a result, male sexual differentiation and hyperactivation of X-linked genes occurs. If the X/A ratio is 1.0, as in a female, then *Sxl* is set in its functionally "on" state, where it acts to elicit female sexual differentiation, and it also causes genes on the X to be expressed at the lower level characteristic of females. The setting of *Sxl* in its active state requires other factors in addition to the signal provided by the X/A ratio = 1.0. These other factors include the maternally-contributed product of the daughterless gene (Cline 1976; 1980; 1985; 1989), the zygotically-acting *sis-a* and *sis-b* genes (Cline 1986; 1988; 1989), and the product of the recently identified *sans fille* (*snf*) locus (Oliver et al. 1988), also known as *liz* (Steinmann-Zwicky 1988). There is also some evidence that a product of the *Sxl* gene itself, produced very early in embryogenesis, may also play a role in the stable activation of *Sxl* expression (Cline 1988; Salz et al. 1989). Once the *Sxl* gene is set in either the female "on" or male "off" state, it acts continuously to affect the sex-specific expression of the downstream regulatory genes that govern somatic sexual differentiation and dosage compensation throughout the rest of sexual development. Thus, the *Sxl* gene can be thought of as a sex-specific binary genetic switch that mediates the choice between the male and female sexual states.

The ability of *Sxl* to act as a stable genetic switch is thought to be related to a positive autoregulatory property exhibited by the *Sxl* gene as revealed by genetic analysis (Cline 1984). The hypothesis that *Sxl* regulates its own expression is based on the observation that some *Sxl* alleles can act in *trans* to activate other *Sxl* alleles that normally would remain inactive in the absence of *Sxl* function. Because of this autoregulatory property, once the *Sxl* gene is activated in females, it remains in its active, functional state, and is then independent of the signals that lie upstream of *Sxl* in the regulatory pathway, such as that provided by the X/A ratio. Recent molecular observations have suggested a possible mechanism by which *Sxl* might act to maintain its continued expression once it is activated early in female development (see below).

The effect of *Sxl* on somatic sexual differentiation and dosage compensation is brought about through the action of sets of regulatory genes that lie downstream of *Sxl* in the regulatory hierarchy and are specific to each of those processes. The *maleless*, *male-specific lethal-1*, *male-specific lethal-2*, and *mle(3)132* (also known as *msl-3*) genes are among the downstream components of the pathway controlling dosage compensation (Fukunaga et al. 1975; Belote and Lucchesi 1980a; Uchida et al. 1981; Skripsky and Lucchesi 1982). These genes are thought to respond to *Sxl*, and their activities are required throughout much of development for the X chromosome hyperactivation that occurs in males. The male-specific lethal phenotype of loss-of-function mutations of these genes is a consequence of the failure of X chromosome hyperactivation in males (Belote and Lucchesi 1980a; Belote 1983; Okuno et al. 1984; Breen and Lucchesi 1986). The observation that the *mle*, *msl-1*, and *msl-2* mutants fail to affect the expression of the X-linked, early acting gene, *runt*, while the Sxl^{F1} mutant does affect *runt* expression, has led to the proposal that there may be additional regulatory genes controlled by *Sxl* and specifically involved in dosage compensation in the early embryo (Gergen 1987; see also Cline 1984). There is also a requirement of *Sxl* for female germline development and function (Cline 1983b; Schupbach 1985; Salz et al. 1987), although little is known of the downstream genes in this developmental pathway that respond to the state of *Sxl* expression. The *ovo* gene (Oliver et al. 1987), which is required for germline development in females but not in males, is one candidate for such a gene.

2.6. Downstream Genes Controlling Somatic Sexual Differentiation

The control of somatic sexual differentiation by *Sxl* is mediated through the functioning of at least four genes that lie downstream of *Sxl* in the regulatory pathway. These genes, *transformer* (*tra*), *intersex* (*ix*), *double sex* (*dsx*), and *tranformer-2* (*tra-2*), were originally identified as mutants that disrupted normal sexual development (Sturtevant 1945; Meyer and Edmondson 1951; Hildreth 1965; Watanabe 1975). The observation that all aspects of somatic sexual development (i.e., morphological, biochemical, behavioral) are affected by these mutants suggests that they represent regulatory functions. All four of these genes are autosomally-linked, and they map to well-separated sites in the genome. Loss-of-function alleles at these loci are recessive and none appear to affect viability. The *tra* and *tra-2* mutants have no effect on sexual differentiation in XY; AA flies but cause chromosomally female individuals (i.e., XX; AA) to develop as males with respect to all aspects of somatic sexual development. The germline cells of these flies are not transformed to maleness, so these XX males are not fertile. The *ix* mutant also has no effect on XY males but causes chromosomally female individuals to develop with a sexually intermediate phenotype. The *dsx* mutant affects both XY and XX individuals, causing them to develop as intersexes. The intersexes arising as the result of mutation at *ix* or *dsx* are different from the triploid intersexes described earlier in that these mutants are not mosaics of male and female tissue. Rather, they exhibit intermediate sexual phenotypes at the level of individual cells, as if individual cells were simultaneously trying to follow both male and female pathways of differentiation. While the *tra-2* gene does play some role in spermatogenesis (Belote and Baker 1983), none of these mutants is capable of transforming the germ cells of either sex (Marsh and Wieschaus 1978; Schupbach 1982), consistent with the abovementioned notion that somatic and germline sex are under separate genetic controls.

By characterizing the genetic properties of various mutant alleles at *tra*, *tra-2*, *dsx*, and *ix*, and by defining the epistatic interactions among these genes in double-mutant combinations, Baker and Ridge (1980) put forth a model of how these four genes might be arranged in a single regulatory pathway controlling sex differentiation of the soma (for reviews, see Baker and Belote 1983; Cline 1985; Nöthiger and Steinmann-Zwicky 1985). One central feature of this model is that the *dsx* locus has the unusual property that it is capable of being actively expressed in either of two mutually exclusive modes. According to this model (Figure 8.3), in chromosomally female individuals (XX; AA, X/A ratio = 1.0), the *tra* and *tra-2* genes are functionally "on" in response to the *Sxl* "on" condition, and their products are required to set the bifunctional *dsx* gene into its female mode of expression. When *dsx* is expressed in the female mode (dsx^F), it acts to repress all male-specific differentiation, and as a result, normal female development occurs. The functional product of the *ix* gene is needed in females for the correct repression of male differentiation. In males (XY; AA, X/A ratio = 0.5), the *tra* and *tra-2* genes are functionally "off" as a result of the *Sxl* "off" condition. Consequently, *dsx* is expressed in its male mode (dsx^M). When *dsx* is expressed in this way, female-specific differentiation is repressed and normal male development ensues. The *ix* gene does not appear to be required for male differentiation. It should be noted that this genetic model does not specify at what molecular level these on/off or male/female binary genetic switches are being controlled.

Genetic analyses have also been useful for determining when during development these regulatory genes must be acting. Studies using temperature-sensitive alleles of the *tra-2* gene have shown that the functioning of this gene, and by inference, that of the other genes in this regulatory hierarchy, is first required by at least the second larval instar stage, and then at many times throughout development for proper sexual development (Belote and Baker 1982; Epper and Bryant 1983; Chapman and Wolfner 1988). The functioning of this regulatory pathway is required even in the adult fly for female-specific yolk protein gene expression (Belote et al. 1985b) and repression of male-specific courtship behavior (Belote and Baker 1987).

3. Molecular Studies of the Genes Controlling Sexual Differentiation

The genetic studies described in the preceding sections resulted in the identification of a number of specific genes that play important roles in controlling somatic sexual differentiation and led to a model, shown in Figure 8.3, of how these genes might interact as parts of a single regulatory pathway. The isolation of some of these genes by molecular cloning techniques has made it possible to test various aspects of this model and to extend the description of these gene interactions to the molecular level. In this section I will describe what has been learned

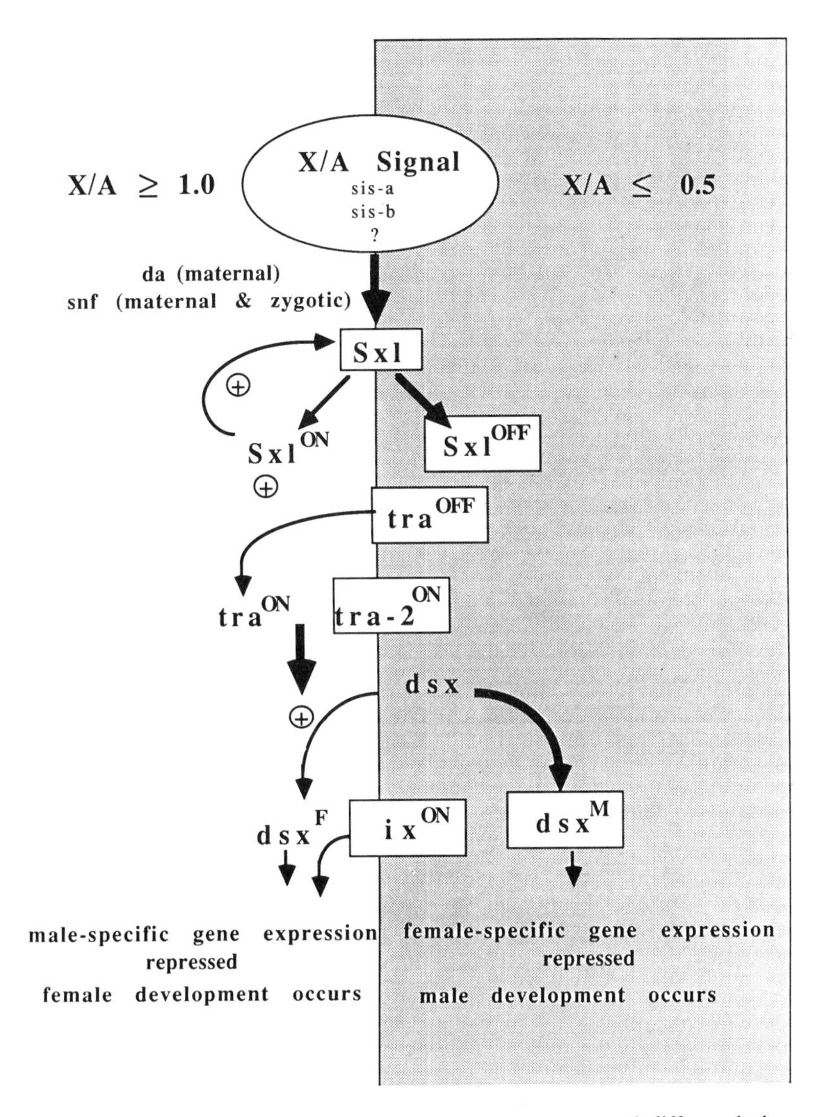

Figure 8.3. The regulatory pathway controlling somatic sexual differentiation, as determined by genetic studies. The *Sxl* gene responds to the X/A ratio such that it is functionally on in females and functionally off in males. The *sis-a* and *sis-b* genes appear to play a major role in specifying or assessing the X/A ratio signal. The activation of *Sxl* requires the maternally-contributed *da* gene product and the zygotically acting *snf* gene. If *Sxl* is off, the *tra* gene fails to be activated and *dsx* remains in its male mode of expression, where it acts to repress female-specific differentiation. If *Sxl* is on, it acts to turn on the *tra* gene. The *tra* gene acts together with the *tra-2* gene to switch *dsx* into its female mode of expression, where it acts to repress male-specific differentiation. The *ix* gene is required for the proper repression of male-specific differentiation in females. In addition to activating *tra*, the *Sxl* gene also exhibits a positive autoregulatory property, which allows it to remain on once it is activated early in development.

about the molecular genetics of sex differentiation and how it relates to the above genetic model. To date, five of the genes involved in controlling somatic sexual differentiation have been cloned and analyzed at the molecular level: *Sex-lethal* (Maine et al. 1985a; 1985b; Salz et al. 1987; Bell et al. 1988; Salz et al. 1989), *daughterless* (Caudy et al. 1988a; 1988b; Cronmiller et al. 1988), *transformer* (Belote et al. 1985b; 1989; Butler et al. 1986; McKeown et al. 1987; 1988; Boggs et al. 1987; Sosnowski et al. 1989), *transformer-2* (Amrein et al. 1988; Goralski et al. 1989), and *double sex* (Belote et al. 1985a; Baker and Wolfner 1988; Burtis and Baker 1989).

3.1. *Sex-lethal* and *daughterless*

The *Sxl* locus was cloned using the technique of P-element transposon tagging and chromosome walking starting with the P-element insertion mutant Sxl^{fPb} (Maine et al. 1985a, 1985b). As the first step toward defining the size and organization of the *Sxl* gene, whole-genome Southern blot analysis, using DNA isolated from a variety of *Sxl* mutant alleles, was carried out to see if any of them were associated with rearrangements within the cloned region (Maine et al. 1985a, 1985b). The original analysis found that 10 of the tested *Sxl* mutant alleles were associated with DNA rearrangements, and that these alterations in the DNA were localized to within an 11 kb region, suggesting this as a minimum size of the *Sxl* functional unit. As another approach toward defining the size and structure of the *Sxl* gene, Northern blot analyses, using various restriction fragments from the *Sxl* region as probes, were carried out (Maine et al. 1985b; Salz et al. 1989). As shown in Figure 8.4, the pattern of transcripts arising from the *Sxl* locus is complex, with at least 10 overlapping RNA species identified by this method (Salz et al. 1989). That these RNAs correspond to *Sxl* transcripts is supported by the observations that they all overlap the region containing the 11 kb interval which is disrupted by *Sxl* mutations; and the expression of this complex pattern of RNAs is affected by several *Sxl* mutant alleles which have been examined. Like several other regulatory genes that have been characterized in *Drosophila*, the *Sxl* transcription unit is large, approximately 23 kb. The transcription units flanking *Sxl* represent genes of unknown function, and they do not show any sex- or stage-specific pattern of expression (Salz et al. 1989). They do not appear to correspond to the two essential genes flanking *Sxl* that had been previously identified by mutational analysis (Nicklas and Cline 1983). Examination of the developmental profile of the *Sxl* transcripts revealed that they belong to two distinct sets of RNAs: early transcripts found up until the blastoderm stage of embryogenesis and then disappearing, and late transcripts appearing at, or just after, the blastoderm stage and then persisting throughout development.

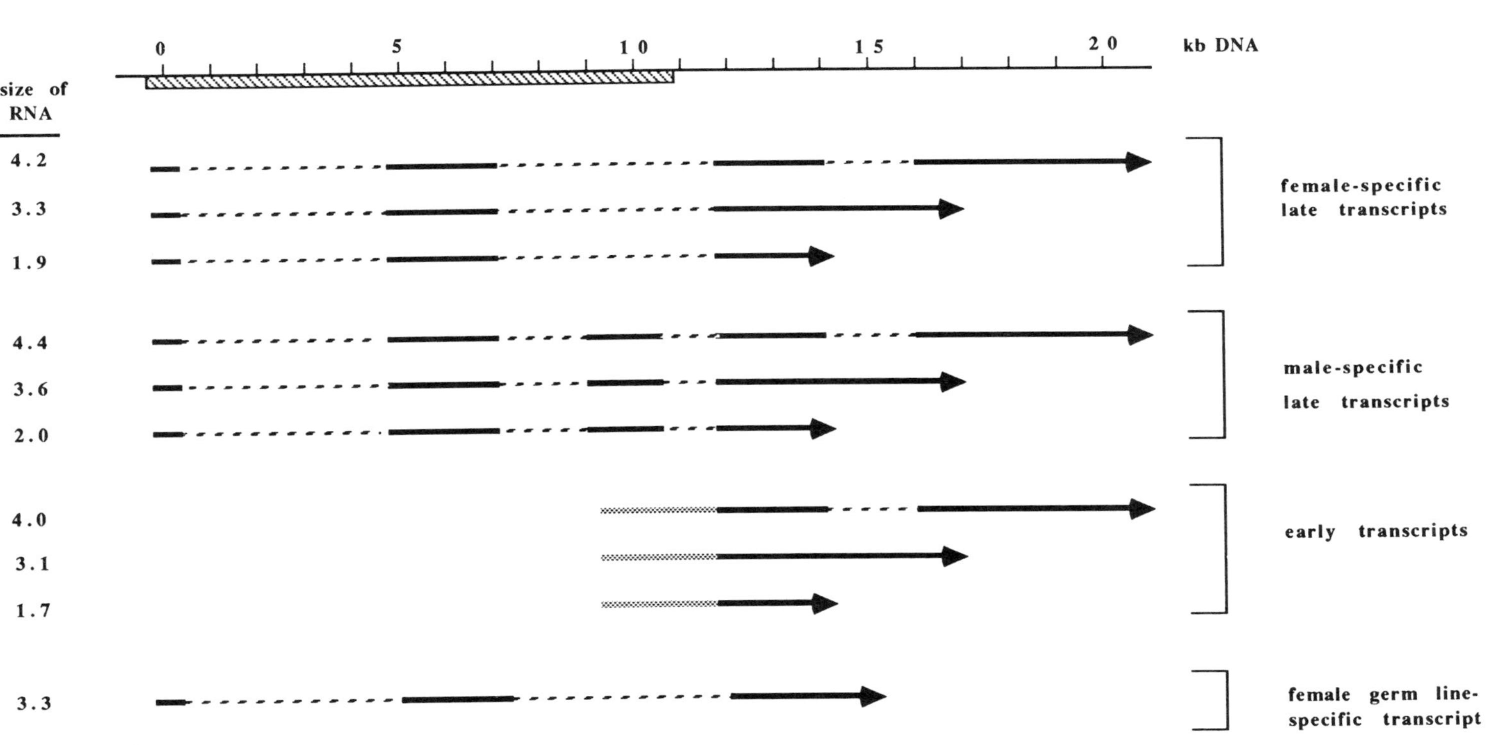

Figure 8.4. Transcription pattern of the *Sxl* gene. The crosshatched bar, just below the scale, represents the 11 kb region of the *Sxl* transcription unit where *Sxl* mutants have been molecularly mapped. The dark lines and arrows represent the approximate organization and structure of the 10 different *Sxl* transcripts as determined by Northern blot analysis. The shaded lines in the early transcripts indicate the uncertainty of assigning the transcription start sites for the transcripts. This figure is derived from the results of Salz et al. (1989).

According to the genetic model, the *Sxl* gene is functional (i.e., "on") in females but nonfunctional ("off") in males. While the sex-specificity of the early transcripts is not known, late transcripts are observed in both sexes. Since it had been previously shown that males completely lacking the *Sxl* gene were perfectly viable and fertile (Salz et al. 1987), the appearance of *Sxl* RNAs in males was not anticipated. What this finding implies is that the on/off switch at *Sxl* is not based simply on a sex-specific transcription mechanism. However, the late transcripts do exhibit a sex-specific pattern, in that there are different size classes of RNAs seen in males as compared to females (Salz et al. 1989). In females, three late transcripts are seen, of sizes 4.2, 3.3, and 1.9 kb. The major differences among these RNAs are at the 3′ ends of the transcripts. In males there are three slightly larger transcripts, of sizes 4.4, 3.6, and 2.0 kb. These transcripts are very similar in structure to the three female transcripts, with one important difference: they all include a small internal exon that is absent from the female transcripts.

Analysis of *Sxl* cDNAs has provided more detail on the differences between the male and female specific *Sxl* transcripts (Bell et al. 1988). All of the male-specific *Sxl* RNAs contain an exon of ~180 bp that is downstream of the exon containing the presumptive translation initiation codon for the *Sxl* protein (Figure 8.5). This exon contains stop codons that interrupt the long open reading frame present in the female specific transcripts. Thus, the male-specific *Sxl* transcripts appear to encode a truncated version of the *Sxl* protein that is presumably nonfunctional. In females, because of the alternative splicing event, the *Sxl* transcripts do contain long open reading frames that potentially encode a polypeptide of 354 amino acids (~39,000 daltons). This presumably represents the functional *Sxl* gene product. This sex-specific alternative RNA splicing of *Sxl* pre-mRNA could thus provide the mechanism by which *Sxl* function is expressed in a female-specific manner.

Another finding to come out of the analysis of *Sxl* cDNAs concerns the sequence of the predicted protein product of the female-specific transcripts. Conceptual translation of the female-specific *Sxl* RNAs reveals that the presumptive protein contains a pair of sequence motifs having significant sequence similarity to a conserved domain, the ribonucleoprotein consensus sequence (see Figure 8.6) found in several proteins implicated in binding to RNA (Bandziulis et al. 1989; Query et al. 1989). This suggests that at least one of the functions of the *Sxl* gene product involves a direct interaction with RNA. As mentioned earlier, the *Sxl* gene product acts in a positive autoregulatory manner to promote its own continued expression. Given the molecular observations, one reasonable hypothesis to explain this autoregulation is that the female-specific *Sxl* protein binds to the *Sxl* pre-mRNA and blocks splicing at the male-specific exon. In this event, splicing occurs by default at the

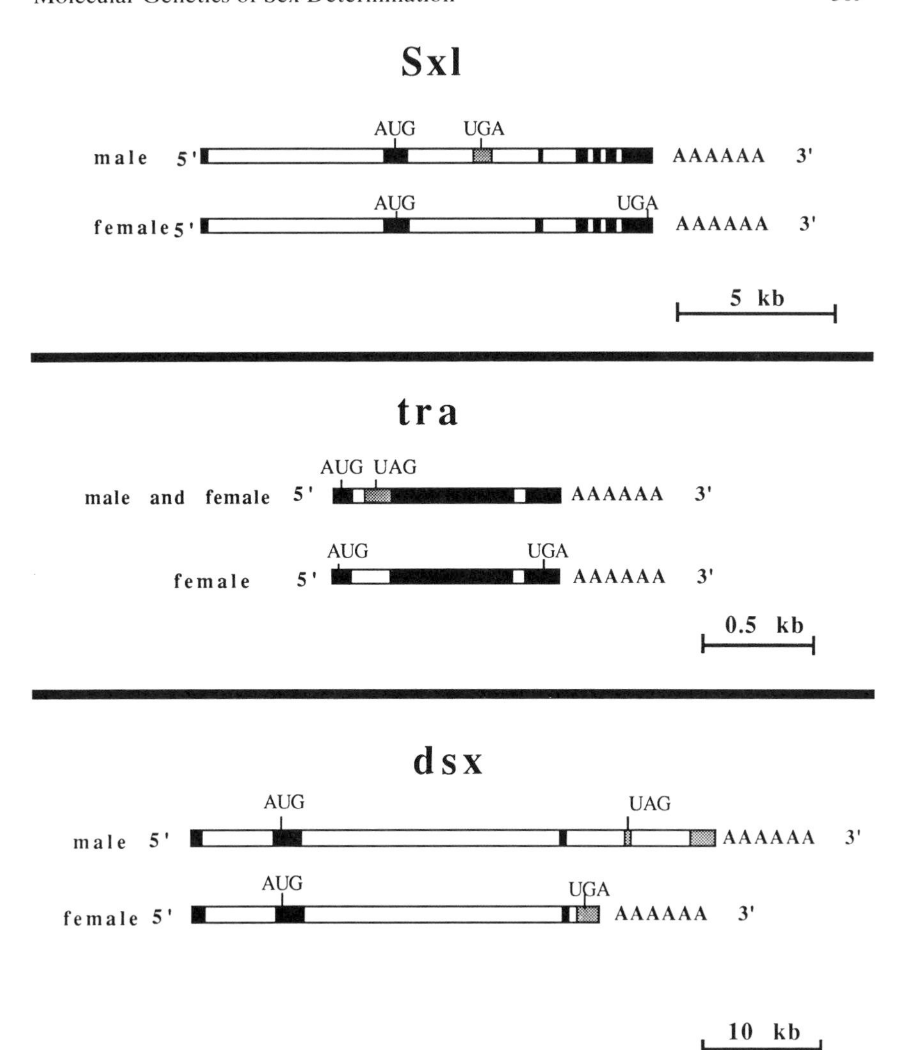

Figure 8.5. Structure of sex-specific transcripts at *Sxl*, *tra*, and *dsx*. The open boxes are introns, the dark boxes are exons, and the shaded areas are sex-specific exons. See text for details. These diagrams are based on the results presented in Bell et al. 1988 (*Sxl*), Boggs et al. 1987 (*tra*), and Burtis and Baker 1989 (*dsx*).

downstream female 3′ splice site, resulting in the production of the female-specific RNAs. Thus, once functional *Sxl* protein is produced early in development, it acts to cause continued production of the female-specific, fuctional *Sxl* protein. As discussed below, the same molecular mechanism is also sufficient to explain the effect of *Sxl* on somatic sexual differentiation.

```
RNP  Consensus
   Sequence    ......L F V G N L ......E. . L . . . F . . .F G . I .............K · · K G F G F V X F.........A...........L . G
                     I Y I K G          D                   Y     V             R       R   Y A       Y                     I

transformer-2· · · · ·I G V F G L ......Q· · V· · · F· · ·Y G . I .............R · · R G F C F I Y F.........A··········E . G

Sex lethal  1 ·····L I V N Y L......D· ·L· · · F. . F G . V.............N . . K T G Y S F G Y.........M··········L . G

Sex lethal  2 .....L Y V T N L......D  L. . . F. . I G . I ............R. . R G V A F V R Y.........A...........P . G
```

Figure 8.6. Comparison between the ribonucleoprotein consensus sequence (RNP consensus) and regions of the presumptive *Sxl* and *tra-2* proteins. *Sxl 1* and *Sxl 2* refer to the two copies of the RNP domain present in the predicted polypeptide. The standard one letter amino acid code is used. The sequence of *tra-2* is from Goralski et al. (1989), and the sequence of *Sxl* is from Bell et al. (1988). The RNP consensus is from Bandziulis et al. (1989).

While the on/off regulation of *Sxl* function throughout most of development can be explained as the result of a mechanism involving alternative RNA splicing of *Sxl* pre-mRNA, the regulation of *Sxl* early in development might be quite different. For example, under the simplest form of the above model, the initial activation of *Sxl* would have to occur only in females, otherwise the autoregulatory property of the *Sxl* product would impose the female sexual state on every embryo. The observation that there are *Sxl* transcripts unique to the early embryo is consistent with the initial activation of *Sxl* being regulated in a different way from that of late *Sxl* expression. As shown in Figure 8.4, these early transcripts are produced from a different promoter than the one used for transcription of the late RNAs. Salz et al. (1989) have proposed that the initial activation of *Sxl* might be controlled at the transcriptional level, with the early *Sxl* RNAs being produced only in females. This female-specific transcriptional activation would be under the control of the signal provided by the X/A ratio and also require the maternally-contributed *da* gene product.

The *daughterless* gene was cloned by chromosome walking and was localized on the molecular map of the region by whole-genome Southern blot analysis of several *da* mutant alleles (Caudy et al. 1988b; Cronmiller et al. 1988). The *da* gene is represented by a transcription unit of approximately 5.3 kb. This gene specifies two transcripts, 3.2 and 3.4 (or possibly 3.7), that are present in both sexes and at all stages of development. The molecular characterization of *da* cDNAs indicates that the *da* gene potentially encodes a protein with significant sequence similarity to the human immunoglobulin gene enhancer-binding proteins, E12 and E47, thought to act as transcriptional activators (Murre et al. 1989). Given this, it is not unreasonable to suppose that the gene product of the *da* gene also acts as a transcriptional activator (Caudy et al. 1988a; Cronmiller et al. 1988). Genetic evidence has suggested that the *da* gene has separate maternal and zygotic activities, both of which are required for normal development (Cline 1976; 1980; Cronmiller and Cline 1987). In addition to its role in the initial activation of *Sxl* (Cline 1980; 1983a), the *da* gene has also been implicated in the determination of the peripheral nervous system (Caudy et al. 1988a; 1988b). This function requires the expression of the *da* gene in the zygote and does not involve the activation of *Sxl*. One reasonable hypothesis is that the *da* protein is capable of forming heteromeric complexes with other factors, and that the different complexes would act as transcriptional activators for different genes, depending on the particular complex formed (Cline 1989).

According to a model suggested by Cline and his collaborators (Cronmiller et al. 1988; Salz et al. 1989), the *da* gene product, present in the egg as a maternally-supplied factor, may interact with some component(s) specified by an X/A ratio of 1.0, possibly including the products of the *sis-a* and *sis-b* genes, to transcriptionally activate the *Sxl* gene from its early promoter. This would lead to the production of a functional form of the *Sxl* protein only in females. Soon after, at about four hours post-fertilization, transcription of the *Sxl* gene switches to using the late promoter. Transcription from this promoter does not require *da* protein and occurs in both males and females. In females, the *Sxl* protein produced from the early expression of *Sxl* interacts directly with the *Sxl* pre-mRNA to inhibit splicing at the male-specific exon, and the female-specific processed RNA is produced. Translation of this message yields more functional *Sxl* protein which can likewise promote continued production of additional female-specific RNAs. Thus, once the *Sxl* protein is produced, the *Sxl* gene is stably set for the remainder of development in its female mode of expression. In males, there is no *Sxl* protein present, so the *Sxl* pre-mRNA is always spliced to yield RNAs containing the male-specific exon. These RNAs are not functional messages in that they are not translated into functional *Sxl* product.

3.2. *transformer*

The *Sxl* gene controls somatic sexual differentiation through its effect on the sex-specific expression of the *transformer* gene (Cline 1979; Nagoshi et al. 1988; McKeown et al. 1988). The *tra* gene was cloned by carrying out chromsomal walks through the cytological region where the gene was known to map (Belote et al. 1985a; Butler et al. 1986; McKeown et al. 1987). Whole-genome Southern blot analysis of flies carrying cytologically visible chromosome rearrangements, and deletion mapping of the known complementation groups in the region, made it possible to correlate the molecular map of this region with the genetic and cytogenetic maps. Once the position of the *tra* gene had been delimited in this way, P-element mediated germ line transformation methods were used to introduce cloned DNA fragments back into the *Drosophila* genome to see which ones could complement the *tra* mutation. In this way, the *tra* gene was found to be contained within a 2.0 kb fragment of DNA (Butler et al. 1986; McKeown et al. 1987). Once the position of the *tra* gene had been established in this way, detailed analyses of the *tra* transcription unit were carried out, revealing that there are two distinct types of *tra* transcripts, of sizes 1.2 and 1.0 kb. Northern blot analysis showed these transcripts are present from the embryonic through the adult stages of development, and they exhibit a sex-specific pattern of expression. While the 1.2 kb RNA is present at similar levels in both sexes (it is the non-sex-specific *tra* RNA), the 1.0 kb RNA is female-limited (it is the female-specific *tra* RNA).

The isolation and sequencing of *tra* cDNA clones representing both size classes of *tra* transcripts demonstrated that the two types of *tra* RNAs differ from one another by virtue of an alternative splicing event at the first of two introns, diagrammed in Figure 8.5 (Boggs et al. 1987). Sl nuclease protection analysis, using RNA isolated from males and from females, confirmed the existence of this sex-specific alternative splice and failed to reveal any other sex-specific differences in the *tra* transcripts (Boggs et al. 1987). This alternative splicing event is not dependent on the promoter used, since both the non-sex-specific and the female-specific *tra* RNAs have the same 5′ ends (McKeown et al. 1988); and, replacement of the *tra* promoter with the promoter of the heat shock protein gene, *hsp70*, does not prevent the correct sex-specific splicing event from occurring (Boggs et al. 1987).

An examination of the sequences of the two different types of *tra* cDNAs suggests a possible basis for the female-specific function of *tra* that is inferred from the genetic analyses. The female-specific RNA has a single, long open reading frame (ORF), starting with the first AUG and proceeding to within about 90 bases of the polyadenylation site. This ORF potentially encodes a polypeptide of 197 amino acids. This presumptive *tra* protein is highly basic (>30% basic amino acids) and contains a high proportion of interspersed arginine and serine residues. This property is shared by two other molecularly analyzed genes which have been implicated in RNA splicing: the *suppressor of white-apricot* gene of *D. melanogaster* (Zachar et al. 1987), and the human Ul RNA-associated 70 K protein gene (Theissen et al. 1986). As pointed out by Bingham et al. (1988), Arg-Ser domains might be an important feature of genes that play roles in RNA splicing.

In contrast to the female-specific transcript, the non-sex-specific RNA has no long open reading frame. In this case, translation from the first AUG would yield a truncated protein of only 36 amino acids. One simple model suggested by these results is that the polypeptide encoded by the female-specific RNA represents the functional *tra* gene product, and that the non-sex-specific gene product is without function (Boggs et al. 1987). Under this model, the mechanism regulating the on/off expression of *tra* is very similar to that controlling *Sxl* expression. In both cases, the genes are transcribed in both sexes, but the sex-specific alternative splicing of the pre-mRNAs leads to functional gene products only in females.

The small size of the *tra* gene made it possible to use P-element-mediated germline transformation methods to prove that the non-sex-specific transcript is not required for female development, and that the female-specific transcript alone specifies *tra* function (McKeown et al. 1988). In these experiments, a P-element construct containing the female-specific *tra* cDNA sequences placed downstream of the *Drosophila hsp70* gene promoter, called the *hsp-tra-female* gene, was introduced into the genome. It was found that flies, which are XX; tra^{-}/tra^{-} (where tra^{-} indicates a deletion of the wildtype *tra* gene) and which also carry one copy of the *hsp-tra-female* gene, develop as females, even though they completely lack the non-sex-specific *tra* RNA. This indicates that expression of the non-sex-specific *tra* RNA is not necessary for female development, and that expression of the female-specific *tra* RNA is sufficient for causing female differentiation. This, taken together with the observation that the tra^- deletion mutation has no phenotypic effect in XY males, suggests that the non-sex-specific *tra* transcript may be without any function. In terms of the genetic model presented in Figure 8.3, tra^{ON} means that the tra pre-mRNA is spliced to give the female-specific *tra* transcript, which is translated into the functional *tra* protein, and tra^{OFF} means that the tra pre-mRNA is spliced to give the non-sex-specific transcript, which is not translated into functional product.

Another important result from these transformation experiments using *hsp-tra-female* is that chromosomally male individuals (i.e., XY) carrying the *hsp-tra-female* gene show partial, or in some cases complete, somatic female differentiation (McKeown et al. 1988). This result is most consistent with a linear regulatory pathway in which all of the genes that must be functional in order for female development to occur (e.g., *Sxl*, *tra*, *tra-2*, dsx^F and *ix*) either lie upstream of *tra* in the pathway, are induced by the expression of *tra*, or are already expressed but are incapable of causing female development in the absence of *tra* functional product. This result is not easily consistent with models in which there are other genes controlling somatic sexual differentiation, regulated by the X/A ratio or by *Sxl*, but are not in the same linear pathway as *tra*. The observation that the induction of female development by the *hsp-tra-female* gene is not affected by mutations at *Sxl* but does require the wildtype function of *tra-2*, *dsx* and *ix*, is also most consistent with a linear pathway in which the *Sxl* gene functions upstream of *tra*, and *tra-2*, *dsx* and *ix* do not lie upstream of *tra* in the regulatory pathway (McKeown et al. 1988).

3.3. Regulation of the Sex-specific Splicing of *tra* RNA

Genetic and molecular observations have shown that the expression of the *tra* gene is controlled by *Sxl* (Cline 1979; Nagoshi et al. 1988; McKeown et al. 1988; Belote et al. 1989). In the absence of *Sxl* function, as in males or in Sxl^- mutants, only the non-sex-specific *tra* RNA is produced, and *tra* function is "off." In the presence of *Sxl* function, as in females, about half of the *tra* pre-mRNA is spliced in the female specific manner, and functional *tra* gene product is produced. The finding that the presumptive *Sxl* protein has a structure similar to known RNA-binding proteins suggests that the effect of *Sxl* on *tra* splicing might involve a direct interaction of *Sxl* protein with *tra* pre-mRNA, although an indirect effect of *Sxl* on *tra* splicing has not been ruled out.

One important question concerns whether the alternative splicing of *tra* RNA is specified by *cis* acting sequences contained within the intron itself, or whether there are other sequences lying outside of the intron that are essential for sex-specific splicing. To address this question, the sex-specific *tra* intron was excised and inserted into a heterologous gene to see if it still maintained its ability to be spliced in a sex-specific manner (Sosnowski et al. 1989). In these experiments, the *tra* intron was inserted into the Herpes thymidine kinase (TK) cDNA whose expression was driven by the *Drosophila hsp70* promoter. Following transformation of this gene into flies' genomes, examination of the TK transcripts by RNase protection analysis revealed that the TK RNA from this gene exhibits a sex-specific pattern of splicing, with females using primarily the female-specific 3′ splice site, and males using only the non-sex-specific 3′ splice site. This indicates that the *cis*-acting regulatory sequences that are crucial for sex-specific splicing of the *tra* intron are contained within the intron itself (Sosnowski et al. 1989).

There are two basic models for explaining why the non-sex-specific 3′ splice site always wins out over the female-specific 3′ splice site in the competition for splice site selection in males, while the female-specific 3′ splice site frequently, although not exclusively, wins out over the non-sex-specific 3′ splice site in females. The Blockage Model proposes that in the absence of *Sxl* protein, the non-sex-specific splice site is strongly preferred over the female-specific splice site, perhaps because of its proximity. When *Sxl* is expressed (i.e., in females), its product acts, directly or indirectly, to block or retard usage of the non-sex-specific 3′ splice site. As a result, splicing will occur, by default, at the female-specific splice site resulting in the functional *tra* mRNA. The fact that the non-sex-specific splice also occurs in females can be explained if the *Sxl* function is not expressed in all cells, and/or if it is inefficient at blocking the non-sex-specific splice junction. In the alternative model, the Activation Model, it is proposed that *Sxl* function acts in a positive way to activate the female-specific 3′ splice site. In the absence of *Sxl* protein, as in males or *Sxl*$^-$ flies, the female 3′ splice site cannot be used for splicing of the intron.

To help distinguish between these models, *tra* genes carrying small deletions or other in vitro generated mutations of the sex-specific intron, were introduced into the genome by germline transformation, and the in vivo effects of the mutations on sex-specific splicing were monitored (Sosnowski et al. 1989).

In one experiment, sequences from the branch point site to the 3′ splice site of the non-sex-specific splice were deleted, and the mutant gene was put back into the genome, via P-element transformation. The observation that the XY flies carrying this gene exhibit nearly complete female differentiation strongly suggests that the female-specific splice can occur even in the absence of *Sxl* function if the non-sex-specific splicing event is prevented. RNase protection analysis confirms that these flies do utilize the female specific 3′ splice site, as predicted from their sexual phenotype. In those lines where the sex transformation is not complete, the observation that XX; tra^-/tra^- flies are indistinguishable from XY; tra^-/tra^- flies, when both carry the in vitro mutated gene, argues against the Activation Model, since under that model, the presence of *Sxl* function in the XX flies would be expected to cause more efficient splicing at the female-specific splice junction and therefore shift their development toward femaleness.

Another informative experiment is one in which site-directed in vitro mutagenesis was used to alter the non-sex-specific splice site into a form that is still a consensus 3′ splice junction, but one which may no longer be capable of being recognized by *trans*-acting factor(s) (e.g., *Sxl*) that, under the Blockage Model, act to inhibit its use as a splice site in females. Transgenic flies carrying this mutant gene as their only copy of *tra* develop as males, indicating that there is little use of the female splice site in either the presence or absence of *Sxl* activity. RNase protection analysis showed that most of the *tra* RNA in these flies is spliced at the non-sex-specific splice site, but that there is a small amount of female-specific transcript in both XX and XY flies. The use of both the non-sex-specific and the female-specific splice sites in males demonstrates that even small changes in the intron sequence can alter the competition between the two 3′ splice sites, and it also shows that the mutant non-sex-specific 3′ splice site is, if anything, less effective as a splice site than the wild type one. The observation that there are no differences in the splicing patterns of the *tra* RNA in XX and XY flies shows that *Sxl* activity (which is present in XX flies) neither represses use of the non-sex-specific site nor activates the female site in the mutant gene. In other experiments it was shown that most of the sequences that lie between the two 3′ splice sites can be deleted without losing the sex-specific regulation. These results are most easily consistent with a model in which the region of the non-sex-specific 3′ splice site acts as an important site for blockage by *Sxl* or some factor under its control, and that the mutated form no longer is recognized by the *trans*-acting regulator(s) of the splice (Sosnowski et al. 1989).

If the regulation of *tra* splicing by *Sxl* is direct, then the mechanism by which *Sxl* controls splice site choice at *tra* could be the same as that which it uses to control splice site choice in its own pre-mRNA. In both cases, a mechanism involving blockage of one of the 3′ splice sites results in the default usage of a downstream 3′ splice site that is female-specific, and that leads to functional product. Under this hypothesis, one might expect there to be sequence similarity between the proposed sites of blockage in *tra* and *Sxl* pre-mRNAs. In fact, there is a noticeable sequence similarity between the two 3′ splice sites, with the *tra* non-sex-specific 3′ splice site having the sequence TGTTTTTTTTCTAG, and the *Sxl* male-specific 3′ splice site having the sequence TATTTTTTTTCACAG.

3.4. *transformer-2*

The *tra-2* locus was originally identified with the discovery of a second chromosome mutant that had a phenotype very similar to the *tra* mutant; i.e., XX flies homozygous for *tra-2* develop as males with respect to all aspects of their somatic sexual differentiation (Watanabe 1975). One difference between the *tra-2* and *tra* mutants is that, while *tra* null alleles have no detectable effect in XY males, the *tra-2* mutant is a recessive male sterile. The male sterile phenotype is characterized by a block relatively late in spermatogenesis, and it appears to involve a requirement for $tra\text{-}2^+$ activity in the germline (Schüpbach 1982; Belote and Baker 1983). This effect of *tra-2* in the male germ line is paradoxical, since all other sex differentiation genes acting downstream of *Sxl* (i.e., *tra*, *dsx*, and *ix*) have been shown to be dispensable for germ line development and function (Marsh and Wieschaus 1978; Schüpbach 1982).

The *tra-2* gene was cloned by P-element transposon tagging and chromosomal walking (Amrein et al. 1988) and, independently, by microdissection-microcloning methods (Goralski et al. 1989). P-element transformation experiments showed that the *tra-2* gene is contained within a DNA fragment 3.9 kb in length. Northern blot analysis showed that the *tra-2* gene specifies a 1.7 kb transcript. Analysis of *tra-2* cDNA clones revealed there are actually at least three 1.7 kb RNAs, arising from alternative splicing of the *tra-2* precursor RNA. Unlike what is seen with *tra*, there is no evidence that any of these alternative splices at *tra-2* are sex-specific (Goralski et al. 1989).

The genetic studies of sexual differentiation mutants indicate that the *tra-2* gene functions downstream of *Sxl*, and upstream of *dsx* and *ix*, in the regulatory hierarchy (Cline 1979; Baker and Ridge 1980). While the genetic analysis could not order the *tra* and *tra-2* gene functions with respect to each other, the molecular studies do provide information that suggests that these two genes act at the same step in the regulatory pathway, perhaps as parts of the same functional complex. The observation that XX flies carrying the *hsp-tra-female* construct differentiate as males if they are mutant for the *tra-2* gene suggests that the *tra-2* gene does not lie upstream of the *tra* gene in the pathway (McKeown et al. 1988). That the *tra-2* gene does not lie downstream of *tra* in the pathway is suggested by a similar set of experiments. For example, forced expression of a *tra-2* transcript (from an *hsp-tra-2* cDNA construct) is capable of complementing the *tra-2* mutant for its sex transforming phenotype, indicating that this transcript specifies sufficient $tra\text{-}2^+$ function to allow complete female development (T Goralski and B Baker, cited in Baker et al. 1989). However, the observation that forced expression of the *tra-2* cDNA fails to cause either XY flies, or XX flies homozygous for the *tra* mutant, to develop as females argues against a model in which the *tra* gene lies upstream of the *tra-2* gene in a linear pathway and acts merely by turning on the *tra-2* gene. The simplest interpretation of these results is that the *tra* and *tra-2* genes act at the same step in the regulatory pathway, with the *tra-2* gene being expressed in both sexes, but only in the presence of the *tra* gene product acting to promote female development.

The predicted *tra-2* protein, like the presumptive *Sxl* protein, shows a considerable degree of sequence similarity to the ribonucleoprotein consensus sequence (Figure 8.6). The predicted *tra* protein does not have a similar domain, but it does have one exon with a high Arg-Ser content and another exon with a high Pro content. Interestingly, the human 70K U1-RNA associated protein, a protein known to interact with RNA as part of the splicing machinery, has all of these features (RNA binding domain, high Arg-Ser region, high Pro region) in a single polypeptide (Theissen et al. 1986). Given these similarities to a known component of ribonucleoprotein, it is not unreasonable to suppose that *tra* and *tra-2* proteins are associated together and may interact directly with RNA.

In addition to these molecular observations, there is genetic evidence that supports the idea that the *tra* and *tra-2* products physically interact with one another. Specifically, it has been seen that wild-type tra^+ gene dosage has a marked effect on the phenotypes of two temperature-sensitive *tra-2* alleles (Belote and Baker 1982). This kind of dosage sensitive interaction with a temperature-sensitive allele of a second locus might reflect a physical interaction between the two gene products in which the tra^+ protein acts to stabilize the thermolabile $tra\text{-}2^{ts}$ gene product in a multicomponent complex. Such an explanation has been proposed to explain a similar kind of interaction between the SRE genes and a temperature-sensitive allele of the RAD3 gene of *Saccharomyces cerevisiae* (Naumovski and Friedberg 1987).

Taken together, all of the above observations support a model in which the *tra-2* gene product is present in both sexes, but that it is without consequences for sexual differentiation unless it is activated by the presence of the female-specific *tra* protein. When *tra* and *tra-2* are present in the same cell, then they act together to cause the female-specific processing of the *dsx* pre-mRNA, thereby leading to female differentiation. Under the influence of other proteins, or by itself, the *tra-2* protein might be involved in mediating other functions, such as its inferred role in spermatogenesis. This model is formally analogous to the case of the *MATa2* gene product of yeast, whlch alters its DNA binding specificity depending on the presence or absence of the *MATa1* gene product (Goutte and Johnson 1988).

3.5 *double sex*

The *dsx* gene is unique among the known sex differentiation regulatory genes in that its active function is required in both sexes for normal sexual differentiation. Genetic evidence suggests that the *dsx* gene specifies two functional products, one expressed in females that acts to repress male-specific differentiation, and one expressed in males that acts to repress female-specific differentiation (Baker and Ridge 1980). The *dsx* locus was cloned by chromosomal walking, and the position of the *dsx* gene within the cloned region delimited by using whole-genome Southern blot analysis to locate the positions of chromosome rearrangements associated with *dsx* mutants (Baker and Wolfner 1988). The *dsx* locus is large, about 40 kb, and it appears to be organized the same way in males and females (Baker and Wolfner 1988).

The pattern of RNAs derived from the *dsx* gene is complex, with both male-specific and female-specific transcripts seen (Baker and Wolfner 1988). There are no *dsx* transcripts detected in embryos, but they are present in larvae, pupae and adults. The isolation and sequencing of male- and female-specific *dsx* cDNAs showed that these differ from each other by virtue of sex-specific alternative splicing and polyadenylation of a common precursor (Burtis and Baker 1989). Sl nuclease protection analysis confirmed the occurrence of the sex-specific RNA processing that yields the *dsx* mRNAs diagrammed in Figure 8.5. While the male and female transcripts use the same transcription start site and are identical in their first three exons, they differ in their 3′ exons. In females, a small intron is removed by joining a 5′ splice site to a female-specific 3′ splice site. In males, the same 5′ splice site is used, but in this case, a male-specific 3′ splice site that is about 3 kb downstream of the end of the female-specific exon is utilized. Conceptual translation of the male and female transcripts reveals that they could encode polypeptides with the same 397 N-terminal amino acids, but having sex-specific carboxy-terminal regions of 30 and 152 amino acids, respectively. Neither of these presumptive *dsx* proteins shows significant sequence similarity to any other proteins in the protein databases checked (Burtis and Baker 1989). One reasonable hypothesis is that these different polypeptides represent the different active functions specified by the *dsx* gene in the two sexes, i.e., dsx^M and dsx^F in the genetic model.

In order to confirm that these sex-specific cDNAs specify the functional products of the *dsx* gene, Burtis and Baker (1989) placed these sequences downstream of a constitutively expressed *actin* promoter and introduced them into flies otherwise deficient for the *dsx* gene, using P-element transformation methods (Burtis and Baker 1989). Because the synthesis of either the male- or the female-specific *dsx* RNA is predetermined by the cDNA structure, individual transgenic flies could express only the male or only the female *dsx* product, regardless of their chromosomal sex. While neither of these constructs, i.e., *actin-dsx*^{F} and *actin-dsx*^{M}, were capable of supplying sufficient *dsx*$^+$ function to allow normal sexual differentiation, both were able to cause shifts in the sexual phenotype in the direction consistent with their predicted functions. For example, XX and XY flies homozygous for the *dsx* mutation develop with intersexual genitalia, complete pigmentation of the sixth abdominal tergite, and partial pigmentation of the fifth tergite. Homozygous *dsx* flies carrying the *actin-dsx*^{M} construct still have intersexual genitalia, but the pigmentation pattern is shifted toward maleness, with both the fifth and sixth tergites now showing complete pigmentation. Homozygous *dsx* flies carrying the *actin-dsx*^{F} construct also have intersexual genitalia, but in this case the pigmentation pattern is more similar to the female pattern, with the fifth tergite being pigmented only along the posterior border. These molecular results provide strong support for the proposition that the alternatively spliced *dsx* transcripts specify active, but opposite, functions in the two sexes.

As mentioned earlier, the genetic studies suggest that the *tra* and *tra-2* genes function in females to control the sex-specific expression of the *dsx* gene. That they do this by controlling the sex-specific alternative RNA processing of *dsx* pre-mRNA is supported by experiments in which the patterns of transcripts produced by a wild-type *dsx* gene were examined in the presence and absence of mutations at *tra* or *tra-2* . In flies homozygous for either the *tra* or *tra-2* mutant, only the male pattern of *dsx* transcripts is observed, regardless of chromosomal sex, i.e., XX or XY (Nagoshi et al. 1988). Similarly, in XX or XY flies carrying the female-constitutive *tra* gene, *hsp-tra-female*, only the female pattern of *dsx* RNAs is seen (McKeown et al. 1988). These results are most consistent with models in which the *dsx* gene functions downstream of the *tra* and *tra-2* genes in the regulatory pathway. The observation that mutations at *ix* do not affect the sex-specific pattern of *dsx* transcripts further suggests that the *ix* gene lies downstream of, or parallel to, the *dsx* gene, as shown in the genetic model depicted in Figure 8.3.

While there is no experimental evidence that the *tra* and *tra-2* proteins act directly on *dsx* pre-mRNA to control its pattern of splicing and/or polyadenylation, the similarity of the *tra* and *tra-2* presumptive proteins to known ribonucleoproteins suggests that this is a reasonable possibility.

As discussed above in relation to the control of splicing at *tra*, there are two general models for how *trans*-acting factors might regulate an alternative RNA splicing event, one involving blockage of a splice site leading to the default use of an alternative splice site, and another involving the activation of a splice site that normally would not be used in the absence of the *trans*-acting factor. While the regulation of splicing at *Sxl* and *tra* is most easily explained by a blockage model, the regulation of *dsx* might involve a splice site activation mechanism, in which the *tra* and *tra-2* proteins act in a positive way to enhance the use of the female-specific 3′ splice site in the *dsx* precursor RNA. Evidence for this comes from the molecular analysis of four *dsx* mutations in which the mutant alleles appear to constitutively express the male-specific function of *dsx* (R Nagoshi and B Baker, cited in Baker 1989). In all four of these mutants, there are alterations in the DNA that map very close to the site of the 3′ splice junction for the female-specific splice. The *dsx* pre-mRNA from these mutant genes is always spliced at the male splice site. The simplest interpretation of this observation is that normally the *tra* and *tra-2* products act on the *dsx* pre-mRNA at the region near the female-specific 3′ splice site to make it the preferred splice site. If this region is disrupted by mutation, or if *tra* or *tra-2* is absent, then the male-specific splice site is used by default. The observation that the female-specific 3′ splice site is a very poor match to the *Drosophila* 3′ splice site consensus sequence, and that this unusual 3′ splice site sequence is conserved in *D. virilis*, lends support to the idea that the female-specific 3′ splice site is important for the sex-specific regulation of *dsx* (Burtis and Baker 1989).

4. Conclusions

In view of the molecular results detailed in Section 3, the genetic model discussed in Section 2 can now be restated in more specific terms, as diagrammed in Figure 8.7. If the X/A ratio is 1.0, as in a female, and if the maternally-supplied *da* gene product and the zygotically-acting *sis-a* and *sis-b* functions are present, then *Sxl* is transcribed from an embryonic promoter, and the early transcripts give rise to a functional *Sxl* protein. If the X/A ratio is 0.5, as in a male, then these early transcripts are not produced or are present at a much reduced level. Within a few hours after fertilization, *Sxl* transcription is switched to a different, constitutive transcription start site, which produces the *Sxl* late transcripts seen in the post-embryonic stages. In the presence of *Sxl* protein (i.e., in females), the *Sxl* pre-mRNA is spliced to yield the functional *Sxl* message, and more *Sxl* protein is produced. Another role of the female-specific *Sxl* gene product(s) is to control, directly or indirectly, the alternative splicing of *tra* pre-mRNA. In the presence of the *Sxl* female-specific gene product(s), *tra* RNA is spliced to give an mRNA encoding the functional *tra* polypeptide. In the absence of the functional *Sxl* gene product(s), *tra* is spliced in an alternative way that results in a non-functional gene product. The role of the *tra* gene product is to cause, directly or indirectly, the female-specific expression of the *dsx* gene. It does this by acting through, or in conjunction with, the *tra-2* gene. The *tra-2* gene product is present in both sexes, but only when the functional *tra* gene product is

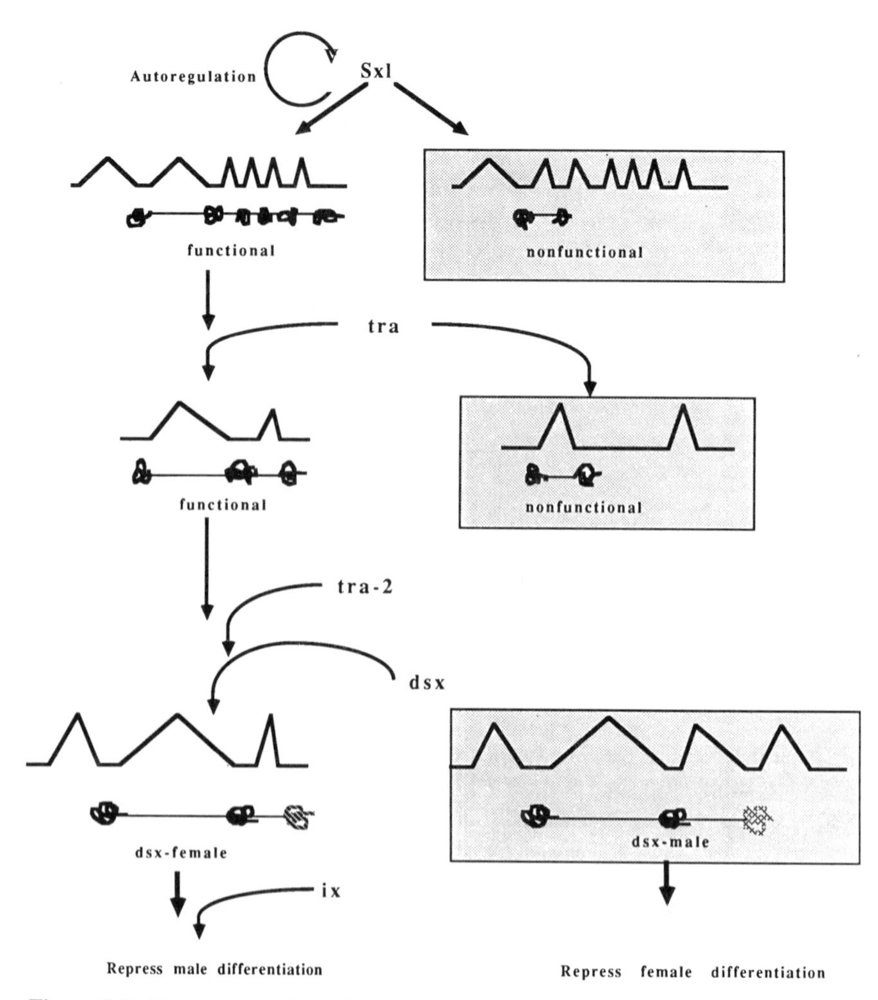

Figure 8.7. Reinterpretation of the genetic model showing the steps in the pathway where alternative RNA processing leads to sex-specific expression of *Sxl*, *tra*, and *dsx*.

present can it act to promote female differentiation. One reasonable possibility is that the *tra-2* and *tra* gene products act together as parts of a complex. In the presence of *tra* and *tra-2* gene products, the *dsx* gene pre-mRNA is processed to yield female-specific transcripts encoding female-specific polypeptides. In the absence of *tra* and/or *tra-2* function, the *dsx* RNA is processed to give male-specific transcripts that encode male-specific polypeptides. The different *dsx* gene products presumably correspond to the dsx^F and dsx^M modes of expression referred to in the genetic model. The role of the *ix* gene product is to act in conjunction with dsx^F to repress expression of the male-specific terminal differentiation functions. Its expression need not be sex specific. It has been shown that, in at least some cases, the downstream target genes are regulated by this hierarchy at the level of their transcription (Krause et al. 1988).

While it is not yet known how widespread this type of sex differentiation regulatory cascade is among insects, mutations affecting sexual development, similar to some of the ones discussed here, have been isolated in the housefly, *Musca domestica* (Inoue and Hiroyoshi 1986). Thus, while the many different insect species are known to have a baffling variety of sex determination mechanisms (e.g., based on a primary determinant of X/A ratio, male-determining Y chromosome, haplo-diploidy system, environmental sex determination signal, etc.), the underlying genetic regulatory pathways may be structured in a similar way (Nöthiger and Steinmann-Zwicky 1985). Now that the molecular tools are available from *Drosophila*, elucidating the genetic pathways controlling sex in other insects may be made easier.

Acknowledgments. I would like to thank James Eberle, Michael O'Neil, and Ken Saville for their comments during the preparation of this chapter. I would also like to thank the National Institutes of Health for their support of my research.

References

Amrein H, Gorman M, Nöthiger R (1988) The sex-determining gene *tra-2* of *Drosophila* encodes a putative RNA binding protein. Cell 55:1025–1035

Baker BS (1989) Sex in flies: the splice of life. Nature 340:521–524

Baker BS, Belote JM (1983) Sex determination and dosage compensation in *Drosophila melanogaster*. Ann Rev Genet 17:345–393

Baker BS, Burtis K, Goralski T, Mattox W, Nagoshi R (1989) Molecular aspects of sex determination in *Drosophila melanogaster*. Genome 31:638–645

Baker BS, Ridge K (1980) Sex and the single cell: on the action of major loci affecting sex determination in *Drosophila melanogaster*. Genetics 94:383–423

Baker BS, Wolfner MF (1988) A molecular analysis of *doublesex*, a bifunctional gene that controls both male and female sexual differentiation in *Drosophila melanogaster*. Genes Devel 2:477–489

Bandziulis RJ, Swanson MS, Dreyfuss G (1989) RNA-binding proteins as developmental regulators. Genes Devel 3:431–437

Bell LR, Maine EM, Schedl P, Cline TW (1988) *Sex-lethal*, a *Drosophila* sex determination switch gene, exhibits sex-specific RNA splicing and sequence similarity to RNA binding proteins. Cell 55:1037–1046

Belote JM (1983) Male-specific lethal mutations of *Drosophila melanogaster*. II. Parameters of gene action during male development. Genetics 105:881–896

Belote JM, Baker BS (1982) Sex determination in *Drosophila melanogaster*: analysis of transformer-2, a sex-transforming locus. Proc Natl Acad Sci USA 79:1568–1572

Belote JM, Baker BS (1983) The dual functions of a sex determination gene in *Drosophila melanogaster*. Devel Biol 95:512–517

Belote JM, Baker BS (1987) Sexual behavior: Its control during development and adulthood in *Drosophila melanogaster*. Proc Natl Acad Sci USA 84: 8026–8030

Belote JM, Handler AH, Wolfner, MF, Livak KJ, Baker BS (1985a) Sex-specific regulation of yolk protein gene expression in *Drosophila*. Cell 40:339–348

Belote JM, Lucchesi JC (1980a) Control of X chromosome transcription by the *maleless* gene in *Drosophila*. Nature 285:573–575

Belote JM, Lucchesi JC (1980b) Male-specific lethal mutations of *Drosophila melanogaster*. Genetics 96:165–186

Belote JM, McKeown M, Andrew DJ, Scott TN, Wolfner MF, Baker BS (1985b) Control of sexual differentiation in *Drosophila melanogaster*. Cold Spring Harbor Symp Quant Biol 50:605–614

Belote JM, McKeown M, Boggs RT, Ohkawa R, Sosnowski BA (1989) Molecular genetics of *transformer*, a genetic switch controlling sexual differentiation in *Drosophila*. Devel Genet 10:143–154

Bingham PM, Chou T-B, Mims I, Zachar Z (1988) On/off regulation of gene expression at the level of splicing. Trends Genet 4:134–138

Birchler JA, Owenby RK, Jacobson KB (1982) Dosage compensation of serine-4 transfer RNA in *Drosophila melanogaster*. Genetics 102:525–537

Boggs RT, Gregor P, Idriss S, Belote JM, McKeown M (1987) Regulation of sexual differentiation in *Drosophila melanogaster* via alternative splicing of RNA from the *transformer* gene. Cell 50:739–747

Bownes M, Nöthiger R (1981) Sex determining genes and vitellogenin synthesis in *Drosophila melanogaster*. Mol Gen Genet 182:222–228

Breen TR, Lucchesi JC (1986) Analysis of the dosage compensation of a specific transcript in *Drosophila melanogaster*. Genetics 112:483–491

Bridges CB (1916) Non-disjunction as proof of the chromosome theory of heredity. Genetics 1:1–52

Bridges CB (1921) Triploid intersexes in *Drosophila melanogaster*. Science 54: 252–254

Bridges CB (1925) Sex in relation to chromosomes and genes. Amer Nat 59: 127–137

Burtis KC, Baker BS (1989) *Drosophila doublesex* gene controls somatic sexual differentiation by producing alternatively spliced mRNAs encoding related sex-specific polypeptides. Cell 56:997–1010

Butler B, Pirrotta V, Irminger-Finger I, Nöthiger R (1986) The sex determining gene *tra* of *Drosophila*: molecular cloning and transformation studies. EMBO J 5:3607–3613

Caudy M, Grell EH, Dambly-Chaudiere C, Ghysen A, Jan LY, Jan YN (1988a) The maternal sex determination gene *daughterless* has zygotic activity necessary for the formation of peripheral neurons in *Drosophila*. Genes Devel 2:843–852

Caudy M, Vässin H, Brand M, Tuma R, Jan LY, Jan YN (1988b) *daughterless*, a *Drosophila* gene essential for both neurogenesis and sex determination, has sequence similarities to *mvc* and the *achaete-scute* complex. Cell 55: 1061–1067

Cavener D, Corbett G, Cox D, Whetten R (1986) Isolation of the eclosion gene cluster and the developmental expression of the *Gld* gene in *Drosophila melanogaster*. EMBO J 5:2939–2948

Chapman KB, Wolfner MF (1988) Determination of male-specific gene expression in *Drosophila* accessory glands. Devel Biol 126:195–202

Cline TW (1976) A sex-specific, temperature-sensitive maternal effect of the *daughterless* mutation in *Drosophila melanogaster*. Genetics 84:723–742

Cline TW (1978) Two closely-linked mutations in *Drosophila melanogaster* that are lethal to opposite sexes and interact with *daughterless*. Genetics 90: 683–698

Cline TW (1979) A male-specific mutation in *Drosophila melanogaster* that transforms sex. Devel Biol 72:266–275

Cline TW (1980) Maternal and zygotic sex-specific gene interactions in *Drosophila melanogaster*. Genetics 96:903–926

Cline TW (1983a) The interaction between *daughterless* and *Sex-lethal* in triploids: A novel sex-transforming maternal effect linking sex determination and dosage compensation in *Drosophila melanogaster*. Devel Biol 95:260–274

Cline TW (1983b) Functioning of the genes *daughterless* and *Sex-lethal* in *Drosophila* germ cells. Genetics 104:s16–sl7

Cline TW (1984) Autoregulatory functioning of a *Drosophila* gene product that establishes and maintains the sexually determined state. Genetics 107: 231–255

Cline TW (1985) Primary events in the determination of sex in *Drosophila melanogaster*. In Halvorson HO, Monroy A (eds) The Origin and Evolution of Sex. AR Liss, Inc, New York, pp 301–327

Cline TW (1986) A female-specific lethal lesion in an X-linked positive regulator of the *Drosophila* sex determination gene *Sex-lethal*. Genetics 113:641–663

Cline TW (1987) Reevaluation of the functional relationship in *Drosophila* between a small region on the X chromosome (3E8-4Fl l) and the sex-determination gene, *Sex lethal*. Genetics 116:sl2

Cline TW (1988) Evidence that *sisterless-a* and *sisterless-b* are two of several discrete "numerator elements" of the X/A sex determination signal in *Drosophila* that switch *Sxl* between two alternative stable expression states. Genetics 119:829–862

Cline TW (1989) The affairs of *daughterless* and the promiscuity of developmental regulators. Cell 59:231–234

Cronmiller C, Cline TW (1987) The *Drosophila* sex determination gene *daughterless* has different functions in the germ line versus the soma. Cell 48:479–487

Cronmiller C, Schedl P, Cline TW (1988) Molecular characterization of *daughterless*, a *Drosophila* sex determination gene with multiple roles in development. Genes Devel 2:1666–1676

Epper F, Bryant PJ (1983) Sex-specific control of growth and differentiation in the *Drosophila* genital disc, studied using a temperature-sensitive *transformer-2* mutation. Devel Biol 100:294–307

Fukunaga A, Tanaka A, Oishi K (1975) *Maleless*, a recessive autosomal mutant of *Drosophila melanogaster* that specifically kills male zygotes. Genetics 81:135–141

Ganguly R, Ganguly N, Manning J (1985) Isolation and characterization of the glucose-6-phosphate dehydrogenase gene of *Drosophila melanogaster*. Gene 35:91–101

Gergen P (1987) Dosage compensation in *Drosophila*: evidence that *daughterless* and *Sex-lethal* control X chromosome activity at the blastoderm stage of embryogenesis. Genetics 117:477–485

Goralski TJ, Edstrom J-E, Baker BS (1989) The sex determination locus *transformer-2* of *Drosophila* encodes a polypeptide with similarity to RNA binding proteins. Cell 56:1011–1018

Goutte C, Johnson AD (1988) *a1* protein alters the DNA binding specificity of *alpha 2* repressor. Cell 52:875–882

Hildreth PE (1965) *doublesex*, a recessive gene that transforms both males and females of *Drosophila* into intersexes. Genetics 51:659–678

Hodgkin J (1989) *Drosophila* sex determination: a cascade of regulated splicing. Cell 56:905–906

Inoue H, Hiroyoshi T (1986) A maternal-effect sex-transformation mutant of the housefly, *Musca domestica*. Genetics 112:469–482

Jowett T, Postlethwait JH (1981) Hormonal regulation of synthesis of yolk proteins and a larval serum protein (LSP2) in *Drosophila*. Nature 292: 633–635

Kraus KW, Lee Y, Lis JT, Wolfner MF (1988) Sex-specific control of *Drosophila melanogaster* yolk protein 1 gene expression is limited to transcription. Mol Cell Biol 8:4756–4764

Lawrence PA, Johnston P (1986) The muscle pattern of a segment of *Drosophila* may be determined by neurons and not by contributing myoblasts. Cell 45:505–513

Lucchesi JC, Manning JE (1987) Gene dosage compensation in *Drosophila melanogaster*. Adv Genet 24:371–430

Lucchesi JC, Skripsky T (1981) The link between dosage compensation and sex differentiation in *Drosophila melanogaster*. Chromosoma 82:217–227

Maine EM, Salz HK, Cline TW, Schedl P (1985a) The *Sex-lethal* gene of *Drosophila*: DNA alterations associated with sex-specific lethal mutations. Cell 43:521–529

Maine EM, Salz HK, Schedl P, Cline TW (1985b) *Sex-lethal*, a link between sex determination and sexual differentiation in *Drosophila melanogaster*. Cold Spring Harbor Symp Quant Biol 50:595–604

Marsh JL, Wieschaus E (1978) Is sex determination in germline and soma controlled by separate mechanisms? Nature 272:249–251

McKeown M, Belote JM, Baker BS (1987) A molecular analysis of *transformer*, a gene in *Drosophila melanogaster* that controls female sexual differentiation. Cell 48:489–499

McKeown M, Belote JM, Boggs RT (1988) Ectopic expression of the female *transformer* gene product leads to female differentiation of chromosomally male *Drosophila*. Cell 53:887–895

Meyer HU, Edmondson M (1951) New mutants. Dros Inform Serv 25:71–74

Morgan TH (1916) Mosaics and gynandromorphs in *Drosophila*. Proc Soc Exper Med 11:171–172

Mukherjee AS, Beermann W (1965) Synthesis of ribonucleic acid by the X chromosomes of *Drosophila melanogaster* and the problem of dosage compensation. Nature 207:785–786

Muller HJ, Zimmering S (1960) A sex-linked lethal without evident effect in *Drosophila* males, but partially dominant in females. Genetics 45:1000–1002

Murre C, McGraw PS, Baltimore D (1989) A new DNA binding and dimerization motif in immunoglobulin enhancer binding, *daughterless*, *MyoD*, and *myc* proteins. Cell 56:777–783

Nagoshi RN, McKeown M, Burtis KC, Belote JM, Baker BS (1988) The control of alternative splicing at genes regulating sexual differentiation in *D. melanogaster*. Cell 53:229–236

Naumovski L, Friedberg EC (1987) The RAD3 gene of *Saccaromyces cerevisiae*: Isolation and characterization of a temperature-sensitive mutant in the essential function and of extragenic suppressors of this mutant. Mol Gen Genet 209:458–466

Nicklas JA, Cline TW (1983) Vital genes that flank *Sex lethal*, an X-linked sex-determining gene of *Drosophila melanogaster*. Genetics 103:617–631

Nöthiger R, Steinmann-Zwicky M (1985) A single principle for sex determination in insects. Cold Spring Harbor Symp Quant Biol 50:615–621

Nöthiger R, Steinmann-Zwicky M (1987) Genetics of sex determination: what can we learn from *Drosophila*? Development 101:17–24

Oakeshott JG, Collet C, Phillis RW, Nielsen KM, Russell RJ, Chambers GK, Ross V, Richmond RC (1987) Molecular cloning and characterization of esterase-6, a serine hydrolase of *Drosophila*. Proc Natl Acad Sci USA 84: 3359–3363

Okuno T, Satou T, Oishi K (1984) Studies on the sex-specific lethals of *Drosophila melanogaster*. VII. Sex-specific lethals that do not affect dosage compensation. Jap J Genet 59:237–247

Oliver B, Perrimon N, Mahowald AP (1987) The *ovo* locus is required for sex-specific germline maintenance in *Drosophila*. Genes Devel 1:913–923

Oliver B, Perrimon N, Mahowald AP (1988) Genetic evidence that the *sans fille* locus is involved in *Drosophila* sex determination. Genetics 120:159–171

Query CC, Bentley RC, Keene JD (1989) A specific 31–nucleotide domain of U1 RNA directly interacts with the 70K small nuclear ribonucleoprotein component. Mol Cell Biol 9:4872–4881

Salz HK, Cline TW, Schedl P (1987) Functional changes associated with structural alterations induced by mobilization of a P element inserted in the *Sex-lethal* gene of *Drosophila*. Genetics 117:221–231

Salz HK, Maine EM, Keyes LN, Samuels ME, Cline TW, Schedl P (1989) The *Drosophila* female-specific sex-determination gene, *Sex-lethal*, has stage-, tissue-, and sex-specific RNAs suggesting multiple modes of regulation. Genes Devel 3:708–719

Sanchez L, Nöthiger R (1982) Clonal analysis of *Sex-lethal*, a gene needed for female sexual development in *Drosophila melanogaster*. Wilhelm Roux Arch Devel Biol 191:211–214

Sanchez L, Nöthiger R (1983) Sex determination and dosage compensation in *Drosophila melanogaster*: production of male clones in XX females. EMBO J 2:485–491

Schüpbach T (1982) Autosomal mutations that interfere with sex determination in somatic cells of *Drosophila* have no direct effect on the germline. Devel Biol 89:117–127

Schüpbach T (1985) Normal female germ cell differentiation requires the female X-chromosome-autosome ratio and expression of *Sex-lethal* in *Drosophila melanogaster*. Genetics 109:529–548

Skripsky T, Lucchesi JC (1982) Intersexuality resulting from the interaction of sex-specific lethal mutations in *Drosophila melanogaster*. Devel Biol 94: 153–162

Smith PD, Lucchesi JC (1969) The role of sexuality in dosage compensation in *Drosophila*. Genetics 61:607–618

Sosnowski BA, Belote JM, McKeown M (1989) Sex-specific alternative splicing of RNA from the *transformer* gene results from sequence dependent splice site blockage. Cell 58:449–459

Steinmann-Zwicky M (1988) Sex determination in *Drosophila*: the X-chromosomal gene *liz* is required for *Sxl* activity. EMBO J 7:3889–3898

Steinmann-Zwicky M, Nöthiger R (1985) A small region of the X chromosome of *Drosophila* regulates a key gene that controls sex determination and dosage compensation. Cell 42:877–887

Steinmann-Zwicky M, Schmid H, Nöthiger R (1989) Cell-autonomous and inductive signals can determine the sex of the germ line of *Drosophila* by regulating the gene *Sxl*. Cell 57:157–166

Stern C (1966) Pigmentation mosaicism in intersex of *Drosophila*. Rev Suisse Zool 73:339–355

Sturtevant AH (1945) A gene in *Drosophila melanogaster* that transforms females into males. Genetics 30:297–299

Taylor BJ (1989) Sexually dimorphic neurons in the terminalia of *Drosophila melanogaster*. I. Development of sensory neurons in the genital disk during metamorphosis. J Neurogenet 5:173–192

Technau GM (1984) Fiber number in the mushroom bodies of adult *Drosophila melanogaster* depends on age, sex, and experience. J Neurogenet 1:113–126

Theissen H, Etzerodt M, Reuter R, Schneider C, Lottspeich F, Argos P, Lührmann R, Philipson L (1986) Cloning of the human cDNA for the U1 RNA-associated 70K protein. EMBO J 5:3209–3217

Uchida S, Uenoyama T, Oishi K (1981) Studies on sex-specific lethals of *Drosophila melanogaster*. III. A third chromosome male-specific lethal mutant. Jap J Genet 56:523–527

Watanabe TK (1975) A new sex-transforming gene on the second chromosome of *Drosophila melanogaster*. Jap J Genet 50:269–271

Wolfner MF (1988) Sex-specific gene expression in somatic tissues of *Drosophila melanogaster*. Trends Genet 4:333–337

Zachar Z, Garza D, Chou T-B, Goland J, Bingham PM (1987) Molecular cloning and genetic analysis of the *Suppressor-of-white-apricot* locus from *Drosophila melanogaster*. Mol Cell Biol 7:2498–2505

Chapter 9

Molecular Genetics of Complex Behavior: Biological Rhythms in *Drosophila*

C.P. Kyriacou

1. Introduction

It is something of a truism to say that behavioral genetics has been a poor relation within the discipline of genetics, even though it can claim an ancestry that dates back to the time of Galton. Whereas biochemical, molecular, population and developmental genetics have taken great strides forward in the past 20–30 years, behavioral genetics, until relatively recently, has attracted little attention among the biological community. When it has made an impact, as it did in the late 1960s and early 1970s, it was for the wrong reasons. The sterile "genes and IQ" debate, which generated lots of heat but little light during that period (for review, see Hay 1985), left geneticists, who were pursuing more mainstream nonbehavioral problems, shrugging their shoulders and offering their sympathies to their behavioral-genetics colleagues who had obviously got out of their depth. "The behavioral phenotype is too difficult to work with," some of my colleagues would (and still do) say.

There is of course an element of truth here. Behavior, even simple behavior, is a very complex phenotype and depends on sensory input and motor output and all the things that go on in between. To produce the integrated motor patterns we call "behavior," many anatomical structures are involved in producing even a very simple response. It is therefore quite understandable that a geneticist, perhaps working on a morphological structure such as a *Drosophila* bristle, might consider any behavioral genetic problem as being intractable. However, in spite of the intrinsic complexities of behavior, various groups of workers have been attempting to get to grips with behavioral phenotypes with varying degrees of success. Behavior, unlike *Drosophila* bristles, varies constantly and frustratingly (see Manning 1961, p 83) with time, requiring a large amount of patience plus a fair knowledge of experimental design in order to study it.

Up until the late 1960s, behavioral geneticists were seemingly content to use the traditional and well tried methods of quantitative genetics, such as selection experiments, diallel crosses, etc., in order to examine the polygenic nature of behavioral variation. Some studies focused on spontaneously occurring single gene mutations—the aim, of course, being to correlate mutant behavior with biochemical changes. The breakthrough came in the late 1960s when Seymour Benzer, working with *Drosophila*, and Sidney Brenner, who studied the nematode *Caenorhabditis*, advocated the use of chemical mutagenesis and mass behavioral screening so as to create mutations in the behavioral systems of choice. In *Drosophila* especially, ingenious screening designs were used to induce mutations affecting visual, flight, locomotor and sexual behaviors. Many of these early mutants such as *drop-dead*, *ether-a-go-go* (see Benzer 1973) showed such gross behavioral abnormalities that they were of little value to the ethologically minded biologist. Nevertheless, work with these mutants showed that it was possible to initiate a controlled dissection of the insect's nervous system, using behavior and genetics as the tools to reveal neural dysfunction. In the nematode, behavioral abnormalities were also used as the first step to isolate mutations which affected the nervous system (Brenner 1973).

With *Drosophila*, in addition to mutations which affected gross locomotor movements, visual behavior or flight, more subtle behavioral screens were applied in successful attempts to isolate mutants that were defective in more "complex" behavior. The two major stories in this field are those concerning biological clocks and learning. In the former, Konopka and Benzer (1971) isolated three *period* (*per*) mutations which dramatically affected the fly's 24-hour circadian behavior. In the latter, Dudai et al. (1976) were able to induce mutations in a gene called *dunce* (*dnc*), which prevented flies from learning a simple classical conditioning procedure. Work on these two behavioral systems has progressed to the point where the relevant genes have been cloned and sequenced (e.g., Bargiello and Young 1984; Reddy et al. 1984; Chen et al. 1986), and in the case of the circadian mutations, much of the "standard" molecular biology has been done. This review will therefore concentrate on the molecular genetic analysis of circadian behavior, which at this level is the most intensively studied eukaryotic behavioral system. It has also provided us with a number of surprises.

2. The *Period* Gene in *Drosophila*

2.1. Isolation

As mentioned earlier, Konopka and Benzer (1971) isolated three X-linked mutations in *D. melanogaster* which dramatically altered two circadian phenotypes, pupal-adult eclosion and locomotor activity. The former of these had been much studied in *D. pseudoobscura* by Pittendrigh and his collaborators (e.g., Pittendrigh 1974). Groups of pupae of different ages are monitored until they emerge as adults. The pupae can be examined under a variety of environmental conditions, constant light (LL) or darkness (DD), or under a 12-hour-light/12-hour-dark cycle (LD1212). Usually, to measure the period of the organism's endogenous clock, pupae may be exposed to light for short periods, but then placed in DD ("free-running" conditions) until eclosion. The adults emerge at characteristic times or "gates" during each successive 24 h period (see Saunders 1982). Most flies that will emerge in a 24 h period will eclose at a time corresponding to "dawn," and consequently, a pattern such as that illustrated in Figure 9.1 will be observed. Approximately every 24 h there will be a burst of adult eclosion and each successive burst will occur approximately 24 h later in normal wild-type flies until all the pupae have emerged.

Konopka and Benzer (1971) chemically mutagenized the X-chromosome and isolated three mutant lines whose free running eclosion cycles were different from the wild-type's. These are also shown in Figure 9.1. The short mutant had a period of 19 h, the long mutant a period of 29 h, and an arrhythmic mutant appeared to have a random pattern of emergence. Studies of individual fly's locomotor activity cycles revealed that these three mutations all had corresponding effects on the fly's "sleep-wake" cycle (also see Figure 9.1). Consequently, the three mutations affected both the group and individual bioassays of the circadian clock. Remarkably, all three mutations mapped to the same point on the X chromosome in a genetically well defined region between the *zeste* and *white* genes (Judd et al. 1972). The three alleles of the *period* (*per*) locus were thus called per^{S} for the short, per^{L1} for the long and per^{01} for the arrhythmic variants.

2.1. Other Phenotypes Affected by the *per* Mutants

2.1.1. Courtship Song Cycles

The *per* mutations have also been observed to affect other characters, some with an obvious temporal element. The best known of these is the male courtship song rhythm (Kyriacou and Hall 1980; 1984; 1985; 1986;

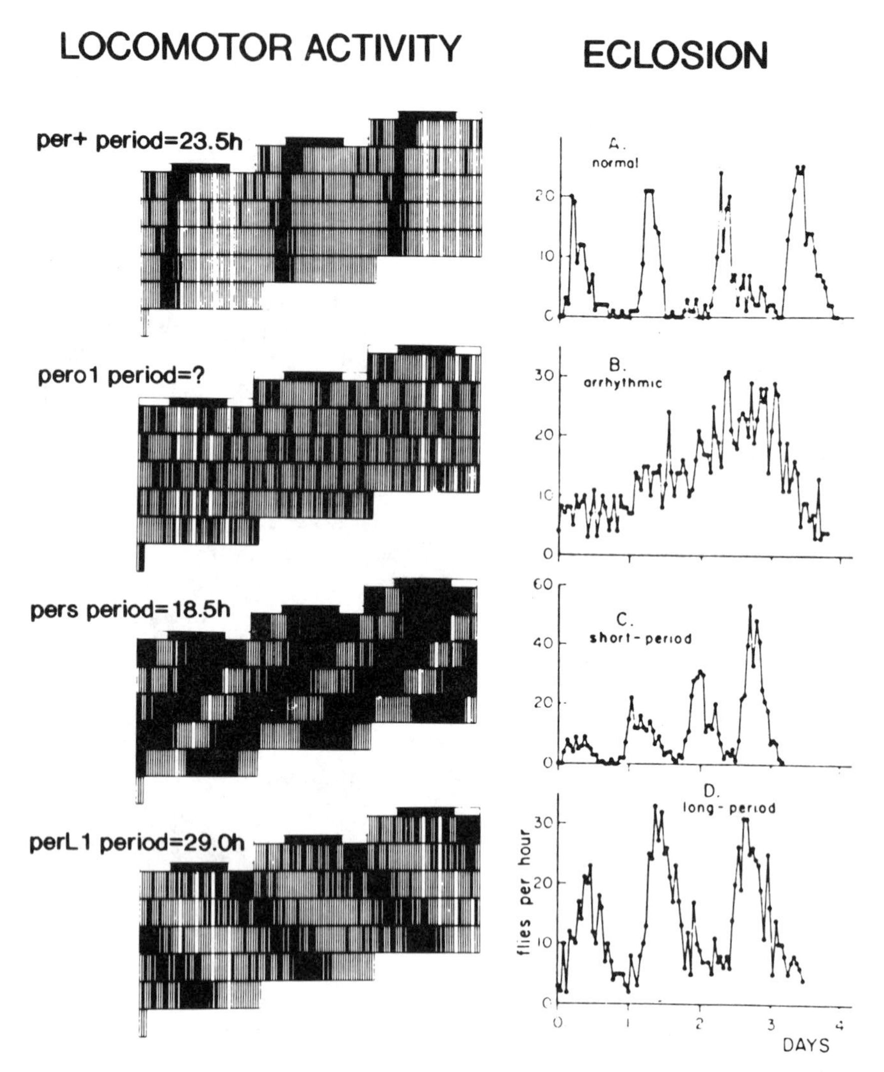

Figure 9.1. Circadian phenotypes in different *per* genotypes. On the left is the circadian free-running data, and on the right the corresponding free-running pupal-adult eclosion cycles (taken from Konopka and Benzer 1971). The activity data are triple plotted, with the first 24 h of data on the top line (the darker the vertical stripe, the more intense the activity; each stripe represents a half hour time bin of activity). Below it, day 1 and day 2 data are plotted; below this, day 1, 2 and 3 data; below this, day 2, 3 and 4 . . . etc. The light and dark bars above the plots represent the entrainment regime (LD 12:12) in which the animals were placed before they were monitored in total darkness (free-run). The per^{+} period is 23.5 h (computed using autocorrelation), the per^{S} period is 18.5 h and the per^{L1} period is 29.0 h. Note in per^{S} activity how the regions of intense activity came about 5.5 h earlier on every successive day, giving the drift to the left. In per^{L1}

1989). Males extend their wings toward the female and vibrate them to produce a "lovesong," which has species-specific characteristics (Cowling and Burnet 1981). A major component of each song burst is a series of pulses (Figure 9.2) separated by an interpulse interval (IPI). These IPIs can vary widely in length during an individual courtship, perhaps between 15 milliseconds (ms) and 80 ms in *D. melanogaster*, and 15 ms and 150 ms in *D. simulans*, and consequently, the IPIs of *D. simulans* are on average slightly longer than those of *D. melanogaster* (Kawanishi and Watanabe 1980; Cowling and Burnet 1981; Kyriacou and Hall 1986). Kyriacou and Hall (1980) observed that the variation in IPI production by the male was not random, and that, by calculating an average from all the IPIs which fell into successive 10 s time frames, a cyclical pattern of mean IPIs was observed (Figures 9.2 and 9.3). The period of this oscillation was between 50–65s for different wild-type strains of *D. melanogaster* (Figure 9.3), and 35–40 s for different strains of *D. simulans* (Kyriacou and Hall 1980; 1986). The per^{S} mutant, however, sang with 40 s IPI cycles, whereas per^{L1} males sang with 80 s IPI cycles (Kyriacou and Hall 1980). The per^{01} mutant revealed no obvious IPI rhythm. Examples of mutant song rhythms are shown in Figure 9.3. Thus, an ultradian cycle (period less than 24 h) also appeared to be affected by the *per* mutations (Figure 9.3).

One should add here that the statistical analysis of courtship song rhythms is rather difficult, and Kyriacou's and Hall's (1980) interpretation of their data has been criticized by Ewing (1988) and Crossley (1988), who suggested that 60 s cycles were not observed in wild-type song. However, by reanalyzing their 1980 data with appropriate spectral methods designed to cope with the particular idiosynchrasies of song data (Figure 9.3), Kyriacou and Hall (1989) demonstrated that their initial interpretations were indeed correct. Furthermore, they were able to show that in the data of Crossley (1988), 60 s periodicites were observed at a level which could not have occurred by chance. I refer the interested reader to the articles above and to reviews of some of the methodological and statistical issues in this debate by Schilcher (1989) and Hall and Kyriacou (1989).

The song cycle shows some interesting differences from circadian cycles. Circadian cycles, under conditions of bright light, became

these periods of intense activity came 5 h later on every successive day, giving a drift to the right. The periods of intense activity in per^{+} is at approximately the same time each day. In per^{01}, no obvious periodicity can be observed (but see Dowse et al. 1987; Dowse and Ringo 1987). The eclosion data show the number of adults emerging every hour over 4 days. Note how arrhythmic (per^{01}) mutants have no obvious eclosion cycle, whereas the wild-type, long (per^{L1}) and short (per^{S}) variants have approximately 24 h, 29 h and 19 h cycles, respectively.

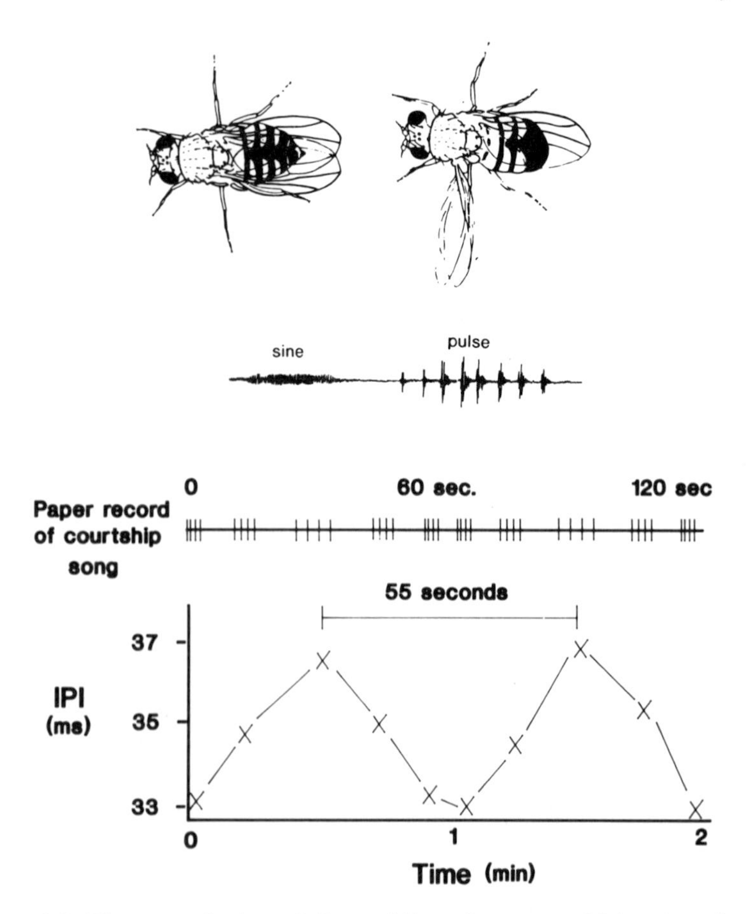

Figure 9.2. The song rhythm of *Drosophila melanogaster*. Male wing vibration gives a burst of song, which may have two components, "sine" song or pulse song (Cowling and Burnet 1981). The pulses are transcribed onto paper records, and the mean interpulse interval (IPI) for every 10 s of real time is computed. A cycle of mean IPI lengths is obtained, which in wild-type songs has a period of about 55 s (Kyriacou and Hall 1980).

"damped" as the clock moves into a "phase-less" state, and eclosion and activity cycles soon become arrhythmic under these conditions (Saunders 1982; Winfree 1987). However, male flies reared in bright light have a perfectly good song cycle (Kyriacou and Hall 1980). Furthermore, mosaic studies using the per^S mutation performed by R Konopka, J Hall and CP Kyriacou (preliminary results described in Hall 1984) reveal that a mosaic with a completely per^S-like cephalic, but per^+ thoracic ganglion, has a short locomotor activity cycle but a wild-type song rhythm. The reciprocal

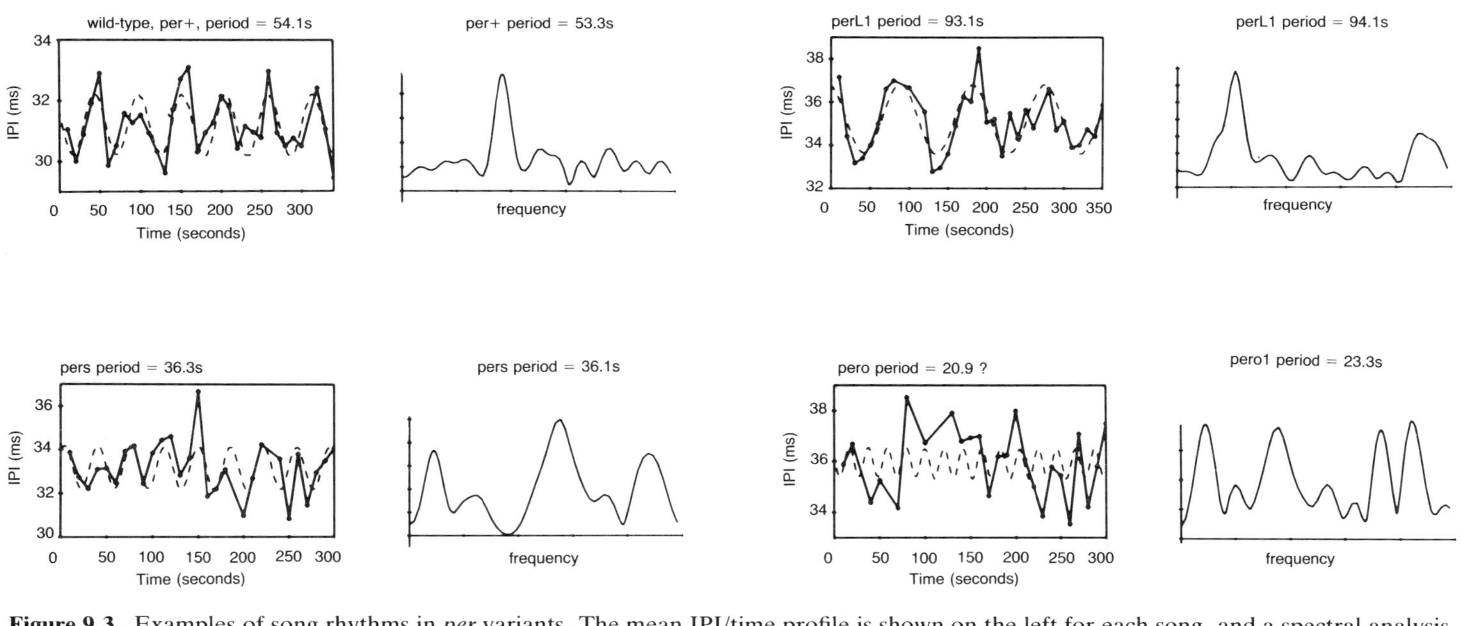

Figure 9.3. Examples of song rhythms in *per* variants. The mean IPI/time profile is shown on the left for each song, and a spectral analysis of this pattern is shown immediately to the right (Kyriacou and Hall 1989; Hall and Kyriacou 1990). The highest point in the spectrogram can be converted to a period. Note how for per^{01} there are a number of spectrogram peaks all of about the same height, whereas with the other genotypes one peak is prominent. The mean IPI/time plot has a dotted sine wave superimposed on the data, which represents the best sinusoidal fit and is computed by regression (Kyriacou and Hall 1989). In the per^{01} song the best fitting sine wave had a very short ≈ 21 s period (see text).

type of mosaic gives the opposite behavioral profile, demonstrating the anatomical independence of the activity and song "oscillators." This mosaic study augments an earlier study published by Konopka et al. (1983), which revealed a cephalic focus for the per^S mutation in the control of circadian activity cycles.

2.1.2. Development Time

Kyriacou et al. (1990) demonstrated that the *per* mutants also affect a nonbehavioral character, namely development time. Under conditions of very bright light, where the insects' circadian eclosion clock is arrested (see Saunders 1982; Kyriacou and Hall 1980; Kyriacou et al. 1990), per^S and per^{01} flies develop faster than per^+, and per^{L1} flies take considerably longer than per^+ to complete development. It is perhaps not surprising that mutants might take longer to develop, as they are generally less viable, but it is surprising that per^S and per^{01} mutants are *faster* to develop than the wild-type. The faster development for per^S seems to mirror its shorter circadian and ultradian cycles. Furthermore, there is evidence that per^{01} may not be truly arrhythmic, either in locomotor activity cycles or courtship song rhythms, because sophisticated spectral analyses of both of these per^{01} phenotypes suggest the presence of fast but weak cycles (Dowse et al. 1987; Dowse and Ringo 1987; Hamblen-Coyle et al. 1989; Kyriacou and Hall 1989). Consequently, the faster development of per^{01} could be viewed as a reflection of a fast temporal process.

It is important to re-emphasize that this developmental experiment is performed under conditions where all the genotypes are made arrhythmic because of the bright light conditions. Thus the effects of the *per* mutations on development time are independent of the circadian machinery.

2.1.3. Learning

Jackson et al. (1983) have reported that per^{L1} mutants and another long period *per* mutant, per^{L2}, fail to learn an experience-dependent modification of courtship behavior (e.g., Siegel and Hall 1979; Gailey et al. 1982). Other autosomal mutants with longer than normal cycles (Jackson 1983) were also impaired in their learning performance. However, per^S and per^{01} mutants were essentially normal. This intriguing result makes an argument for endogenous rhythms having a role in these simple conditioning procedures. However, more recent data suggest that the learning defect in per^{L1} mutants is restricted to a very specific conditioning test only, and furthermore, that it is only this long circadian period mutation which causes any learning defect (Gailey et al. 1991).

2.1.4. Other Phenotypes

The per^{01} mutants have been observed to have altered cellular circadian cycles in salivary gland cells (Weitzel and Rensing 1981). Furthermore, an often cited but very brief report that per^{01} pupae have arrhythmic heartbeats (Livingstone 1981) has yet to be confirmed. In fact, investigators working in the same laboratory have been known to come to opposite conclusions about the effects of the per^{01} mutation on pupal heartbeat (J Wood and H Fillery in the author's laboratory). It is presently premature to conclude from Livingstone's (1981) result that per^{01} mutants have an altered heartbeat "clock."

Whatever the case regarding heartbeats, it is clear from the previous sections that *per* has several effects on behavior and development. In the case of circadian rhythms, song rhythms and development time, the phenotypes have in common the property of an easily recognized temporal program. The learning defects in per^{L1} are more difficult to understand. Consequently we might imagine that, given this pleiotropy of effects, the *per* protein might not be uniquely localized in a small set of neural (or other) cells. Indeed, the mosaic results would suggest that *per* should be expressed in both thoracic and cephalic regions. This turns out to be the case.

3. Molecular Structure of the *Period* Gene

3.1. Cloning and Transformation

The *per* gene was cloned by two groups, one at Rockefeller University (Bargiello and Young 1984), and the other at Brandeis University (Reddy et al. 1984). Cloning the gene was facilitated by the fact that a number of chromosomal deletions, and one specific translocation, removed or disrupted the *per* gene (Smith and Konopka 1981). Several different transcripts were found to hybridize to DNA around the *per* region, and one of these transcripts, about 900 bases long, was initially described as oscillating in abundance during the circadian cycle (Reddy et al. 1984). This finding has turned out to be an artifact (Lorenz et al. 1989), but it caused some excitement at the time. The probability of finding a "cycling" transcript in the close vicinity of a clock gene tended to focus attention on this particular transcript as being the candidate for the *per* mRNA. However, P-element mediation transformation experiments using various overlapping DNA restriction fragments from this region soon pinpointed a 4.5 kb transcript, encoded by DNA located distally (to the left) of the DNA encoding the 900 base mRNA as being the key factor (Figure 9.4). Only those DNA fragments which encoded part or all

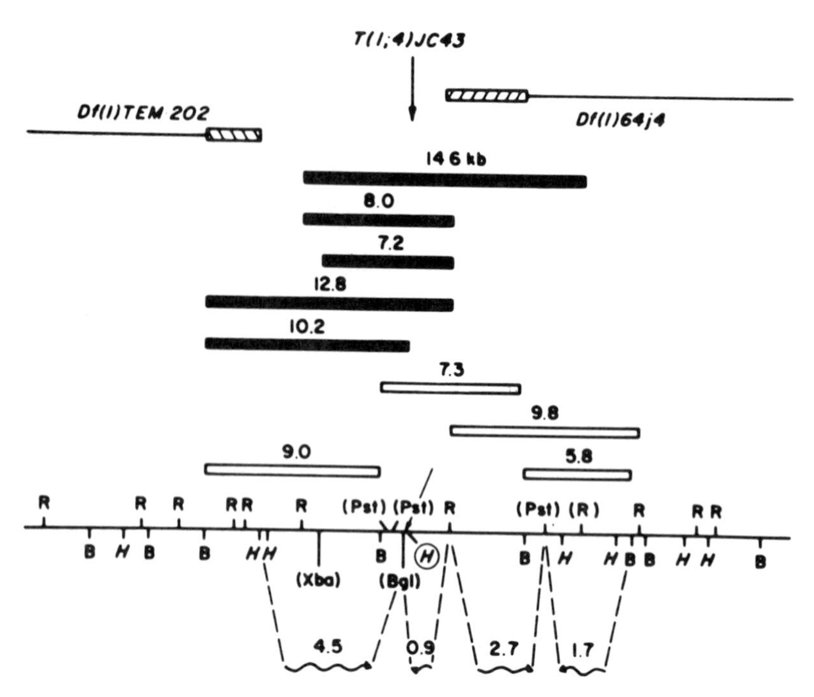

Figure 9.4. Molecular genetics of the *period* gene (from Hamblen et al. 1986). Df(1)TEM202 and Df(1)64j4 are two deletions, whose breakpoints are shown as hatched lines and which are missing material to the right and left, respectively. T(1:4)JC43 is a translocation that disrupts the *period* gene (see text). Solid bars are DNA fragments that rescue rhythmicity, both for circadian and song phenotypes in per^{01} or per^{-} mutants. Open bars are DNA fragments that fail to restore circadian or ultradian rhythmicity. Transcripts and their direction of transcription are shown under the restriction map, and they are positioned so that they correspond to their respective encoding DNA fragments. Note that rescuing fragments cover most of the 4.5 kB mRNA.

of the 4.5 kb transcript were able to restore, at least to some degree, both the circadian activity and lovesong cycles to per^{01} mutants (Bargiello et al. 1984; Zehring et al. 1984; Hamblen et al. 1986). Interestingly, the rescue was not complete in that most of the early transformants showed incomplete penetrance and longer than normal circadian and lovesong periods (Bargiello et al. 1984; Zehring et al. 1984; Hamblen et al. 1986). When a DNA fragment was used which carried more information at the 5′ end of the gene, free running activity periods of about 24.5 h, and song cycles of 60 s were observed in transformants (Citri et al. 1987; Yu et al. 1987b). Baylies et al. (1987) analyzed the levels of the 4.5 kb RNA in lines transformed with a 7.1 kb DNA fragment missing some of the

3′ noncoding sequences and correlated these levels with the average circadian period observed in each strain. An inverse correlation between circadian period and RNA titre was observed, so that the higher the level of RNA, the shorter the period. Periods in the per^L range (up to 40 h) were obtained in these strains, but no periods shorter than 24 h were observed. The *per* mutants themselves (including per^{01}) had normal levels of the 4.5 kb transcript, and therefore, the differences in periods between the mutants cannot be due to differences in gross levels of this RNA species (Bargiello and Young 1984; Young et al. 1985; Hamblen et al. 1986).

3.2. The DNA Sequence

per has been sequenced in a number of different *Drosophila* species (Reddy et al. 1986; Jackson et al. 1986; Citri et al. 1987; Colot et al. 1988; Thackeray and Kyriacou 1990). In *D. melanogaster, per* has eight exons (see Figure 9.5), the first of which is noncoding followed by a long intron, and then the coding sequence, which begins in exon 2 (see Citri et al. 1987). There are believed to be at least two alternative splicing pathways, which result in two other minor mRNA species differing at the 3′ end from the one illustrated in Figure 9.5 (Citri et al. 1987). The result is that slightly different proteins are encoded at the carboxy terminus. All three cDNAs are extremely poor in rescuing per^{01} locomotor and song arrhythmicity in transformation experiments (Citri et al. 1987; J Rutila and M Hamblen-Coyle pers comm; CP Kyriacon unpublished). This

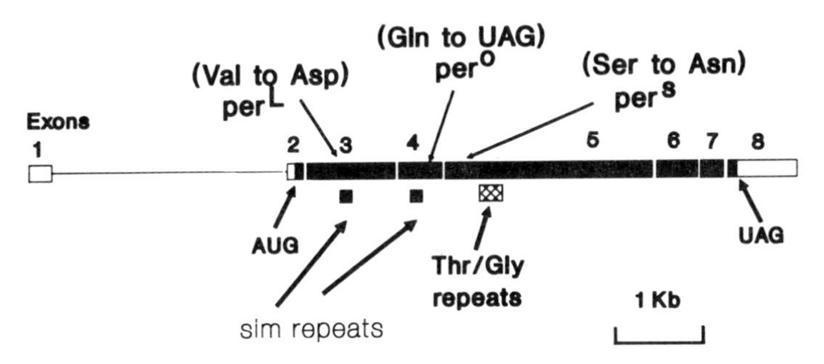

Figure 9.5. Intron/exon map of the *period* mRNA. This corresponds to the major species of transcript (Citri et al. 1987). Note the long first intron depicted as a horizontal line. Untranslated exons are represented as open blocks, and coding sequence as filled blocks. The positions and amino acid alterations of each of the *per* mutations are shown, as are the positions of the Thr-Gly encoding repeats and the two 51 amino acid *sim* repeats (see text). The initiating codon AUG and termination codon UAG are given.

suggests that the long first intron may provide an important regulatory function, something recently confirmed by Liu et al. (1991).

3.3. The *per* Protein

The primary translation product of the *per* locus has 1,200 or so amino acids, a molecular weight of 127,000 (Citri et al. 1987), and bears similarities to certain other proteins. The most spectacular, but limited, similarity is with the conceptual product of the *frequency* (*frq*) locus in *Neurospora* (McClung et al. 1989). Mutants of the *frq* gene shorten or lengthen the organism's circadian cycle of growth and spore formation (Dunlap 1990 for review). Therefore, *frq* mutations, at least phenotypically, show similar effects to the *per* mutations. The finding that both *per* and *frq* have a run of Threonine-Glycine pairs (Figure 9.5) is almost miraculous (see Dunlap 1990). However, this limited similarity between them tells us little about the possible mechanisms of action of these proteins. The Thr-Gly repeat bears some resemblance to Ser-Gly repeats found in mammalian proteoglycans (Bourdon et al. 1987). A *Drosophila per* probe has also detected a gene carrying a Thr-Gly encoding region in the mouse (Shin et al. 1985). Threonine and serine are similar amino acids and could potentially act as sites for the attachment of glycosaminoglycan (GAG) molecules. Mammalian proteoglycans are glycosylated in this way (Ruoslahti 1988), and preliminary biochemical analysis of the *per* product suggests that it, too, is glycosylated (Reddy et al. 1986, Bargiello et al. 1987).

The *per* gene also encodes some Ser-Gly pairs, so these could also act as attachment sites for GAG molecules. Yu et al. (1987b) deleted the *per* gene of the uninterrupted Thr-Gly encoding repeat and then transformed per^{01} mutants with this *in vitro*- mutated *per* gene. The behavioral results suggested that at 25°C, circadian rhythms were unaffected. The surprising result from this investigation was that deleting the Thr-Gly encoding motif resulted in a shortening of the normal 60 s song cycle to 40 s, suggesting that the Thr-Gly region played a prominent role in the song oscillator mechanism. The secondary structure of the conceptual *per* protein reveals that the Thr-Gly region may serve to separate the *per* protein into two domains (Costa et al. 1991). How the deletion of this apparent "spacer" region can affect one "clock" phenotype so drastically, but not the other, is a mystery!

Another similarity exists between the *per* protein and the protein encoded by a gene called *single-minded* (*sim*) that has also been isolated from *D. melanogaster*. Mutations of this gene disrupt the development of the embryonic ventral midline cells (Thomas et al. 1988). The *sim* protein has two 51 amino acid repeats which are also found in the *per* protein (Crews et al. 1988, and see Figure 9.5). The *sim* protein appears to be found in the nucleus (Crews et al. 1988), and it has been suggested that perhaps the *sim* protein may be a transcriptional regulatory factor (Nambu et al. 1990). More recently, a mammalian gene encoding a protein which translocates the aryl hydrocarbon receptor from the cytosol to the nucleus has been cloned (Hofman et al. 1991). The *arnt* gene as it is called, also encodes the two 51 amino acid repeats found in the *per* and *sim* proteins. Furthermore, *arnt* encodes a basic helix-loop-helix motif found in proteins that bind DNA as homo- or heterodimers. Thus *per* and *sim* have similarities to a protein that may act as a transcriptional regulator. This is particularly interesting in view of recent work showing that the *per* mRNA cycles in abundance and that the *per* protein can regulate its own mRNA cycling (Hardin et al. 1990; and see Section 3.4 below). Therefore, the exciting possibility exists that the *per* protein, perhaps like *arnt*, may be involved in a transcriptional cascade which may "turn on" any genes whose expression is required in a circadian manner.

Thus the most important clues to *per*'s mode of action is the limited but exciting similarity to *frq*, its resemblance to mammalian proteoglycans and its possible transcriptional role. Other similarities include the finding of DNA encoding Thr-Gly repeats in *Acetabularia* (Li-Weber et al. 1987) and the isolation of another gene in *D. melanogaster* that has a Thr-Gly repeat during a low stringency screen with a *Shaker* gene probe (Butler et al. 1989). *Shaker* is a gene encoding a putative potassium channel protein (Papazian et al. 1987; Tempel et al. 1987), so the possible but unlikely significance here might be that *per* might also have something to do with membrane channels.

3.4. A Model for *per* Action

However speculative this latter suggestion might be, it does lead us, if somewhat indirectly, into the work which at present provides us with a "global" model of how the *per* protein might function. This model, the "coupling model," depends on the *per* protein being found at the cell surface. Several apparently unrelated pieces of information are required to put this theory together.

1. Proteoglycans can modulate gap junctional communication in primary liver cell cultures (Spray et al. 1987)—in other words, proteoglycans can alter the way cells communicate with each other.
2. *per* transcripts and products can be found near cell membranes (Bargiello et al. 1987; Saez and Young 1988), as well as in the cytoplasm of many tissues (Reddy et al. 1984; James et al. 1986; Siwicki et al. 1988, Liu et al. 1988).
3. per^{01} mutants are not arrhythmic, but reveal weak, though significant short periodicities (5–12 h) in locomotor activity (Dowse et al. 1987;

Dowse and Ringo 1989) and perhaps also in song cycles (Kyriacou and Hall 1989).

4. Mathematical models of circadian cycles (Pavlidis 1973) have suggested that, by coupling together oscillators which themselves have high frequency periods, a longer circadian cycle can emerge.

Taking (3) and (4) together, it was suggested that what one might be seeing in *per*^{01} mutants is the "decoupling" of the component fast-frequency oscillators, giving rise to a phenotype which shows the fast frequencies (albeit weak) of the component oscillators (Dowse et al. 1987; Dowse and Ringo 1989; Bargiello et al. 1987). If one adds (1) and (2), the *per* protein could conceivably serve to couple together the hypothetical tonically-beating cells. Thus, in *per*^{01} mutants, coupling between cells would be removed, whereas in *per*^{S} mutants, the protein would be more active, providing "tighter" coupling between cells. By analogy the *per*^{L} mutants would have weak coupling.

There is some experimental evidence for this hypothesis. Bargiello et al. (1987) showed that dye injected into a salivary gland cell would move faster into a neighboring cell if both cells were *per*^{S} compared to when they were both *per*$^{+}$. Dye movement between adjacent *per*^{01} cells was very weak. Electrical coupling between *per*^{S} cells was also stronger than that between wild-type cells and was very poor between *per*^{01} cells (Bargiello et al. 1987). Thus, the coupling hypothesis has some support, at least using the large larval salivary gland cells as models. Whether such coupling can ever be demonstrated in nerve cells is open to question. Therefore, it may be that *per* provides a coupling function, coordinating the oscillations among different oscillating cells. This, however, begs the question, "What are the genes that provide the primary signals for the cells involved to oscillate?", and this can only be answered by further mutagenesis.

The attractiveness of the "coupling" theory is that it may explain why so many different temporally-regulated phenotypes are affected by *per* mutations. One can readily see that coupling between the neural oscillators determining the song rhythm might be altered in *per* mutants, even though the component oscillators are still "ticking" as normal. In development, gap junctions are believed to play an important role during cell determination and differentiation (e.g., Warner 1985). Altering the pace of gap junction communication in *per* mutants might explain the effects seen in their development times (Kyriacou et al. 1990).

One must not get too carried away with the coupling hypothesis, and some of the difficulties with it have been summarized by Hall and Kyriacou (1990). Problems with the proteoglycan component of the hypothesis (1 above) include the fact that when Ser-Gly encoding repeats are altered to Thr-Gly repeats in vertebrates, glycosylation can be disrupted (Bourdon

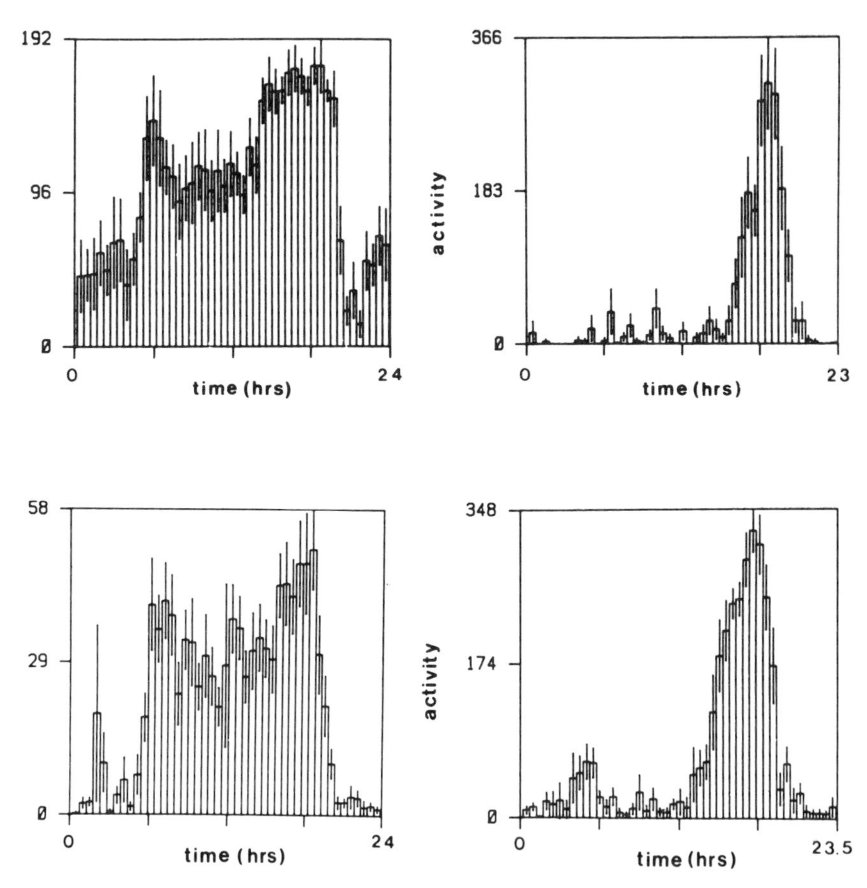

Figure 9.6. The effects of transforming a *D. melanogaster/D. pseudoobscura per* fusion gene on the activity of *per*$^+$ *D. melanogaster* flies (from Petersen et al. 1988, Figures 9.2 and 9.4). Each figure represents the "eduction profile" of one individual fly. The bars represent the average (±sem) number of locomotor activity events in half-hour time bins during a free-run (DD) after prior light entrainment (LL). The top left hand corner gives a profile for wild-type *D. melanogaster*, and the top right hand corner the data for wild-type *D. pseudoobscura*. The data are collected over several days and "folded" over a 24 h period for *D. melanogaster* (this represents the free-running period) and 23 h for *D. pseudoobscura* (representing the slightly shorter period of this species). The bottom left hand figure represents the eduction profile for *per*$^+$ *D. melanogaster* transformed with a 13.2 kb DNA fragment that covers the complete *D. melanogaster per* transcription unit. The bottom right hand figure gives the data from *per*$^+$ *D. melanogaster* transformed with a *per* gene containing the *D. pseudoobscura* coding sequence. Note the strikingly similar activity profile of this "partly hybrid" fly compared to the wild-type *D. pseudoobscura* pattern above it.

et al. 1987). Also the uninterrupted Thr-Gly encoding repeat can be deleted with little effect on the circadian rhythm, although some Ser-Gly and Thr-Gly pairs do remain in this Thr-Gly deleted gene (Yu et al. 1987b). Finally, a more general problem with the hypothesis is suggested by the results of Siwicki et al. (1988), who, using an anti-*per* antibody, found the protein to cycle in abundance in the brain and visual system. This finding led to a reevaluation of the previously reported "steady state" levels of the *per* gene's 4.5 kb transcript (Bargiello and Young 1984; Reddy et al. 1984; and see Section 3.1 above). A careful re-analysis revealed that, in fact, the *per* mRNA also showed a circadian cycle in abundance when head extracts were analyzed (Hardin et al. 1990). Furthermore, the *per* protein appears to be involved in a feedback cycle so it can regulate its own mRNA cycling (Hardin et al. 1990; Zweibel et al. 1991; Section 3.3 above). This finding creates some problems for the coupling hypothesis since it proposes that the period of the oscillation depends upon the "tightness" of the cellular interactions. If this intercellular association is cycling in strength *during* the circadian period, this makes the coupling hypothesis difficult to entertain, at least in its simplest form. However, it is still a useful model for understanding *per*'s possible mode of action. We must await the genetic analysis of gap junctions themselves in *Drosophila* in order to scrutinize it more closely.

3.5. Molecular Analysis of *per* Mutants

There are a number of *per* mutants, the original three, per^{S}, per^{L1}, and per^{01}, and per^{01}, per^{03} (Smith and Konopka 1982; Hamblen et al. 1986), per^{04} (Hamblen-Coyle et al. 1989), per^{L2} (Konopka 1987), and per^{LVar} (Hall and Kyriacou 1990 for a review). The per^{02} and per^{03} mutations are identical to per^{01} and all encode a "stop" codon at amino acid 464 (see Figure 9.5) (Yu et al. 1987a; Baylies et al. 1987; Hamblen-Coyle et al. 1989). per^{04} appears to be a genuinely new per^{0}-type allele and causes interesting behavioral differences from per^{01} (Hamblen-Coyle et al. 1989). The available molecular data on this mutation from Northern blots tentatively suggest the presence of a truncated 3 kb transcript rather than the normal 4.5 kb message (Hamblen-Coyle et al. 1989).

The per^{S} mutation causes a serine-to-asparagine substitution and is found downstream of the per^{01} site (Yu et al. 1987a; Baylies et al. 1987). The per^{L1} mutation encodes an aspartate rather than a valine upstream of per^{01}(Baylies et al. 1987; and see Figure 9.5).

The amino acids around and including the serine that is replaced in the per^{S} mutant meet the criteria of a site for phosphorylation (Yu et al. 1987a) or glycosylation (Section 3.3). The absence of such modifications at that site in the per^{S} mutant could cause a clock to "speed up." It must be stressed, however, that there is no direct evidence yet as to whether such post-translational modifications are indeed exerted at this site *in vivo*. Interestingly, the antibody used by Siwicki et al. (1988; 1989) and Zerr et al. (1990) was based on an epitope located close to the per^S site. We can conclude that this region may represent an active and accessible site which other molecules can recognize and bind.

The per^{L1} mutation falls just upstream of one of the repeated 51 amino acid encoding motifs (Figure 9.5) which are also found in the conceptual *sim* protein (Crews et al. 1988). However, neither this nor the nature of the amino acid substitution itself give us much idea as to how this mutation might disrupt the *per* protein. Of the other per^L-like variants, per^{L2} has a similar phenotype to per^{L1}, but per^{LVar} is difficult to assess in any comprehensible genetic terms (Hall and Kyriacou 1990).

3.6. Other Clock Mutations

We must also bear in mind that there are other mutations that affect circadian behavior which are not allelic to *per*. On the autosomes, *psi-2*, *psi-3* and *gat* alter the phasing of eclosion and have minor effects on the free-running period (Jackson 1983). On the X chromosome, *Clk* leads to a shortened circadian periodicity of 22 h but leaves the song cycle with a normal period (Dushay et al. 1990). *Clk* maps very closely to *per* and could possibly be a mutation in a 3′ flanking region of *per* (Dushay et al. 1990). Finally, the X-linked *Andante* (*And*), mutant has a long, free-running period of approximately 26 h (Konopka et al. 1991) and maps to the *miniature-dusky* (m-dy) region (Newby et al. 1991). The molecular analyses of *Clk* and *And* (now known as dy^{And}) are awaited with great interest.

There also exist other mutations which disrupt circadian behavior. The most dramatic of these is *disconnected* (*disco*), in which the eyes are anatomically disconnected from the optic lobes (Steller et al. 1987). *disco* adults are arrhythmic in locomotor behavior and show very weak eclosion rhythms (Dushay et al. 1989). These "clock" defects might be understood in the following terms. The central "pacemaker" may be located in the fly's optic lobes, as it is in other insects (Saunders 1982). For the pacemaker to be entrained, it must receive stimulation from the eyes. If light is prevented from reaching the pacemaker in *disco* mutants, then entrainment is impossible. Perhaps also the pacemaker requires a light signal via the eyes to set it in motion. Thus, in *disco* mutants the endogenous 24 h circadian period, which is a property of the pacemaker, may never be established. Consistent with this view, flies reared completely in the dark are also arrhythmic (Dowse and Ringo 1989). However recent evidence clearly shows that in *disco* mutants the *per* mRNA cycling is intact (Hardin et al. 1992). Therefore the oscillator appears to be functioning, and consequently the arrhythmicity observed in *disco* mutants must be due to a defect in the way the output from the oscillator is transferred to the effector organs.

Other mutations also create mild disturbances of the circadian period, particularly those with defects in the optic lobes (Helfrich 1986). Finally, a blind mutant *no-receptor-potential-A* (*norpA*) (Pak 1975) has a slightly faster than normal, free-running period (Dushay et al. 1989). The *norpA* transcript is observed both in the compound eye and in the brain of adults (Bloomquist et al. 1988). The DNA sequence from *norpA* suggests a phospholipase-C translation product (Bloomquist et al. 1988), so perhaps this enzyme is involved in pacemaker function via its effects in intracellular signaling pathways (Bloomquist et al. 1988; Dushay et al. 1989).

3.7. Where Is the Clock?

The molecular work with the mutants described above aims to provide information about either the location of the pacemaker or the biochemistry of circadian mechanisms. Probably the most specific evidence on the location of the pacemaker cells comes from the *per*-antibody study of Siwicki et al. (1988), and Zerr et al. (1990). These workers found *per* expression in the photoreceptors of the eyes and ocelli, in cell bodies of the lateral brain, posterior brain and optic lobes, as well as in other (probably glial) cells through the central nervous system. The lateral neurons stained positively with the antibody during the day, although the other brain structures stained very prominently during the night. The day-night oscillation in the staining pattern "free-ran" in constant darkness. The lateral neurons were the only neurons to stain in both adults and mature pupae. Siwicki et al. (1988) were thus tempted to speculate that these cells should be an important component, if not the cardinal component, of the pacemaker. Their reasoning was that both adults and pupae must express *per* in order to modulate their locomotor activity and pupal-adult eclosion phenotypes; thus any cells which stained at both developmental stages were key candidates for being the "pacemaker." In more detailed studies with the same *per* antibody, staining was also observed to be generally weaker than wild-type in per^{L1} mutants, and the amplitude of the rhythm in per^{L1}'s staining intensity appeared to be rather shallow under a light-dark cycle (Zerr et al. 1990). Under a light-dark cycle, the intensity of staining in the *per* mutants also changed with a different cyclical trajectory compared to per^{+}. Monitoring the intensity under free-running conditions revealed a circadian period in per^{S} of about 20 h (Zerr et al. 1990).

Thus it appears that cycles of immunoreactivity are altered in the per^{S} and per^{L1} mutants both in light-dark cycles and in constant darkness (for per^{S} at least). The overall signal strengths obtained with the antibody also differ between per^{L1} and the wild-type, giving added credence to the description of *per*^{L} as a hypomorph (Smith and Konopka 1982; Coté and Brody 1986; Baylies et al. 1987). The *per*^{01} mutant does not stain because the epitope recognized by this antibody is encoded downstream of the translational "stop" codon. The pattern of *per*-antibody staining in *disco* mutants was also studied and appeared to be normal for the photoreceptors and glial cells (Zerr et al. 1990). However, the lateral neurons were not stained in these mutants. Either the *per* protein was absent from these cells or, more likely, the mutation had severely disturbed the development of these neurons. Could it be, then, that the *disco* mutants are behaviorally arrhythmic because their pacemaker cells have been ablated? Again the recent work of Hardin et al. (1992) with *disco* mutants might suggest that because the *disco* mutant pacemaker appears to be functioning, perhaps the lateral neurons which are not staining in the mutants are not absolutely critical to the pacemaker.

It also became clear from these investigations that many non-neural tissue types express *per*. The staining in larval salivary gland (Bargiello et al. 1987) has already been discussed, although some disagreement remains as to whether these signals are "real" or an artifact (see Hall and Kyriacou 1990 for discussion). However, Saez and Young (1988) have obtained positive signals from Northern blots of hundreds of dissected larval salivary glands, so it is possible that any differences between laboratories are due to differences in techniques. In a similar vein, Bargiello et al. (1987) could not detect the presence of *per* transcript in the embryonic nervous system observed by James et al. (1986). However, this discrepancy has been resolved, and the Rockefeller group now find embryonic brain signals (M Young pers comm).

Other tissues such as ring glands, male and female reproductive tracts, gut, and Malpighian tubules also express *per* products. Some of these tissues, such as female ovaries and male testes, are known to express circadian rhythms (Giebultowicz et al. 1988), so perhaps *per* is mediating these non-neural rhythms. However, it may also be the case that the *per* protein functions in a manner unrelated to rhythmicity in these tissues. A detailed review of the expression patterns of *per* can be found in Hall and Kyriacou (1990).

3.8. Heat Shock Promotor—*per* Fusion

Related to the spatial and developmental expression of *per* are two elegant studies by Ewer et al. (1988, 1990). By fusing a heat-shock promoter (hsp) to the *per* coding sequence and transforming *per*^{01} mutants with this construct, they were able to determine when transcription of the *per* gene was necessary in order to obtain rhythmicity in adult circadian locomotor behavior. The results suggested that the gene had only to be turned "on" with a heat shock during the adult phase in order to obtain rhythmicity.

Having the gene "on" during earlier development, but then "off" at the adult stage led to arrhythmicity. The simple conclusion was that the *per* gene does not need to program adult circadian behavior by "building" a cycling neuronal network during ontogeny. The mode of action of *per* was "physiological" and not "developmental" because reducing the temperature in adulthood caused previously rhythmic flies to become arrhythmic quickly. This transient nature of the action of the *per* protein can perhaps be understood if, as suggested in section 3.4 above, it is a coupling protein which, when "on," quickly alters gap junctional communication between cells and gives a rhythmic behavioral output. The protein might act as a hormone, being secreted into the haemolymph and coupling together various neural oscillator cells. Indeed, there is some evidence, both from anatomical studies (Konopka and Wells 1980) and transplantation experiments (Handler and Konopka 1979), that hormonal factors might play a role in circadian behavior.

However, we might expect that if *per* is secreted in this way, then a peptide signal sequence would be obligatory in the primary amino acid structure; no such signal peptide is evident from conceptual translation of the *per* gene sequence. Also, the conclusions based on these results with the hsp-*per* transformants must be tempered in that the construct may have been "leaky" and therefore allowed some *per* protein to be made under the restrictive low temperature. This may have been enough to "build" the structures necessary for later "clock" output. Also, because the rhythms in these transformants were weak when the gene was turned "on" in the adult, it could be argued that this was because the developmental function of *per* was not adequate at the embryonic stages. Perhaps this could have been tested by raising flies at high temperature for their early developmental stages, maintaining the high temperature into adulthood, and comparing the strength of the rhythmicity in these individuals, against those whose gene had been turned "on" only in adulthood.

3.9. Evolutionary Studies

The *per* gene coding sequence has been determined in four *Drosophila* species, *D. melanogaster* (Reddy et al. 1986; Jackson et al. 1986; Citri et al. 1987; Yu et al. 1987a; 1987b), *D. yakuba* (Thackeray and Kyriacou 1990), *D. virilis* and *D. pseudoobscura* (Colot et al. 1988). Some sequence data also exist for *D. simulans* (Wheeler et al. 1991). Comparison of the *per* protein sequences of the two closely related species *D. melanogaster* and *D. yakuba* reveals that there are large stretches of highly conserved amino acid residues interwoven with three or four short "variable" regions. One of these is the Thr-Gly repeat region, which in *D. yakuba* encodes only 15 Thr-Gly pairs (Thackeray and Kyriacou 1990), as opposed to

17–23 in *D. melanogaster* (Yu et al. 1987b). The region immediately downstream of the Thr-Gly repeat also shows differences between the eight species of the *D. melanogaster* subgroup (Peixoto et al. 1992). These species-specific differences are functionally important in the case of *D. melanogaster* and *D. simulans*, as the amino acid changes between the two species determine the cycle lengths in the songs of the males (Wheeler et al. 1991). These are 55 s for *D. melanogaster* versus 35–40 s for *D. simulans* (Kyriacou and Hall 1980, 1986), and depend on only *four* amino acid substitutions. Similarly the *D. pseudoobscura per* gene can impose upon a *D. melanogaster* host, a *pseudoobscura*-like circadian cycle in locomotor activity as illustrated in Figure 9.6 (Petersen et al. 1988). Consequently, for songs and locomotor behavior, the species-variable regions within *per* carry species-specific functional information. As circadian activity cycles and song rhythms could be important isolating mechanisms, the possibility exists that *per* is a key "speciation' gene (Coyne 1992). *D. melanogaster* strains are polymorphic for the length of the uninterrupted Thr-Gly repeat (Yu et al. 1987b), and comparison of the corresponding sequences from two strains of *D. simulans* also reveals a polymorphism for either 23 or 24 Thr-Gly pairs. Within *D. melanogaster*, Costa et al. (1991) have studied the evolution of different natural Thr-Gly length alleles. These appear to evolve quickly by deletion or duplication of complete Thr-Gly encoding "cassettes." Furthermore, the pattern of length polymorphism shows significant clinal variation in that populations from southern Europe carry high frequencies of a 17 Thr-Gly allele, whereas populations from northern Europe carry high levels of the 20 Thr-Gly allele (Costa et al. submitted). Consequently, it appears that temperature may be acting as a selective agent, favoring shorter Thr-Gly alleles at higher temperatures.

The antibody to *per* made by Siwicki et al. (1988) has also been used to examine whether *per*-like proteins are present in other organisms. In the marine gastropods, *Aplysia* and *Bulla*, the antibody detects neurons and fibers in the eyes (Siwicki et al. 1989). Remarkably, these labeled structures represent the pacemaker cells which determine the circadian periodicity in the frequency of action potentials recorded from the optic nerve (for review, see Jacklet 1989). Therefore, a *per* relative is conserved in the circadian machinery of such widely divergent species as *Drosophila*, *Aplysia* and *Bulla*. The antibody was also used in Western blots and detected proteins of 55 kD and 70 kD in the eyes and central ganglia of *Aplysia*, but a larger protein of 170 kD was found in neural extracts from *Bulla* (Siwicki et al. 1989). The *per* antibody should therefore prove useful in the screening of gastropod "expression" libraries in order to isolate the genes encoding these proteins. The neurophysiological and cellular implications of such future studies will be discussed in the final section.

4. Conclusions

It is clear from the above discussions that the biological rhythms of *Drosophila* have become a model system for the study of the molecular genetics of complex behavior in eukaryotes. Only for egg-laying behavior in *Aplysia* is there a comparable body of molecular information (Scheller et al. 1984, for review). However, work on bacterial chemotaxis is far ahead of either of these systems, and a large number of bacterial genes influencing chemotaxis have been sequenced and manipulated (see Stock 1987, for review). With biological rhythms, the most obvious requirement at present is for further genetic and molecular analyses of other genes which contribute to behavioral rhythmicity. Only a few other loci besides *per* have been identified, such as *Clk* (Dushay et al. 1990), dy^{And} (Newby et al. 1991) and *norpA* (Dushay et al. 1989). In this respect it is most encouraging that *per* appears to belong to a "family" of genes whose products are implicated in the circadian control of other species. The limited similarity of *Neurospora's frq* gene to *per*, and the staining of nerve cells in the circadian pacemakers of *Aplysia* and *Bulla* by the *Drosophila per* antibody, are particularly significant. These results suggest that two different regions of the *per* protein (the Thr-Gly region shared by *frq* and the 14 amino acid epitope recognized by the *per* antibody) have been conserved through evolution in the circadian systems of distantly related oganisms. The discovery of *per*-like proteins in *Bulla* and *Aplysia* should also greatly facilitate cellular and neurophysiological analysis of biological clocks, which would be extremely difficult to undertake on *Drosophila's* small nerve cells. The possibility of cloning the *per*-like relatives of these larger invertebrates, either by using antibody or DNA probes and then reintroducing the manipulated genes back into the relevant cells, is on the horizon. Unravelling more of the mysteries of biological timing will depend on this kind of cellular-molecular strategy, and perhaps the answers will come by working with the circadian systems of these larger, more accessible organisms. However, I suspect that the fruit-fly will continue to play a significant leading role for many years to come.

Acknowledgments. I would like to thank Jeff Hall for his comments on the manuscript and my colleagues and collaborators at Leicester and Brandeis Universities who have contributed to much of this work. My modest contribution to the work has been underwritten by SERC grant GRD 79982 and a CEC grant.

References

Bargiello TA, Jackson FR, Young MW (1984) Restoration of circadian behavioural rhythms by gene transfer in *Drosophila*. Nature 312:752–754

Bargiello TA, Young MW (1984) Molecular genetics of a biological clock in *Drosophila*. Proc Natl Acad Sci USA 81:2142–2146

Bargiello TA, Saez TA, Baylies MK, Gasic G, Young MW, Spray DC (1987) The *Drosophila* gene *per* affects intercellular junctional communication. Nature 328:686–691

Baylies MK, Bargiello TA, Jackson FR, Young MW (1987) Changes in abundance or structure of the *per* gene product can alter periodicity of the *Drosophila* clock. Nature 326:390–392

Benzer S (1973) Genetic dissection of behaviour. Scient Amer 229(6):24–37

Bloomquist BT, Shortridge RO, Schneuwly S, Perdew M, Montell C, Steller H, Rubin GM, Pak WL (1988) Isolation of a putative phospholipase C gene of *Drosophila*, *norpA*, and its role in phototransduction. Cell 54:723–733

Bourdon MA, Shiga M, Rouslahti E (1987) Gene expression of the chondroitin sulfate proteoglycan protein PGl9. Mol Cell Biol 7:33–40

Brenner S (1973) Genetics of behaviour. Brit Med Bull 29:269–271

Butler A, Wei A, Baker K, Salkoff L (1989) A family of putative potassium channel genes in *Drosophila*. Science 243:943–947

Chen C-N, Denome S, Davis RL (1986) Molecular analysis of cDNA clones and the corresponding genomic coding sequences of the *Drosophila dunce*$^+$ gene, the structural gene for cAMP phosphodiesterase. Proc Natl Acad Sci USA 83:9313–9317

Citri Y, Colot HV, Jacquier AC, Yu Q, Hall JC, Baltimore D, Rosbash M (1987) A family of unusually spliced and biologically active transcripts is encoded by a *Drosophila* clock gene. Nature 326:42–47

Cogne JA (1992) Genetics and speciation. Nature 355:511–515

Colot HV, Hall JC, Rosbash M (1988) Interspecific comparisons of the *period* gene of *Drosophila*. EMBO J 7:3929–3937

Costa R, Peixoto AA, Thackeray JR, Dalgleish R, Kyriacou CP (1991) Length polymorphism in the Threonine-Glycine encoding repeat region of the *period* gene in *Drosophila*. J Mol Evol 32:238–246

Costa R, Peixoto AA, Barbujani G, Kyriacou CP. A cline in a *Drosophila* clock gene. Manuscript submitted

Coté GG, Brody S (1986) Circadian rhythms in *Drosophila melanogaster*: Analysis of period as a function of gene dosage at the *per* (*period*) locus. J Theor Biol 121:487–503

Cowling DE, Burnet B (1981) Courtship songs and genetic control of their acoustic characteristics in sibling species of the *D. melanogaster* subgroup. Anim Behav 29:924–935

Crews ST, Thomas JB, Goodman CS (1988) The *Drosophila single-minded* gene encodes a nuclear protein with sequence similarity to the *per* gene product. Cell 52:143–151

Crossley SA (1988) Failure to confirm rhythms in *Drosophila* courtship song. Anim Behav 36:1098–1109

Dowse HB, Hall JC, Ringo JM (1987) Circadian and ultradian rhythms in *period* mutants of *Drosophila melanogaster*. Behav Genet 17:19–35

Dowse HB, Ringo JM (1987) Further evidence that circadian rhythms in *Drosophila* are a population of coupled ultradian oscillators. J Biol Rhythms 2:65–76

Dowse HB, Ringo JM (1989) Rearing *Drosophila* in constant darkness produces phenocopies of *period* clock mutants. Physiol Entomol 62:785–803

Dudai Y, Jan Y-N, Byers D, Quinn WC, Benzer S (1976) *dunce*, a mutant of *Drosophila* deficient in learning. Proc Natl Acad Sci USA 73:1684–1688

Dunlap J (1990) Closely watched clocks: molecular analysis of circadian rhythms in *Neurospora* and *Drosophila*. Trends Genet 6:159–165

Dushay MS, Rosbash M, Hall JC (1989) The *disconnected* visual system mutations in *Drosophila melanogaster* drastically disrupt circadian rhythms. J Biol Rhythms 4:1–27

Dushay MS, Konopka RJ, Orr D, Greenacre M, Kyriacou CP, Rosbash M, Hall JC (1990) Phenotypic and genetic analysis of *Clock*, a new circadian rhythm mutant in *Drosophila melanogaster*. Genetics 125:557–578

Ewer J, Hamblen-Coyle M, Rosbash M, Hall JC (1990) Requirement for *period* gene expression in the adult and not during development for locomotor activity rhythms of imaginal *Drosophila melanogaster*. J Neurogen 7:31–73

Ewer J, Rosbash M, Hall JC (1988) An inducible promotor fused to the *period* gene in *Drosophila* conditionally rescues adult *per*-mutant arrhythmicity. Nature 333:82–84

Ewing AW (1988) Cycles in the courtship song of male *Drosophila melanogaster* have not been detected. Anim Behav 36:1091–1097

Gailey DA, Jackson FR, Siegel RW (1982) Male courtship in *Drosophila*: the conditioned response to immature males and its genetic control. Genetics 102:771–782

Gailey DA, Villella A, Tully T. 1991 Reassessment of the effect of biological rhythm mutations on learning in *Drosophila melanogaster*. J Comp Physiol. 169:685–697

Giebultowicz JM, Bell RA, Imberski RB (1988) Circadian rhythm of sperm movement in the male reproductive tract of the gypsy moth, *Lymantria despar*. J Insect Physiol 34:527–532

Hall JC (1984) Complex brain and behavioral functions disrupted by mutations in *Drosophila*. Devel Genet 4:355–378

Hall JC, Kyriacou CP (1990) Genetics of biological rhythms in *Drosophila*. Adv Insect Physiol 22:221–298

Hamblen M, Zehring WA, Kyriacou CP, Reddy P, Yu Q, Wheeler DA, Zwiebel LJ, Konopka RJ, Rosbash M, Hall JC (1986). Germ-line transformation involving DNA from the *period* locus in *Drosophila melanogaster*. Overlapping genomic fragments that restore circadian and ultradian rhythmicity to *per*^{0} and *per* mutants. J Neurogenet 3:249–291

Hamblen-Coyle M, Konopka RJ, Zwiebel LJ, Colot HV, Dowse HB, Rosbash M, Hall JC (1989) A new mutation at the *period* locus with some novel effects on circadian rhythms. J Neurogenet 3:249–291

Handler AM, Konopka RJ (1979) Transplantation of a circadian pacemaker in *Drosophila*. Nature 279:236–238

Hardin PE, Hall JC, Rosbash M (1990) Feedback of the *Drosophila period* gene product on the circadian cycling of its messenger RNA levels. Nature 343: 536–540

Hardin PE, Hall JC, Rosbash M (1992) Behavioral and molecular analyses suggest that circadian output is disrupted by *disconnected* mutants in *D. melanogaster*. EMBO J 11:1–6

Hay DA (1985) Essentials of Behaviour-Genetics. Blackwell, Palo Alto, 359 pp

Helfrich C (1986) The role of the optic lobes in the regulation of the locomotor activity rhythms in *Drosophila melanogaster*: behavioural analysis of neural mutants. J Neurogenet 3:321–343

Hoffman EC, Reyes H, Chu F, Sander F, Conley LH, Brooks BA, Hankinson O (1991) Cloning of a factor required for the activity of the Ah (dioxin) receptor. Science 252:954–958

Jacklet JW (1989) Cellular neuronal oscillators. In Jacklet JW (ed) Neuronal and Cellular Oscillators. Marcel Dekker, New York, pp 482–527

Jackson FR (1983) The isolation of biological rhythm mutations on the autosomes of *Drosophila melanogaster*. J Neurogenet 1:3–15

Jackson FR, Gailey DA, Siegel RW (1983) Biological rhythm mutations affect an experience dependent modification of male courtship behavior in *Drosophila melanogaster*. J Comp Physiol A 151:545–552

Jackson FR, Bargiello TA, Yun SH, Young MW (1986) Product of *per* of *Drosophila* shares homology with proteoglycans. Nature 320:185–188

James AA, Ewer J, Reddy P, Hall JC, Rosbash M (1986) Embryonic expression of the *period* clock gene of *Drosophila melanogaster*. EMBO J 5:2313–2320

Judd BH, Shan MW, Kaufman TC (1972) The anatomy and function of a segment of the X chromosome in *Drosophila melanogaster*. Genetics 71:139–156

Kawanishi M, Watanabe TK (1980) Genetic variations of courtship song of *Drosophila melanogaster* and *Drosophila simulans*. Jap J Genet 55:235–240

Konopka RJ (1987) Neurogenetics of *Drosophila* circadian rhythms. In Huettel MD (ed) Evolutionary Genetics of Invertebrate Behaviour. Plenum, New York, pp 215–221

Konopka RJ, Benzer S (1971) Clock mutants of *Drosophila melanogaster*. Proc Natl Acad Sci USA 68:2112–2116

Konopka RJ, Wells S (1980) *Drosophila* clock mutations affect the morphology of a brain neurosecretory cell group. J Neurobiol 11:411–415

Konopka RJ, Wells S, Lee T (1983) Mosaic analysis of a *Drosophila* clock mutant. Mol Gen Genet 190:284–288

Konopka RJ, Smith RF, Orr D (1991) Characterisation of *Andante*, a new *Drosophila* clock mutant, and its interaction with other clock mutants. J Neurogen 7:103–114.

Kyriacou CP, Hall JC (1980) Circadian rhythm mutations in *Drosophila* affect short-term fluctuations in the male's courtship song. Proc Natl Acad Sci USA 77:6929–6933

Kyriacou CP, Hall JC (1982) The function of courtship song rhythms in *Drosophila*. Anim Behav 30:794–801

Kyriacou CP, Hall JC (1984) Learning and memory mutations impair acoustic priming of mating behaviour in *Drosophila*. Nature 308:62–65

Kyriacou CP, Hall JC (1985) Action potential mutations stop a biological clock in *Drosophila*. Nature 314:171–173

Kyriacou CP, Hall JC (1986) Inter-specific genetic control of courtship song production and reception in *Drosophila*. Science 232:494–497

Kyriacou CP, Hall JC (1989) Spectral analysis of *Drosophila* courtship song rhythms. Anim Behav 37:850–859

Kyriacou CP, Oldroyd M, Wood J, Sharp M, Hill M (1990) Clock mutations alter development time in *Drosophila*. Heredity 64:395–401

Liu X, Lorenz L, Yu Q, Hall JC, Rosbash M (1988) Spatial and temporal expression of the *period* gene in *Drosophila melanogaster*. Genes Devel 2:228–238

Liu X, Yu Qiang, Huang Z, Zwiebel L, Hall JC, Rosbash M (1991) The strength and periodicity of *D. melanogaster* circadian rhythms are differentially affected by alterations in *period* gene expression. Neuron 6:753–766

Livingstone MS (1981) Two mutations in *Drosophila* affect the synthesis of octopamine, dopamine and serotonin by altering the activities of two different amino-acid decarboxylases. Neurosci Abstr 8:384

Li-Weber M, de Groot GJ, Schweiger HG (1987) Sequence homology to the *Drosophila per* locus in higher plant nuclear DNA and in *Acetabularia* chloroplast DNA. Mol Genet 209:1–7

Lorenz LJ, Hall JC, Rosbash M (1989) Expression of a *Drosophila* mRNA is under circadian clock control during pupation. Development 107:869–880

Manning A (1961) The effects of artificial selection of mating speed in *Drosophila melanogaster*. Anim Behav 9:82–91

McClung CR, Fox BA, Dunlap JC (1989) *Frequency*, a clock gene in *Neurospora* shares a sequence element with the *Drosophila* clock genes *period*. Nature 339:558–562

Nambu JR, Franks RG, Hu S, Crews S (1990) The *single-minded* gene is required for the expression of genes important for the development of CNS midline cells. Cell 63:63–75

Newby LM, White L, DiBartolomeis SM, Walker BJ, Dowse HB, Ringo JM, Khuda N, Jackson FR (1991) Mutational analysis of the *Drosophila miniature-dusky* (*m-dy*) locus: effects on cell size and circadian rhythms. Genetics 128:571–582

Pak WL (1975) Mutants affecting the vision of *Drosophila melanogaster*. In King RC (ed) Handbook of Genetics Vol 3. Plenum, New York, pp 723–733

Papazian DM, Schwartz TL, Tempel BL, Jan YN, Jan LY (1987) Cloning of genomic and complementary DNA from *Shaker*, a putative potassium channel gene in *Drosophila*. Science 237:749–753

Pavlidis T (1973) Biological Oscillators: Their Mathematical Analysis. Academic Press, New York, 207 pp

Peixoto AA, Costa R, Wheeler DA, Hall JC, Kyriacou CP (1992) Evolution of the Threonine-Glycine repeat region of the *period* gene in the *melanogaster* species subgroup. J Mol Evol. In press

Petersen G, Hall JC, Rosbash M (1988) The *period* gene of *Drosophila* carries species-specific behavioural instructions. EMBO J 7:3939–3947

Pittendrigh CS (1974) Circadian oscillations in cells and the circadian organization of multicellular systems. In Schmitt FO, Worden FG (eds) The Neurosciences. Third Study Program Cambridge, MIT Press, Massachussets, pp 437–458

Reddy P, Zehring WA, Wheeler DA, Pirrotta V, Hadfield C, Hall JC, Rosbash M (1984) Molecular analysis of the *period* locus in *Drosophila melanogaster* and identification of a transcript involved in biological rhythms. Cell 38: 701–710

Reddy P, Jacquier AC, Abovich N, Petersen G, Rosbash M (1986) The *period* clock locus of *D. melanogaster* codes for a proteoglycan. Cell 46:53–61

Ruoslahti E (1988) Structure and biology of proteoglycans. Ann Rev Cell Biol 4:229–255

Saez L, Young MW (1988) In situ localisation of the *per* clock protein during development of *Drosophila melanogaster*. Mol Cell Biol 8:5378–5385

Saunders DS (1982) Insect Clocks. Pergamon Press, Oxford, 409 pp

Scheller RH, Kaldany RR, Kreiner T, Mahon AC, Nambu J, Schaefer M, Taussig R (1984) Neuropeptides: mediators of behaviour in Aplysia. Science 225:1300–1308

Schilcher F von (1989) Have cycles in the courtship song of *Drosophila* been detected? Trends Neuroscience 12:311–313

Shin HS, Bargiello TA, Clark BT, Jackson FR, Young MW (1985) An unusual coding sequence from a *Drosophila* clock gene is conserved in vertebrates. Nature 317:445–448

Siegel RW, Hall JC (1979) Conditioned responses in courtship behaviour of normal and mutant *Drosophila*. Proc Natl Acad Sci USA 76:3430–3434

Siwicki KK, Eastman C, Petersen G, Rosbash M, Hall JC (1988) Antibodies to the *period* gene product of *Drosophila* reveal diverse tissue distribution and rhythmic changes in the visual system. Neuron 1:141–150

Siwicki KK, Strack S, Rosbach M, Hall JC, Jacklet JW (1989) An antibody to the *Drosophila period* protein recognises circadian pacemaker neurons in *Aplysia* and *Bulla*. Neuron 3:51–58

Smith RF, Konopka RJ (1981) Circadian clock phenotypes of chromosome aberrations with a breakpoint at the *per* locus. Mol Gen Genet 183:243–51

Smith RF, Konopka RJ (1982) Effects of dosage alterations at the *per* locus on the circadian clock of *Drosophila*. Mol Gen Genet 185:30–36

Spray DC, Fujita M, Saez JC, Choi H, Watanabe T, Heitzberg L, Rosenberg C, Reid LM (1987) Proteoglycans and glycosaminoglycans induce gap junction synthesis and function in primary liver cell cultures. J Cell Biol 105:541–551

Steller H, Fischbach KF, Rubin GM (1987) *disconnected*: a locus required for neural pathway formation in the visual system of *Drosophila*. Cell 50: 1139–1153

Stock J (1987) Mechanisms of receptor function and the molecular biology of information processing in bacteria. Bioessays 6:199–203

Tempel BL, Papazian DM, Schwartz TL, Jan YN, Jan LY (1987) Sequence of a probable potassium channel component encoded at the *Shaker* locus of *Drosophila*. Science 237:770–775

Thackeray JR, Kyriacou CP (1990) Molecular evolution in the *Drosophila yakuba period* locus. J Mol Evol 31:389–401

Thomas JB, Crews ST, Goodman CS (1988) Molecular genetics of the *single-minded* locus; a gene involved in the development of the *Drosophila* nervous system. Cell 52:133–141

Warner A (1985) The role of gap junctions in amphibian development. J Embryol Exper Morph Suppl 89:365–380

Weitzel G, Rensing L (1981) Evidence for cellular circadian rhythms in isolated fluorescent dye-labeled salivary glands of wild type and an arrhythmic mutant of *Drosophila melanogaster*. J Comp Physiol 143:229–235

Wheeler DA, Kyriacou CP, Greenacre ML, Yu Q, Rutila JE, Rosbash M, Hall JC (1991) Molecular transfer of a species-specific behavior from *Drosophila simulans* to *Drosophila melanogaster*. Science 251:1082–1085

Winfree AT (1987) The Timing of Biological Clocks. Scientific American Books, New York, 199 pp

Young MW, Jackson FR, Shin HS, Bargiello TA (1985) A biological clock in *Drosophila*. Cold Spring Harbor Symp Quant Biol 50:865–875

Yu Q, Jacquier AC, Citri Y, Colot HM (1987a) Molecular mapping of point mutations in the *period* gene that stop or speed up biological clocks in *Drosophila melanogaster*. Proc Natl Acad Sci USA 84:784–788

Yu Q, Colot HV, Kyriacou CP, Hall JC, Rosbash M (1987b) Behaviour modification by *in vitro* mutagenesis of a variable region within the *period* gene of *Drosophila*. Nature 326:765–769

Zehring WA, Wheeler DA, Reddy P, Konopka RJ, Kyriacou CP, Rosbash M, Hall JC (1984) P-element transformation with *period* locus DNA restores rhythmicity to mutant, arrhythmic *Drosophila melanogaster*. Cell 39:369–376

Zerr DN, Rosbash M, Hall JC, Siwicki KK Circadian rhythms of *period* protein immunoreactivity in the CNS and the visual system of *Drosophila*. J Neurosci 10:2749–2762

Zweibel LJ, Hardin PE, Liu X, Hall JC, Rosbash M (1991) A post-transcriptional mechanism contributes to circadian cycling of a *per*-β-galactosidase fusion protein. Proc Natl Acad Sci USA 88:3882–3886

Chapter 10

Molecular Analysis of Insect Meiosis and Sex Ratio Distortion

Terrence W. Lyttle, Chung-I Wu, and R. Scott Hawley

1. Introduction: Gametogenesis as a Developmental System

It is arguable that gametogenesis is, if not preeminent, at least among the most critical processes in development. Errors in any other developmental system may result at worst in the genetic death of the individual organism affected, while errors in gametogenesis may lead to the propagation of deleterious genetic variants over many generations, potentially resulting in far reaching evolutionary effects. As our understanding of the general machinery of meiosis and gametogenesis improves, so do our opportunities for utilizing this information to co-opt the gametogenic process to alter the genetic makeup of individuals or even whole populations. A perfect example of this is the way in which our understanding of the meiotic disturbances arising from hybrid dysgenesis led to the development of genetic transformation technology in *Drosophila* (Spradling and Rubin 1981; Spradling 1986).

Here we will be reviewing current molecular genetic approaches to the study of insect meiosis. Even in a model organism such as *Drosophila*, which is the focus of much of this work, there have been distinct differences in the approaches to the study of spermatogenesis and oogenesis. Oogenesis in *Drosophila* has in recent years become a model developmental system, now understood in exquisite molecular detail (cf. Manseau and Schupbach 1989). Further, because meiosis in *Drosophila* females exhibits exquisite systems of pairing, recombination and segregation, it has been the subject of intense laboratory mutational analysis. The existence of a wide array of point mutations defining classical genetic loci has made it possible in recent years to begin the molecular analysis of female meiosis (cf. Carpenter 1988; Eisen and Camerini-Otero 1988; Hawley 1988; Yamamoto et al. 1989; Endow et al. 1990; McDonald and Goldstein 1990, for summaries). In contrast, studies of spermatogenesis in *Drosophila* and other insects have been largely restricted to ultrastructural analyses of developing spermatids. Indeed, because natural populations or hybrids between species are the source of most data on genetic disturbances of male reproduction, research interest has often focused on the evolutionary consequences of abnormalities of spermatogenesis, particularly those involving the sex chromosomes.

In this review, meiosis in both sexes will be considered. For male insects, our particular emphasis will be on the genetic basis of sex ratio distortion arising from defects in the normal developmental blueprint followed during meiosis and subsequent developmental stages of spermatogenesis. Since the most important function of the meiotic process is to ensure Mendelian segregation of homologous chromosomes, we will first briefly discuss the general evolutionary consequences of non-Mendelian segregation. Subsequent sections will focus on sex ratio distortion arising from non-Mendelian segregation of the sex chromosomes, both in naturally occurring and laboratory systems, and our current knowledge concerning their molecular basis. Finally, we will turn to meiosis in *Drosophila* females to review the current directions in molecular genetic analysis of those genes involved in governing the general processes of chromosome pairing, exchange and segregation.

2. Evolutionary Consequences of Gamete Competition and Meiotic Drive

Higher plants and animals exhibit an alternation between haploid and diploid nuclear phases during the life cycle. Population genetic theory for these organisms tends to focus primarily on the role of diploid deterministic selection in directing adaptive evolution. This is perhaps not surprising in light of the fact that the diploid stage of the life cycle is longer and more highly developed than the corresponding haploid stage. If theorists consider the latter stage at all, it is viewed simply as a source of background noise arising from the stochastic sampling of gametes, interfering with the fidelity of Darwinian natural selection, but producing no net advantage to one genetic variant over another. The prevalent assumption is that the meiotic process and the ensuing haploid gametic phase basically serve to faithfully translate, from one generation to the next, the diploid selective advantage accruing to a given allelic form. This would seem especially appropriate for higher animals, where the gametic stage is generally so underdeveloped as to result in gametic function being largely independent of genetic content.

However, when defective meiosis alters the transmission frequency of an allele, the result may be the modification or even reversal of its expected evolutionary fate. For example, genetic disturbances of meiosis or gametogenesis may lead to an allele or chromosome gaining a transmission advantage, such that one member of a pair of segregating alleles or chromosomes is regularly recovered in the gametes in excess of the expected 50% proportion. The term *meiotic drive* has been coined to describe this form of non-Mendelian segregation (Sandler and Novitski 1957) or distorted transmission ratio; the genetic variant recovered in excess is said to be "driven." *Genic* meiotic drive is initially limited in its impact to the population dynamics of the drive locus itself and those loci which happen to be in close linkage. Alleles at these latter loci may enjoy indirect drive through genetic hitchhiking, leading eventually to the establishment of drive haplotypes (Hedrick 1988). The haplotype may be extended by the incorporation of chromosome rearrangements which reduce recombination and promote further linkage disequilibrium between the drive locus and more distant modifier loci (Prout et al. 1973; Thomson and Feldman 1974). In the extreme the haplotype may come to include the whole chromosome, leading to a form of *chromosomal* meiotic drive.

In general, chromosomal meiotic drive usually arises as a direct consequence of intrinsic chromosome structure rather than through the evolution of a drive locus and its linked modifiers. For example, this is seen for certain B chromosomes (the largely heterochromatic supernumerary chromosomes found in some plant, orthopteran and coccid species, among others; Jones and Rees 1982 for a review) which undergo extra rounds of replication to increase in number within the genome. It would be expected that chromosomal meiotic drive would be more common in female gametogenesis, where there is an inherent asymmetry based on the fact that only one of the four products of meiosis is regularly functional. For example, in several fly species it has been observed that the smaller member of asymmetric chromosome pairs is oriented on the female meiotic spindle in such a way as to enhance its inclusion in the egg (Novitski 1967; Foster and Whitten 1974; 1991). Other cases of chromosomal drive include the preferential segregation of heterochromatic knobs in maize (Rhoades and Dempsey 1966).

As we proceed, it is important to keep in mind a perhaps subtle, but important distinction between *true meiotic drive* (such as the examples of chromosomal drive just discussed) and cases of *gamete competition*, which may result incidentally in meiotic drive. A strict definition of meiotic drive requires that abnormalities of the meiotic divisions themselves be responsible for the nonrandom segregation; that is, a chromosome or allele is recovered in excess because it undergoes extra replications, or because it is recovered preferentially in those meiotic products destined to become gametes. In this case, the number of gametes carrying the driven allele increase in absolute number, although the total number of functional gametes remains constant. Conversely, several of the best studied cases of genic drive (e.g., Segregation distorter (*SD*) and Sex ratio (*SR*) in *Drosophila*, male drive (*MD*) in mosquito, *t*-haplotype transmission distortion in mouse) essentially represent a form of haploid selection resulting from post-meiotic gamete competition. In each of these cases, while it is during male meiosis that the drive allele induces a lesion in those chromosomes carrying its homolog, it is actually the subsequent dysfunction of sperm carrying the affected homolog that leads to the transmission advantage. Thus, the *absolute* number of successful gametes carrying the favored element is not necessarily increased, a prerequisite for effective meiotic drive. In fact, an effective segregation advantage at the population level can arise from such gamete competition only when: (1) competition occurs among gametes from single individuals (rather than in gamete pools of mixed parental origin, e.g., pollen competition on the stigma), and (2) gamete loss results in a less than proportional loss of individual fecundity (Mulcahy 1975). If these conditions hold, then gamete competition can be modeled as mathematically equivalent to true meiotic drive. On the other hand, a loss of fecundity strictly proportional to the loss of gametes results in no effective drive in the population, even with complete drive in the individual (cf. Hartl 1972). Not surprisingly, meiotic drive systems depending on gamete competition are more likely to be found in males of monogamous species, where dysfunction of a portion of the sperm has the smallest impact on fecundity.

To summarize, true meiotic drive will usually be chromosomal and occur in females, while genic drive arising from gamete competition will generally be restricted to males.

3. Sex Ratio Evolution

Fisher (1930) first demonstrated why the 1:1 sex ratio is evolutionarily stable for most organisms (assuming equal parental costs for rearing males and females, see Hamilton 1967; Uyenoyama and Feldman 1978; Lyttle 1981, for more recent discussions). His argument can be loosely paraphrased as follows. The total contribution of each sex to the genetic pool of the next generation must be equal. In populations with a sex ratio that is biased overall, an individual of the rarer sex will therefore, on average, make a greater genetic contribution to ensuing generations than does a member of the common sex. All else being equal, any individual with an excess of the rarer sex among his or her progeny will consequently tend to have more descendants. Any autosomal genetic variant contributing to this advantageous individual sex ratio bias will be selected to increase in frequency in the population, ultimately causing the overall sex ratio to return to 1:1. At that point, all individual sex ratios have equal fitness, and further sex ratio evolution ceases.

The existence in many organisms of heteromorphic sex chromosomes showing unbiased transmission through meiosis of the heterogametic sex (XY males in mammals and many insects, ZW females in birds, butterflies and moths) would seem to offer an obvious mechanism for ensuring this outcome. However, it is important to note that the Fisherian argument for maintaining the optimal 1:1 sex ratio does not apply to the sex chromosomes themselves. X- or Y-linked genes that alter the sex ratio in their own favor (perhaps by meiotic drive) have a large gametic selective advantage operating at the chromosome level, irrespective of their general negative effect on diploid population fitness (Hamilton 1967). While sex ratio distortion can arise from other biological causes (including cytoplasmic factors; see Uyenoyama and Feldman 1978), this review will focus particularly on sex ratio distortion arising from meiotic drive favoring one of the sex chromosomes. As this must occur in the heterogametic sex, which is usually the male for most insects (Bull 1983), it follows that we will be concerned primarily with genic forms of meiotic drive.

4. Sex Chromosome Drive

Most forms of genic drive that depend on gamete dysfunction cause some reduction in male fecundity. This is true for cases of both sex chromosomal and autosomal drive systems (cf. Hartl and Hiraizumi 1976; Youngson et al. 1981; Wu 1983a; 1983b; Silver 1985). However, here the similarity ends. The relative fecundity of drive homozygotes to heterozygotes is critical in determining the fate of autosomal drive systems; it clearly has little impact on sex chromosome drive, which of necessity is never homozygous in the heterogametic sex. Strong autosomal drive systems can become fixed without any major detrimental effect on the population; conversely, a sufficiently strong sex ratio distortion may lead to population extinction (Figure 10.1). The effect of sex chromosome drive on population structure and individual fitness is of a much higher magnitude than is the case for autosomal drive systems. For example, Hamilton (1967) demonstrated that Y drive chromosomes spread faster through a population, and X drive chromosomes slower, than does an autosomal element of similar drive strength restricted to acting in male meiosis (Figures 10.1 and 10.2). This is a consequence of

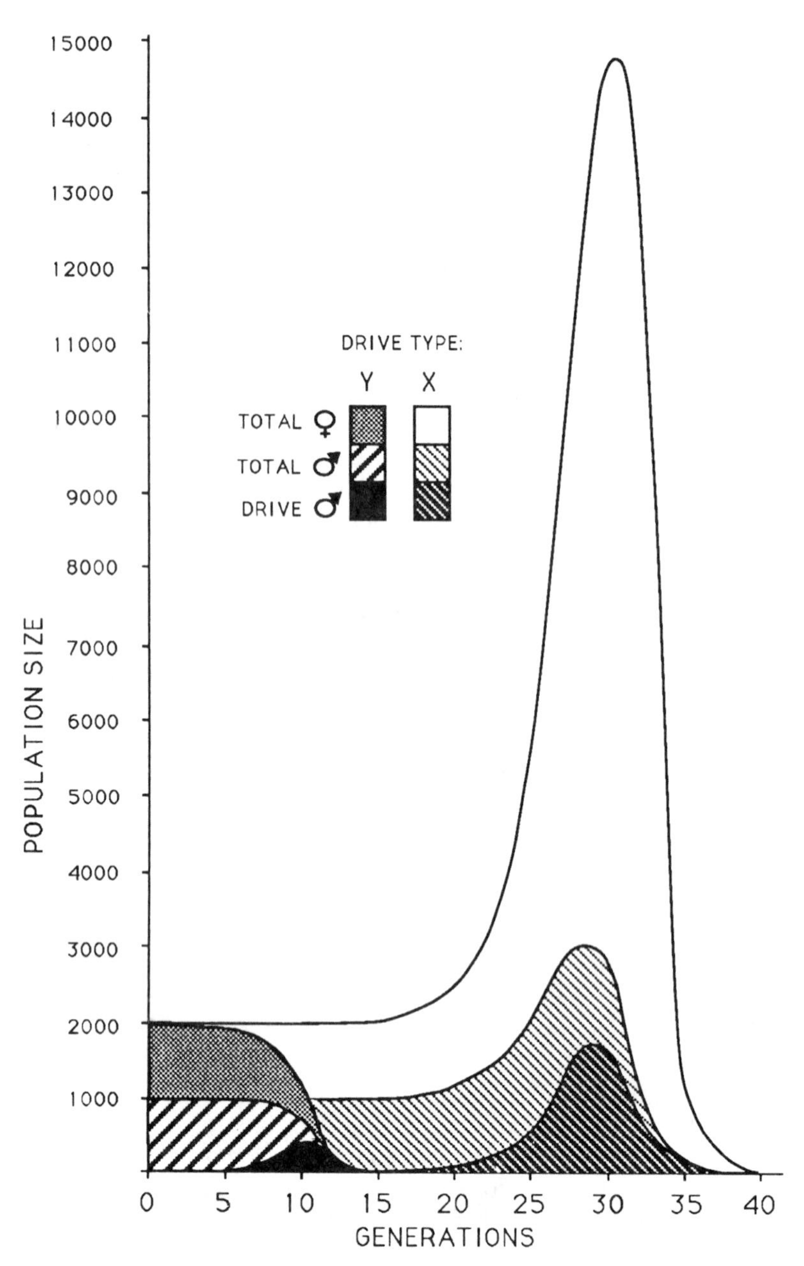

Figure 10.1. Change in population size during invasion of sex chromosome drive. With complete Y drive (left side of figure), the population size begins to drop immediately, reaching zero in 10–15 generations as the drive Y replaces the nondriving form to give a unisexual male population. Conversely, invasion of an X chromosome showing complete drive in males (right side of figure) initially leads to an increase in population size, as females increase in frequency. Population extinction is not reached until 40 generations, when the population consists solely of females. (Modified from Hamilton 1967, using his parameterization.)

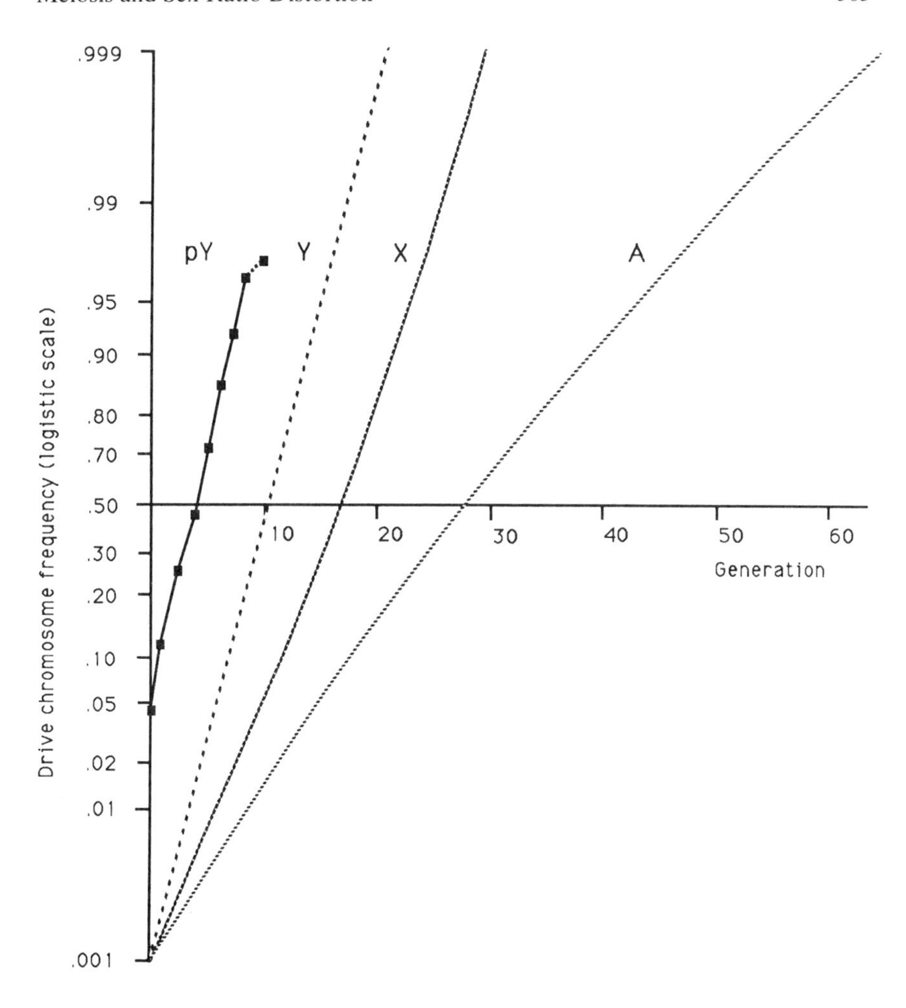

Figure 10.2. Change in frequency of drive chromosomes over time. Curves are generated assuming no fitness differences among chromosomes other than those arising from meiotic drive in heterozygotes. Paths A, X and Y indicate increase in frequency of an invading autosome, X chromosome or Y chromosome, respectively, showing complete meiotic drive (based on simulations of Hamilton 1967). Path pY represents the average results obtained from several experimental population cages of *D. melanogaster* challenged by pY drive (Lyttle 1977, see text). The experimental cages were started with higher initial frequencies of drive males than were used in the simulations (0.05 rather than 001); otherwise the slope of the experimental curve for pY drive agrees closely with the simulation for true Y drive.

the fact that a driving Y chromosome is expressed in males in every generation, while a driving autosome finds itself in males only half the time, and a driving X chromosome only one-third of the time. Since the limiting factor for population fecundity is usually egg, not sperm, production, strong X chromosome drive may actually initially foster an increase in population size. In contrast, strong Y drive almost immediately causes a decrease in population size concomitant with the decrease in the proportion of females (Figure 10.1), producing an immediate detrimental effect on the population. Strong Y drive is of necessity transitory, since either drive suppression must evolve very quickly or the population may be pushed to extinction. This may explain why few Y drive systems have been observed in nature (see below). Further, it has been suggested (Hamilton 1967) that the heterochromatinization of the Y chromosome itself, or of autosomal segments that have become attached to the Y through translocation (cf. *Drosophila miranda*, Steinemann 1982), may in part be a relic of the suppression of ancient Y chromosome drive systems.

A distorted sex ratio exerts such a great effect on fitness that modifiers of sex chromosome drive should be strongly selected. There are at least two distinct ways in which a deviant sex ratio arising from meiotic drive can be returned to 1:1; either autosomal suppressors or sex chromosomes less sensitive to the distorting effects of the driving chromosome, or both, will accumulate. Essentially, the fitness of the driving sex chromosome is enhanced by the accumulation of linked modifiers that increase the strength of distortion. Conversely, the general negative effects of a distorted progeny sex ratio on organismal fitness lead to the evolution of drive suppressors on all other chromosomes, as Fisher would have predicted. This is particularly true for the non-driving sex chromosome. Thus, a system of sex chromosome drive is most likely to evolve into a two-locus polymorphism with linkage disequilibrium. The drive allele will tend to be found in coupling with enhancer and in repulsion with suppressor alleles at the same linked modifier locus. Linkage disequilibrium is further promoted by selection for genetic variants, such as chromosome inversions, which reduce recombination between the drive and modifier loci. This maintenance of tight linkage is extremely important because gene complexes, rather than single gene loci, are the genetic basis of virtually all meiotic drive systems so far analyzed. Of course, recombination between the sex chromosomes of heteromorphic males is essentially eliminated already, although there is still an opportunity for recombination in XX females. Since drive systems generally involve multiple loci, it might be expected that the tight linkage inherent to heteromorphic chromosomes would make sex chromosome drive more common than autosomal drive, as appears to be the case in nature (Hamilton 1967).

There are at present four known naturally occurring cases of sex ratio distortion caused by meiotic drive. One of these is a very complex system in the wood lemming *Myopus schisticolor* (Fredga et al. 1977) and will not be considered here. The other three, however, are in insects and are discussed below.

4.1. W Chromosome Drive in Butterflies

Several African butterfly species (where females are ZW, males ZZ) of the genera *Danaus* and *Acraea* show W-chromosome drive, which is often strong enough to produce unisexual female broods (Owen 1974; Smith 1975). In *Danaus chrysippus*, Smith (1975) identified an autosomal locus, linked to loci controlling wing color and pattern, that seems partially to suppress W drive by saving some of the Z-bearing gametes that also carry the proper suppressor allele. There is some evidence that this suppressor variability may permit *Danaus* to use morph patterns as a cue for adjusting the sex ratio to track cyclical changes in the environment. Unfortunately, there is no further information available on the genetic, much less the molecular, basis of this system of sex ratio distortion.

4.2. Male Drive in Mosquitoes

In mosquitoes, sex is determined by a single locus on a largely homomorphic pair of sex chromosomes, with *M/m* and *m/m* representing the male and female genotypes, respectively. Both *Culex quinquefasciatus* and *Aedes aegypti* harbor male distorter chromosomes that are probably the result of different mutations at the same genetic locus (*D*). In the former species, the drive allele is a rare recessive, and only homozygous *d/d* males produce an excess of sons (Sweeney and Barr 1978). In *A. aegypti*, distorter chromosomes carry a more common dominant allelic form, such that *MD* chromosomes act as distorters in males, while *md* chromosomes are the sensitive targets. *D* and *M* are closely linked (Hickey and Craig 1966; map distance 1.2%, Newton et al. 1978). In contrast to the *trans*-acting nature and quantitative action of many genic drive systems (cf. the *SD* system discussed below), *D* apparently acts to cause distortion of *m*-bearing sperm only in *MD/md* genotypes (Hickey and Craig 1966); that is, *Md/mD*, *MD/mD* and *Md/md* males are nondistorters.

MD/md males produce a sex ratio (proportion of males) as high as 0.99 in selected laboratory crosses (Wood and Newton 1991). In nature, however, the effective sex ratio measured in 14 lines known to carry *MD* was in the range 0.499–0.606, as opposed to a control range of 0.486–0.550 obtained from 18 lines lacking the drive chromosome (Wood and Newton 1991). This illustrates the point made earlier that, if they are to avoid extinction when challenged by Y chromosome drive, populations must rapidly evolve some mechanism of genetic suppression. It also suggests that inactive Y chromosome drive systems may be revealed when

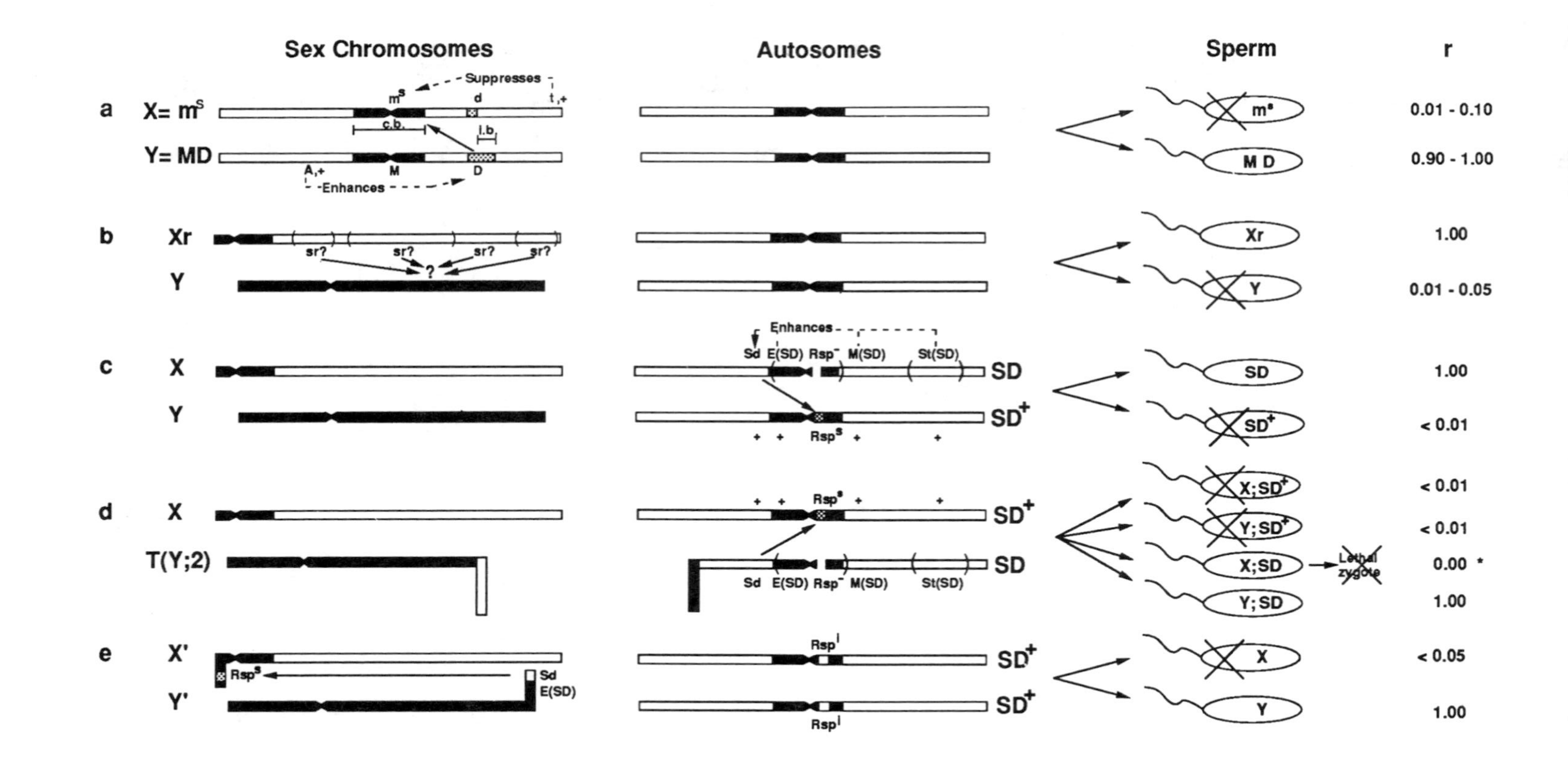

Sex Chromosomes
Autosomes
Sperm
r
a
X= m^S
Y= MD
Suppresses
Enhances
m^s
d
t,+
c.b.
l.b
A,+
M
D
m^s
M D
0.01 - 0.10
0.90 - 1.00
b
Xr
Y
sr?
?
Xr
Y
1.00
0.01 - 0.05
c
X
Y
Enhances
Sd
E(SD)
Rsp^-
M(SD)
St(SD)
SD
SD^+
Rsp^s
SD
SD^+
1.00
< 0.01
d
X
T(Y;2)
Rsp^s
SD^+
SD
Sd
E(SD)
Rsp^-
M(SD)
St(SD)
X;SD^+
Y;SD^+
X;SD
Y;SD
Lethal zygote
< 0.01
< 0.01
0.00 *
1.00
e
X'
Y'
Rsp^s
Sd
E(SD)
Rsp^l
SD^+
SD^+
X
Y
< 0.05
1.00

individuals taken from natural populations are outcrossed to laboratory strains. In the case of *A. aegypti*, the suppression is a consequence of the presence of a range of *md* chromosomes of varying sensitivity to segregation distortion in many wild populations (Wood 1976; Suguna et al. 1977). Some of these insensitive chromosomes carry an allele at a linked modifier locus (*t*) which apparently reduces the susceptibility of the X chromosome to *MD*-induced gamete dysfunction. Conversely, some *MD* chromosomes carry an enhancer (*A*) of drive located in the opposite arm from the suppressor. Figure 3a summarizes the genetic map of these several loci.

Sperm dysfunction apparently arises as a consequence of the preferential breakage of the X (= *md*) chromosome during male meiosis, in association with chiasma formation (Newton et al. 1976; Wood and Newton 1977; Sweeney and Barr 1978). Most breaks occur at four specific sites (Wood and Newton 1991). Those X-bearing spermatozoa that are produced often have organelles in an abnormal, disintegrated state, and are presumably mostly nonfunctional (Wood and Newton 1991).

←

Figure 10.3. Meiotic drive mechanisms in male insects. In all cases, *R* refers to the recovery proportion (i.e., the probability of survival) of a class of sperm in males showing meiotic drive. Typically, sperm carrying an insensitive allele at the target locus are unaffected ($R = 1.00$), while those carrying a sensitive target allele are usually rendered dysfunctional ($0 \leqslant R \leqslant 0.05$). Parentheses represent chromosome inversions, solid bars represent heterochromatin, and constriction of the bar represents the centromere. A solid line with an arrowhead indicates action of a drive allele at a sensitive target locus, dashed lines represent modification of drive action by secondary loci. Stippling represents polymorphism between drive and nondrive chromosomes for staining with the Hoechst 33258 fluorochrome. Five types of drive are represented and discussed in detail in the text. **A.** Male drive in mosquito (i.b. = intercalary C band present on both chromosomes, c.b. = centromeric C band present only on X). **B.** Sex-Ratio in the obscura group of *Drosophila*. **C.** Segregation distorter in *D. melanogaster*. *SD* chromosomes have apparently lost a chromosomal band (shown as a gap) present in SD^+ chromosomes. This band is Hoechst-positive in chromosomes sensitive to meiotic drive (see c and d), Hoechst-negative in insensitive chromosomes (see e below). **D.** pY drive. A translocation between the Y and *SD* chromosomes leads to the production of genetically unbalanced gametes. SD^+ gametes are rendered dysfunctional by *SD*, while X:*SD* gametes lead to lethal (see asterisk) aneuploid zygotes deficient for a large piece of the second chromosome. Only euploid *SD* sons survive, and these are karyotypically identical to their fathers. **E.** Using the *SD* system to produce true Y drive. Drive elements are translocated to the Y (denoted Y'), and the *Rsp* target sequences translocated to the X (denoted X'). When these are the only target sequences present in the genome, virtually all X-bearing sperm are rendered dysfunctional.

The biological mechanism for male drive is not yet established. Some cytological data concerning the gross DNA structure of *MD* and *md* chromosomes are available, however. While the sex determining locus (*M* vs. *m*) has been placed near the centromere, the drive allele (*D*) has been localized to a region midway in one arm of the Y chromosome, which often carries a large intercalary C band (see Figure 10.3a, and Newton et al. 1978). This band is also found on the X, but on the Y it is particularly polymorphic in terms of both its presence and size. Compared to the X in this region, the Y chromosome also shows more extensive staining by both quinacrine and Hoechst-33258 fluorochromes (Wallace and Newton 1987), suggesting the possibility that the DNA base composition for this region differs significantly between the two chromosomes, with the Y more AT rich. At present, it is not possible to determine whether this association of the *D* allele with cytogenetic polymorphism is purely fortuitous, represents a functional relationship, or indicates the nascent heterochromatinization of part of the Y chromosome.

4.3. X Chromosome Drive in *Drosophila*

In contrast to the cases of W or Y chromosome drive, which are known to occur only among some butterfly or mosquito species, X chromosome drive occurs over a wide range of the *Drosophila* phylogeny. *Sex-Ratio* (*SR*) chromosomes, as driving X chromosomes are termed (cf. Morgan et al. 1925, Sturtevant and Dobzhansky 1936), have been found in *D. melanica* and *paramelanica* (Stalker 1961), *D. mediopunctata* (De Carvalho et al. 1989), *D. quinaria* and *D. testacea* (J Jaenike pers comm) of the subgenus Drosophila, and in many species in the subgenus Sophophora, including at least three species in the affinis subgroup (notably *D. affinis* and *D. athabasca*) and several species of the obscura group (*D. azteca*, *D. obscura*, *D. pseudoobscura* and *D. persimilis*; see Novitski 1947 for a review). In most of these species, *SR* chromosomes carry inversions distinguishing them from the standard X chromosome (Wu and Beckenbach 1983). These most likely owe their evolution to their effect on reducing recombination between the as yet unidentified drive locus and linked drive enhancers.

Males with the *Sex-Ratio* chromosome (*Xr*) transmit predominantly X-bearing sperm, in extreme cases leading to all female progeny. Direct cytological examination of developing spermatogonial cysts indicates that *SR* males have approximately half as many normal sperm per cyst. Abnormal sperm are eliminated during individualization or appear to degenerate soon after (Hauschteck-Jungen and Maurer 1976). While the molecular basis of *SR* is completely unknown, its widespread occurrence in the *Drosophila* radiation makes the condition of great biological interest. A brief discussion of *SR* will not only help in better understanding the *Segregation distorter* system of *Drosophila*, to be discussed below, but may serve the larger purpose of emphasizing the potential reward from molecular evolutionary studies of the *Sex-Ratio* phenomenon.

Xr can be defined as an ultraselfish chromosome (Wu and Hammer 1990), as it causes Y-bearing sperm to degenerate. As a consequence, virtually all the sperm of *Xr/Y* males carry *Xr*, whereas only half of the sperm in normal males are X-bearing. Provided that sperm are not in limiting supply, *Xr* will therefore be transmitted at twice the rate of a normal X, leading to its rapid fixation in the population. As discussed above, the evolutionary limitation on such sex-linked meiotic drive is that, if it is too successful, it will eventually cause the host population to become extinct, owing ultimately to a deficiency of males. Obviously, if *Xr* normally caused population extinction, we would not be in a position to have ever isolated it from nature. We must then ask what has happened to prevent its fixation in natural populations. There are two possibilities: *Xr* incurs a hidden fitness disadvantage roughly balancing its segregation advantage, such that there is no net selection in its favor; or genetic suppressors of *Sex-Ratio* distortion have evolved.

4.3.1. Components of Selection Against SR

The theoretical conditions for the maintenance of the *Xr*-X-Y polymorphism have been established by Edwards (1961). Many experiments have since been carried out to measure selection components (including male virility, female virility and larval viability) acting against *Xr* in *D. pseudoobscura* (Curtsinger and Feldman 1980; Beckenbach 1983). The most interesting (and controversial) issue is whether *SR* males have lowered virility (=fecundity, in more general population genetic literature). Because half of the sperm in these males degenerate, reduced fecundity may be expected. In such a case, selection against *Xr* is necessarily linked to meiotic drive itself. However, as fecund males normally ejaculate more sperm than females are capable of storing, a reduction in absolute numbers of sperm does not usually affect the realized fecundity of *Xr* males (Beckenbach 1978; Wu 1983a). This raises the question of why *Drosophila* males ejaculate more sperm than needed for insemination. The evidence suggests that a larger ejaculate is more effective in displacing sperm of a previous male stored in the receptacles of the female; thus *SR* males are less fecund as a consequence of weaker sperm displacement (Wu 1983a; 1983b). In order to properly assess the role of reduced virility in controlling the evolution of the *SR* trait, it will ultimately be necessary to measure components of selection in natural populations.

4.3.2. *Genetic Suppressors of* SR

In general, both autosomal and Y-linked genetic variants that suppress *SR* will have an advantage. The advantage of the latter is self-evident. Similarly, autosomal suppressors will increase in the population by virtue of their effect in rectifying a biased sex ratio (Eshel 1985). In the progeny, such a suppressor allele is distributed evenly between the two sexes, whereas its normal (i.e., nonsuppressing) alternative allele tends to be recovered primarily in daughters, which have lowered fitness as a consequence of their excess in the population. Such suppressors have been discovered in some species (Stalker 1961), and it is conceivable that *SR* might have disappeared in others because of suppression. Why, then, are suppressors absent in species like *D. pseudoobscura* (Policansky and Dempsey 1978; Beckenbach et al. 1982)? A possible explanation is that selection may act against suppressors because gametic phase disequilibrium often results in suppressors ending up in genotypes with reduced fitness (Wu 1983c). These results suggest that strong selection against *SR* males, rather than selection against females carrying *Xr*, may be effective in preventing the emergence of autosomal suppressors.

4.3.3. *Genetics of* Sex-Ratio

As is the case for most other meiotic drive systems, *SR* is associated with various inversions. The location of loci responsible for *SR* varies with the species in which the *Xr* chromosome is found. Thus, while *SR* loci are associated with XR (the right arm of the X) in the obscura group, in *D. testacea SR* maps to the telocentric X, homologous to XL in the obscura group. This suggests that *SR* has evolved at least twice independently. While unique inversions are associated with the *SR* system in each species, they are not themselves required for the expression of *Sex-Ratio*. For example, between the sibling species *D. pseudoobscura* and *D. persimilis*, *Xr* (*D. persimilis*) and X (*D. pseudoobscura*) are homosequential, yet only one of the two is a distorter. Such variable patterns of rearrangements illustrate the point that these inversions presumably evolve to reduce recombination between complexes of *Sex-Ratio* genes on *Xr* and can serve this same purpose irrespective of whether they arise on X or *Xr*. Within each species, such inversions restrict the efficacy of genetic analysis by recombination. Fortunately, the fertility of hybrid females between these two sibling species make it feasible to introgress *Xr* (*D. persimilis*) into *D. pseudoobscura* and, reciprocally, X (*D. pseudoobscura*) into *D. persimilis* (Wu and Beckenbach 1983). Males carrying recombinants between the homosequential *Xr* (*D. persimilis*) and X (*D. pseudoobscura*) chromosomes were tested for the presence of the *SR* trait. Surprisingly, no fewer than four *SR* factors were identified on the *Xr* chromosome, suggesting very extensive differentiation between the two different X chromosome forms. These data are summarized in Figure 10.3b.

4.3.4. Molecular Studies of SR*—a Perspective*

Many questions about the evolution of *SR* can conceivably be approached through the examination of nucleotide sequence variations across the various segments of the *Xr* chromosome, as has been done for other meiotic drive systems (cf. the *SD* system below). Between the older inversions separating X and *Xr*, DNA sequences are likely to be more divergent. It should be feasible to use these differences to reconstruct the history of the system of inversions. We would expect the most interesting DNA measurement to be the level of interspecific divergence between *Xr* chromosomes. It is possible that the *D. pseudoobscura Xr* was derived from *D. persimilis* by introgression, by analogy to the way *t*-haplotypes have apparently spread in mice (see Wu and Hammer 1990), and acquired its inversions after introgression. In fact, there is some direct indication of hybridization between these species in nature (Powell 1983). Investigation of the molecular evolution of *SD* chromosomes (see below) may provide important guidelines for the appropriate approach to the molecular population genetic analysis of *Sex-Ratio*.

It may also eventually be feasible to clone the sequence on the Y chromosome that serves as the target of *SR* distortion. The first step would be to select for Y-borne deletions that do not respond to *Sex-Ratio* distortion. Because mass matings of *Xr*/Y males will yield F_2 progeny only when one of the males carries an insensitive Y, a large screening for deletions is quite feasible. The cloning of sequences corresponding to the deleted region can then be accomplished using the same strategy employed in cloning the *Responder* locus of the *Segregation distorter* system (see below, and Wu et al. 1988). The repeated references to the *SD* system (discussed below) suggest that a comparative study between *SR* and *SD* will be of great value in understanding the underlying molecular genetics of meiotic drive.

5. Mimicking Sex Chromosome Drive with Autosomal Systems

Unfortunately, as the above descriptions suggest, the level of detailed genetic or molecular information on the structure and function of the several naturally occurring cases of sex chromosome drive is very limited. Considerably more progress has been made toward understanding the genetic basis of autosomal meiotic drive systems, such as the *Segregation distorter* (*SD*) system in *Drosophila melanogaster* (reviews in Hartl and Hiraizumi 1976; Crow 1979; Sandler and Golic 1985; Temin et al. 1991) and *t*-haplotype distorted transmission ratios in mice (reviewed in Silver 1985). Because it occurs in an organism for which an extensive array of genetic, cytological and molecular techniques are now available, the *SD* system has proved to be of particular value as a model for understanding the general effects of meiotic drive. Further, through the use of a number of cytogenetic tricks, it is possible to convert the autosomal *SD* system, at least in the laboratory, into a sex chromosome drive system (Lyttle 1977; 1979; 1989). By comparing the population dynamics of *SD* acting in these two chromosomal forms, it has been possible both to prove the experimental applicability of Hamilton's (1967) theoretical conclusions concerning sex ratio distortion, and to measure the differential impact on the evolution of drive suppressors and enhancers occurring when meiotic drive causes sex ratio distortion. Simultaneous with, and to a certain extend depending on, studies of the evolution of the *SD* system both in the laboratory and in nature, there has been a concerted attack on the genetic and molecular structure of the two main elements in the *SD* system, the *Sd* and *Rsp* alleles.

5.1. The Genetic Structure of *SD* Chromosomes

SD second chromosomes have at least two major genetic loci: *Segregation distorter* (*Sd*, located at 37D2–6 in the euchromatin of the autosomal left arm; see Brittnacher and Ganetzky 1983) and *Responder* (*Rsp*, located in the proximal centromeric heterochromatin of the right arm; Lyttle 1989; Brittnacher and Ganetzky 1989). In addition, *SD* chromosomes carry several known linked modifiers of segregation distortion, including particularly the major allele *E(SD)* (located in the distal left arm heterochromatin, less than a single map unit from *Sd*; Brittnacher and Ganetzky 1984; Temin et al. 1991), *M(SD)* (located in the proximal right arm; Hiraizumi et al. 1980), and a number of minor modifiers in the distal right arm (*St(SD)*) held in tight linkage disequilibrium by the inversion *In(2R)NS*. *SD* chromosomes recovered from nature are not only often associated with this cosmopolitan inversion, but may also carry one or more of a suite of unique inversions as well (except in the Mediterranean basin, where *SD* chromosomes are generally inversion free; Nicoletti and Trippa 1967, and discussion below). It is interesting to note that some of the unique inversions may be relatively recent additions to the genome arising from the action of *hobo* transposable elements (TW Lyttle and D Haymer, in press).

Sd exists in two allelic forms, corresponding to either the presence (*Sd*) or the absence (Sd^+) of segregation distortion activity, while *Rsp* alleles range more or less continuously (Martin and Hiraizumi 1979; Hiraizumi and Thomas 1984; Temin and Marthas 1984; Lyttle et al. 1986) from supersensitivity (*Rsp*^{ss}) through standard sensitivity (*Rsp*^{s}) to insensitivity (*Rsp*^{i}) in their response to *Sd*, depending on the number of DNA copies of a 120 bp repeating unit present at the *Rsp* site in the 2R heterochromatin (Wu et al. 1988, see below). The *Rsp* locus appears to be genetically quite large; on larval neuroblast mitotic chromosome spreads it can be visualized as a chromosomal band with a strong affinity for binding the fluorochrome Hoechst S33258 (Pimpinelli and Dimitri 1989). The size and staining intensity of the band for a given *Rsp* allele show a strong positive correlation, both with the sensitivity of the allele to segregation distortion, and to the number of *Rsp* DNA repeat copies present (Pimpinelli and Dimitri, in Temin et al. 1991). Genetic studies have demonstrated that sensitivity to segregation distortion cosegregates with this Hoechst-positive region when it is translocated to either the Y (Lyttle 1989) or X (Walker et al. 1989) chromosomes.

The *Sd* allele, perhaps in conjunction with *E(SD)* or other modifier loci, causes the dysfunction of *Rsp*^{s}- or *Rsp*^{ss}-bearing sperm by interfering with proper chromatin condensation during spermiogenesis. The visible defects are a failure of defective sperm to undergo the normal lysine-arginine rich histone transition (Kettaneh and Hartl 1976; Hauschteck-Jungen and Hartl 1982) and, ultimately, a failure to achieve proper sperm individualization (Tokuyasu et al. 1972; 1977—analogous to the defect described for *SR* above), although temperature shift experiments suggest that *Sd* may actually be acting in Meiosis I (Mange 1968). Because *Sd* apparently has no effect on *Rsp*^{i} alleles, a male of genotype *Rsp*^{i}/*Rsp*^{s} which carries a copy of *Sd* as well usually transmits more than 95% (and up to 100%) *Rsp*^{i}-bearing sperm. This proportion is defined as the statistic k. Segregation in females is normal ($k = 0.5$). Since *Sd* and *Rsp*^{i} are normally linked, the k value is usually considered to be the property of a whole *SD* chromosome. From a different perspective, the strength of drive can also be measured as the probability of survival (R) of an *Rsp*^{s} bearing sperm (Lyttle 1986). R values are therefore associated with *SD*$^{+}$ chromosomes. In a normal cross involving an *SD*/*SD*$^{+}$ male, k and R are related by the transformation:

$$R = \frac{1 - k}{k}$$

A chromosome may be designated as *SD* (*Sd Rsp*^{i}), sensitive *SD*$^{+}$ (*Sd*$^{+}$ *Rsp*^{s}), insensitive *SD*$^{+}$ (*SD*$^{+}$ *Rsp*^{i}) or suicidal (*Sd Rsp*^{s}). *Sd*, *E(SD)* and *Rsp*^{s} are all classified as neomorphic mutations; that is, genetic deletions of these alleles yield phenotypes indistinguishable from wild-type as defined by the *Sd*$^{+}$, *E(SD)*$^{+}$ and *Rsp*^{i} alleles, respectively (Temin et al. 1991). The structure and genetic interactions of the loci in *SD* and sensitive *SD*$^{+}$ chromosomes are summarized in Figure 10.3c.

5.2. The Molecular Stucture of *Sd* and *Rsp*

5.2.1. Molecular Cloning of the Rsp *Locus*

In the absence of alternative molecular approaches, the successful cloning of *Rsp* actually relied on the major assumption that the Rsp^s allele consists of a large array of repetitive DNA sequences which either are missing or exist in a much lower copy number on Rsp^i chromosomes (Wu, in Temin et al. 1991). By screening a λ-phage library of *D. melanogaster* DNA with total genomic DNA from fly lines differing only in the presence or absence of a small deletion for Rsp^s, Wu et al. (1988) were able to identify several λ-clones which hybridized only to Rsp^s DNA. From one of these they obtained an *Rsp* clone containing an array of DNA repeats, each about 120 base-pairs (bp) long and rich in AT content, with short runs of As and Ts (see Figure 10.4). The repeats are organized mainly as dimers consisting of two 120 bp repeats (Wu et al. 1988). A dimer is delineated by TCTAGA (an *Xba*I restriction site) at one end (position 1–6 of the left hand monomer in Figure 10.4), while the corrresponding site in the middle of the dimer has changed to TCTACA (position 1–6 of the right hand monomer). Within a dimer, the left and right hand repeats show approximately 20% divergence at the nucleotide level, significantly higher than the 4.2 to 4.3% divergence obtained when left or right hand repeats, respectively, are compared as a class among themselves across dimers. This repeat structure has most of the characteristics of satellite DNAs (Miklos 1985). These are a class of tandem repeats with very simple sequences, usually AT or GC rich, which form a satellite band on a CsCl gradient. Satellite DNAs are both ubiquitous and abundant in higher eukaryotes. This represents, perhaps, the first direct evidence for a phenotype associated with such satellite sequences.

There are three lines of evidence that this array of repeats represent the *Rsp* locus. The first is that the *Rsp* sensitivity of chromosomes from nature correlates with repeat copy number. Among 35 tested chromosomes, there is a nearly perfect correlation between the copy number of the repeat and sensitivity to the sperm killing effect of *SD* (cf. Figure 10.5), with typical sensitive and supersensitive chromosomes having about 700 and 2,500 copies, respectively. On the other hand, insensitive chromosomes usually have between 100–200 repeat copies, with *SD* chromosomes having fewer than 20 copies. In fact, *SD* chromosomes appear to be missing the whole genetic region that normally harbors the repeats (Pimpinelli and Dimitri 1989), a fact indicated by the Rsp^- label in Figure 10.3. This observation that only *SD* chromosomes have essentially eliminated the *Rsp* repeats is compatible with the hypothesis that the latter have a function, albeit a dispensable one (Wu et al. 1989). Given

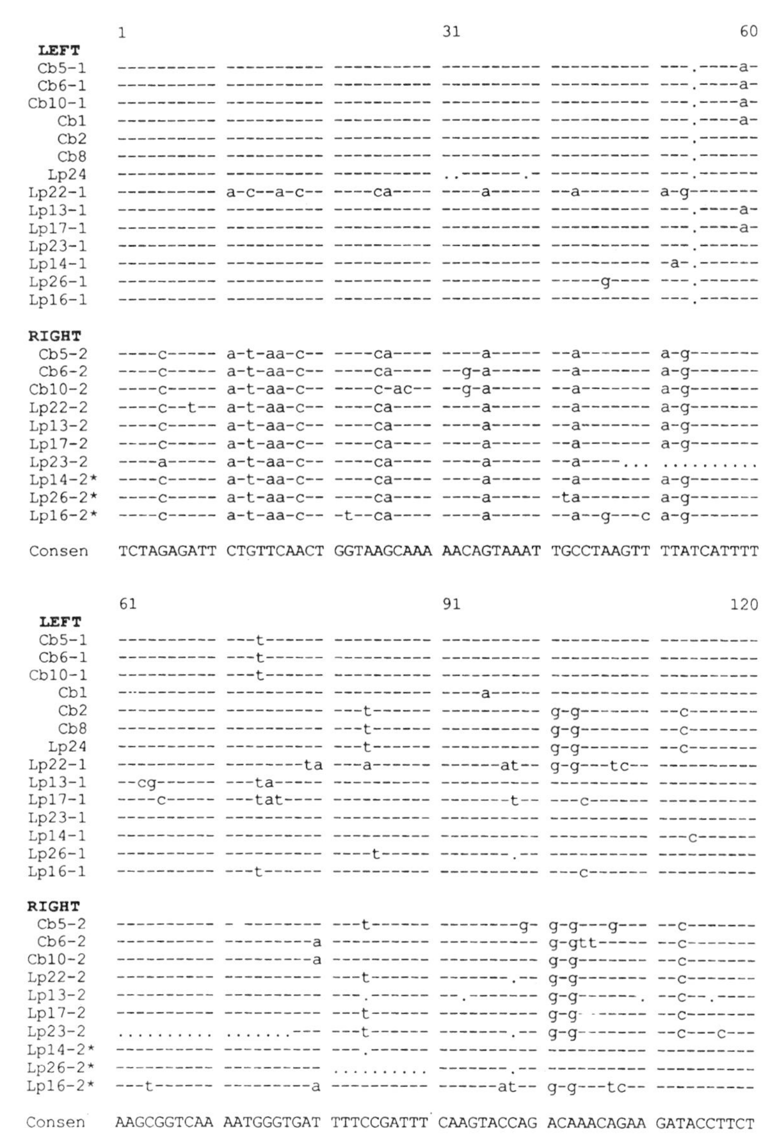

Figure 10.4. DNA sequence variation in the *Rsp*^{s}-associated satellite repeat. Each line represents an independently isolated copy of the basic repeating unit, which is 120 bp, delimited by X*ba*I restriction sites and arranged as a dimer into a larger 240 bp repeating unit. The left hand repeat of the dimer begins with the X*ba*I site (TCTAGA) at position 1, the right hand repeat begins with the same sequence, except with a C at position 5. Left and right repeats with the same prefix (e.g., Cb5–1, CbS-2) were extracted together as part of the same dimer. The consensus sequence for the left repeat is presented at the bottom, with dashes representing identity with this sequence.

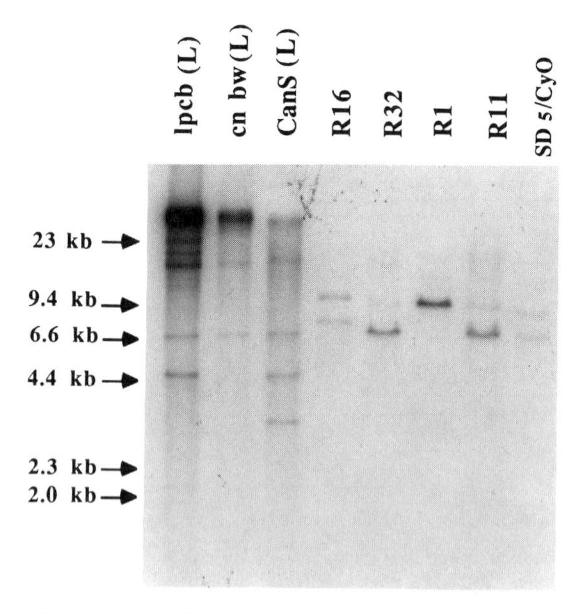

Figure 10.5. Variation of X*ba*I repeat copy number with sensitivity to *SD*. For these gels, genomic DNA, from stocks whith *Rsp* alleles that are supersensitive (lane 1), sensitive (lane 2), or partially sensitive (lane 3), are digested with *Eco*RI and hybridized to *Xba*I repeat DNA. Lanes 4–7: digests of stocks homozygous for chromosomal deletions of *Rsp* (derived from the *cn bw* stock of lane 2), which simultaneously remove most of the *Xba*I repeats. Lane 8: a completely insensitive *SD* balanced stock. (Modified from Figure 1 of Wu et al. 1988.)

the low frequency of *SD* in most natural populations (1–5%), non-*SD* chromosomes may be selectively favored to maintain at least a small array of repeats, thus balancing a requirement for some minimal level of the normal repeat function against a conflicting need to avoid *SD*-mediated sperm dysfunction. The advantage for SD chromosomes in avoiding autocidal distortion presumably promotes their evolution to complete insensitivity.

The second line of evidence that the cloned repeats represent the *Rsp* locus is that the repeat array is deleted whenever *Rsp* is deleted. From an Rsp^s second chromosome, Ganetzky (1977) constructed five lines deleting *Rsp* sensitivity. Four of these remain insensitive; these simultaneously retain fewer than 30 repeat copies, compared to 700 copies in the parental chromosome (Figure 10.5).

The final line of evidence for the congruence of the cloned repeats and *Rsp* is that repeat array and *Rsp* sensitivity are simultaneously translocated to the Y chromosome. Of a group of 17 translocations between the Y and second chromosomes constructed by Lyttle (1989), 10 move the Rsp^s allele to the Y (as judged by the acquired sensitivity of the Y chromosomome to *SD*), while the remaining seven are broken adjacent to Rsp^s, but leave it segregating with the second chromosome. The repeat array cosegregates with Rsp^s in every line. In one case, where both the resulting Y and second chromosomes retain partial sensitivity to *SD*, molecular data also indicate translocation of a partial array of repeats (Wu et al. 1988; see also Figure 10.6).

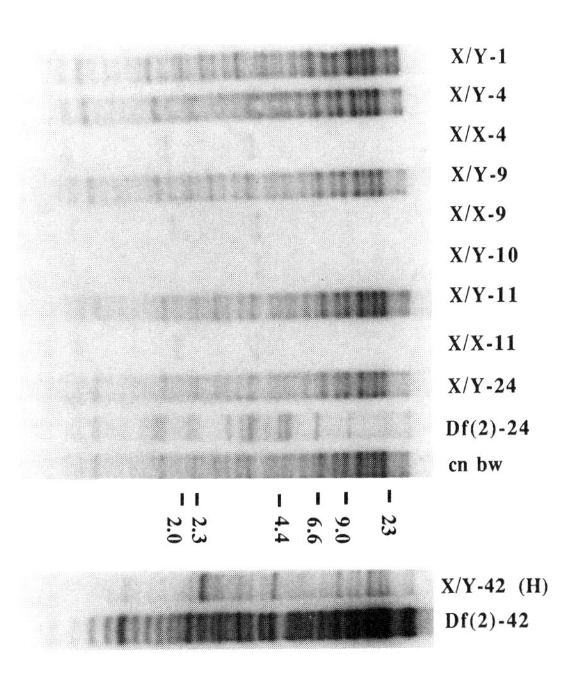

Figure 10.6. Location of *Xba*I repeats in translocations moving *Rsp* to the Y. Several translocations (denoted by number) move a portion of the *Rsp* region of a *cn bw* second chromosome (a standard SD^+ line) to the Y. Total genomic DNA is recovered from flies carrying either the resulting Y chromosome (e.g., *Y-24*) or the corresponding second chromosome deficiency (e.g., *Df(2)-24*) alone. Total genomic DNA is digested with *Msp*I, and probed with the *Xba*I repeat as in Figure 5, with the *cn bw* stock of lane 11 serving as a control. *Y-1*, *Y-4*, *Y-9* and *Y-11* carry many repeat copies and are sensitive to *SD*-mediated sperm dysfunction, while *Y-10* is insensitive and carries no repeats. *Y-24* is highly sensitive to *SD*, while *Df(2)-24* is weakly sensitive; conversely, *Y-42* is weakly sensitive, while *Df(2)-42* is highly sensitive. Thus, there is a complete correspondence between sensitivity to *SD* and *Xba*I repeat copy number. (Modified from Figure 4 of Wu et al. 1988.)

5.2.2. *Molecular Evolution of the* Responder *Satellite DNA*

Since none of several second chromosomes from *D. simulans* examined so far has the *Sd* allele, we might expect this sibling species of *D. melanogaster* to carry a large array of the *Rsp* repeat. Surprisingly, every one of the *D. simulans* strains tested exhibited a very low (<10) repeat copy number. This is in sharp contrast with the 100 to 2,500 copies commonly observed for naturally-occurring SD^+ chromosomes in *D. melanogaster*. Similar, low repeat numbers were obtained from two other sibling species, *D. mauritiana* and *D. sechellia*. The DNA of any of 12 other species from either the *Drosophila* or the *Sophophora* subgenus shows no detectable hybridization to the *Rsp* repeat. The presence of the *Rsp* repeat in sibling species of *D. melanogaster*, albeit in very low copy number, is expected given the observation that the nucleotide divergence of these species from *D. melanogaster* is no more than 5% (Zwiebel et al. 1982), while the age of the Rsp^s dimeric structure (at 20% divergence) appears to be much greater. Since the DNA divergence among *Rsp* dimer repeats (4.2–4.3%, see above) is comparable to the divergence among the *D. melanogaster* sibling species, we conclude that the expansion of the array took place after the divergence of *D. melanogaster* from its sibling species. Such an expansion provided a target for segregation distortion and may have been a precondition for the evolution of *SD*. The absence of *Rsp* in *D. pseudoobscura* suggests that *SR* and *SD* are not likely to share DNA homology at their respective target loci.

A striking feature of the *Rsp* locus in *D. melanogaster* is the enormous variation in repeat structure among chromosomes. Among 60 second chromosomes, 59 distinct restriction patterns were observed from genomic southerns probed with repeat DNA; only two chromosomes isolated from flies caught in the same bucket of bait appeared to be identical. One possibility is that this high level of variability arises as a consequence of unequal crossing over between mismatched repeats, either on sister chromatids or homologous chromosomes, similar to the explanation put forth for the rapid changes in copy number observed for ribosomal RNA genes in *D. melanogaster* (Tartof 1974; Endow and Komma 1986). The occurrence of unequal exchange can be inferred directly from the occurrence of "hybrid" 120 bp repeat units, that is, those simultaneously sharing DNA homology with both the left and right units of the 240 bp repeat dimer (Wu and Hammer 1990). Such hybrid repeats are presumably the result of exchange between chromosomes (or chromatids) in which the left and right hand 120 bp repeats are paired out of register. Recently, mathematical models for the evolution of the *SD* system have been developed (N Takahata and C-I Wu unpublished) incorporating unequal sister chromatid exchange (USCE, as modeled by Takahata

1981) at the *Rsp* locus into the mathematical model for *SD* developed by Charlesworth and Hartl (1978). The preliminary results suggest that allowing a high rate of USCE significantly improves the fit of these models to the observed patterns of polymorphism observed in nature for the loci making up the *SD* system.

5.2.3. Molecular Structure of the Sd *Locus*

Brittnacher and Ganetzky (1983) used deletion analysis of *SD* chromosomes to localize the *Sd* allele cytologically to polytene bands 37D2–6 of the *D. melanogaster* salivary gland chromosome map. A recombinant bacteriophage library produced from a microdissection of this polytene region (V Pirrotta pers comm) was screened by Powers (see Temin et al. 1991) to identify a single DNA clone hybridizing to band 37D2. This, in turn, was used to isolate a series of overlapping bacteriophage clones from a second library constructed from an SD^+ line. These overlapping SD^+ clones were then used to characterize the restriction map for a stretch of about 220 kb of DNA spanning 37D2–6. This DNA restriction fragment map was then compared to the corresponding chromosomal region from an *SD* chromosome line to identify *SD*-specific differences. Only a few restriction fragment length polymorphisms (RFLPs) were observed between the two lines, and Powers found only one of these showed complete linkage to the SD/SD^+ difference. It appears that all *SD* chromosomes, and only these, carry a tandem duplication of a 5 kb segment of genomic DNA, localized to band 37D5. Moreover, every mutation which leads to loss of *SD* activity also causes the loss or alteration of this 5 kb duplication. This argues convincingly, albeit indirectly, that this duplication event represents the *Sd* mutation. Further, cDNA analysis indicates that there are a number of novel mRNA transcripts produced in this region by *SD* flies which are missing in SD^+. Since none of these seem to be both *SD* and male specific, there is as yet no clear indication as to which, if any, of these transcripts might be involved in the $Sd\text{-}Rsp^s$ interaction. These results are discussed in detail by Powers, in Temin et al. (1991) and Powers and Ganetzky (1991).

5.2.4. Molecular Evolution of the Sd *Region*

Further examination of DNA polymorphisms around *Sd*, within Powers' original walk of 220 kb, have revealed intriguing patterns of DNA divergence (Wu and Hammer 1990). The evolution of DNA sequences linked to the *Sd* allele are interesting for at least two reasons. First, such data will reveal the level of divergence between *SD* and SD^+ chromosomes, shedding light on the evolution of *SD*. Second, owing to the linkage relationship between the two major drive loci, the *Sd* region is expected to exhibit a very distinct linkage relationship among polymorphic sites. Referring to Figure 10.3, we can see that in SD/SD^+ heterozygotes recombination events proximal to *Sd* (i.e., falling between *Sd* and Rsp^s) will generate unfavorable recombinants (such as suicide chromosomes), whereas recombination distal to *Sd* should yield no deleterious products. The *Sd* mutation is thus expected to represent a sharp demarcation point between two regions with contrasting levels of effective recombination (high distal and low proximal) and linkage disequilibrium (low distal and high proximal). Examining groups of both *SD* and SD^+ chromosomes, Wu and Hammer (1990, summarized in Table 10.1) have confirmed this prediction for restriction fragment length polymorphisms (RFLPs) in two DNA regions flanking *Sd*. Based on these RFLP patterns, they have been able to identify two types of *SD* chromosomes. Type I chromosomes are restricted to the Mediterranean basin and are usually inversion free. On the other hand, Type II chromosomes are found worldwide (including parts of the Mediterranean region), usually have one or more of the suite of inversions discussed above, and have several unique RFLP differences in the DNA region from 28 to 43 kb immediately proximal to *Sd*, resulting in strong linkage disequilibrium (note the differences between the two Spanish *SD* chromosomes). Since the distance between *Sd* and the RFLP sites making up this Type II "haplotype" is greater than 30 kb, a distance sufficient to bring linkage equilibrium in many regions of the *D. melanogaster* genome (Aquadro et al. 1986; Langley and Aquadro 1987), the observed disequilibrium can best be explained as arising from the effects of reduced recombination combined with strong selection against recombinant *Sd* Rsp^s suicide chromosomes. On the other hand, little linkage disequilibrium is detected in the DNA region distal to *Sd*. This is expected, since exchanges in this region presumably occur freely (*SD* inversions are all in 2R or the 2L pericentric heterochromatin), and the resulting recombinant products are not maladaptive. Further, the low and stable frequency of *SD* in nature (1–5%) insures that it exists primarily in SD/SD^+ heterozygotes, increasing the opportunity for recombination and the shuffling of polymorphic sites. For example, the Australian *SD* and SD^+ chromosomes of Table 10.1 share a common *Hae*II restriction site, which presumably marks a recent distal exchange event.

A simple interpretation is that Type I chromosomes represent an ancestral or prototype form, similar to SD^+, while Type II's are derived forms that have subsequently acquired both *SD*-specific DNA changes and inversions, which act to close much of the rest of the *SD* chromosome to recombination. The distinctive pattern of linkage associations seen in *SD* chromosomes must surely reflect a long history of interaction between meiotic drive and selection in this system, suggesting that the SD-SD^+ polymorphism is old. The fact that the weakest meiotic drive is exhibited

Table 10.1. RFLP Patterns in the Sd Region*

		Left (20–35 kb Distal to Sd)						Sd	Right (28–43 kb Proximal to Sd)					
		*Eco*RI	*Hind*III	*Cla*I	*Xho*I	*Bam*HI	*Hae*II		*Eco*RI	*Hind*III	*Cla*I	*Xho*I	*Bam*HI	*Hae*II
SD	Roma	A	A	A	A	A	A	+	A	B	A	A	A	A
	SPNI	A	A	A	C	A	A	+	A	B	A	A	A	A
	SPN II	A	A	A	A	A	A	+	C	D	B	B	C	A
	USA (5)	A	A	A	A	A	A,E	+	C	D,D′	B	B	C	A
	JPN	A	A	A	A	K	A	+	C	D	B	B	C	A
	AUS	A	F	B	C′	M	B′	+	C	D′	B	B	C	A
SD+	CantonS	A	A	A	A	B	A	–	A	A	A	A	A	B
	AUS (2)	A	A	A	A	L	B′	–	A	B,E	A	A	A	A
	CAL (9)	A	A,C,E	A	A	A,B,J	A	–	A	A,B	A	A	A	A,D
	WIS (3)	A	A	A,G	A,E	A,B	A	–	A	A,C	A	A,B′	A	B,A
	NYS (7)	B,A,F	B,D	A,E	A,E	B,F,G	B,A	–	A	A,B	A	A	A	B,A
	EUR (4)	A,D	A	A,C,E′	A	A,C	A,F	–	A,B	B	A	A,D	A	A

* Letters designate restriction patterns, where pattern A is always the consensus SD+ type. Patterns which cannot be unambiguously distinguished are marked with a prime. Chromosomes are classified by their geographic origin: SPN (Spain), JPN (Japan), AUS (Australia), CAL (California), WIS (Wisconsin), NYS (New York State), EUR (Europe). The number in parenthesis is the number of chromosomes surveyed. Roman numeral after the SPN SD chromosomes (I, II) refer to type of RFLP pattern, with pattern I typical of SD Roma, pattern II typical of all other SD chromosomes found worldwide. Most pattern II SD chromosomes also carry a small pericentric inversion (sex text).

by those *SD* chromosomes which are inversion-free is consistent with this being the ancestral form.

The divergence among Type II *SD* chromosomes for inversion karyotype (Hartl and Hiraizumi 1976), is in marked contrast to the high level of nucleotide similarity (Wu and Hammer 1990) among *SD* chromosomes within each type. It is quite plausible that the inversion polymorphism involving Type II forms is recent and only transient. The observation of the gradual replacement during the last 30 years in Madison, Wisconsin, of a local *SD* inversion karyotype by a form found virtually worldwide, is compatible with this interpretation (Temin and Marthas 1984).

5.2.5. Molecular Mechanism of Segregation Distortion

Because the molecular structures of key components of the *SD* system have been determined only recently, it is to be expected that the molecular mechanism underlying segregation distortion is still unknown. However, the cloning of *Rsp*^{s} does help in evaluating several competing explanations for *SD*-mediated sperm dysfunction. First, as many higher organisms (*Drosophila* included, McCloskey 1966) normally show no transcriptional activity during spermiogenesis, it has been often hypothesized that spermiogenic failure may result from unscheduled *Rsp*^{s} transcription catalyzed by *Sd*. However, no detectable transcription from *Rsp*^{s} can be observed in any stage of development of the whole organism, with or without *Sd* in the background genome (P Doshi and C-I Wu unpublished results). It is still possible that the aberrant transcription is restricted to the germ cells and might have escaped detection. In situ hybridization of *Rsp* repeat DNA to testis tissues may resolve the issue. Second, we can also rule out the possibility that failure of *Rsp*^{s}-bearing sperm is due to nuclease activities that specificially cleave *Rsp*^{s} repeats. Genomic DNA isolated from the testes of *SD* males does not appear to be cleaved at the *Rsp* locus, nor is there any difference in the level of methylation of the *Rsp* repeats from testes of *SD* and *SD*$^+$ males. This can be shown by restriction digestion with enzymes that recognize the same sequence but have different sensitivity to methylated bases.

Genetic (Ganetzky 1977) and EM (Tokuyasu et al. 1972; 1977) data indicate that *Sd* interferes with the maturation of sperm carrying *Rsp*^{s}, but these results do not tell us if such an interference is direct or indirect at the molecular level, although several molecular models based entirely on genetic observations have been proposed (Hiraizumi et al. 1980; Brittnacher and Ganetzky 1983; Lyttle et al. 1986; Lyttle 1989). Some of these postulate a direct interaction between *Sd* and *Rsp*, while others invoke a product from a third locus mediating that interaction. Further, the consequence of the *Sd-Rsp* interaction, whether direct or indirect, has been suggested by some (U Nur and C-I Wu unpublished) to most likely be a delay (as opposed to disruption) in the maturation of *Rsp*^{s}-bearing sperm. Finally, as there are many enhancers and suppressors of segregation distortion (see below), it is likely that one or more of them (in particular, *E(SD)*) also mediate the *Sd-Rsp* interaction. Thanks to the cloning of these two major elements, assays for the ability of the *Sd* product(s) to form complexes with *Rsp* repeats are now being carried out directly (P Powers, pers comm). Such studies may provide the much needed clue to the nature of the molecular interaction.

5.3. Suppressors of *SD*

Mutations of Rsp^s to Rsp^i (probably deletions, based on the known DNA structure of Rsp^s) can instantly convert SD^+ chromosomes from sensitivity into insensitivity. One of the ongoing evolutionary puzzles of the *SD* system is why, in the absence of any obvious normal function for Rsp^s, many *Drosophila* populations should simultaneously maintain high frequencies of both Rsp^s (usually 50–80%) and *SD* (1–3%) (Kataoka 1967; Hartl 1970; Hihara 1974; Hartl and Hartung 1975; Trippa et al. 1980; Temin and Marthas 1984). One possible explanation is that the effect on fitness of Rsp^s loss is too small to measure under normal conditions. To test this notion, Wu et al. (1989) took advantage of the existence of Rsp^i chromosomes, viable as homozygotes, which have been produced as the result of small X-ray induced heterochromatic deletions of an otherwise genetically identical Rsp^s parent chromosome (Ganetzky 1977). They montitored laboratory populations (size = 5,000) of *D. melanogaster* segregating for such Rsp^i and Rsp^s second chromosomes. Because Rsp^s is made up of so many copies of the 120 bp repeating unit (anywhere from 700–2400), the *Rsp* genotype of individual flies from population cages (or indeed, of flies collected directly from nature) can be determined directly by squashing flies onto nitrocellulose paper and probing the residual hemolymph with labeled *Rsp* repeat DNA. Over a period of a year, the frequency of Rsp^i/Rsp^i genotypes in the cages dropped to 10% from an initial value of nearly 40%. This suggests that Rsp^s alleles do have a fitness advantage over their Rsp^i counterparts over the long term; however, it is impossible to eliminate the possibility that other sequences deleted along with Rsp^s are primarily responsible for these differences. Further, it is hard to reconcile the possibility of a normal function for Rsp^s satellite DNA sequences with the fact that, as discussed above, these repeats have so far been seen at most only in very low copy number in other species examined. One possibility, of course, is that Rsp^s-like repeat copies are present in these species, but have diverged in DNA sequence to such an extent that they no longer hybridize to the *D. melanogaster* probes.

In addition to the obvious source of suppression of segregation distortion arising from deletion of Rsp^s, a number of studies have identified additional X-linked and autosomal suppressors of *SD* activity arising in natural populations (Kataoka 1967; Hartl 1970; Hartl and Hartung 1975; Trippa and Loverre 1975). Very little is known about these suppressors. In most cases they are not even precisely mapped, although a major X suppressor, present at frequencies as high as 85% in some populations, has been localized to the vicinity of the *vermilion* (*v*) gene (Kataoka 1967). Because *SD* is an autosomal drive system generating relatively weak selection for suppressors, their slow rate of evolution for the latter is difficult to observe directly in experiments using laboratory populations. However, by manipulating the *SD* system to cause distorted sex ratios, the rate of evolution of these same suppressors can be significantly enhanced. To this end, Lyttle (1979; 1981; 1984) translocated a portion of an *SD* autosome onto the Y chromosome to produce a form of Y chromosome drive (called pseudo-Y, or pY drive, see Figure 10.3d). This was not true Y chromosome drive, because segregation distortion was still directed against SD^+ and not the X chromosome (Figure 10.3d), but its impact on the population and the evolution of genetic suppression was the same as if it had been a Y-drive system. This is discussed in the next section.

6. Extinction of Laboratory Populations by *SD*-Mediated Sex Ratio Distortion

Well-documented cases of sex chromosome drive in nature are either characterized by relatively weak sex ratio distortion (male drive in mosquito) or detrimental genotype-specific fitness effects (Sex-Ratio in *Drosophila*). These drawbacks make it difficult to use these systems to study the predicted demographic consequences of extreme sex ratio distortion, as depicted in Figures 10.1 and 10.2. However, some analysis of the potential role of sex ratio distortion for population suppression has been carried out in mosquito, where Curtis et al. (1976) demonstrated that a combination of male drive linked by translocation to selectively favored markers was sufficient to suppress target populations in outdoor field trials. In general, the lack of detailed genetic mapping information in these organisms makes it difficult to examine the general dynamics of the evolution of genetic suppression of sex chromosome drive, much less actually isolate, or even map, the loci in question. It is reasonable to presume that *de novo* creation of a strong sex ratio drive system in the laboratory offers the best chance to observe the evolution of drive suppression mechanisms.

The *SD* system of drive operates in an organism easily amenable to genetic manipulation. Further, because *SD* is autosomal, and therefore normally slow to evolve suppressors, it is easy to find *SD* chromosomes that give complete drive (k = 1.00, R = 0.00). As discussed above, however, *SD* can be converted from autosomal drive to Y-drive by simply involving the *SD* autosome in a reciprocal translocation with the Y chromosome (giving pY-drive, Figure 10.3d). In a series of studies by Lyttle (1977; 1979; 1981), several laboratory populations with a mixture of pY-drive and normal males were established. Not only did such experimental populations provide a test of the impact of sex ratio distortion on population demographics (Figure 10.2), but the strong selective pressure favoring suppression of male drive acted as a kind of evolutionary hothouse in forcing a more rapid accumulation of drive suppressors than occurs for *SD* in its normal autosomal form. The strength of pY-drive in the lines used to establish these populations varied from $k = 0.94$ to $k = 1.00$; that is, from 94% to exclusive production of sons from pY-drive fathers. Since the absolute number of such sons exceeded the number produced by normal males from the same population, pY-drive continued to increase in frequency among males, and the overall population sex ratio was pushed further away from the optimal 1:1 value. Eventually it reached the intrinsic sex ratio of the pY-drive line itself. For strong pY-drive, this led to rapid population extinction owing to lack of females, usually within seven to eight generations. Included in Figure 10.2 is a summary of the population dynamics for a number of replicated populations (Lyttle 1977) established with different initial frequencies of males showing strong pY-drive ($k = 1.00$). It is clear that the dynamics of population extinction follow a path for Y chromosome drive very close to that predicted by Hamilton's (1967) theory.

Normal males were eventually eliminated even in cages harboring weaker pY-drive lines (initial $k = 0.94$); however, the persistence of low frequencies of females (≈6%) prevented population extinction. These populations were monitored for more than 40 generations (≈480 days) and assessed for changes in sex ratio distortion. In most populations, the strength of drive declined slowly and linearly with time to a low value of $k = 0.83$, accompanied by a drop in sex ratio. For these populations, the observed decrease in drive strength could be attributed to polygenic, quantitatively acting suppressors, each of small effect, distributed across both the X and non-*SD* autosomes (Lyttle 1979). In one population, however, sex ratio distortion was rapidly neutralized by the accumulation of sex chromosome aneuploids (XXY females and XYY males), with the sex ratio stabilizing at ≈60% male (Lyttle 1981). Since pY drive causes sex ratio distortion through the elimination of zygotes lacking a Y chromosome (Figure 10.3e), aneuploids carrying extra Y chromosomes are a source of significant numbers of surviving XXY daughters. For this cage, further accumulation of drive suppressors ceased.

In general, the cage results suggest that conditions favoring the accumulation of drive suppressors (e.g., weak distortion, slow population extinction) are insufficient for maintaining aneuploidy, while conditions favoring aneuploidy (e.g., strong distortion, very low production of females) lead to population extinction before polygenic drive suppressors can accumulate. Thus, the different mechanisms for neutralizing sex-ratio distortion are complementary in action.

This illustrates the importance of direct experimental analysis in testing population genetic models. The potential role for aneuploidy in suppressing sex ratio distortion would likely have been missed had the analysis of sex ratio distortion been approached by theoretical mathematics or computer simulation alone. Now that both X and Y chromosomes exist that carry the *Rsp*^{s} target DNA sequences (see above), it is possible to use the *SD* system to produce true X and Y chromosome drive in the laboratory, rather than pY-drive. Figure 10.3e illustrates SD-mediated Y chromosome drive, using these new chromosomal constructs. Cage studies with true Y and X drive will undoubtedly lead to a better understanding of the population dynamics of sex ratio distortion, as well as to a more detailed genetic mapping, perhaps at the DNA level, of specific drive suppressor genes.

7. The Biochemical Basis of Meiosis

So far, we have focused on the developmental and population consequences of defects in insert spermatogenesis leading to gamete competition and effective meiotic drive. Strictly speaking, none of these defects can yet be attributed to a failure in the normal processes of male meiosis. This is perhaps not surprising; since so little is known about these normal processes, it would be difficult to monitor any changes in their expression. This is not a limitation in female meiosis, where the well-defined stages of meiosis are an integral part of the intensely studied developmental stages of oogenesis, and where it is easier to measure the effect of mutational perturbation on meiotic processes. In this section, we review the status of our understanding of the biochemical events which underlie female meiosis. Because the most comprehensive study has been in *Drosophila melanogaster*, we will focus on systems in that organism.

The basic events of the meiotic process may be stated as pairing, recombination and segregation. The function of pairing is to identify and properly align homologous chromosomes. Reciprocal recombination (also known as exchange and crossing-over) serves the vital function of aligning chromosomes on the metaphase plate by linking homologous chromosomes together to form a bivalent. This physical linkage, called a chiasma, holds the bivalent in position by balancing the poleward forces acting at the centromeres. Chiasma resolution at the beginning of anaphase then allows the segregation of homologous chromosomes to opposite poles.

Our long-term objective must be to define meiosis in terms of the basic molecular events that allow pairing, recombination and chromosome segregation. We would, for example, like to discuss recombination in terms of well-defined enzymes such as ligases, topoisomerases, nucleases, etc. To some extent, this goal has been achieved by straightforward biochemical approaches. For example, Eisen and Camerini-Otero (1988) have reported the partial purification and characterization of a recombinase from *Drosophila* embryos. This enzyme exhibits a homology-dependent, strand-exchange activity that is characteristic of RecA-like proteins. Similarly, a variety of topoisomerases, nucleases and a ligase (Rabin et al. 1986) have also been isolated and characterized. However, because the enzymes were isolated either from large scale preparations of embryos or adults, most, if not all, probably represent functions required for general replication and/or repair processes and not meiosis-specific functions.

Given the small number of cells undergoing meiosis in *Drosophila* females, the use of biochemical methods to purify meiosis-specific proteins or to analyze the meiotic function of a given protein is likely to prove extremely arduous. It is perhaps, then, not surprising that most of our understanding of the meiotic process in *Drosophila* is derived from the molecular analysis of genes that were initially defined by meiotic mutations. Moreover, most of the progress in this area has come from the study of genes required for recombination and for segregation. A summary of our progress along those lines is provided below.

7.1. The Molecular Biology of Meiotic Recombination in *Drosophila*

Meiotic recombination occurs between paired homologs in early prophase and results in the physical interchange of genetic material between two nonsister chromatids. Although the precise molecular basis for meiotic recombination in *Drosophila* remains obscure, it is clear that the general mechanism will be similar to those elucidated in *E. coli* and the fungi (Szostak 1983; Carpenter 1987, for reviews). Basically, exchange is initiated by the formation of a DNA-DNA heteroduplex created by the pairing of complementary single strands of DNA from each of the two chromatids. At sites where the base sequences of the two chromatids differ, mismatches will be formed. Moreover, the exchange of single strands creates a so-called Holiday junction. It is the enzymatic cleavage of this structure that results in the actual recombination of two chromatids.

In early meiotic prophase (zygotene) synapsis occurs and the synaptonemal complex is laid down. In *D. melanogaster* females this step is defined by the *c(3)G* mutation (chromosome 3—map location 57.4, Gowen and Gowen 1922). No normal synaptonemal complex is visible in meiocytes of females homozygous for *c(3)G* (Rasmussen 1975). This suggests that the wild-type form of *c(3)G* may encode either some function necessary for normal synapsis or some component of the synaptonemal complex itself.

Reciprocal meiotic recombination events are marked at pachytene by the appearance of the so-called late recombinational nodules (for review, see Carpenter 1988). These nodules appear during pachytene with both a frequency and distribution paralleling that of crossover events (Carpenter 1979). Moreover, as detailed below, several recombination deficient mutations affect the number and morphology of recombination nodules.

A second class of recombination nodules, known as early or ellipsoidal nodules, are present earlier in prophase (generally during zygotene), the time at which synapsis occurs and the synaptonemal complex is laid down. These early nodules have been suggested to play a role in gene conversion events serving to facilitate homolog recognition and thus synapsis (for review, see Carpenter 1989).

In *D. melanogaster* females, the precise biochemical structures of the synaptonemal complex, the early recombination nodule, and the late recombination nodule remain an open question. What data are available have been obtained from studies of mutations which inhibit recombination. There have been several large scale searches for mutations that disrupt female meiosis in *D. melanogaster* (for review, see Baker and Hall 1976). Each of these screens was based on detecting mutations that increase nondisjunction. Since exchange is the primary mechanism for ensuring disjunction, it is perhaps not surprising that for many of these mutations the underlying defect is in the exchange process rather than in disjunction. Several of these mutations are considered in detail below.

7.1.1. mei-9 *(*meiotic mutant 9*; chromosome 1–map position 6.5)*

Females homozygous for *mei-9* show greatly reduced levels of meiotic exchange. However, the pattern of residual exchanges along the chromosomes resembles wild-type. *mei-9* is thus considered to be defective in the exchange process itself, rather than in the establishment of the preconditions that are essential to determining the normal number and distribution of exchanges (Baker and Hall 1976). Two alleles of *mei-9* have also been assayed with respect to their effects on gene conversion (i.e., heteroduplex formation and mismatch repair) at the unrelated *rosy* locus (Carpenter and Baker 1982). Although neither allele reduces the frequency of gene conversion events, both produce post-meiotic segregation events (i.e., mosaic progeny) at high frequency. This result indicates that, during the recombination process, *mei-9*-bearing ova form heteroduplexes but do not repair them. This is understandable in terms of the observation that *mei-9* is defective in excision or mismatch repair (see below). The recombinational phenotype of *mei-9* thus appears to involve two components, namely, a decrease in the frequency of heteroduplex repair and also a decrease in the frequency of reciprocal exchange. This latter phenotype cannot be simply explained in terms of the defect in excision repair.

As a consequence of the decreased frequency of reciprocal exchange, *mei-9* females display greatly elevated frequencies of meiotic nondisjunction and chromosome loss (Baker and Carpenter 1972). This effect is, however, entirely due to the overloading of the exchange-independent, back-up segregation system (known as the distributive system, see below) and not to any direct requirement for the *mei-9*$^+$ gene product in chromosome segregation. Moreover, meiotic chromosome behavior in males (in which meiotic recombination does not occur) is not affected.

The first clue as to the biochemical defect(s) that underlie the *mei-9* phenotype came from the observation that these alleles confer sensitivity to mutagens as a consequence of their defect in excision repair (Boyd et al. 1976b; Nguyen and Boyd 1977). This defect is also manifested by a high frequency of mitotic chromosome breakage and instability (Baker et al. 1978; Gatti 1979). For example, larval neuroblasts of *mei-9*/Y males display a high frequency of spontaneous chromosome breaks in both the eu- and heterochromatin (Gatti 1979).

Crude extracts of *mei-9* cultured cells show a reduction in apurinic endonuclease activity (Osgood and Boyd 1982). Venugopal et al. (1990) have also shown that the specific activity of an apurinic endonuclease is reduced 98% in ovaries from *mei-9* females when compared to wild-type. However, several lines of evidence suggest that although the *mei-9* gene product alters and influences the levels of this apurinic endonuclease, this nuclease is not the direct product of the *mei-9*$^+$ gene (for review, see Venugopal et al. 1990). Several laboratories are now attempting to clone the *mei-9*$^+$ locus, and significant progress in understanding this vital exchange and repair function can be expected in the near future.

7.1.2. mei-41 *(*meiotic mutant 41*; 1– 54.2)*

Many alleles of *mei-41* are female-sterile or nearly so. Females homozygous for the more fertile alleles of *mei-41* exhibit reduced levels of meiotic exchange. The observed reductions in exchange are not uniform, but rather are most extreme in distal regions. Chiasma interference is also diminished (Baker and Hall 1976). Thus, unlike *mei-9*, *mei-41* is believed to be defective in both the establishment of the number and position of exchange sites and in the mechanism of exchange itself. As was the case for *mei-9* bearing females, these reduced levels of exchange allow for high frequencies of meiotic loss and nondisjunction (see Baker and Hall 1976).

Ultrastructural analysis of pachytene in *mei-41* mutant females demonstrates a reduced number of late recombination nodules. The residual nodules are distributed in a fashion that parallels residual exchange events (Carpenter 1979). Over half of those nodules observed are morphologically abnormal and are associated with unusual regions of relatively uncondensed chromatin.

mei-41 alleles also confer sensitivity to mutagens. This presumably results from a defect in a caffeine-sensitive, postreplication repair pathway; that is, such mutants suffer an inability to synthesize normal length DNA on a UV-damaged template (Boyd and Setlow 1976; Boyd et al. 1976a; Boyd and Shaw 1982). This defect in DNA repair is also manifested by a high frequency of mitotic chromosome breakage and instability (Baker et al. 1978; Gatti 1979). The relationship between this defect in postreplication repair and the accompanying defects in meiotic recombination remains unclear. However, in prokaryotes, postreplication repair is dependent on sister-chromatid recombination. Thus, the presence of both defects in *mei-41* flies may indicate a requirement for the *mei-41*$^+$ gene product in the execution of normal recombination events. Transposon-tagged alleles of *mei-41* are now available, and we can anticipate that DNA clones of this locus will be available in the near future.

7.1.3. mei-218 *(meiotic mutant 218; 1– 56.2)*

Females homozygous for *mei-218* exhibit reduced levels of meiotic exchange. The residual exchanges are distributed such that the probability of euchromatic exchange becomes more nearly proportional to the polytene-chromosome length (Baker and Carpenter 1972; Baker and Hall 1976). This relaxation of the normal constraints on the distribution of euchromatic exchange is clearly demonstrated by the fact that *mei-218* females allow exchange between the normally achiasmate fourth chromosomes (Sandler and Szauter 1978).

Although *mei-218* alleles reduce the frequency of reciprocal meiotic exchange, the absolute frequency of gene conversion at the unrelated *rosy* locus is twofold elevated relative to wild-type controls (Carpenter and Baker 1982; Carpenter 1984). Moreover, co-conversion distances are shorter than those observed in either controls or *mei-9* females (Carpenter 1984). Thus, the function of this locus is required for the generation of reciprocal exchanges and not for the formation of heteroduplex intermediates or gene conversion (Carpenter 1984).

Ultrastructural analysis of pachytene cells in *mei-218* and *mei-218*$^{6-7}$ females demonstrates a reduced number of late recombination nodules (about 8% of normal), which are distributed in a fashion that parallels the residual exchange events (Carpenter 1979; 1989). Many of the nodules observed are morphologically abnormal. There is also some evidence that early recombination nodules may be either fewer in number or more ephemeral in *mei-218* females (Carpenter 1989).

Recently, Wendy Whyte (in the laboratory of RS Hawley) has mapped *mei-218* to a small genetic interval for which a large chromosome walk has been completed. There are multiple alleles of *mei-218*, including at least one spontaneous allele, which probably arose from the insertion of a transposon (Ashburner 1990). It seems clear that a molecular approach to *mei-218* will be feasible in the near future.

Thus, for several genes identified so far only by meiotic mutants, a molecular characterization seems imminent. Such studies will greatly enhance our understanding of both recombination and repair. We now turn our attention to those functions required for chromosome segregation.

7.2. The Molecular Biology of Chromosome Segregation

As noted above, exchanges serve the vital function of assuring the disjunction of homologous centromeres. During prometaphase, each centromere of the bivalent attaches to the spindle and begins to move toward one of the two poles. Once the two centromeres have oriented toward opposite poles, the progression of the two centromeres toward the poles is halted at the metaphase plate by the chiasma. This represents a stable position in which the bivalent will remain until anaphase I.

Thus, the ability of bivalents to orient their centromeres toward opposite poles is achieved by balancing the tension between the bivalent and the two spindle poles. Meiotic segregation must then involve at least three separate components or structures, namely the chiasma itself, the spindle, and sites on the chromosomes that generate poleward traction.

7.2.1. Genes Affecting Chiasma Structure and Function

Numerous studies in *Drosophila* have shown that chiasmate exchanges between chromosomes are fully sufficient to ensure conjunction, and that precise homologous disjunction is dependent on such crossover events (see Baker and Hall 1976). There is an average of one exchange event (or chiasma) per chromosome arm, and these exchanges are limited to euchromatic intervals (Weinstein 1936; Baker and Hall 1976).

When chromosomes fail to undergo exchange, they then enter the distributive pairing system (Grell 1976). Unfortunately, this system cannot faithfully assure the segregation of more than one or two pairs of chromosomes; hence, normal chiasma frequencies are necessary to prevent disjunctional errors arising when too many chromosomes enter the distributive system.

Most of the meiotic mutants obtained so far in *Drosophila* act by simply reducing exchange (Baker and Hall 1976; and Section 7.1 above). In order to study chiasma function, we would like to obtain mutants that allow exchange to occur at normal or near normal frequencies but prevent those exchanges from ensuring proper disjunction. In order to exclude mutants that define some general component of chromosome disjunction (such as spindle anomalies), we would also require that the distributive system function normally in such mutant meioses. Of the more than fifty meiotic mutants analyzed so far, only two, *ald* and *ord* possess this phenotype.

7.2.1.1. ald *(*altered disjunction*; 3–61).* This mutation was characterized by O'Tousa (1982). In *ald* females chiasmata frequently fail to commit chromosomes to segregating from each other. Nevertheless, chromosomes that have undergone exchange still participate in distributive disjunctions.

This phenotype is indeed unique. Exchange defective mutations (such as *mei-9*) have been shown to increase secondary nondisjunction, but only as a consequence of increasing the fraction of nonexchange tetrads (Baker and Hall 1976). On the other hand, mutants such as *nod* that disrupt pairing or segregation in the distributive system (see below), do not allow exchange chromosomes to enter the distributive system (Carpenter 1973; Zitron and Hawley 1989).

Although *ald* does have some effect on the choice of partners for nondisjunction in the distributive system, it does not affect the fidelity of distributive disjunction. Thus, *ald* does not appear to affect general processes or structures for meiotic disjunction (such as spindle structure or centromere function), but rather appears specifically to affect the ability of an exchange to ensure disjunction.

7.2.1.2. ord *(*orientation disruptor*; 2–13.5).* This mutant was characterized by Mason (1976) and Goldstein (1980). Although *ord* affects meiosis in both sexes, our attention here is focused on the observation that exchange does not ensure disjunction in *ord* females (Mason 1976). In males, *ord* clearly allows precocious sister chromatid separation at anaphase I (Goldstein 1980), yielding a high frequency of reductional and equational nondisjunction events. *ord* also reduces the total amount of exchange in females. Mapping data of Mason (1976) suggest that all three phenotypes of *ord* (failure of exchange, failure of chiasmata to ensure disjunction, and a defect in sister chromatid adhesion) are the consequence of a single mutational defect. However, Mason (1976) also showed that the distributive system in *ord* is still capable of ensuring the segregation of those nonexchange chromosomes in which sister chromatid cohesion is maintained. Thus, *ord* is not likely to define a general disjunctional defect, but rather a specific defect in sister chromatid adhesion and chiasma function. Indeed, a preliminary ultrastructural analysis of pachytene morphology in *ord* females suggests a possible defect in chromatin condensation and the morphology of the lateral element of the synaptonemal complex (ATC Carpenter, cited in Hawley 1988).

The phenotypes of chiasma failure and precocious sister chromatid separation are consistent with Maguire's (1978) model which postulates that both processes depend on proper function of the lateral elements of the synaptonemal complex. Thus, a detailed genetic and morphological analysis of a number of *ord* alleles is a promising route to understanding relationships between the continuity of the synaptonemal complex, chiasma function, and sister chromatid cohesion.

7.2.2. Genes Affecting the Distribution System

Mutations of two loci, *ncd* and *nod* allow normal pairing and exchange, but subsequently impair segregation.

7.2.2.1. ncd *(*non-claret disjunctional*; 3–100.7).* The *ncd* locus was originally identified by the mutation ca^{nd} (claret nondisjunctional of Lewis and Gencarella 1952). However, more recent genetic and molecular studies have shown that the original ca^{nd} mutation disrupts two adjacent genes: *ncd*, which is required for chromosome segregation, and *ca*, which is required for normal eye color (O'Tousa and Szauter 1980; Sequiera et al. 1989; Yamamoto et al. 1989).

ncd alleles induce chromosomal nondisjunction and loss at meiosis I and mitotic chromosome loss during early cleavage divisions. Although *ncd* mutations induce both exchange and nonexchange X chromosomes to undergo nondisjunction at high frequency (Davis 1969; Sequiera et al. 1989), nonexchange chromosomes nondisjoin and are lost much more frequently than are exchange chromosomes (Carpenter 1973). Thus, as pointed out by Baker and Hall (1976), "the wild type allele of *ncd* specifies a function required for the regular disjunction of nonexchange, as well as exchange chromosomes, but an exchange may serve as the partial equivalent of the function—it can facilitate regular disjunction of chromosomes even in the absence of the ncd^+ function."

Several studies have demonstrated that the physical basis of the *ncd* defect is the abnormal segregation of chromosomes on disorganized spindles with abnormal numbers of poles. This defect is manifested both at meiosis and during early cleavage divisions (Wald 1936; Kimble and Church 1983). Davis (1969) has argued that mutations at the *ncd* locus result in aberrant spindle fibers, while others have suggested that the defect is in the centromeres (Baker and Hall 1976) or in the microtubule organizing centers (Kimble and Church 1983).

Subsequently, the *ncd* locus has been shown to encode a protein with good homology to kinesin heavy chain in the motor domain (Endow et al. 1990; McDonald and Goldstein 1990). Kinesin was originally identified as a microtubule motor required for organelle transport along microtubule arrays in squid giant axons and subsequently has been found in a variety of organisms and tissues including neurons, sea urchin eggs, yeast, and *Drosophila* (Vale et al. 1985; Scholey et al. 1985; Saxton et al. 1988; Meluh and Rose 1990). Several studies have shown that kinesin is a microtubule-activated ATPase, whose activity is directly linked to kinesin-driven motility (for review, see Saxton et al. 1988). Evidence from a variety of systems indicates that kinesin is a heterotetramer with two heavy and two light chains (Bloom et al. 1988; Saxton et al. 1988).

7.2.2.2. nod *(*no distributive disjunction*; 1–36).* Mutations at the *nod* locus specifically induce the nondisjunction and loss of nonexchange chromosomes, i.e., they disrupt distributive segregation (Carpenter 1973; Zhang and Hawley 1990). These mutations have no effect on either the frequency of exchange or on the segregation of exchange bivalents. Since the other autosomes will usually form exchange bivalents, the most obvious effect of *nod* mutations on female meiosis is therefore on the disjunction of the fourth chromosomes. Fourth chromosome nondisjunction frequencies approach 85% in *nod/nod* females, with the vast majority of the exceptional gametes having lost the fourth chromosome. Mutations at the *nod* locus also induce random X chromosome disjunction in those meiocytes containing nonexchange X chromosomes (Carpenter 1973; Zhang and Hawley 1990).

It has been shown recently that the *nod* locus encodes a protein that, like the *ncd* protein, is a member of the kinesin heavy chain superfamily (Zhang et al. 1990). Since kinesin is a microtubule motor, the suggestion is that the *nod* locus encodes a spindle motor that is required to hold distributively "paired" chromosomes at the metaphase plate until anaphase. Ordinarily, the presence of chiasmata serve this purpose by balancing the poleward tension on homologous kinetochores, holding the bivalent at the plate (Nicklas 1974). However, in the absence of chiasmata, homologues would require other forces to maintain them properly at the metaphase plate.

Zhang et al. (1990) have proposed a mechanism by which the *nod* kinesin-like protein might play a role in this process (see Figure 10.7). They argue that the *nod* protein is a component of the so-called polar ejection force, which apparently acts along the whole length of the chromosomes to propel them away from the spindle pole (Rieder et al. 1986; Salmon 1989). In this case, the *nod* protein would facilitate the movement of chromosomes away from the pole along microtubule tracks. Chromosomes using the distributive system would then be held at the metaphase plate, because that would be the position where opposing antipoleward forces were balanced. In this model, the *nod* protein serves a function similar to that of chiasmata, which probably hold homologous chromosomes at the metaphase plate by balancing the poleward pulling forces generated at the centromere (Nicklas 1974).

Two lines of evidence argue that the *nod* gene product is involved in mitosis as well as in meiosis. First, *nod* transcription is easily detected throughout development. Second, we have recently shown that a dosage- and cold-sensitive mitotic lethal mutation, known as *l(1)TW-6^{cs}* (Wright 1974), is in fact an antimorphic allele of *nod* (RS Hawley, C New, R Rasooly and BS Baker unpublished data). Death at the restrictive

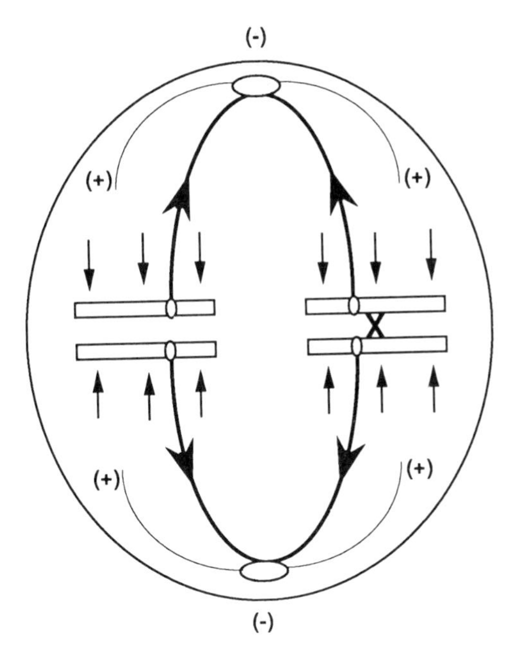

Figure 10.7. A speculative model for *nod* function. Note that the *nod*$^{+}$ protein generates a balancing force.

temperature apparently occurs as a consequence of frequent chromosome breakage and loss during mitosis (Baker et al. 1978). It seems likely that this defect arises as a consequence of the production of an aberrant (poisonous) protein rather than the ectopic expression of the normal *nod* protein (P Zhang and RS Hawley unpublished data). Thus, both genetic and molecular data argue that the *nod* protein is expressed both during meiosis and mitosis and that, while the *nod* protein is only required in meiosis, a mutation that encodes a poisonous product can interfere with both processes.

In summary, there appears to be a family of kinesin-like proteins in *Drosophila*, whose functions are required for chromosome segregation and other microtubule-mediated motility processes (Endow et al. 1990; McDonald and Goldstein 1990). It can be argued that those functions are sufficiently overlapping that the loss of any one member (such as *ncd* or *nod*) results in a very restricted phenotype. Moreoever, alleles of *nod* and *ncd* show strong interaction (Knowles and RS Hawley unpublished data), as would be expected if these two proteins play distinct but overlapping roles in the process of chromosome segregation.

Clearly, there is much more to be done. However, there have been several large recent screens for meiotic and mitotic mutations, and rapid means for identifying gene products are now available.

8. Summary and Conclusions

Both spermatogenesis and oogenesis in insects are developmentally complex. At least in *Drosophila*, oogenesis has in recent years become a model developmental system, now becoming understood at the molecular level (Manseau and Schupbach 1989). In comparison, studies of spermatogenesis have been largely restricted to cytological and ultrastructural analyses, with few molecular studies. This disparity arises partly because researchers have focused on the great importance of egg development in contributing to early embryogenesis. The importance of oogenesis is that it is a developmental process, contributing both to the making of embryos as well as the making of eggs. On the other hand, spermatogenesis seems to be a much more labile system. For many organisms (particularly *Drosophila*), it is usually male, rather than female, sterility which results from species hybridization. Also, backcross experiments reveal spermatogenesis to be highly vulnerable to genetic perturbations, whereas oogenesis appears quite robust (Dobzhansky 1936; Coyne 1985; Orr 1989). This lability is perhaps understandable for species in which the male is the heterogametic sex, and the phenotypes of new Y-linked mutations are exposed to selection immediately. As we have pointed out, the lability of spermatogenesis is perhaps best demonstrated by the fact that naturally occurring systems of genic meiotic drive are also associated only with spermatogenic aberrations (Crow 1988; Charlesworth 1988).

The approach to the molecular genetic analysis of female meiosis has sometimes been technically complicated, but straightforward in terms of interpreting the evolutionary impact of mutations defective in meiosis. In *Drosophila*, genes have been identified that play a major role in recombination/DNA repair (e.g., *mei-218*, *mei-9* and *mei-41*, the latter two of which play roles in excision and post-replication repair, respectively), chiasma structure and function (e.g., *ald* and *ord*), and in segregation (e.g., the kinesin heavy chain genes *ncd* and *nod*). Defective alleles for one or more of these genes generally lead to abnormal meioses and would be expected to be selectively eliminated from populations harboring them (although some meiotic mutants have indeed been isolated from nature; Baker and Hall 1976). We can expect continued rapid progress in the identification of genes controlling single steps in the stages of female meiosis in *Drosophila*, followed quickly by the isolation of the corresponding DNA and the identification of the protein product. Such sequences offer a starting point to identify analogous genes in other insects. On the other hand, mutations exhibiting meiotic drive in males that have only been isolated from natural populations are rarely due to simple genetic mutations at single loci and may have profound impact at the population or even evolutionary level, especially when the sex chromosomes are involved. These results suggest that it may be difficult to mutagenically induce meiotic drive in the laboratory. Further, compared to female meiosis, the molecular mechanisms of the various known cases of drive in males may prove more refractory to analysis.

Are there any generalities which can be gleaned from the several drive systems discussed here? The most striking observation is that they are always multilocus systems. That is, the drive element always seems to operate against a secondary locus, where the distorted segregation ratios are actually observed. There have been no cases observed where the same DNA sequence can act as both drive and target loci. Thus, a meiotic drive system can only become established if the drive element becomes linked to an insensitive allelic form at the target locus. This can happen initially by pure chance, or weak initial linkage can be tightened by the accumulation of chromosome rearrangement or other suppressors of recombination. These requirements favor X chromosome over autosomal drive forms.

There are important similarities among the three cases of meiotic drive discussed. In each case, drive arises because the high levels of gametic wastage do not result in comparable reductions in male fecundity. For each system, sperm dysfunction is *cis*-acting; i.e., it depends on the presence in the sperm of a target DNA sequence, which has somehow been affected by the action of the drive allele. This immediately suggests that the product of the drive allele binds at specific target DNA sites to cause dysfunction. In the case of *SD*, the drive product is probably a protein, and the target DNA may be the Responder satellite repeat sequence. In the case of male drive in mosquito, the drive allele itself may be associated with AT rich satellite DNA sequences, while the target sequences are as yet unknown. The action of both *SD* and *SR* in *Drosophila* results in a similar phenotype of sperm dysfunction—the failure to achieve proper individualization of sperm carrying sensitive target alleles. All three systems are susceptible to suppression by modifier loci, and each drive chromosome comes to be associated with changes in chromosome structure, which probably act to tighten linkage among drive and target loci.

Nevertheless, the similarities among these drive systems probably are the result of convergent evolution, rather than homologous mechanism of action. For instance, there is no evidence that *SD* or *SR* in *Drosophila* involve chromosome breakage, while this is clearly the mode of action of *MD* in mosquito. *SR* appears to involve multiple drive loci, while *MD* involves multiple target sites for chromosome breakage; *SD* falls in between. Further, the drive alleles of *SD* and *SR* are *trans*-acting and probably (at least in the case of *SD*) additive in effect. Conversely, in the *MD* system the drive allele apparently can cause drive only when linked to *M* and only when present in a single copy.

Even considering these three systems alone, there would seem to be several stages in spermatogenesis, each susceptible of being co-opted to produce meiotic drive; indeed, any step in spermatogenesis, which involves binding of a regulatory or structural gene product to specific chromosomal DNA sequences, may offer a point open to exploitation by meiotic drive systems. In some ways it is perhaps surprising that Mendelian segregation is so persistent in evolution, given the number of ways it may be open to exploitation by such ultraselfish DNA elements (Crow 1988).

For meiotic drive to be useful as a tool for population suppression of pest insects, it is necessary that the population contain only sensitive allelic forms at the target locus, and that drive be very strong. Otherwise, both laboratory and field experiments make it clear that target populations will quickly evolve drive suppression, either by the accumulation of insensitive alleles at the target locus, or by the fixation of unlinked secondary suppressors of drive. If a drive element can act only at specific DNA sequences at a secondary target locus, this may make it difficult to transfer meiotic drive systems from one species to another by means of genetic transformation.

A more realistic approach will be to attempt to isolate native drive systems directly from the species of interest. While this is currently not practical for many pest organisms, the possibilites should change radically with the increase in our understanding of the molecular genetic basis, both of normal gametogenesis and of the abnormal spermatogenesis exhibited in meiotic drive systems. On the one hand, once the molecular basis of a known abnormality (such as meiotic drive) in one species is established, it may be possible to screen for mutations (or directly alter the homologous DNA sequences) affecting the analogous step in a second pest species. Conversely, a knowledge of the general genetic control of meiosis and gamete development may allow us to engineer, *de novo*, forms of meiotic drive currently not seen in natural populations. In either case, it is clear that a better understanding of the genetic control of gametogenesis will be a critical step in the advancement of both basic and applied insect genetic studies.

Acknowledgments. Production of this contribution was supported by NSF Grant DCB-8815749 and USDA Grant HAW00763-G to TW Lyttle, NSF Grant DCB-8517504 to RS Hawley, and Research Career Development Award NIH GM39902, KO4 HG00005 and a National Down's Syndrome Society Fellowship to C-I Wu.

References

Ashburner M (1990) *Drosophila*. A Laboratory Handbook. Cold Spring Harbor Laboratory Press, New York, 1331 pp

Aquadro CF, Susan SF, Bland ML, Langley CH, Laurie CC (1986) Molecular population genetics of the Alchohol dehydrogenase gene region of *Drosophila melanogaster*. Genetics 114:1165–1190

Baker BS, Carpenter ATC (1972) Genetic analysis of sex chromosomal meiotic mutants in *Drosophila melanogaster*. Genetics 71:255–286

Baker BS, Carpenter ATC, Ripoll R (1978) The utilization during mitotic cell division of loci controlling meiotic recombination and disjunction in *Drosophila melanogaster*. Genetics 90:531–578

Baker BS, Hall JC (1976) Meiotic mutants: genetic control of meiotic recombination and chromosome segregation. In Ashburner MA, Novitski E (ed) Genetics and Biology of *Drosophila*, Vol 1a Academic Press, New York, pp 351–434

Beckenbach AT (1978) The Sex-Ratio trait in *Drosophila pseudoobscura*: fertility relations of males and meiotic drive. Amer Nat 112:97–117

Beckenbach AT (1983) Fitness analysis of the Sex-Ratio polymorphism in experimental populations of *Drosophila pseudoobscura*. Amer Nat 121:630–648

Beckenbach AT, Curtsinger CW, Policansky D (1982) Fruitless experiments with fruitflies: The sex-ratio chromosomes of *Drosophila pseudoobscura*. Dros Inform Serv 58:22

Bloom GS, Wagner MC, Pfister KK, Brady ST (1988) Native structure and physical properties of bovine brain kinesin and identification of the ATP binding subunit polypeptide. Biochemistry 27:3409–3416

Boyd JB, Golino MD, Nguyen TD, Green MM (1976a) Isolation and characterization of X-linked mutants of *Drosophila melanogaster* which are sensitive to mutagens. Genetics 84:485–506

Boyd JB, Golino MD, Setlow RB (1976b) The *mei-9^a^* mutant of *Drosophila melanogaster* increases mutagen sensitivity and decreases excision repair. Genetics 84:527–544

Boyd JB, Setlow RB (1976) Characterization of postreplication repair in mutagen-sensitive strains of *Drosophila melanogaster*. Genetics 84:507–526

Boyd JB, Shaw KES (1982) Postreplication repair defects in mutants of *Drosophila melanogaster*. Mol Gen Genet 186:289–294

Brittnacher JG, Ganetzky B (1983) On the components of Segregation Distortion in *Drosophila melanogaster*. II. Deletion mapping and dosage analysis of the *SD* locus. Genetics 103:659–673

Brittnacher JG, Ganetzky B (1984) On the components of Segregation Distortion in *Drosophila melanogaster*. III. Nature of Enhancer of SD. Genetics 107: 423–434

Brittnacher JG, Ganetzky B (1989) On the components of Segregation Distortion in *Drosophila melanogaster*. IV. Construction and analysis of free duplications for the *Rsp* locus. Genetics 121:739–750

Bull JJ (1983) Evolution of Sex Determining Mechanisms. Benjamin/Cummings Publishing, Menlo Park

Carpenter ATC (1973) A mutant defective in distributive disjunction in *Drosophila melanogaster*. Genetics 73:393–428

Carpenter ATC (1979) Recombination nodules in recombination-defective females of *Drosophila melanogaster*. Chromosoma 83:59–80

Carpenter ATC (1984) Meiotic roles of crossing over and gene conversion. Cold Spring Harbor Symp. Quant Biol 48:23–30

Carpenter ATC (1987) Gene conversion, recombination nodules, and the initiation of meiotic synapsis. Bioessays 6:232–236

Carpenter ATC (1988) Thoughts on recombination nodules, meiotic recombination and chiasmata. In Kucherlapati R, Smith GR (eds) Genetic Recombination ASM Press, Washington DC, pp 497–525

Carpenter ATC (1989) Are there morphologically abnormal early recombination nodules in the *Drosophila melanogaster* meiotic mutant *mei-218*? Genome 31:74–80

Carpenter ATC, Baker BS (1982) On the control of the distribution of meiotic exchange in *Drosophila melanogaster*. Genetics 101:81–89

Charlesworth B (1988) Driving genes and chromosomes. Nature 332:394–395

Charlesworth B, Hartl DL (1978) Population dynamics of the *Segregation Distorter* polymorphism of *Drosophila melanogaster*. Genetics 89:171–192

Coyne JA (1985) The genetic basis of Haldane's rule. Evolution 43:362–381.

Crow JF (1979) Genes that violate Mendel's rules. Sci Amer 240(2):134–146

Crow JF (1988) The ultraselfish gene. Genetics 118:389–391

Curtis CF, Grover KK, Suguna SG, Uppal DK, Dietz K, Agarwal HV, Kazmi SJ (1976) Comparative field cage tests of the population suppressing efficiency of three genetic control systems for *Aedes aegypti*. Heredity 36:11–29

Curtsinger JW, Feldman MW (1980) Experimental and theoretical analyses of the "sex-ratio" polymorphism in *Drosophila pseudoobscura*. Genetics 94: 445–466

Davis DG (1969) Chromosome behavior under the influence of claret-nondisjunctional in *Drosophila melanogaster*. Genetics 61:577–594

De Carvalho AB, Peixoto AA, Klaczko LB (1989) Sex ratio in *Drosophila mediopunctata*. Heredity 62:425–428

Dobzhansky T (1936) Studies on hybrid sterility. II. Localization of sterility factors in *Drosophila pseudoobscura* hybrids. Genetics 21:113–135

Edwards AWF (1961) The population genetics of sex ratio in *Drosophila pseudoobscura*. Heredity 16:291–304

Eisen A, Camerini-Otero RD (1988) A recombinase from *Drosophila melanogaster* embryos. Proc Natl Acad Sci USA 85:7481–7485

Endow S, Komma DJ (1986) One-step and stepwise magnification of a *bobbed* lethal chromosome in *Drosophila melanogaster*. Genetics 114:511–523

Endow SA, Henikoff S, Soler-Niedziela L (1990) A kinesin-related protein mediates meiotic and early mitotic chromosome segregation in *Drosophila*. Nature 345:81–83

Eshel I (1985) Evolutionary genetic stability of Mendelian segregation and the role of free recombination in the chromosomal system. Amer Nat 125: 412–420

Fisher RA (1930) The Genetical Theory of Natural Selection. Oxford University Press, Oxford

Foster GG, Whitten MJ (1974) Unequal segregation in *Lucilia cuprina*. Genetics 77:s22–s23

Foster GG, Whitten, MJ (1991) Meiotic drive in *Lucilia cuprina* and chromosome evolution. Amer Nat 137:403–415

Fredga K, Gropp A, Winking H, Frank F (1977) A hypothesis explaining the exceptional sex ratio in the wood lemming (*Myopus schisticolor*). Hereditas 85:101–104

Ganetzky B (1977) On the components of Segregation Distortion in *Drosophila melanogaster*. Genetics 86:321–355

Gatti M (1979) Genetic control of chromosome breakage and rejoining in *Drosophila melanogaster*. Proc Natl Acad Sci USA 76:1377–1381

Goldstein LSB (1980) Mechanisms of chromosome orientation revealed by two meiotic mutants in *Drosophila melanogaster*. Chromosoma 78:79–111

Gowen MS, Gowen JW (1922) Complete linkage in *Drosophila melanogaster*. Amer Nat 56:286–288

Grell RF (1976) Distributive pairing. In Ashbumer MA, Novitski E (eds) Genetics and Biology of *Drosophila*, Vol 1a. Academic Press, New York, pp 435–486

Hamilton WD (1967) Extraordinary sex ratios. Science 156:477–488

Hartl DL (1970) Meiotic drive in natural populations of *Drosophila melanogaster*. IX. Suppressors of *Segregation Distorter* in wild populations. Canad J Genet Cytol 12:594–600

Hartl DL (1972) Population dynamics of sperm and pollen killers. Theor Appl Genet 42:81–88

Hartl DL, Hartung N (1975) High frequency of one element of *Segregation Distorter* in natural populations of *Drosophila melanogaster*. Evolution 29:512–518

Hartl DL, Hiraizumi Y (1976) Segregation Distortion after fifteen years. In Ashburner M, Novitski E (eds) The Genetics and Biology of *Drosophila*, Vol 1b. Academic Press, New York, pp 615–666

Hauschteck-Jungen E, Hartl DL (1982) Defective histone transition during spermiogenesis in heterozygous *Segregation-Distorter* males of *Drosophila melanogaster*. Genetics 101:57–69

Hauschteck-Jungen E, Maurer B (1976) Sperm dysfunction in sex ratio males of *Drosophila subobscura*. Genetica 46:459–477

Hawley RS (1988) Exchange and chromosome segregation in eukaryotes. In Kucherlapati R, Smith GR (eds) Genetic Recombination ASM Press, Washington DC, pp 497–523

Hedrick PW (1988) Can segregation distortion influence gametic disequilibrium?. Genet Res 52:237–242

Hickey WA, Craig GB (1966) Genetic distortion of sex ratio in mosquito, *Aedes aegypti*. Genetics 53:1177–1196

Hihara YK (1974) Genetic analysis of modifying system of Segregation Distortion in *Drosophila melanogaster*. II. Two modifiers for *SD* system on the second chromosome of *D. melanogaster*. Jap J Genet 49:209–222

Hilliker AJ (1976) Genetic analysis of the centromeric heterochromatin of chromosome 2 of *Drosophila melanogaster*: Deficiency mapping of EMS-induced lethal complementation groups. Genetics 83:765–782

Hiraizumi Y, Martin DW, Eckstrand IA (1980) A modified model of Segregation Distortion in *Drosophila melanogaster*. Genetics 95:693–706

Hiraizumi Y, Thomas AM (1984) Suppressor systems of *Segregation Distorter* (*SD*) chromosomes in natural populations of *Drosophila melanogaster*. Genetics 106:279–292

Jones RN, Rees H (1982) B Chromosomes. Academic Press, New York

Kataoka Y (1967) A genetic system modifying Segregation Distortion in a natural population of *Drosophila melanogaster* in Japan. Jap J Genet 42:327–337

Kettaneh NP, Hartl DL (1976) Histone transition during spermiogenesis is absent in *Segregation Distorter* males of *Drosophila melanogaster*. Science 193:1020–1021

Kimble M, Church K (1983) Meiosis and early cleavage in *Drosophila melanogaster* eggs. J Cell Sci 62:301–318

Langley CH, Aquadro CF (1987) Restriction map variation in natural populations of *Drosophila melanogaster*: white locus region. Mol Biol Evol 4:651–663

Lewis EB, Gencarella W (1952) Claret and nondisjunction in *Drosophila melanogaster*. Genetics 37:600–601

Lyttle TW (1977) Experimental population genetics of meiotic drive systems. I. Pseudo-Y chromosomal drive as a means of eliminating cage populations of *Drosophila melanogaster*. Genetics 86:413–445

Lyttle TW (1979) Experimental population genetics of meiotic drive systems. II. Accumulation of genetic modifiers of *Segregation Distorter* (*SD*) in laboratory populations. Genetics 91:339–357

Lyttle TW (1981) Experimental population genetics of meiotic drive systems. III. Neutralization of sex-ratio distortion in *Drosophila* through sex chromosome aneuploidy. Genetics 98:317–334

Lyttle TW (1984) Chromosomal control of fertility in *Drosophila melanogaster*. I. Rescue of $T(Y{:}A)/bb^{l158}$ male sterility by chromosome rearrangement. Genetics 106:423–434

Lyttle TW (1986) Additive effects of multiple *Segregation Distorter* (*SD*) chromosomes on sperm dysfunction in *Drosophila melanogaster*. Genetics 114:203–216

Lyttle TW (1989) The effect of novel chromosome position and variable dose on the genetic behavior of the Responder (*Rsp*) element of the *Segregation Distorter* (*SD*) system of *Drosophila melanogaster*. Genetics 121:751–763

Lyttle TW, Brittnacher JG, Ganetzky B (1986) Detection of *Rsp* and modifier variation in the meiotic drive system *Segregation Distorter* (*SD*) of *Drosophila melanogaster*. Genetics 114:183–202

Lyttle TW, Haymer DS (1992) The role of the transposable element *hobo* in the origin of endemic inversions in wild populations of *D. melanogaster*. Genetica (in press)

Maguire MP (1978) A possible role for the synaptonemal complex in chiasma maintenance. Exper Cell Res 112:297–308

Mange EJ (1968) Temperature sensitivity of Segregation Distortion in *Drosophila melanogaster*. Genetics 58:399–413

Manseau LJ, Schupbach T (1989) The egg came first of course! Anterior-posterior pattern formation in *Drosophila* embryogenesis and oogenesis. Trends Genet 4:400–405

Martin DW, Hiraizumi Y (1979) On the models of Segregation Distortion in *Drosophila melanogaster*. Genetics 93:423–435

Mason JM (1976) Orientation disrupter (*ord*): a recombination-defective and disjunction-defective meiotic mutant in *Drosophila melanogaster*. Genetics 84:545–572

McCloskey JD (1966) The problem of gene activity in the sperm of *Drosophila melanogaster*. Amer Nat 100:211–218

McDonald HB, Goldstein LSB (1990) Identification and characterization of a gene encoding a kinesin-like protein in *Drosophila*. Cell 61:991–1000

Meluh PB, Rose MD (1990) KAR3: a kinesin-related gene required for yeast nuclear fusion. Cell 60:1029–1034

Miklos GLG (1985) Localized highly repetitive DNA sequences in vertebrate and invertebrate genomes. In MacIntyre RJ (ed) Molecular Evolutionary Genetics. Plenum Press, New York

Morgan TH, Bridges CB, Sturtevant AH (1925) The genetics of *Drosophila*. Bibliogr Genet 2:1–262

Mulcahy DL (1975) The biological significance of gamete competition. In Mulcahy DL (ed) Gamete Competition in Plants and Animals. North-Holland Publishing Company, Amsterdam, pp 1–3

Muller HJ, Settles F (1927) The nonfunctioning of genes in spermatozoa. Z Induk Abstam-Vererb 43:285–312

Newton ME, Wood RJ, Southem DI (1976) A cytogenetic analysis of meiotic drive in the mosquito, *Aedes aegypti* (L.). Genetica 45:297–318

Newton ME, Wood RJ, Southern DI (1978) Cytological mapping of the *M* and *D* loci in the mosquito, *Aedes aegypti* (L.). Genetica 48:137–143

Nguyen TD, Boyd JB (1977) The meiotic-9 (*mei-9*) mutants of *Drosophila melanogaster* are deficient in repair replications of DNA. Mol Gen Genet 158:141–147

Nicklas RB (1974) Chromosome segregation mechanisms. Genetics 78:205–213

Nicoletti B, Trippa G (1967) Osservazioni citologiche su di un nuovo caso di "Segregation Distortion" (*SD*) in una popolazione naturale di *Drosophila melanogaster*. Atti Assoc Genet Ital 12:361–365

Novitski E (1947) Genetic analysis of an anomalous sex ratio condition in *Drosophila affinis*. Genetics 32:526–534

Novitski E (1967) Nonrandom disjunction in *Drosophila*. Ann Rev Gen 1:71–86

Orr HA (1989) Genetics of sterility in hybrids between two subspecies of *Drosophila*. Evolution 43:180–189

Osgood CJ, Boyd JB (1982) Apurinic endonuclease activity from *Drosophila melanogaster*. Reduced enzymatic activity excision defective mutants of the *mei-9* and *mus-201* loci. Mol Gen Genet 184:135–139

O'Tousa J (1982) Meiotic chromosome behavior influenced by mutation-altered disjunction in *Drosophila melanogaster* females. Genetics 102:503–524

O'Tousa J, Szauter P (1980) The initial characterization of *non-claret disjunctional* (*ncd*): evidence that ca^{nd} is the double mutant, *ca ncd*. Dros Inform Serv 55:119

Owen DF (1974) Seasonal change in sex ratio in *Acraea quirina* (F.) (Lep. Nymphalidae), and notes on the factors causing distortions of sex ratio in butterflies. Entomol Scand 5:110–114

Pimpinelli S, Dimitri P (1989) Cytogenetic analysis of the Responder (*Rsp*) locus in *Drosophila melanogaster*. Genetics 121:765–772

Policansky D, Dempsey BB (1978) Modifiers and Sex-ratio in *Drosophila pseudoobscura*. Evolution 32:922–924

Powell JR (1983) Interspecific cytoplasmic gene flow in the absence of nuclear gene flow: Evidence from *Drosophila*. Proc Natl Acad Sci USA 80:492–495

Powers PA, Ganetzky B (1991) On the components of segregation distortion in *Drosophila melanogaster*. *V* Moleculae Analysis of the *SD* Locus. Genetics 129:133–144

Prout T, Bundgaard J, Bryant S (1973) Population genetics of modifiers of meiotic drive. I. The solution of a special case and some general implications. Theor Pop Biol 4:446–465

Puro J, Nokkala S (1977) Meiotic segregation of chromosomes in *Drosophila melanogaster* oocytes. Chromosoma 63:273–286

Rabin BA, Hawley RS, Chase JW (1986) DNA ligase from *Drosophila melanogaster* embryos: purification and physical characterization. J Biol Chem 261:10637–10645

Rasmussen SW (1975) Ultra-structural studies of meiosis in males and females of the $c(3)G^{17}$ mutant of *Drosophila melanogaster* Meigen. C R Trav Lab Carlsberg 40:163–173

Rhoades MM, Dempsey E (1966) The effect of abnormal chromosome 10 on preferential segregation and crossing over in maize. Genetics 53:989–1020

Rieder LL, Davison EA, Jensen LCW, Cassimeris L, Salmon ED (1986) Oscillatory movements of monooriented chromosomes and their position relative to the spindle pole result from the ejection properties of the aster and half-spindle. J Cell Biol 103:581–591

Salmon ED (1989) Microtubule dynamics and chromosome movement. In Hyams JS, Brinkley BR (eds) Mitosis. Academic Press, New York, pp 119–181

Sandler L, Golic K (1985) Segregation Distortion in *Drosophila*. Trends Genet 1:181–185

Sandler L, Novitski E (1957) Meiotic drive as an evolutionary force. Amer Nat 41:105–110

Sandler L, Szauter P (1978) The effect of recombination defective mutations on fourth-chromosome recombination in *Drosophila melanogaster*. Genetics 90:699–712

Saxton WM, Porter ME, Cohn SA, Scholey JM, Raff EC, McIntosh JR (1988) *Drosophila* kinesin: characterization of microtubule motility and ATPase. Proc Natl Acad Sci USA 85:1109–1113

Scholey JM, Porter ME, Grisson PM, McIntosh JR (1985) Identification of kinesin in sea urchin eggs, and evidence for its localization in the mitotic spindle. Nature 318:483–486

Sequiera W, Nelson CR, Szauter P (1989) Genetic analysis of the claret locus of *Drosophila melanogaster*. Genetics 123:511–524

Silver LM (1985) Mouse *t* haplotypes. Ann Rev Genet 19:179–208

Smith DAS (1975) All-female broods in the polymorphic butterfly *Danaus chrysippus* and their ecological significance. Heredity 34:363–371

Spradling AC (1986) P element mediated transformation. In Roberts DB (ed) *Drosophila*: a Practical Approach. IRL Press, Oxford, pp 175–197

Spradling AC, Rubin GM (1981) *Drosophila* genome organization: conserved and dynamic aspects. Ann Rev Genet 15:219–264

Stalker HD (1961) The genetic systems modifying meiotic drive in *Drosophila paramelanica*. Genetics 46:177–202

Steinemann M (1982) Multiple sex chromosomes in *Drosophila miranda*: a system to study the degeneration of a chromosome. Chromosoma 86:59–76

Sturtevant AH, Dobzhansky T (1936) Geographical distribution and cytology of "sex-ratio" in *Drosophila pseudoobscura* and related species. Genetics 21:473–490

Suguna SG, Wood RJ, Curtis CF, Whitelaw A, Kazmi SJ (1977) Resistance to meiotic drive at the M^D locus in an Indian wild population of *Aedes aegypti*. Genet Res 29:123–132

Sweeney TL, Barr AR (1978) Sex ratio distortion caused by meiotic drive in a mosquito, *Culex pipiens* (L.). Genetics 88:427–446

Szostak JW, Orr-Weaver TL, Rothstein RJ, Stahl FW (1983) The double-strand track repair model for recombination. Cell 33:25–35

Takahata N (1981) A mathematical study on the distribution of the number of repeated genes per chromosome. Genet Res 38:97–102

Tartof KD (1974) Unequal mitotic sister chromatid exchange as the mechanism of ribosomal RNA gene magnification. Proc Natl Acad Sci USA 71:1272–1276

Temin RG, Ganetzky B, Powers PA, Lyttle TW, Pimpinelli S, Dimitri P, Wu C-I, Hiraizumi Y (1991) *Segregation Distorter* (*SD*) in *Drosophila melanogaster*: Genetic and molecular analyses. Amer Nat 137:287–331

Temin RG, Marthas M (1984) Factors influencing the effect of Segregation Distortion in natural populations of *Drosophila melanogaster*. Genetics 107:375–393

Thomson GJ, Feldman MW (1974) Population genetics of modifiers of meiotic drive. II. Linkage modification in the Segregation Distortion system. Theor Pop Biol 5:155–162

Tokuyasu KT, Peacock WJ, Hardy RW (1972) Dynamics of spermiogenesis in *Drosophila melanogaster*. I. Individualization process. Z Zellforsch 124: 479–506

Tokuyasu KT, Peacock WJ, Hardy RW (1977) Dynamics of spermiogenesis in *Drosophila melanogaster*. VII. Effects of *Segregation Distorter* (*SD*) chromosome. J Ultrastruct Res 58:96–107

Trippa G, Loverre A (1975) A factor on a wild third chromosome (IIIRa) that modifies the Segregation Distortion phenomenon in *Drosophila melanogaster*. Genet Res 26:113–125

Trippa G, Loverre A, Ciccetti R (1980) Cytogenetic analysis of an *SD* chromosome from a natural population of *Drosophila melanogaster*. Genetics 95: 399–412

Uyenoyama MK, Feldman MW (1978) The genetics of sex ratio distortion by cytoplasmic infection under maternal and contagious transmission: an epidemiological study. Theor Pop Biol 14:471–497

Vale RD, Reese TS, Sheetz MP (1985) Identification of a novel force-generating protein, kinesin, involved in microtubule-based motility. Cell 42:39–50

Venugopal S, Guzdar SN, Deutsch WA (1990) Apurinic endonuclease activity from wild-type and repair-deficient *mei-9 Drosophila* ovaries. Mol Gen Genet 221:427–433

Wald H (1936) Cytological studies on the claret mutant type of *Drosophila simulans*. Genetics 21:264–279

Walker ES, Lyttle TW, Lucchesi JC (1989) Transposition of the Responder element (*Rsp*) of the Segregation Distorter system (*SD*) to the X chromosome in *Drosophila melanogaster*. Genetics 122:81–86

Wallace AJ, Newton ME (1987) Heterochromatin diversity and cyclic responses to selective silver staining in *Aedes aegypti* (L.). Chromosoma 95:89–93

Weinstein A (1936) The theory of multiple-strand crossing-over. Genetics 21: 55–199

Wood RJ (1976) Between-family variation in sex ratio in the Trinidad (T30) strain of *Aedes aegypti* (L.) indicating differences in sensitivity to the meiotic drive gene M^D. Genetica 46:345–361

Wood RJ, Newton ME (1977) Meiotic drive and sex ratio distortion in the mosquito *Aedes aegypti*. Proc Plen Symp Genetic Control of Insects, XVth Int Cong Entomol, 97–105

Wood RJ, Newton ME (1991) Meiotic drive causing sex ratio distortion in mosquitoes. Amer Nat 137:379–391

Wright TRF (1974) A cold-sensitive zygotic lethal causing high frequencies of nondisjunction during meiosis I in *Drosophila melanogaster* females. Genetics 76:511–536

Wu C-I (1983a) Virility selection and the Sex-Ratio trait in *Drosophila pseudoobscura*: I. Sperm displacement and sexual selection. Genetics 105: 651–652

Wu C-I (1983b) Virility selection and the Sex-Ratio trait in *Drosophila pseudoobscura*: II. Multiple insemination and overall virility selection. Genetics 105:663–679

Wu C-I (1983c) The fate of autosomal modifiers of the Sex-ratio trait in *Drosophila* and other sex-linked meiotic drive systems. Theor Pop Biol 24:107–120

Wu C-I, Beckenbach AT (1983) Evidence for extensive genetic differentiation between the Sex-Ratio and the Standard arrangement of *Drosophila pseudoobscura* and D. *persimilis* and identification of hybrid sterility factors. Genetics 105:71–86

Wu C-I, Hammer MF (1990) Molecular evolution of ultraselfish genes of meiotic drive systems. In Selander RK, Whittam T, Clark A (eds) Evolution at the Molecular Level. Sinauer, Massachusetts

Wu C-I, Lyttle TW, Wu M-L, Lin G-F (1988) Association between a satellite DNA sequence and the Responder of *Segregation Distorter* in *D. melanogaster*. Cell 54:179–189

Wu C-I, True JR, Johnson N (1989) Fitness reduction associated with the deletion of a satellite DNA array. Nature 341:248–251

Yamamoto AH, Komma DJ, Shaffer CD, Pirotta V, Endow SA (1989) The claret locus in *Drosophila* encodes products required for eye color and for meiotic chromosome segregation. EMBO J 8:3543–3552

Youngson JHAM, Welch HM, Wood RJ (1981) Meiotic drive at the D (M^D) locus and fertility in the mosquito, *Aedes aegypti* (L.). Genetica 54:335–340

Zhang P, Hawley RS (1990) The genetic analysis of distributive segregation in *Drosophila melanogaster*. II. Further genetic analysis of the *nod* locus. Genetics 125:115–127

Zhang P, Knowles BA, Goldstein LSB, Hawley RS (1990) A kinesin-like protein required for distributive chromosome segregation in *Drosophila*. Cell 62: 1053–1062

Zitron AE, Hawley RS (1989) The genetic analysis of distributive segregation in *Drosophila melanogaster*. I. Isolation and characterization of Aberrant X segregation (*Axs*), a mutation defective in chromosome partner choice. Genetics 122:801–821

Zweibel LJ, Cohn VH, Wright DR, Moore GP (1982) Evolution of single-copy DNA and the ADH gene in seven Drosophilids. J Mol Evol 19:62–71

Chapter 11

Strategies and Tools for Extending Molecular Entomology Beyond *Drosophila*

Stephen C. Trowell and Peter D. East

1. Introduction

The future of insect control is widely regarded as depending on biotechnology to ameliorate the twin problems of chemical residues and insecticide resistance. In this context however, biotechnology should not be viewed narrowly as a set of techniques for manipulation of nucleic acids. Two biotechnological approaches to insect pest management, which are becoming increasingly important, lean heavily on protein biochemistry and immunochemistry. The first approach entails isolation of genes from nondrosophilid insects, often as the starting point for creating genetically engineered insecticides. The second approach is exploitation of antibody-based diagnostic kits to assist insect control programs, for example by allowing identification of insect species or estimation of their resistance status.

Regarding the first approach, interest in applying the techniques of molecular biology to insects other than *Drosophila melanogaster* is clearly growing. In many instances interest is motivated by a desire to develop novel insect control measures through the medium of genetically engineered crops, domestic animals or microbial vectors. In some cases the potential practical application of a research project may be a medium- to long-term goal. Nevertheless, isolation and characterization of relevant genes is an extremely powerful way to begin to elucidate fundamental aspects of insect biochemistry. Whatever the purposes of the research, however, moving beyond *D. melanogaster* almost invariably entails the need to isolate and manipulate a gene from an insect about whose genetics and physiology little is known. In this Chapter we will describe biochemical techniques being developed that will reduce or obviate the need to perform a sophisticated genetic analysis before undertaking molecular genetic studies in an insect species.

The second approach to insect control that relies heavily on protein chemistry and immunochemistry falls under the general heading of "identification" or "diagnostic" technologies. A number of entomological problems revolve around the ability to perform rapid and simple species, strain, or entomopathogen identification. Identifications can be performed on the basis of characteristic protein or DNA differences—the benefits and limitations of each differ. As with the first approach, research in this area is directed at insects for which we have little or no genetic information, so many of the methods and tools coincide.

It is our brief in this Chapter to present an overview of these methods and tools. The Chapter is divided into two main Sections: the first (2.) dealing with the problem of isolating a specified target gene from a non-drosophilid insect, and the second (3.) discussing the options for generating and using molecular diagnostic kits to solve identification problems. Protein-specific probes, which may be antibodies or any other highly specific protein-binding molecules (ligands), are central to several of the strategies in both Sections because they are often the best option in the absence of genetically based strategies.

2. Cloning an Insect Gene from a Non-drosophilid Insect

Our aim in this Section is to discuss the problem of cloning the gene for a low abundance, nonenzymic protein from a genetically unexplored insect. The stipulations that the target protein be scarce and nonenzymic relate to the likely interests of molecular entomologists. An essential component of most biological signaling systems is the high affinity binding of a small molecule, or ligand, to a specific receptor. Most current and proposed strategies for insect control rely on using or disrupting this type of interaction. Current strategies rely heavily on chemical disruption of neural signaling (Hutson and Roberts 1985). It is hoped that future biorational control methods, based on genetically engineered insect genes, will allow us to expand our range of targets to include hormonal (e.g., Hammock et al. 1990) and neurohormonal (Menn and Bořkovec 1989) signaling systems. Cloned receptor genes therefore represent a potentially valuable resource for developing novel genetically engineered insecticides (e.g., Trowell et al. 1990; and Section 2.2.2.3 below).

Receptors for insect steroid hormones (Cherbas et al. 1988) and juvenile hormones (Palli et al. 1991) are present at very low concentrations in target tissues. The lack of quantitative data from insects on the density of receptors for peptide hormones is itself evidence of their scarcity. The evidence from mammals indicates that receptors for neurotransmitters and peptide hormones indeed represent a very minor proportion of tissue protein (Haga et al. 1990). Finally, the targets of many current insecticides include various receptors and ion-channels (Hutson and Roberts 1985), which are also present at very low concentrations.

Within the first part of this Section (2.1), we present an overview of the various strategies for cloning genes from an insect which does not have a sophisticated body of genetic research as a background. Two techniques, subtractive mRNA screening (2.1.1) and cloning by homology (2.1.2), are used at the DNA level only and require no knowledge of the properties of the gene's product. The alternative approach, with which we are more concerned, is to start at the protein level and work toward the cognate gene. To do this, one can either use protein-specific probes to probe an "expression library" (2.1.3) or use short segments of synthetic DNA, "oligonucleotide probes," with sequences reverse-engineered from the amino-acid sequence of a region of the target protein (2.1.4) to probe a DNA library.

The importance of protein-specific probes in cloning a gene from the protein end leads us to devote the whole of Section 2.2 to discussion of the types of probes available, the principles underlying them and the ways they may be generated. Antibody probes are dealt with in Section 2.2.1, but discussion of conventionally produced monoclonal and polyclonal antibodies is brief because both are already widely familiar. However, some less familiar variations on these techniques, which are particularly applicable to the problems encountered in molecular entomology, are dealt with in more detail. These include anti-idiotypic antibodies (2.2.1.1), complementary peptides as immunogens (2.2.1.2), prokaryotic antibody expression libraries (2.2.1.3) and differential immunization (2.2.1.4). Specific binding molecules, or ligands, other than antibodies (2.2.2) are discussed in terms of radioactively labeled ligands (2.2.2.1), biotinylated ligands (2.2.2.2) and peptide substitutes for non-peptide ligands (2.2.2.3).

Finally in this Section, we go into some detail on techniques for using protein-specific probes in association with gel electrophoresis to obtain partial amino-acid sequence data from a few tens or hundreds of picomoles of a target protein (2.3.1).

2.1. General Strategies for Cloning Insect Genes

Drosophila genetics allows the cloning of genes by transposon tagging, chromosome walking and a number of other techniques based on isolating the nucleic acid sequence that harbors a characterized mutation (see chapters by ffrench-Constant et al. and Tearle et al. in this volume). In other insects the suite of available techniques is more limited. Among them are isolation of a target gene by differential mRNA screening or by homology of the target gene to one that has already been cloned. The first of these methods is limited to cloning genes that are specific to a particular tissue. It cannot be used with any certainty to clone the gene encoding a defined target protein. The second method is limited to identifying genes that have already been cloned in another species, or very closely related genes.

Because of these limitations, we will devote more space to alternative techniques that can be applied when one wishes to isolate a protein with a particular function but without precedents to guide the isolation. The principal techniques available in these cases are: screening of a protein expression library with a protein-specific probe and screening a DNA library with a degenerate oligonucleotide probe based on some amino-acid sequence data from the protein of interest.

2.1.1. Subtractive or Selective mRNA Screening

Where the protein of interest is a major or predominant product of a particular tissue or life-stage, it is possible to use a differential screening procedure. To do this, cDNA, prepared from a tissue expressing the protein of interest, is screened for fragments that hybridize to mRNA from the same tissue, but not to mRNA from a tissue known not to express the protein. The problem then is to identify the functions of the proteins encoded by tissue-specific cDNAs.

A shotgun version of the method has been used extensively in investigations of proteins specific to the insect compound eye (Hyde et al. 1990; Smith et al. 1990). This was facilitated by the availability of an eyes absent (*eya*) mutant of *D. melanogaster*. Table 11.1 lists some examples of cloning by subtractive mRNA screening.

2.1.2. Cloning by Homology

Progress may be greatly expedited if the gene of interest has already been cloned in another species. A clone or oligonucleotide derived from the first species may be used to screen for the gene in a second species, with the chances of success being higher the more closely the species are related.

The use of a heterologous probe to clone an insect gene by its homology to a vertebrate gene with the same function is well exemplified by the isolation of the *D. melanogaster* R1–6 opsin gene by its homology to bovine cDNA for opsin (O'Tousa et al. 1985; Zuker et al. 1985). In this case the work grew out of years of painstaking work in obtaining the total amino-acid sequences of vertebrate opsins (Hargrave et al. 1980;

Table 11.1. Some examples of insect genes cloned by differential screening with tissue- or stage-specific mRNAs

Gene(s) Cloned†	References
Galleria mellonella two larval hemolymph proteins	Ray et al. (1987)
Dacus oleae larval cuticle proteins	Souliotis and Dimitriadis (1989)
Drosophila melanogaster arrestin homologue	Hyde et al. (1990) Smith et al. (1990)
D. melanogaster cyclosporin A-binding protein	Shieh et al. (1989)

† Used loosely to include isolation of either a genomic clone or a full-length cDNA

Ovchinnikov et al. 1982; Pappin et al. 1984), which meant that the cDNA of the bovine opsin gene was already available.

Obviously, this method cannot be used unless one has information on the gene of interest in at least one other species. Also, use of this strategy is generally limited by the requirement that nucleotide similarities between the probe and the target gene are of the order of 60% or greater. The *D. melanogaster* R1–6 gene above was identified using a bovine opsin cDNA probe on the basis of 36% amino-acid similarity (O'Tousa et al. 1985). Since the degree of similarity between the probe and target genes cannot be known until after the target gene has been cloned and sequenced, it is usual to restrict use of the method to gene families that are known to be highly conserved, or to species that are closely related. Table 11.2 lists a number of cases where the method has been applied to the isolation of genes from *D. melanogaster*, but it is certainly applicable in other insect species.

If the gene of interest belongs to a multi-gene family, it may also be isolated by homology to another member of the family for which a cDNA

Table 11.2. Some examples of insect genes cloned by their homologies to genes previously isolated in other species

Gene(s) Cloned†	References
Drosophila melanogaster FMRFamide prohormone gene	Schneider and Taghert (1988)
D. melanogaster serotonin receptor	Witz et al. (1990)
D. melanogaster kinesin-like protein	McDonald and Goldstein (1990)
D. melanogaster steroid hormone receptor-related proteins	Henrich et al. (1990)
D. melanogaster cecropin genes	Kylsten et al. (1990)
D. melanogaster ocellar opsin	Cowman et al. (1986) Feiler et al. (1988)

† Used loosely to include isolation of either a genomic clone or a full-length cDNA

is available. This technique has been used successfully to clone many members of the steroid and thyroid hormone receptor superfamily. In *D. melanogaster* five genes of unknown function have been isolated by this strategy using their homology to vertebrate steroid hormones (Henrich et al. 1990). A variation on this technique utilizes the polymerase chain reaction (PCR), (Erlich 1989; ffrench-Constant et al. this volume) to clone members of multi-gene families using primers derived from highly conserved sequences (e.g., McDonald and Goldstein 1990).

As alluded to above, a disadvantage of the approach is that sometimes genes with unknown functions (Evans 1988; Henrich et al. 1990) will be isolated, or the function of a gene that has been isolated may initially be misassigned (Feiler et al. 1988). Also, in some cases a number of pseudogenes may be isolated. While this may not be a problem if related functional and pseudogenes are present in a gene cluster (Kylsten et al. 1990), it may prove disastrous in other circumstances. A full discussion of the information needed to verify the identity of a cloned gene is presented by Kimmel (1987).

A more daunting situation arises when attempting to clone a gene with a particular function in the absence of any relevant molecular data for homologous genes. In these cases the principal techniques are either to isolate the gene from an expression library (one capable of producing protein products from the inserted cDNA fragments) using a probe specific for the target protein, or to probe a DNA library with an oligonucleotide designed on the basis of some amino acid sequence from the protein. The two methods are not mutually exclusive and can be used sequentially to increase the probability of obtaining good full length clones for the protein of interest (Hanzlik et al. 1989). Even where only one of the two methods is used to isolate the target gene, it will usually be essential to use the other to confirm the identity of the cloned gene (e.g., Lerro and Prestwich 1990).

2.1.3. Cloning from an Expression Library

In this approach a cDNA library is constructed from a tissue known to express the protein of interest. Ideally this process would be guided by a knowledge of the protein's mRNA transcription profile. Since this is unlikely to be known, intelligent guess-work based on timing of the protein's expression is probably the best that can be done. The Maniatis manual (Sambrook et al. 1989) and protocols supplied by molecular biology companies provide detailed instructions concerning the construction and screening of expression libraries.

The screening process usually relies on some binding property of the target protein, most generally using antibodies to the protein of interest. The types of ligands that can assist with purifying proteins extracted from insect tissues (Section 2.2) are also applicable, in principle, to the screening of expression libraries. Several different detection methods are available, including those based on colorimetric and radiographic visualization, or alternatively, selective enrichment of the target protein.

A variety of different vectors has been developed for expression of foreign DNA sequences. In many instances the choice of vector/expression system will be determined by the characteristics of the ligand to be used for library screening. The robustness of the ligand's interaction with the target protein is an important limiting factor. The two major options are to express a cDNA library in a prokaryote or a eukaryote system.

The prokaryotic system of choice is usually *E. coli*, but one may use either plasmid or bacteriophage vectors. Bacteriophage lambda cloning vectors (Sambrook et al. 1989), such as λ gt11 (and its derivatives) and λ ZAP, offer the advantages of high cloning efficiency and technical simplicity of library screening but place greatest constraint on the type of ligand-target interaction likely to be detectable. Fusion proteins are exposed to harsh conditions when the cells are lysed to liberate the phage, and therefore screening will probably be most successful where the ligand binds a relatively short stretch of contiguous amino acids.

Plasmid expression vectors are relatively less efficient than bacteriophage, but they offer the advantage of comparatively mild lysis conditions and consequently the potential for preservation of complex surfaces for ligand binding. The most commonly used plasmid vectors for construction of cDNA libraries belong to the pUC, pUR and pEX series (Sambrook et al. 1989). These vectors share the features of a multiple-cloning site, replication at high copy number, and a tightly regulated, inducible promoter. The latter is essential to minimize expression of potentially toxic fusion proteins during plasmid amplification.

Plasmid vectors have the advantage that they can efficiently express quite large quantities of fusion proteins. However, they suffer from the disadvantage that making colony replicas for immunological or ligand screening is very labor-intensive. In practice therefore, plasmid vectors may be preferred when screening libraries for moderately abundant messages (for example, up to 50,000 clones to be screened), but for rarer messages where more than 100,000 recombinants must be tested, phage vectors are more commonly used (Kimmel and Berger 1987).

For insect proteins, cultured eukaryote cells and an appropriate eukaryote expression vector offer the greatest likelihood of preserving proteins' native structures. The expression vectors used are often either bacterial plasmids, modified by the addition of eukaryote promoters and polyadenylation signals (e.g., Uhler and McKnight 1987), or eukaryote viruses (e.g., Seed and Aruffo 1987), with baculoviruses being a popular choice (reviewed by Luckow and Summers 1988). Compared to the prokaryote systems above, eukaryote cell expression systems suffer the combined disadvantages of more complicated and expensive culture and screening requirements. However, they probably provide the only realistic option where the target for ligand recognition involves specific post-translational modifications or integration of the protein into a cell membrane (for examples, see Seed and Aruffo 1987; Gearing et al. 1989). This is likely to apply to cell-surface receptors of the type involved in peptide hormone and neurotransmitter signaling.

Expression of cDNA libraries in heterologous systems is subject to a number of constraints. Recognition of a protein by a specific probe can be compromised by poor utilization of promoters and enhancers, low levels of translation, possible truncation of the product or failure to perform any one of a spectrum of possible post-translation modifications normally. Despite the possible problems, these techniques have been successfully applied to *D. melanogaster* and a number of other insects. Table 11.3 details some recent uses of expression cloning techniques in molecular entomology.

2.1.4. Cloning Using an Oligonucleotide Probe

The fourth possible approach is to isolate enough of the protein of interest to obtain some N-terminal and/or internal amino acid sequence data, and Section 2.3 describes some recent developments in electrophoretic techniques particularly suitable for this purpose. The amino acid sequence is used to guide the synthesis of oligonucleotide probes. Subsequently, a cDNA or genomic library can be screened for clones hybridizing to the oligonucleotide probe.

One important technical problem is that knowledge of an amino-acid sequence does not specify a unique nucleotide sequence because of the degeneracy in the third base of many codons. Most species have a characteristic bias in their use of nucleotides in the third position of the codon, and tables of codon bias can therefore be used to reduce the number of alternative versions of an oligonucleotide probe that must be synthesized. For most nondrosophilid insects the codon bias is unknown due to insufficient nucleotide sequence data.

A more sophisticated approach (Gautam et al. 1989) utilizes two or more oligonucleotide probes, derived from some internal amino acid sequence of the protein of interest, as polymerase chain reaction (PCR) primers to amplify a portion of cDNA encoding the protein. The probe amplified by PCR may then be used to screen a cDNA library in order to obtain the full coding sequence for the gene of interest. The method possesses the advantages of employing a probe that does not include any degenerate bases and is longer than an oligonucleotide. Hybridization to the gene of interest can therefore be performed at high stringencies with the prospect of good signal-to-noise ratios.

Table 11.3. Some examples of expression-cloning of insect genes

Gene Cloned[†]	References
Manduca sexta hemolymph juvenile hormone carrier protein	Lerro and Prestwich (1990)
Drosophila melanogaster neuroglian	Bieber et al. (1989)
Musca domestica cytochrome P-450	Feyereisen et al. (1989)
Heliothis virescens juvenile hormone esterase	Hanzlik et al. (1989)
D. melanogaster arrestin homolog	Yamada et al. (1990)
Bombyx mori prothoracicotropic hormone	Kawakami et al. (1990)

[†] Used loosely to include isolation of either a genomic clone or a full-length cDNA

In fact, the properties of PCR amplification are especially relevant for entomological problems where small size and difficulty in isolating adequate quantities of individual tissues have been major impediments to the preparation of cDNA. Table 11.4 provides examples in which PCR amplification has been used to overcome problems of isolating rare cDNAs or to generate species-specific probes starting with heterologous, degenerate primers.

Technically, screening a cDNA library with nucleic acid probes is usually easier than screening expression libraries with antibodies or other ligands. In the past the requirement to purify sufficient protein for amino acid sequencing before screening with an oligonucleotide may commence has been a major obstacle. However, the increasing sophistication of the technologies available for small scale purification of scarce proteins, especially utilizing SDS-PAGE and electroblotting, are making this an increasingly attractive strategy. Oligonucleotide screening has already been used quite extensively to isolate insect genes (Table 11.5 shows some examples), and its use will probably increase.

Table 11.4. Some examples of PCR technology with potential application in molecular entomology

Application	References
Subtractive cNDA cloning	Timblin et al. (1990)
cDNA library construction from limited amounts of total RNA	Belyavsky et al. (1989) Domec et al. (1990)
Amplification of gene-specific cDNA and genomic DNA fragments with heterologous primers	Kamb et al. (1989) Váradi et al. (1989)
Construction of cDNA probes using degenerate oligonucleotide primers based upon amino acid sequence information	Gautam et al. (1989)

[†] Used loosely to include isolation of either a genomic clone or a full-length cDNA

Table 11.5. Some examples of insect genes cloned after partial amino-acid sequencing of the protein the encode and construction of synthetic oligonucleotides

Gene Cloned[†]	References
Drosophila melanogaster esterase 6	Oakeshott et al. (1987)
Manduca sexta apolipophorin III	Cole et al. (1987)
Heliothis virescens juvenile hormone esterase	Hanzlik et al. (1989)
M. sexta eclosion hormone gene	Horodyski et al. (1989)
Bombyx mori bombyxin	Adachi et al. (1989)

[†] Used loosely to include isolation of either a genomic clone or a full-length cDNA

For non-enzymic proteins, a high affinity ligand or an antibody specific for the protein of interest is a crucial requirement in order to monitor the protein purification. Even if the protein of interest exhibits an enzymic activity that is readily assayed, a specific antibody can greatly facilitate the final stages of a purification, especially where these involve polyacrylamide gel electrophoresis.

From the above, it should be clear that unless one's gene of interest, or a closely related gene, has already been cloned in another organism, a specific ligand or antibody is an important or essential requirement. Even where the target protein displays enzymic activity, acetylcholinesterase for example, specific probes are far from superfluous since they open up avenues such as in situ localization, screening of expression libraries and affinity purification (ffrench-Constant et al. this volume). The rest of this Section will therefore consider the types of protein-specific probes that are available, how they may be generated, the benefits and drawbacks of each, and some ways in which they may be used.

2.2. Generating Protein-Specific Probes

The range of commercially produced antibodies and radiolabeled or biotinylated ligands available for research on insects is very limited. One or other of these probes is an essential requirement for the cloning strategy outlined in Section 2.1.3 and is highly desirable for the protein-purification phase of an oligonucleotide screening strategy (Section 2.1.4). This Section therefore describes a number of strategies for the molecular entomologist to generate a protein-specific probe appropriate to his or her own research problem.

2.2.1. Antibody Probes

Antibodies are some of the more versatile and useful probes available. A good antibody or antiserum to a protein of interest will, in principle, allow the researcher to identify a receptor, localize it in situ, purify it and clone the cognate gene.

Polyclonal antibodies are often obtained by immunization of rabbits. Ideally, they require the isolation of more than a hundred micrograms of highly purified antigen to allow several immunizations of a pair of rabbits. They have the advantages that they are cheap, fast and easy to prepare, and tend to be robust and of high affinity. Their major disadvantages are their relative lack of specificity, including the danger of reaction against contaminants in the immunizing antigen. The antibody response may also vary significantly from rabbit to rabbit and with successive bleeds. The reader is referred to Harlow and Lane (1988) for a comprehensive treatment of polyclonal antibody production.

The alternative to raising polyclonal antisera is to generate monoclonal antibodies (MAbs). This involves cloning and immortalizing the antibody-producing cells of a mouse in cell culture. Lymphoblasts, which are dividing progenitors of the antibody-secreting cells (or lymphocytes), are fused with tumor cells of the lymphocyte lineage (myeloma cells), usually by means of a chemical agent. Some of the progeny of the fusion (hybridomas) take on the immortality of their myeloma parent, as well as the antibody-secreting properties and specificity of their single lymphoblast parent. Stable hybridoma cell lines can be grown indefinitely in cell culture or frozen and stored in liquid nitrogen, and each line secretes a single class of antibodies of fixed specificity.

The time and cost involved in generating MAbs is significant, and some may be of low affinity or be very sensitive to conditions under which they are used. However, often MAbs are characterized by very high specificity and a good MAb is, in principle, available in perpetuity. Even though impure immunizing antigen may be used, the chances of successfully obtaining antibodies of the desired specificity increases with increasing antigen purity. It is ideal if relatively pure antigen is available in 10–100 μg quantities. Where sufficient antigen is available, standard procedures for obtaining antibodies may be followed with confidence (Goding 1983; Campbell 1984; Harlow and Lane 1988). In future, the emerging technology of expressing the variable domains of antibodies in bacterial expression libraries may make generating antibody probes a routine procedure in any molecular biology laboratory (2.2.1.3).

Despite these emerging technologies, the scarcity of receptors will always make it difficult to raise antibodies against them. It is often extraordinarily difficult to obtain enough purified receptor to immunize an animal or screen an expression library. As an alternative, however, it is frequently possible to obtain milligram quantities of an unlabeled ligand that binds to the protein, whether it is a peptide or any other small organic molecule. If this is the case, an anti-idiotypic approach may be contemplated, as outlined below.

2.2.1.1. Anti-idiotypic Antibodies. This technique currently involves generating MAbs and entails significant investment in equipment, consumables and manpower. Also, some proficiency in synthetic chemistry will often be required. Nevertheless, anti-idiotypic antibodies have found growing acceptance in recent years, since they have been used successfully to obtain antibodies to a number of mammalian steroid hormone receptors (e.g., Cayanis et al. 1986; Lombes et al. 1989), as well as some cell-surface receptors, including the acetylcholine receptor (Cleveland et al. 1983), putative vasopressin receptors (Knigge et al. 1988), the substance-P receptor (Couraud et al. 1990) and opioid receptors (Coscia et al. 1991). (For more comprehensive lists, see Schick and Kennedy 1989; Couraud and Strosberg 1991.)

Despite the fact that, initially, anti-idiotypic antibodies were generated against receptors that had already been well characterized, the method is also applicable in the absence of any molecular information concerning the receptor. Examples include the identification of receptors responsible for the targeting and retention of proteins within specific subcellular compartments, such as endoplasmic reticulum (Vaux et al. 1990) and mitochondria (Pain et al. 1990). To date, there are no reports of the technique being used for molecular studies of insects. However, it may prove to be extremely useful in cases where unlabeled ligand (e.g., insecticide, hormone, neurotransmitter) can be obtained in milligram quantities and linked to a carrier protein.

The method relies on the network of interacting antibodies that arise in the course of the mammal's normal humoral response to an antigen. The ramifications of the network are extensive and complex (Köhler et al. 1989), but it appears that the technique can be surprisingly simple to use, at least for laboratories generating MAbs routinely.

The first step in generating anti-idiotypic antibodies is to link multiple copies of the ligand covalently to a carrier protein. This process, termed haptenization, is essential to obtain a good immune response against a small molecule. When a mammal is immunized with a haptenized hormone, for example (Figure 11.1), the initial class of antibodies that arises (the Ab1 class) includes some that share topographical features with the hormone-binding region of the receptor. This is not surprising since both the receptor's hormone binding site and the antibody's antigen binding sites are constrained by the shape of the hormone. The shared topography between parts of the receptor and the Ab1 means that the Ab1 can act as a surrogate receptor.

A second class of antibodies (the Ab2 class) arises as a normal autoimmune response to the variable regions of the Ab1 (its idiotypic determinant, hence "anti-idiotypic" antibody). Among this second class are some antibodies whose antigen-binding site is the region of the Ab1 that mimics the receptor's hormone binding site. This subset of the Ab2

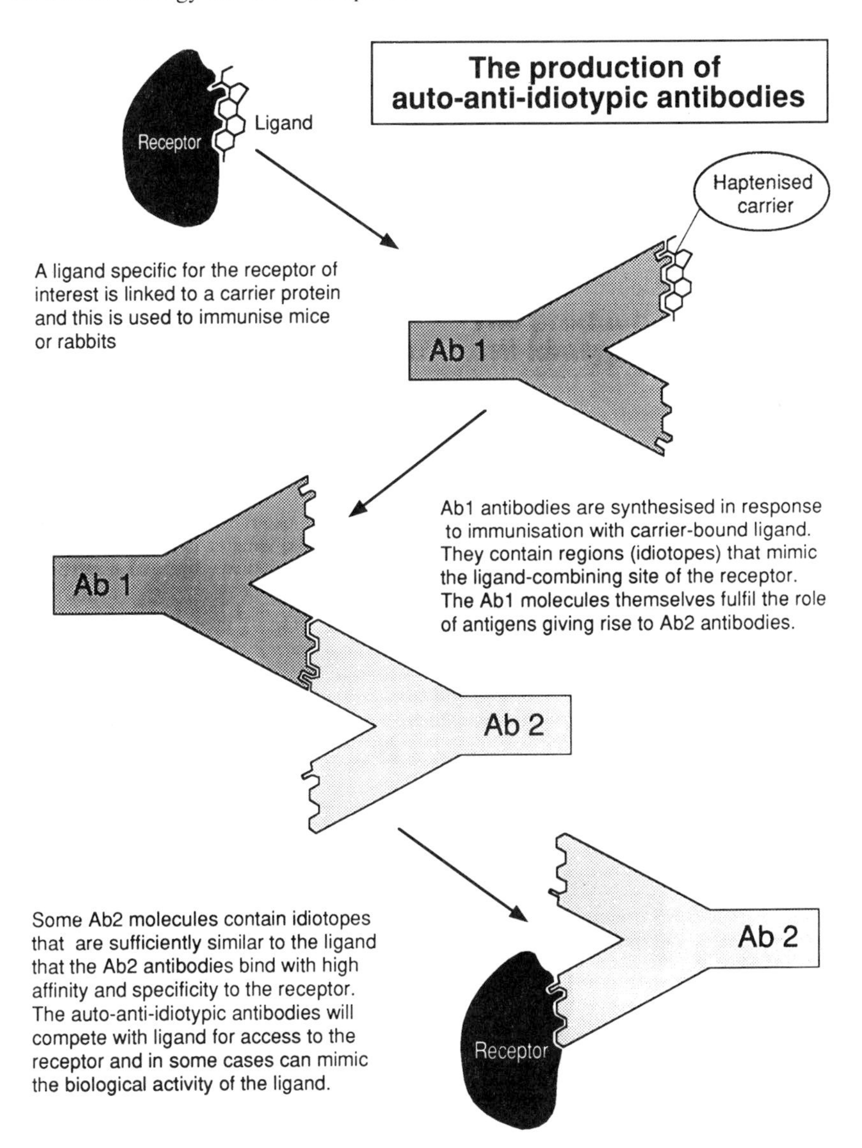

Figure 11.1. Making anti-receptor antibodies when the receptor is not available. A ligand specific for the receptor is conjugated to a carrier protein (a process known as haptenization), and the conjugate is used for immunization in place of the receptor.

class cross-reacts with the hormone's receptor and will generally compete with hormone for access to its binding site. These Ab2 antibodies are powerful research tools.

Traditionally, the route to obtaining anti-idiotypic antibodies involved generating and purifying the Ab1, followed by reimmunization of a second animal with the Ab1 in order to obtain Ab2. However, in several cases this has been shown not to be necessary (Cleveland and Erlanger 1986), because, as stated above, the production of both Ab1 and Ab2 is a normal part of the mammalian immune response. For example, Ab2-secreting clones (called "auto-anti-idiotypic" because they are produced in response to the mouse's own Ab1) have been obtained at approximately half the frequency of Ab1 clones after immunization with a haptenized ligand (Cleveland et al. 1983), and this does not seem to be atypical of what can be achieved.

To obtain anti-receptor antibodies using this technique, the most demanding requirement is a convenient assay with which to identify the anti-idiotypic antibodies. In the absence of a pure receptor preparation, the most reliable way to screen for monoclonal anti-idiotypic antibodies involves two steps. The first step is to generate an Ab1 in a species other than a mouse. The second step is to purify this antibody and use it as a surrogate receptor against which to screen mouse monoclonal antibodies using an enzyme-linked immunosorbent assay (ELISA) or a related assay (e.g., Cayanis et al. 1986; Cleveland and Erlanger 1986; Couraud et al. 1990). The reason for using both rabbits and mice is that, in practise, regions in the variable region of Ab1 that are tightly constrained by the immunizing ligand tend to be conserved among different species, whereas idiosyncratic regions of Ab1 that are not constrained by the ligand are not conserved. Therefore an Ab1, generated in a rabbit and purified by ligand-affinity chromatography, is a much better choice for identifying murine Ab2 monoclonals, raised in response to the same ligand, than an Ab1 from a mouse.

The two-species strategy described above is not necessary if purified receptors are available for screening (e.g., Cleveland et al. 1983), but this is unlikely to be the case for any insect receptor. Conceivably, unpurified preparations containing the receptors of interest could be used in an ELISA assay, but such a strategy is only likely to succeed if the receptor is unusually abundant and we are unaware of any examples. However, indirect immunofluorescence microscopy has been used to screen for anti-receptor antibodies in a case where the subcellular distribution of the receptor could be inferred (Vaux et al. 1990). An alternative class of screening assays involves assaying for competition between anti-idiotypic antibody and natural ligand in a ligand-receptor radiobinding assay (Bennett and Yamamura 1985) or a functional assay of ligand-receptor binding (e.g., Pain et al. 1990).

It is generally inadequate to use the cloning of a gene through an anti-idiotypic antibody as the sole proof of the identity of that gene (Meyer 1990). Indeed, this is a general comment that is relevant to all of the cloning strategies discussed (Kimmel 1987), but it is a particularly pertinent when a novel receptor has been identified using the anti-idiotypic approach.

2.2.1.2. Complementary Peptides as Immunogens. Use of complementary peptides is a technique conceptually related to the auto-anti-idiotypic approach for making antireceptor antibodies. It rests on the surprising observation that for any given positive strand of mRNA encoding a peptide "A," the negative strand of mRNA appears to encode a peptide "B" that is complementary to A (Bost et al. 1985). The complementarity is manifested in specific binding of peptide A to its negative strand-encoded counterpart B. The interaction closely mimics binding of A to its protein receptor. The implication is that a complementary sequence of amino acids is constrained by some unexplained set of rules to fit its positive strand peptide, just as the combining site of an Ab1 (Figure 11.1) is constrained topologically to fit the haptenized ligand that gave rise to it. The complementary peptide is therefore an alternative to an Ab1 as a receptor substitute for the generation of anti-receptor antibodies.

This property has been used in a number of instances to generate antibodies to peptide hormone receptors and other protein binding molecules. A complementary peptide, synthesized according to the sequence coded by negative strand mRNA for a hormone of interest, is used as the immunogen to produce anti-receptor antibodies. It should be emphasized that the technique is controversial (Goldstein and Brutlag 1989; Blalock 1990), partly because of the lack of a convincing theoretical basis for the complementarity and also because a number of reports of negative results have appeared (reviewed in Blalock 1990). On the other hand, the technique appears to have been used successfully on a number of occasions to raise antibodies to peptide combining sites (see Table 11.6).

Should this approach be chosen, the first step would be to synthesize the complementary peptide of interest and then link it to a carrier protein. It would therefore be wise to include a convenient residue for the coupling procedure, a lysine for example, if there is none present in the sequence of the complementary peptide. Any such additions should obviously be placed well away from regions of the parent peptide that are required for biological activity.

As a primary screen for antibodies, the immunizing peptide, ideally linked to a different protein from that used in the immunization, can be used as the antigen in any appropriate immunoassay. However, it is also crucial for efficient and critical evaluation of the method to have

Table 11.6. Examples of use of the complementary peptide approach to generate antibodies to peptide receptors

Peptide of Interest	References
Bovine adrenocorticotropic hormone (ACTH, corticotropin)	Bost et al. (1985)
Human luteinising hormone releasing hormone (LHRH)	Mulchahey et al. (1986)
Human fibronectin	Brentani et al. (1988) Pasqualini et al. (1989)
Rat angiotensin II	Elton et al. (1988)
Bovine γ-endorphin	Carr et al. (1986)

a secondary screening assay to identify antibodies which bind to the target receptor. The screening assay would ideally be based on biological activity of the antibodies. Alternatively, specific competition with the ligand for binding to its receptor could be tested in a ligand-binding assay (e.g., Bennett and Yamamura 1985).

2.2.1.3. Prokaryotic Antibody Expression Libraries. As indicated at the beginning of this Section, obtaining antibodies of the required specificity and properties can be a time-consuming and expensive process. In the absence of a workable alternative, the auto-anti-idiotypic route may provide the researcher with antibodies for the purification and characterization of receptors or for expression cloning of a target gene, even though the effort may be considerable. Nevertheless, a faster and cheaper route to obtaining antibodies of the required specificity would be highly desirable. Two solutions can be envisaged. The first aims to increase the number of independent clones that can be sampled from a conventional fusion of spleen and myeloma cells to produce MAbs. The second and more radical solution is to eliminate immunization with purified receptors and instead screen a preexisting antibody library, which contains a range of specificities, for receptor-specific antibodies.

The first of these solutions involves membrane screening of monoclonals, based on techniques developed for screening bacterial expression libraries (Gherardi et al. 1990). In this technique, the hybridoma products of a fusion between antibody-secreting lymphoblasts and immortal myeloma cells are plated out on agar in dishes, rather than in the wells of tissue culture plates. A nitrocellulose membrane is coated with the antigen of interest and placed over the clones so that any secreted antibodies can react with the antigen. The membrane is then processed to detect the presence of any specific antibodies. The advantages of the method are that manipulation of the hybridomas and their supernatants is almost eliminated, while a much larger number of clones and specificities may be examined than is possible conventionally.

In the long run the second solution, expression of functional antibody fragments in bacterial libraries, seems to have more potential to increase the number of antibody specificities that may be screened at one time. The principle underlying this new method is that while the antigen-binding regions of antibodies are highly variable, with millions of different sequences represented in the mature immune system, other regions of the antibodies fall into only a few subtypes. Using PCR primers based on the invariant regions, it is therefore possible to amplify a family of DNA molecules that encode a complete range of antibody specificities. When cloned into a prokaryote expression system, these molecules form the basis of an antibody library. The library can be screened, using standard techniques, for antibodies of the desired specificity.

The technique may suffer from some limitations. Natural antibodies are actually complex proteins with four or more polypeptide subunits, which are of two different classes, called heavy and light chains. Antibody libraries only contain cut-down versions of natural antibodies. Furthermore, the natural associations of particular heavy and light chains are lost in the process of making a library. The same range of specificities cannot therefore be expected from this technique as those obtained *in vivo*. In particular, it is not clear whether bacterial expression libraries will consistently provide very high affinity variable regions comparable to those generated *in vivo* by somatic mutation after antigen stimulation (Winter and Milstein 1991).

However, it remains to be seen whether these will be serious drawbacks in practice. The cut-down antibodies should have fewer problems of access to binding sites than their more bulky natural counterparts, and recombination in a library can generate a vast range of composite binding sites formed by novel associations of heavy and light chains.

In one study (Huse et al. 1989), a library of Fab (antigen-binding fragment, the monomeric antigen-binding region of an antibody) fragments, containing randomly assorted heavy and light chains, was cloned from an immunized mouse. The library was expressed in a bacteriophage vector and screened for the specificity of interest. Specific Fabs were obtained which had very high affinities for the test hapten (represented by dissociation constants in the nanomolar range). Using this method, a large number of independent clones (10^6 to 10^7 per day) were screened, which implies that a highly diverse range of antibody specificities can be made available. Commercial kits based on this method (Alting-Mees et al. 1990) are available for researchers to make and screen their own mouse or human antibody libraries.

Using PCR it is also possible to make antibody libraries that express smaller fragments of antibodies than in the example cited above, which nevertheless retain significant binding capacity. One approach involves the production of a library of individual variable heavy chain genes (dAbs) in *E. coli*. Antibodies of this type have also been isolated with high affinities (in the 20 nanomolar range; Ward et al. 1989). In this case, immunization of the donor mouse was found to be beneficial but not essential. The reader is referred to some recent reviews (Hudson 1990; Plückthun 1990; Winter and Milstein 1991) on antibody library construction and related techniques for more information.

Despite all the potential advantages of antibody libraries, there remains the problem that some form of receptor preparation, preferably relatively pure, is still required for screening the library. A combination of the antibody library and the anti-idiotypic approaches might be used to overcome this constraint in a two-step process. The first step would be to probe the library with a radiolabeled ligand. Ligand-specific antibodies isolated from the first screen would themselves be used to probe the whole library again, this time for anti-idiotypic antibodies capable of binding to receptors specific for the ligand of interest.

2.2.1.4. Differential Immunization. Differential immunization is a technique that may prove useful for expediting antibody production where pure antigen is not available and no specific ligands are known. The principle of the technique involves generating an immune response by immunizing with unwanted antigens, followed by destruction of the growing lymphoblasts responsible for the response, using a dose of 40–65 mg/kg cyclophosphamide. Subsequently, a second combination of antigens is injected containing, in addition to the unwanted antigens, the antigen of interest. The immunized animal now contains only lymphoblasts capable of responding to the antigen of interest, and the proportion of specific antibodies is increased as a consequence. The technique can be employed when one has two sets of impure antigens available that differ only in the target antigen. An example would be where a null strain exists for the protein of interest.

This technique was first used to select antibodies to biologically active determinants of a heparan sulfate proteoglycan (Matthew and Patterson 1983). Using conventional immunization techniques, only antibodies that bound to biologically inactive regions of heparan sulfate were recovered. (This phenomenon, where antibodies to more potent antigenic regions dominate the immune response to the exclusion of others, is termed "immunodominance"). Cyclophosphamide was therefore used to suppress the immunodominant response and unmask the weaker, biologically interesting antigens. Subsequently, the method was successfully used to select antibodies to particular tubulin isotypes (Lewis et al. 1987).

Our laboratory is currently employing the technique to raise species-specific monoclonal antibodies to two closely related *Helicoverpa* species (see Section 3.). The strategy is to immunize, with a relatively crude mixture of proteins from one species, and eliminate the ensuing immune response with cyclophosphamide. Immunization with proteins from the second species engenders a much weaker immune response than normal, but it is dominated by antigens that are unique to the second species. This use of the differential immunization technique involves a much wider range of antigens in the immunizing mixture than those previously published. In this case, we have found that two rounds of immunization and cyclophosphamide treatment are required to achieve complete immunosuppression before immunization with the group of immunogens containing the antigen of interest can begin.

Use of cyclophosphamide is not the only way in which tolerization against unwanted antigens may be achieved. Indeed, classical tolerization with small but increasing amounts of unwanted immunogen is a way to achieve this. However, it is unlikely to be successful where a mixture of unwanted antigens is present in widely varying concentrations. A novel possibility, so far untried outside the strictly biomedical area, involves blocking the action of helper T-lymphocytes at the time of primary immunization. Monoclonal antibodies directed against the $L3T4^{+}$ bearing class of T-lymphocytes are injected at the time of immunization with the unwanted antigens. This method was shown to tolerize the humoral response (Benjamin and Waldmann 1986; Gutstein et al. 1986; Gutstein and Wofsy 1986) and has been proposed as a tolerizing technique with general applicability (Benjamin and Waldmann 1986).

An equivalent effect to the use of *in vivo* tolerizing treatments is possible at the stage when lymphocytes have been removed from the mouse preparatory to generating hybridomas. In this case the mouse will have been immunized with the complete mixture of antigens, including the desired target protein. Prior to the fusion, the spleen is removed and unwanted antibody specificities removed by incubating a suspension of spleen cells with magnetic beads that have been coated with the unwanted antigen mixture only. This has the effect of removing inappropriate lymphocyte specificities. Beads suitable for this purpose are now available from a number of commercial suppliers (Amersham, Dynal and Promega to name only three), and some reports of their use in protocols analogous to the one suggested have been published (Horton et al. 1989; Janssen and Rios 1989).

2.2.2. Ligands

The major types of ligands include hormones, peptide neuromodulators, enzyme cofactors, inhibitors and transition state analogs of enzymes. To be useful as tools for identification they must carry some reporter function. The most common form of reporter is a radioisotopic label, but over recent years nonradioactively labeled compounds, especially those exploiting the biotin-avidin reaction, have gained increasing usage.

2.2.2.1. Radioactively Labeled Ligands. Radiolabeled ligands may be divided into two classes. The first includes simple radioisotopically labeled versions of the natural ligand, such as tritiated juvenile hormone III (JHIII). Chemically these are identical to the natural molecule. This class also includes radiolabeled analogs that, while not occurring in insects, bind in an identical manner to the *in vivo* ligand. Tritiated ponasterone A, which binds to the same receptors and in the same fashion as 20-hydroxyecdysone, is a good example. In the second class are radiolabeled photoaffinity analogs of the natural ligands. These molecules are generally related to natural ligands but have been chemically modified to include a group transformed to a highly reactive state by absorption of ultraviolet light. They are therefore suitable for covalent labeling of their respective binding proteins and receptors.

Radiolabeled ligands for receptor research must be labeled at a far higher specific activity than for metabolic or radiolytic enzyme assays, because receptor molecules are generally present in tissues in vanishingly small amounts. This constraint, together with the need for a relatively simple labeling method, effectively limits the choice of isotopes to tritium (approx. 29 Ci per milligram atom) and 125-iodine (2176 Ci per milligram atom). The costs of commercial labeling of compounds by tritium exchange may run into several thousands of dollars and generally cannot be used for unsaturated molecules or those carrying reactive groups for fear of destroying biological activity. The method is therefore beyond the reach of most researchers.

However [^{125}I]iodination of polypeptide hormones (Walker and Burgess 1985; Soutar and Wade 1990) is a simple technique that is possible in most biochemical laboratories with access to suitable facilities. In theory, similar radioiodination techniques are also applicable to small ligands carrying hydroxylated aromatic rings, although the likelihood of disrupting biological activity is probably inversely related to the size of the ligand.

In practice, it is more usual to generate tritiated or radioiodinated ligand analogs using conventional methods of chemical synthesis. Juvenile hormones and their analogs (Prestwich and Wawrzeńczyk 1985; Prestwich et al. 1987a; 1987b; 1988; Boehm and Prestwich 1988a; 1988b; Eng and Prestwich 1988) and ecdysone (Cherbas et al. 1988) have been labeled to high specific activity in this manner. Recently a radioiodinated juvenile hormone analog has been used to identify and purify the JH-binding protein from the hemolymph of *Manduca sexta* (Kulcsár and Prestwich 1988). However, nonradioactive methods such as biotin-labeling can also be considered for small ligands (see Section 2.2.2.2 below) and, where applicable, will probably gain growing acceptance.

A wide variety of reagents is available for the introduction of photolabels into peptides and proteins (Eberle and de Graan 1985; Prestwich

1991). For other ligands, generating suitable photoaffinity analogs is beyond the capabilities of most biologists. However, excellent work has followed when the time and trouble has been invested to synthesize them, for example, in the case of the photoaffinity analogs of the juvenile hormones (Prestwich et al. 1982; 1984; 1985; 1987a; Koeppe et al. 1984; Stupp and Peter 1984; 1987).

2.2.2.2. Non-radioactively Labeled Ligands. Biotin-labeling is a convenient and versatile alternative to radioisotopic labeling of compounds. A biotin label represents an extremely effective reporter group because a number of detecting agents based on avidin or streptavidin are available commercially. Each molecule of avidin carries four binding sites for biotin, each of which has a dissociation constant of approximately 10^{-15} M. Biotin-labeling reactions may be relatively easily performed in any laboratory.

The detection systems that can be used with biotin may be as simple as a colorigenic enzyme (horseradish peroxidase or alkaline phosphatase) linked directly to avidin. However, for enhanced sensitivity, additional steps can be interposed, so that each biotinylated ligand gives rise to a tree of reporter molecules. The range of reporter and effector molecules available allows the biotin-avidin/streptavidin system to be used in immunocytochemistry, immunoblotting, ELISA, affinity chromatography and, in the case of cell-surface targeted ligands, affinity panning (i.e., selective enrichment of cells expressing a cloned gene from a large population of nonexpressing cells, Harada et al. 1990).

Some mammalian hormones and drugs linked to biotin are available commercially. For example, Sigma offers dexamethasone, testosterone, propranalol and insulin linked to biotin. While these particular compounds may be irrelevant for purification of insect hormone receptors, the methodologies used to generate them (Atlas et al. 1978; Manz et al. 1983) are equally applicable to insect studies.

A number of commercial suppliers (Boehringer-Mannheim, Calbiochem, ICN, Pierce, Sigma) also provide biotin linked to various reactive groups for conjugation to molecules containing primary amines, thiols, aldehydes and alcohols, as well as nucleic acids. Spacer arms minimize the likelihood of steric hindrance interfering with the recognition of either the ligand by its receptor or biotin by avidin/streptavidin.

Biological activity is likely to be lost if a significant amount of the ligand couples to biotin through a site that participates in receptor binding. This becomes an increasingly serious consideration as the size of the ligand decreases. Small non-peptide ligands present the greatest challenge because the potential problems of steric hindrance may be compounded by the absence of convenient coupling groups. Some skill in chemical synthesis, often beyond the molecular biologist, may be required to incorporate the biotin group.

Nevertheless, the power of the technique is illustrated by a recent report (Redeuilh et al. 1990) of the biotinylation of oestradiol and use of the resulting probe to affinity purify oestrogen receptors from calf uterine tissue 1590-fold, with a yield of 25%. Such a probe would probably be an equally effective tool for detecting oestrogen receptors on protein blots or even cytochemically within tissue sections.

2.2.2.3. Substitution of Peptides for Non-peptide Biologically-active Ligands. A third option for generating protein-specific probes is based on the recent development of very large random peptide libraries that may be screened for molecules with a particular binding specificity (Cwirla et al. 1990; Devlin et al. 1990; Scott and Smith 1990). Functionally, the products of this process are equivalent to the combining sites of anti-idiotypic antibodies (2.2.1.1) or antibodies raised against a peptide complementary to a natural hormone (2.2.1.2), but they are generated in a shotgun procedure, akin to the way that the antibody libraries discussed in Section 2.2.1.3 are generated.

The basis of the method is the chemical synthesis of a partly randomized mixture of oligonucleotides. In the published accounts, complex mixtures of oligonucleotides, encoding all the possible sequence permutations of a six amino acid peptide (Cwirla et al. 1990; Scott and Smith 1990) or a 15 amino acid peptide (Devlin et al. 1990) were synthesized. The oligonucleotide mixtures were cloned into bacteriophage, and a library of between 2×10^7 and 3×10^8 phage was screened using two or more selective enrichment steps. Target specificities were peptides that would bind to streptavidin or to anti-myohaemerythrin or anti-β-endorphin antibodies. In all cases, peptides with the required specificity were recovered, the highest published dissociation constant being $0.35\,\mu M$ (Cwirla et al. 1990), representing moderately high affinity.

While this is a very exciting new technology, with the potential to replace non-peptide ligands with high affinity peptide substitutes (Devlin et al. 1990) by itself, it still requires bulk purification or cloning and expression of a receptor gene for use as a peptide screening agent.

As a refinement, we wish to propose that it may be conveniently combined with a preliminary step in which antibodies are raised against the ligand of interest, whether it be a steroid or isoprenoid insect hormone, an insecticide, a pheromone, or any other biologically active molecule. The antibodies would then be used as a receptor substitute (cf. Cwirla et al. 1990; Scott and Smith 1990) to screen a peptide library for peptides of the required specificity. The DNA inserts from phage that react positively would be excised and sequenced. Following this, the peptides they encode could be synthesized chemically and compared with natural ligands in assays of their biological activity and in radiobinding assays. Those peptides with sufficiently high affinity can be used as affinity purification reagents or be labeled with biotin or radioisotopes for use in isolating and cloning the receptor gene. Such an approach would be valid even where a radiolabeled natural ligand could not be obtained and would require only the ability to haptenize a suitable specific ligand for initial antibody production.

In the long term, the greatest potential of this approach is the promise it holds for the generation of short synthetic genes encoding peptides that are agonists or antagonists for almost any insect receptor (certainly any for which a natural or synthetic ligand is available) without the need ever to purify the receptor itself or clone its gene. Thus the technology could be used to mimic non-peptide hormones, insecticides, toxins, pheromones or indeed any other class of biologically potent molecules that bind to protein receptors.

Furthermore, the synthetic genes will be short sequences that require no post-translational modification for activity. They may perhaps be designed in tandem arrays with proteinase recognition sites interspersed between them or even be endowed with recognition sites allowing for their uptake by particular cell types. Such peptide genes seem to us to offer significant potential as genetically engineered insect control agents, whether for dissemination via entomopathogenic viruses or insertion into genetically manipulated crops or domestic animals. They may be more attractive than full length natural genes for these purposes.

In the case of non-peptide hormones produced by multi-enzyme pathways, peptides that mimic the natural product would have enormous advantages. They might bypass the need to purify receptors or other proteins involved in transduction and may expedite the whole process of developing novel genetically engineered insect control agents. Such an approach would offer the opportunity to revisit whole classes of insecticidal agents in genetically engineered form without the problems of environmental persistence associated with the original agents. One might also predict that peptide mimics of non-peptide insecticides, while they may be susceptible to the same target site resistance mechanisms as the original agents, will be immune to pre-existing penetration and metabolic types of resistance.

2.3. Use of Protein-Specific Probes in Isolating a Gene

2.3.1. Protein Purification

As indicated above, two relevant areas in which protein-specific probes may be required are identification of a protein during its purification from tissue extracts or in screening a cDNA expression library. Of course, protein-specific ligands may also be immobilized and used in affinity chromatography for purifying proteins, but this is a discrete field in its own right. Interested readers are referred to some general references on affinity chromatography (Turkova 1978; Pharmacia 1979; Dean et al. 1985).

Almost any probe that is specific for a protein of interest may be used to detect it in some adaptation of a standard ligand-binding assay or ELISA. Such a ligand could therefore be used in test-tube- or microplate-based assays. Equally, antibodies and radioactive or biotin-labeled ligands can be used in cytochemical studies of whole tissues or tissue-sections. However, when attempting to purify microgram quantities of protein for amino acid sequence determination, electrophoresis holds a special place.

Isoelectric focusing (denaturing or nondenaturing) and native or SDS-polyacrylamide gel electrophoresis (PAGE), either individually or combined as a two dimensional system, are routinely used as high resolution analytical techniques. They are also quite capable of yielding microgram quantities of a target protein of the purity required for automated sequence analysis if used in combination with electroblotting. Thus electrophoresis is often applied as the final step in a purification, and its high resolving power reduces the number of other purification steps that may be required. Electrotransfer of the proteins onto membranes has been performed using glass fiber sheets under acidic conditions (Aebersold et al. 1986) or modified glass fiber or polyvinylidene difluoride (PVDF) under mildly alkaline conditions (Matsudaira 1987; Aebersold and Leavitt 1990). Polybrene coating of the glass fiber is one of the easier and more common modifications used (Vandekerckhove et al. 1985; Bergman and Jörnvall 1987; Bergman and Jörnvall 1990). Either glass fiber or PVDF may be submitted directly to the cup of a sequenator because they are sufficiently inert to withstand the chemical conditions required during sequencing. A polypropylene membrane has also recently been made available commercially (Schleicher and Schuell) for the same purpose.

Unfortunately, a high proportion of proteins are found to be N-terminally blocked after purification and sequencing by these methods. Artifactual N-terminal blocking may arise at several steps during the electrophoresis, and a number of precautions may be introduced to minimize this possibility (Moos et al. 1988; Choli and Wittmann-Liebold 1990). Nevertheless, there remain an unknown proportion of proteins that may carry N-terminal modifications in vivo. In these cases, and where some internal amino acid sequence is desirable (e.g., for the generation of more than one oligonucleotide sequence), peptides may be generated from the purified protein (Carrey 1990; Kawasaki et al. 1990) and separated by high performance liquid chromatography (HPLC) (Ward et al. 1990) or another electrophoresis step (Huang and Matthews

1990; Promega Probe-Design^{TM} kit). The main drawback is that extra losses are incurred with each additional processing step. At least one company, FMC Corporation, supplies an agarose gel matrix that, it is claimed, can be substituted directly for acrylamide with no loss of resolution. It has the advantage that proteins may be extracted with near 100% yield for sequencing or further processing. Gel matrices of this type may help overcome some of the problems of low recovery traditionally experienced when eluting proteins or peptides from polyacrylamide gel electrophoresis.

It is a major advantage to be able to detect the target protein unequivocally on a membrane after electroblotting. Ideally, the labeled protein band would then be cut out and sequenced. Two-dimensional gels using SDS have the most resolving power but suffer from the major disadvantage that many ligand and antibody binding reactions are disrupted by the denaturation of the protein (Tovey 1990; Beisiegel 1986). In the case of photoaffinity ligands that have been covalently linked to the target protein prior to electrophoresis, denaturation is not an obstacle, but for other ligands it may be a serious one.

A few proteins survive SDS-electrophoresis and electroblotting with their ligand-binding properties intact. An example is the low density lipoprotein receptor (Daniel et al. 1983). However, even if the capacity to bind a ligand is lost after SDS-PAGE, a number of proteins may be renatured to the point that they recover functionally (enzyme activity or ligand binding ability) following SDS-PAGE (Beisiegel 1986; and Table 11.7).

A variety of protocols have been used to renature proteins after denaturation by SDS, including incubation of the gel in 6 M urea and washing with a solution containing casein. Perhaps the simplest technique is the use of a number of washes with a buffer containing glycerol, followed by electrotransfer at high pH. Use of high purity SDS has also been reported to assist subsequent attempts at renaturation. Maintenance of nonreducing conditions during electrophoresis (Daniel et al. 1983) may be important in preserving protein activity, but reduction of disulphide bridges prior to electrophoresis is not always detrimental (Lacks and Springhorn 1980). Another possible route to reconstituting ligand-binding activity is the addition of a redox buffer during the renaturation step to allow correct reformation of disulphide bridges. The latter technique has proven useful in some instances where a chemically synthesized peptide is folded into its native state (Tam 1987). Table 11.7 summarizes some of the techniques for protein renaturation that have been used.

If SDS is incompatible with use of a particular ligand-protein combination, even after attempting renaturation, sufficient resolution may be obtained by performing two-dimensional electrophoresis under non-

Table 11.7. A summary of techniques used for renaturing proteins after SDS-polyacrylamide electrophoresis

Proteins Renatured (Detection Method)	Renaturation Method	Reference
ATCase, Lac repressor, aldolase, RNA polymerase (enzyme activity)	Elute protein in 0.1% SDS, dialyse vs 6 M urea, pass over Dowex 1-X2, dialyse versus pH 8.0 buffer containing 20% glycerol	Weber and Kuter (1971)
DNase I, proteases, amylases, dehydrogenases, esterases (enzyme activity)	Rinse gel with water, incubate for three 1 h washes in 0.04 M Tris-HCl pH 7.6	Lacks and Springhorn (1980)
Sperm antigens (radio-immunobinding)	Incubate gel 2 × 5 h in 0.01 M Tris-HCl pH 7.6 containing 4 M urea	Lee et al. (1982)
neurophysin, insulin, calcitonin, vasopressin, β-endorphin (avidin-biotin immunobinding)	Wash 3 × 10 min in 0.05 M Tris-HCl pH 7.4 containing 20% glycerol, electroblot in carbonate/ bicarbonate buffer pH 10	Rosenbaum et al. (1989)
transforming growth factor-a (mitogenic activity and receptor-binding) N.b. this was a 50 amino acid peptide synthesized chemically rather than denatured by SDS-PAGE	Dissolve peptide in 8 M urea, dialyse sequentially versus 8 M, 4 M, 2 M urea containing 0.1 M dithiothreitol, 1 M urea containing glutathione redox buffer, 0.1 M Tris-HCl pH 8.0 and finally 1 M acetic acid, lyophilise	Tam (1987)

denaturing conditions. There is no obstacle to electroblotting proteins from native gels to nitrocellulose or PVDF membranes. However, in the absence of denaturing detergents, it is also possible to electroblot proteins to anion- or cation-exchange membranes. For example, ion-exchange membranes have been used successfully to detect JH-binding proteins from several insect species after electroblotting (Wiśniewski 1989; SC Trowell unpublished results). These have the benefits of maintaining the protein's native conformation and allowing easy elution of the proteins, if required. After the target protein has been identified by ligand or antibody binding, a parallel or duplicate blot can be eluted with SDS sample buffer for final clean-up by SDS-PAGE prior to transfer to an inert membrane for sequencing.

Detection of radioactive compounds on blots can be achieved with special tritium counters, autoradiography using special film, or by liquid scintillation counting of fractionated strips of the blot. Spray-on fluorophores, such as En^{3}Hance from NEN (Jefferies and Roberts 1990) and Amplify from Amersham (SC Trowell unpublished observations), allow fluorography of the blot in a very convenient and sensitive manner.

3. Diagnostic Applications

In this Section we turn to diagnostic applications of protein-specific probes. This is preceded by a discussion of the relative advantages and disadvantages of antibody-based versus DNA-based tests (3.1). We discuss the use of diagnostic technologies in identifying pathogens (3.2) both of beneficial insects (3.2.1) and genetically engineered pathogens (3.2.2). Preparation of pathogen-specific antibodies is also covered briefly (3.2.3). A second major class of uses for diagnostic technologies is in species identification (3.3). This is illustrated using three case studies from the area of applied entomology (3.3.1). We also discuss the potential of diagnostics for assisting in the development and monitoring of biological control programs (3.3.2). The particular problems of raising antibodies for these uses are considered in Section 3.3.3. Finally, we discuss insect strain identification (3.4), particularly in relation to insecticide resistance status (3.4.1). This is perhaps the most demanding challenge for diagnostic technologies.

There are a number of commercial problems that require rapid and easy identification or discrimination of insect species and strains (for example, resistant versus susceptible insects) as well as their pathogens. There is indeed already a significant demand for diagnostic probes, assays and kits in applied entomology. Species identification might be required for quarantine purposes in order to facilitate the process of screening potential biological control agents, to assess the biological differences among sibling members of a species complex or to guide insect control strategies in the field. At a more detailed level, the ability to distinguish insecticide-susceptible and -resistant strains in the field is the goal of many working in the area of resistance management.

DNA-based probes, either based on RFLP patterns or direct recognition of sequence homologies by nonradioactive means, have their place, as do antibodies and immunochemically-based assays. We will discuss a number of factors that may influence the choice between DNA-based or antibody-based assays. These include the number and type of samples and their likely stability, the level of training of those performing the tests, and the type of information one requires.

3.1. DNA-Versus Antibody-Based Diagnostic Tests

DNA sequence differences, detected as restriction fragment length polymorphisms (RFLPs), can be used to characterize organisms down to the individual level. DNA-based tests can therefore readily be adapted to species or strain identification. Aquadro (this volume) gives an excellent review of the population genetics of RFLPs. Traditionally, detection methods relied on the use of a ^{32}P-labeled probe, which has limited their usefulness, but a number of colorimetric methods can now be substituted (Leary et al. 1983; Renz and Kurz 1984; Forster et al. 1985; Kemp 1989; Kemp et al. 1989; Triglia et al. 1990) with little or no loss in sensitivity.

RFLP assays generally require an electrophoresis step to separate DNA molecules of different length. This may be appropriate as a routine procedure in a clinical biochemistry or forensic pathology laboratory but is poorly suited to use by nonspecialists away from central laboratories. With suitable facilities, one technician could use these techniques to survey a few thousand individuals over a year. However, new developments may make DNA-based tests more amenable for widespread use.

In one new development, termed AP-PCR for arbitrarily primed PCR, random mixtures of PCR primers can be used to recover strain-specific markers from small samples without the need to use an electrophoretic step (Welsh and McClelland 1990; Williams et al. 1990). On the other hand, ELISA-like assays have been developed that employ a double-stranded DNA-binding protein to retain the DNA of interest in a microtitre well (Kemp 1989; Kemp et al. 1989; Triglia et al. 1990) in conjunction with a PCR step and a specific DNA probe linked to a colorigenic enzyme.

These innovations offer the prospect of much more efficient assays and the capacity to accommodate far more samples than electrophoretic methods. The ultimate goal would be to deliver the user friendliness of antibody-based assays with the supreme discrimination of DNA-based methods. It may well be that this can only be achieved in cases where a specific DNA sequence is overexpressed in one strain and absent in others, thereby avoiding the necessity for a PCR step. Such a favorable case is exemplified by the development of DNA probes specific for members of the *Anopheles gambiae* species complex (Gale and Crampton 1987; 1988) that can be used to species-type individual mosquitoes.

Antibody tests, on the other hand, already routinely deliver the capacity to handle thousands of samples and can easily be used to assay a hundred samples in a day without access to laboratory facilities. A single worker with laboratory facilities could quite easily test one or two thousand insects in a day, allowing information from a wide geographical region to be processed rapidly. In principle, antibodies can detect even the smallest differences between the macromolecules of different strains or species. In practice, however, the finer the distinction to be made, the more time and effort will be expended in obtaining suitable antibodies. Even though specific antibodies could be obtained, there may be certain instances, for example target-site resistance involving a point mutation (ffrench-Constant et al. this volume), where a practical antibody-based field assay could not be developed. In such cases RFLP or other DNA-based laboratory analysis may be the only option.

3.2. Pathogen Identification

3.2.1. Identifying Pathogens of Beneficial Insects

Entomopathogens are likely targets for immunoasssays from two distinct points of view. In the case of economically important beneficial insects, principally bees and silkworms, there is a requirement to be able to diagnose diseases rapidly in the field. This has already been achieved for *Bacillus larvae* (Olsen et al. 1990) and *Nosema bombycin* (Ke et al. 1990), but other insect pathogens are obvious future targets. In the case of microbial pathogens, there are often biochemical or microscopical tests which allow diagnosis. However, these may be slow or require the involvement of technical personnel with special training and experience.

3.2.2. Identifying Genetically Engineered Pathogens

A potentially much larger area of applications comprises the identification and typing of microorganisms used in biological control. A survey of the literature reveals an increasing number of papers reporting immunochemical methods for the detection and assay of pesticides, toxins, herbicides and their breakdown products (Dreher and Podratzki 1988; Jung et al. 1989; Stanker et al. 1989; Schmidt et al. 1990). While there are instrumental methods for the detection of all of these compounds, they require major capital equipment and often extensive sample preparation before they may be submitted for analysis. The administrative and transport costs of moving samples to a central laboratory may far outweigh the actual costs of performing the assays. In addition, some samples may be unstable over the time period needed to transport them. Immunochemical assays are therefore seen as having distinct benefits for routine testing away from central laboratories.

Given the absence of competing instrumental methods for monitoring genetically engineered insecticides, there is likely to be a proportionately greater demand to develop immunochemical and DNA-based probes for them. These tools will be used for initial registration studies, to establish proprietary rights over a particular strain of microorganism developed for insect control and to monitor for spread or change of that organism in the environment during field use.

3.2.3. Generation of Diagnostic Probes for Entomopathogens

Entomopathogens may generally be cultured in large quantities, either in tissue culture or in insect hosts, and purified by standard biochemical procedures. Antigenically they are likely to bear little relationship to their hosts. One can therefore generally expect to be able to prepare a range of antibodies (both polyclonal and monoclonal) that are highly specific for the pathogen over its host. Strain-specific recognition of the pathogen may also be an important requirement, but it will generally be more difficult to achieve. Only testing under a variety of conditions can establish which are the most specific and sensitive antibodies and the least affected by interfering substances that may be present in samples taken outside the laboratory.

3.3. Species Identification and Differentiation

In a small proportion of cases, immunochemical assays or DNA-based probes may provide significant advantages over conventional morphological or isozyme analysis for distinguishing similar sympatric species. Immunochemical assays may be beneficial where the identification must be made by nonexperts, away from central laboratories, very rapidly or without access to major equipment. Immunochemical assays also have the advantage that they can readily be applied to very large numbers of samples. There are a number of potential applications of immunochemical assays for identification of closely related insect species. Our background and interests lead us to identify two major ones: those cases where there is a need to discriminate between two species which differ in pest status or insecticide resistance, and those cases where a quick and easy species test is required to set up, monitor or apply a biological control problem.

3.3.1. Distinguishing Species Which Differ in Pest Status or Insecticide Resistance

We will use the example of *Helicoverpa armigera* and *Helicoverpa punctigera* (formerly known as *Heliothis armigera* and *Heliothis punctigera*, respectively), two polyphagous insects which are significant pests of Australian cotton crops, to illustrate this potential application of molecular diagnostic technology. Morphologically, eggs and early instars of the two species are indistinguishable. While later stages are morphologically distinct, it requires an experienced eye to perform the discrimination and mistakes are easily made. *H. armigera* has a considerable propensity to develop resistance to commonly used insecticides while, at least in the cotton cropping ecosystem, *H. punctigera* does not.

Pyrethroids are among the more cost-effective insecticides, and the cotton industry aims to preserve their usefulness for as long as possible (Forrester 1990a). It is therefore highly desirable to minimize pyrethroid selection on populations that are predominantly composed of *H. armigera.* Currently, this is best achieved by a management strategy (Anonymous 1983; Daly and Fitt 1990) that limits the periods during which particular classes of insecticides may be employed. While generally successful, the current strategy suffers from two disadvantages. First, it is rather inflexible in that it cannot take account of significant seasonal and geographical fluctuations in the ratios of *H. armigera* and *H. punctigera.* The second problem is that the basal level of pyrethroid resistance is increasing slowly from year to year (Forrester 1990b).

Any measures which can reduce the selection for resistance still further, while achieving cost-effective control, would be highly desirable. The cotton industry is therefore funding our laboratory to generate monoclonal antibodies that will discriminate between the two species (Trowell and Daly 1990). We will then incorporate the antibodies into a kit which can be used routinely by cotton growers and their consultants to guide decisions on insecticide usage.

An example from the field of medical entomology will further serve to illustrate the potential utility of these methods in routine differentiation of species that are impossible to distinguish by morphological methods. It has been argued (Service 1985) that the success or failure of malaria eradication programs in tropical Africa will, to some extent, depend on the ability of entomologists to discriminate sibling species of the *Anopheles gambiae* complex which differ markedly in their behavior and ecology. Two different DNA-based tests have been developed for this purpose (Gale and Crampton 1987; 1988; Paskewitz and Collins 1990), although the tests would currently be restricted to laboratory use.

A third example of the potential applications of molecular diagnostics in this area is of the rapid screening of specimens in quarantine to identify particularly important agricultural pests. For example, the Old World screw-worm fly (*Chrysomia bezziana*) is not currently present in Australia but if introduced would cause serious economic problems for pastoral agriculture. A species-specific molecular probe could be used by non-experts to identify samples and distinguish them from larvae of endemic parasitic Diptera. In this case an antibody test could complement and eventually replace morphological identification currently being used for this purpose.

3.3.2. Assistance in Biological Control Strategies

Traditionally, identification of arthropod predators as possible biological control agents has relied on painstaking and labor-intensive field observations. As an alternative to this approach, Greenstone (Greenstone 1989; Greenstone and Morgan 1989) has developed antibodies specific for *Helicoverpa zea.* The antibodies were used to screen the stomach contents of potentially useful predators for evidence that they had fed on *H. zea.* It may be possible to use similar cost-effective strategies generally to identify potential biological control agents (Greenstone 1989). The technique could also be adapted to allow determination of the frequencies of particular parasitoids in a pest population. Where the species involved in biological control programs are particularly small, for example mites or nematodes, antibodies could allow agronomists to distinguish routinely between pests and beneficials.

3.3.3. Generation of Probes

In aiming for an antibody capable of discriminating two closely related species, one may choose a random approach—immunizing with antigens from one of the species and screening sequentially against antigens of each target species. Conventional monospecific or even monoclonal antibodies are not well suited to the task because of the relatively small number of specificities that may be screened and the tendency largely to recover monoclonals directed against the immunodominant antigens. However, in the future antibody fragments may be generated from prokaryote libraries (cf. Section 2.2.1.3. above). The advantages of being able to survey a much wider range of specificities would translate into the potential to obtain discriminating antibodies at much lower cost than is possible with conventional techniques. Alternatively, antigens peculiar to one of the species may be allowed to engender antibodies. Such an immunization strategy could use a purified protein known to differ between the two species (e.g., An et al. 1990) or differential immunization as described in Section 2.2.1.4.

3.4. Strain Identification

3.4.1. Insecticide Resistance Status

There are instances where one would wish to be able to identify insecticide resistant strains of insects and distinguish them from their susceptible conspecifics. There are two levels at which this is relevant. The first is the level of determining, then monitoring the number, severity and frequencies of resistant alleles in field populations. This can only practically be done at the DNA level, particularly because some insects can repress their resistance phenotype while retaining the resistance genes (Field et al. 1989). The information thus obtained is vital for development of informed resistance management strategies. However, at the second level one would want to follow the actual proportion of individuals expressing the most significant resistance alleles in field populations on a routine basis. This information would be used to guide week-to-week decisions on usage of insecticides within the guidelines established by one's resistance management strategy. For statistical accuracy, relatively large numbers of insects should be sampled within each discrete area, and antibody tests, if available, may well be preferable over DNA tests for this purpose because of their greater ease of use.

3.4.2. Generation of Probes

RFLP analysis is quite capable of detecting strain-specific DNA polymorphisms associated with insecticide resistance. For example, Raymond et al. (1991) were able to go further than simple diagnosis of resistance. They demonstrated, using RFLP analysis, that a particular biochemical variant of organophosphate-resistant *Culex pipiens* had achieved its worldwide distribution by geographical spread after a single mutational event.

On the other hand, the use of antibodies to detect differences as subtle as those that may be associated with the development of insecticide resistance seems the most daunting of tasks. Nevertheless, if the basis of the resistance is an overexpression of a protein, then it may also be feasible to raise specific antibodies that will identify the resistant phenotype. This has been achieved for a resistance-specific cytochrome P-450 in the case of *D. melanogaster* (Sundseth et al. 1989; ffrench-Constant et al. this volume). Antibodies of this type could certainly be used as a rapid and effective way of monitoring the type and extent of resistance in pest populations as long as they were backed up by occasional monitoring of the genetic status of the resistant population.

4. Conclusion

Much of the justification behind the funding of insect molecular biology is the promise that this discipline holds for solving problems in agricultural and medical entomology. Many of the practical benefits of molecular entomology will flow from engineering of insect genes to manipulate the status of pest species and from techniques which will improve our ability to identify pest insects or their pathogens.

Identification of insects and their pathogens using molecular techniques is a narrower field than the engineering of insect genes. It has a special significance, however, in that diagnostic technologies have a much shorter lead time than genetic engineering, and we are therefore likely to see them enter use in economic entomology over the next few years. This difference between the diagnostic and genetic engineering aspects of molecular biology arises partly because of the relative simplicity of the research problems involved in developing diagnostic methods and partly because the complicated registration procedures that will delay the commercial availability of engineered organisms are not required for diagnostic kits.

When it comes to the engineering of insect genes, there can be few biologists who are unaware of the enormous power and usefulness of the genetic and molecular genetic techniques available in *Drosophila melanogaster*. Work on this species has taught us much of what we know about insect molecular biology. However, to achieve the goal of manipulating economically important insects, the focus of attention has already moved to genetic engineering of pest species.

In moving beyond *D. melanogaster* we usually lose the experimental advantages of that species: its small genome, short generation time and ease of culture. Most seriously of all, however, we lose the facility to clone genes using genetic and molecular genetic techniques at the DNA level. The dilemma is that, in the absence of biochemical and physiological background data, molecular biology offers the most efficient and powerful way to study non-drosophilids. The answer to this dilemma, we have argued, is to exploit the range of molecular techniques becoming available to isolate a gene, starting with a knowledge of the binding properties of the protein it encodes.

We have drawn attention to the significant effort already devoted to the purification of insect proteins with a view to gene isolation. Recent advances in probe generation and the purification and partial sequencing of tens to hundreds of picomoles of a protein are making the "protein up" route an increasingly attractive option. In association with the powerful techniques of molecular genetics, this will inevitably lead to the cloning of a growing number of genes from non-drosophilid insects over the next decade. In many cases this research will be driven by utilitarian motives and will have practical benefits, but there should be no doubt that it will also provide profound new insights into a variety of physiological organizations in the insect species concerned.

Acknowledgments. We would like to thank the editors of this volume for the many valuable criticisms and suggestions they made. We are also indebted to Dr JWO Ballard, CSIRO Division of Entomology, for sharing with us his knowledge of PCR and DNA-based species identification and Ms MM Dumancic, also of the Division of Entomology, for critical reading of the manuscript.

References

Adachi T, Takiya S, Suzuki Y, Iwami M, Kawakami A, Takahashi SY, Ishizaki H, Nagasawa H, Suzuki A (1989) cDNA structure and expression of bombyxin, an insulin-like brain secretory peptide of the silkmoth *Bombyx mori*. J Biol Chem 264:7681–7685

Aebersold R, Leavitt J (1990) Sequence analysis of proteins separated by polyacrylamide gel electrophoresis—towards an integrated protein database. Electrophoresis 11:517–527

Aebersold RH, Teplow DB, Hood LE, Kent SBH (1986) Electroblotting onto activated glass. High efficiency preparation of proteins from analytical sodium dodecyl sulfate-polyacrylamide gels for direct sequence analysis. J Biol Chem 261:4229–4238

Alting-Mees M, Amberg J, Ardourel D, Elgin E, Greener A, Gross EA, Kubitz M, Mullinax RL, Short JM, Sorge JA (1990) Monoclonal antibody expression libraries: A rapid alternative to hybridomas. Strategies in Molecular Biology (Stratagene technical publication) 3:1–9

An HJ, Klein PA, Kao KJ, Marshall MR, Otwell WS, Wei CI (1990) Development of monoclonal antibody for rock shrimp identification using enzyme-linked immunosorbent assay. J Agric Food Chem 38:2094–2100

Anonymous (1983) Pyrethroid resistance. The Australian Cotton Grower 4(3): 4–7

Atlas D, Yaffe D, Skutelsky E (1978) Ultrastructural probing of β-adrenoreceptors on cell surfaces. FEBS Letters 95:173–176

Beisiegel U (1986) Protein blotting. Electrophoresis 7:1–18

Belyavsky A, Vinogradova T, Rajewsky K (1989) PCR-based cDNA library construction: general cDNA libraries at the level of a few cells. Nucl Acids Res 17:2919–2932

Benjamin RJ, Waldmann H (1986) Induction of tolerance by monoclonal antibody therapy. Nature 320:449–451

Bennett JP, Yamamura HI (1985) Neurotransmitter, hormone or drug receptor binding methods. In Yamamura HI, Enna SJ, Kuhar MJ (eds) Neurotransmitter Receptor Binding. Raven Press, New York, pp 61–89

Bergman T, Jörnvall H (1987) Electroblotting of individual polypeptides from SDS/polyacrylamide gels for direct sequence analysis. Eur J Biochem 169: 9–12

Bergman T, Jörnvall H (1990) Electroblotting of polypeptides onto glass fiber filters for direct sequence analysis after sodium dodecyl sulfate-polyacrylamide gel electrophoresis. Electrophoresis 11:569–572

Bieber AJ, Snow PM, Hortsch M, Patel NH, Jacobs JR, Traquina ZR, Schilling J, Goodman CS (1989) *Drosophila* neuroglian: A member of the immunoglobulin superfamily with extensive homology to the vertebrate neural adhesion molecule L1. Cell 59:447–460

Blalock JE (1990) Complementarity of peptides specified by "sense" and "antisense" strands of DNA. TIBTECH 8:140–144

Boehm MF, Prestwich GD (1988a) Synthesis of tritium-labeled fenoxycarb and S-31183, two phenoxyphenyl ether insect growth regulators. J Label Comp Radiopharm 25:1007–1015

Boehm MF, Prestwich GD (1988b) Synthesis of [8,9–3H2]-7S-methoprene, a juvenile hormone analog, by selective reduction of a protected trienoate. J Label Comp Radiopharm 25:653–659

Bost KL, Smith EM, Blalock JE (1985) Similarity between the corticotropin (ACTH) receptor and a peptide encoded by an RNA that is complementary to ACTH mRNA. Proc Natl Acad Sci USA 82:1372–1375

Brentani RR, Ribeiro SF, Potocnjak P, Pasqualini R, Lopes JD, Nakaie CR (1988) Characterization of the cellular receptor for fibronectin through a hydropathic complementarity approach. Proc Natl Acad Sci USA 85:364–367

Campbell AM (1984) Monoclonal Antibody Technology. Elsevier, Amsterdam
Carr DJJ, Bost KL, Blalock JE (1986) An antibody to a peptide specified by an RNA that is complementary to γ-endorphin mRNA recognizes an opiate receptor. J Neuroimmunol 12:329–337
Carrey EA (1990) Peptide mapping. In Creighton TE (ed) Protein Structure: A Practical Approach. IRL Press, Oxford, pp 117–144
Cayanis E, Rajagopalan R, Cleveland WL, Edelman IS, Erlanger BF (1986) Generation of an auto-anti-idiotypic antibody that binds to glucocorticoid receptor. J Biol Chem 261:5094–5103
Cherbas P, Cherbas L, Lee S-S, Nakanishi K (1988) 26-[^{125}I]Iodoponasterone A is a potent ecdysone and a sensitive radioligand for ecdysone receptors. Proc Natl Acad Sci USA 85:2096–2100
Choli T, Wittmann-Liebold B (1990) Protein blotting followed by micro-sequencing. Electrophoresis 11:562–568
Cleveland WL, Erlanger BF (1986) The auto-anti-idiotypic strategy for preparing monoclonal antibodies to receptor combining sites. Meth Enzymol 121: 95–107
Cleveland WL, Wasserman NH, Sarangarajan R, Penn AS, Erlanger BF (1983) Monoclonal antibodies to the acetylcholine receptor by a normally functioning auto-anti-idiotypic mechanism. Nature 305:56–57
Cole KD, Fernando-Warnakulasuriya GJP, Boguski MS, Freeman M, Gordon JI, Clark WA, Law JH, Wells MA (1987) Primary structure and comparative sequence analysis of an insect apolipoprotein. Apolipophorin III from *Manduca sexta*. J Biol Chem 262:11794–11800
Coscia CJ, Szücs M, Barg J, Belcheva MM, Bem WT, Khoobehi K, Donnigan TA, Juszczak R, Mchale RJ, Hanley MR, Barnard EA (1991) A monoclonal anti-idiotypic antibody to μ and δ opioid receptors. Mol Brain Res 9: 299–306
Couraud JY, Maillet S, Conrath M, Calvino B, Pradelles P (1990) Use of anti-idiotypic antibodies as probes for in vitro and in vivo identification of substance-P receptor. Mol Chem Neuropathol 12:71–82
Couraud PO, Strosberg AD (1991) Anti-idiotypic antibodies against hormone and neurotransmitter receptors. Biochem Soc Trans 19:147–151
Cowman AF, Zuker CS, Rubin GM (1986) An opsin gene expressed in only one photoreceptor cell type of the *Drosophila* eye. Cell 44:705–710
Cwirla SE, Peters EA, Barrett RW, Dower WJ (1990) Peptide on phage: a vast library of peptides for identifying ligands. Proc Natl Acad Sci USA 87: 6378–6382
Daly JC, Fitt GP (1990) Resistance frequencies in overwintering pupae and the first spring generation of *Helicoverpa armigera* (Lepidoptera: Noctuidae): selective mortality and immigration. J Econ Entomol 83:1682–1688
Daniel TO, Schneider WJ, Goldstein JL, Brown MS (1983) Visualization of lipoprotein receptors by ligand blotting. J Biol Chem 258:4606–4611
Dean PDG, Johnson WS, Middle FM (eds) (1985) Affinity Chromatography: A Practical Approach. IRL Press, Oxford
Devlin JJ, Panganiban LC, Devlin PE (1990) Random peptide libraries: a source of specific protein binding molecules. Science 249:404–406

Domec C, Garbay B, Fournier M, Bonnet J (1990) cDNA library construction from small amounts of unfractionated RNA: association of cDNA synthesis with polymerase chain reaction amplification. Analyt Biochem 188:422–426

Dreher RM, Podratzki B (1988) Development of an enzyme immunoassay for endosulfan and its degradation products. J Agric Food Chem 36:1072–1075

Eberle AN, de Graan PNE (1985) General principles for photoaffinity labeling of peptide hormone receptors. Meth Enzymol 109:129–157

Elton TS, Dion LD, Bost KL, Oparil S, Blalock JE (1988) Purification of an angiotensin II binding protein by using antibodies to a peptide encoded by angiotensin II complementary RNA. Proc Natl Acad Sci USA 85:2518–2522

Eng W-S, Prestwich GD (1988) A short radiosynthesis of natural juvenile hormone III, methyl [12–3H]-10R-10,11–epoxyfarnesoate. J Label Comp Radiopharm 25:627–633

Erlich HA (1989) PCR Technology: Principles and Applications for DNA Amplification. Stockton Press, New York

Evans RM (1988) The steroid and thyroid hormone receptor superfamily. Science 240:889–895

Feiler R, Harris WA, Kirschfeld K, Wehrhahn C, Zuker CS (1988) Targeted misexpression of a *Drosophila* opsin gene leads to altered visual function. Nature 333:737–741

Feyereisen R, Koener JF, Farnsworth DE, Nebert DW (1989) Isolation and sequence of cDNA encoding a cytochrome P-450 from an insecticide-resistant strain of the house fly, *Musca domestica*. Proc Natl Acad Sci USA 86: 1465–1469

Field LM, Devonshire AL, ffrench-Constant RH, Forde BG (1989) Changes in DNA methylation are associated with loss of insecticide resistance in the peach-potato aphid *Myzus persicae* (Sulz.). FEBS Letters 243:323–327

Forrester N (1990a) Insecticide price trends. The Australian Cotton Grower 10(4):42–44

Forrester NW (1990b) Resistance management in Australian cotton. Fifth Australian Cotton Conference, Broadbeach, Queensland, ACGRA.

Forster AC, McInnes JL, Skingle DC, Symons RH (1985) Non-radioactive hybridization probes prepared by the chemical labeling of DNA and RNA with a novel reagent, photobiotin. Nucl Acids Res 13:745–761

Gale KR, Crampton JM (1987) DNA probes for species identification of mosquitoes in the *Anopheles gambiae* complex. Med Vet Entomol 1:127–136

Gale KR, Crampton JM (1988) Use of a male-specific DNA probe to distinguish female mosquitoes of the *Anopheles gambiae* species complex. Med Vet Entomol 2:77–79

Gautam N, Baetscher M, Aebersold R, Simon MI (1989) A G protein gamma subunit shares homology with *ras* proteins. Science 244:971–974

Gearing DP, King JA, Gough NM, Nicola NA (1989) Expression cloning of a receptor for human granulocyte-macrophage colony-stimulating factor. EMBO J 8:3367–3676

Gherardi E, Pannell R, Milstein C (1990) A single-step procedure for cloning and selection of antibody-secreting hybridomas. J Immunol Meth 126:61–68

Goding JW (1983) Monoclonal Antibodies: Principles and Practice. Production and application of monoclonal antibodies in cell biology, biochemistry and molecular biology. Academic Press, London

Goldstein A, Brutlag DL (1989) Is there a relationship between DNA sequences encoding peptide ligands and their receptors? Biochemistry 86:42–45

Greenstone MH (1989) Foreign exploration for predators: A proposed new methodology. Environ Entomol 18:195–200

Greenstone MH, Morgan CE (1989) Predation on *Heliothis zea* (Lepidoptera: Noctuidae): An instar-specific ELISA assay for stomach analysis. Ann Entomol Soc Amer 82:45–49

Gutstein NL, Seaman WE, Scott JH, Wofsy D (1986) Induction of immune tolerance by administration of monoclonal antibody to L3T4. J Immunol 137:1127–1132

Gutstein NL, Wofsy D (1986) Administration of $F(ab')_2$ fragments of monoclonal antibody to L3T4 inhibits humoral immunity in mice without depleting $L3T4^+$ cells. J Immunol 137:3414–3419

Hammock BD, Bonning BC, Possee RD, Hanzlik TN, Maeda S (1990) Expression and effects of the juvenile hormone esterase in a baculovirus vector. Nature 344:458–461

Hanzlik TN, Abdel-Aal YAI, Harshman LG, Hammock BD (1989) Isolation and sequencing of cDNA clones coding for juvenile hormone esterase from *Heliothis virescens*. J Biol Chem 264:12419–12425

Harada N, Castle BE, Gorman DM, Itoh N, Schreurs J, Barrett RL, Howard M, Miyajima A (1990) Expression cloning of a cDNA encoding the murine interleukin 4 receptor based on ligand binding. Proc Natl Acad Sci USA 87:857–861

Hargrave PA, Fong SL, McDowell JA, Mas MT, Curtis PR, Wang JK, Juszcazak E, Smith DP (1980) The partial primary structure of bovine rhodopsin and its topography in the retinal rod disc membrane. Neurochem Int 1:231–244

Harlow E, Lane D (1988) Antibodies: A Laboratory Manual. Cold Spring Harbor Laboratory Press, Cold Spring Harbor

Henrich VC, Maroy P, Sliter TJ, Ren X-J, Gilbert LI (1990) Cloning of genes that encode ecdysteroid regulated DNA-binding proteins in *Drosophila melanogaster*. Invert Reprod Devel 18:117

Horodyski FM, Riddiford LM, Truman JW (1989) Isolation and expression of the eclosion hormone gene from the tobacco hornworm, *Manduca sexta*. Proc Natl Acad Sci USA 86:8123–8127

Horton JK, Evans OM, Swann K, Swinburne S (1989) A new and rapid method for the selection and cloning of antigen-specific hybridomas with magnetic microspheres. J Immunol Meth 124:225–230

Huang JM, Matthews HR (1990) Application of sodium dodecyl sulfate gel electrophoresis to low molecular weight polypeptides. Analyt Biochem 188:114–117

Hudson P (1990) DAbs and fAbs take on mAbs. Today's Life Science 2:38–39

Huse WD, Sastry L, Iverson SA, Kang AS, Alting-Mees M, Burton DR, Benkovic SJ, Lerner RA (1989) Generation of a large combinatorial library of the immunoglobulin repertoire in phage lambda. Science 246:1275–1281

Hutson DH, Roberts TR (1985) Insecticides. In Hutson DH, Roberts TR (eds) Insecticides, John Wiley & Sons, New York, pp 1–34

Hyde DR, Mecklenburg KL, Pollock JA, Vihtelic TS, Benzer S (1990) Twenty *Drosophila* visual system cDNA clones: One is a homolog of human arrestin. Proc Natl Acad Sci USA 87:1008–1012

Janssen WE, Rios AM (1989) Non-specific cell binding characteristics of paramagnetic polystyrene microspheres used for antibody-mediated cell selection. J Immunol Meth 121:289–294

Jefferies LS, Roberts PE (1990) A new method of detecting hormone-binding proteins electroblotted onto glass fiber filter: juvenile hormone-binding proteins from grasshopper hemolymph. J Steroid Biochem 35:449–455

Jung F, Gee SJ, Harrison RO, Goodrow MH, Karu AE, Braun AL, Li QX, Hammock BD (1989) Use of immunochemical techniques for the analysis of pesticides. Pestic Sci 26:303–317

Kamb A, Weir M, Rudy B, Varmus H, Kenyon C (1989) Identification of genes from pattern formation, tyrosine kinase, and potassium channel families by DNA amplification. Proc Natl Acad Sci USA 86:4372–4376

Kawakami A, Kataoka H, Oka T, Mizoguchi A, Kimura-Kawakami M, Adachi T, Iwami M, Nagasawa H, Suzuki A, Ishizaki H (1990) Molecular cloning of the *Bombyx mori* prothoracicotropic hormone. Science 247:1333–1335

Kawasaki H, Emori Y, Suzuki K (1990) Production and separation of peptides from proteins stained with coomassie brilliant blue R-250 after separation by sodium dodecyl sulfate-polyacrylamide gel electrophoresis. Analyt Biochem 191:332–336

Ke ZX, Xie WD, Wang XZ, Long QX, Pu ZL (1990) A monoclonal antibody to *Nosema bombycis* and its use for identification of microsporidian spores. J Invert Pathol 56:395–400

Kemp D (1989) Blue genes: Colorimetric detection of PCR products for diagnostics. Today's Life Science 1:64–73

Kemp DJ, Smith DB, Foote SJ, Samaras N, Peterson MG (1989) Colorimetric detection of specific DNA segments amplified by polymerase chain reactions. Proc Natl Acad Sci USA 86:2423–2427

Kimmel AR (1987) Identification and characterization of specific clones: Strategy for confirming the validity of presumptive clones. Meth Enzymol 152: 507–510

Kimmel AR, Berger SL (1987) Preparation of cDNA and the generation of cDNA libraries: overview. Meth Enzymol 152:307–317

Knigge KM, Piekut DT, Berlove D (1988) Immunocytochemistry of a vasopressin (AVP) receptor with anti-idiotype antibody: inhibition of staining with a peptide (PVA) encoded by an RNA that is complementary to AVP mRNA. Neurosci Letters 86:269–271

Koeppe JK, Kovalick GE, Prestwich GD (1984) A specific photoaffinity label for hemolymph and ovarian juvenile hormone-binding proteins in *Leucophaea maderae*. J Biol Chem 259:3219–3223

Köhler H, Kaveri S, Kieber-Emmons T, Morrow WJW, Müller S, Raychaudhuri S (1989) Idiotypic networks and nature of molecular mimicry: An overview. Meth Enzymol 178:3–35

Kulcsár P, Prestwich GD (1988) Detection and microsequencing of juvenile hormone-binding proteins of an insect by the use of an iodinated juvenile hormone analog. FEBS Letters 228:49–52

Kylsten P, Samakovlis C, Hultmark D (1990) The cecropin locus in *Drosophila*: a compact gene cluster involved in the response to infection. EMBO J 9: 217–224

Lacks SA, Springhorn SS (1980) Renaturation of enzymes after polyacrylamide gel electrophoresis in the presence of sodium dodecyl sulfate. J Biol Chem 255:7467–7473

Leary JJ, Brigati DJ, Ward DC (1983) Rapid and sensitive method for visualizing biotin-labeled DNA probes hybridized to DNA or RNA immobilized on nitrocellulose: bio-blots. Proc Natl Acad Sci USA 80:4045–4049

Lee C-YL, Huang Y-S, Hu PC, Gomel V, Menge AC (1982) Analysis of sperm antigens by sodium dodecyl sulfate gel/protein blot radioimmunobinding method. Analyt Biochem 123:14–22

Lerro KA, Prestwich GD (1990) Cloning and sequencing of a cDNA for the hemolymph juvenile hormone binding protein of larval *Manduca sexta*. J Biol Chem 265:19800–19806

Lewis SA, Gu W, Cowan NJ (1987) Free intermingling of mammalian β-tubulin isotypes among functionally distinct microtubules. Cell 49:539–548

Lombes M, Edelman IS, Erlanger BF (1989) Internal image properties of a monoclonal auto-anti-idiotypic antibody and its binding to aldosterone receptors. J Biol Chem 264:2528–2536

Luckow VA, Summers MD (1988) Trends in the development of baculovirus expression vectors. Bio Technology 6:47

Manz B, Heubner A, Köhler I, Grill H-J, Pollow K (1983) Synthesis of biotin-labeled dexamethasone derivatives. Novel hormone-affinity probes. Eur J Biochem 131:333–358

Matsudaira P (1987) Sequence for picomole quantities of proteins electroblotted onto polyvinylidene difluoride membranes. J Biol Chem 262:10035–10038

Matthew WD, Patterson PH (1983) The production of a monoclonal antibody that blocks the action of a neurite outgrowth-promoting factor. Cold Spring Harbor Symp Quant Biol 48:625–631

McDonald HB, Goldstein LSB (1990) Identification and characterization of a gene encoding a kinesin-like protein in *Drosophila*. Cell 61:991–1000

Menn JJ, Bořkovec AB (1989) Insect neuropeptides: potential new insect control agents. J Agric Food Chem 37:271–278

Meyer DI (1990) Receptor anti-idiotypes—mimics—or gimmicks? Nature 347: 424–425

Moos M, Nguyen NY, Liu T-Y (1988) Reproducible high yield sequencing of proteins electrophoretically separated and transferred to an inert support. J Biol Chem 263:6005–6008

Mulchahey JJ, Neill JD, Dion LD, Bost KL, Blalock JE (1986) Antibodies to the binding site of the receptor for luteinizing hormone-releasing hormone (LHRH): Generation with a decapeptide encoded by an RNA complementary to LHRH mRNA. Proc Natl Acad Sci USA 83:9714–9718

O'Tousa J, Baehr W, Martin R, Hirsh J, Pak WL, Applebury ML (1985) The *Drosophila ninaE* gene encodes an opsin. Cell 40:839–850

Oakeshott JG, Collet C, Phillis RW, Nielsen KM, Russell RJ, Chambers GK, Ross V, Richmond RC (1987) Molecular cloning and characterization of esterase-6, a serine hydrolase of *Drosophila*. Proc Natl Acad Sci USA 84: 3359–3363

Olsen PE, Grant GA, Nelson DL, Rice WA (1990) Detection of American foulbrood disease of the honeybee, using a monoclonal antibody specific to *Bacillus larvae* in an enzyme-linked immunosorbent assay. Canad J Microbiol 36:732–735

Ovchinnikov YA, Abdulaev NG, Feigina MY, Artomonov ID, Zolotarev AS, Moroshnikov AI, Martynow VI, Kostina MB, Kudelin AB, Bogachuk AS (1982) The complete amino acid sequence of visual rhodopsin. Bioorg Khim 8:1424–1427

Pain D, Murakami H, Blobel G (1990) Identification of a receptor for protein import into mitochondria. Nature 347:444–449

Palli SR, Riddiford LM, Hiruma K (1991) Juvenile hormone and "retinoic acid" receptors in *Manduca* epidermis. Insect Biochem 21:7–15

Pappin DJC, Eliopoulos E, Brett M, Findlay JBC (1984) A structural model for ovine rhodopsin. Int J Biol Macromol 6:73–76

Paskewitz SM, Collins FH (1990) Use of the polymerase chain reaction to identify mosquito species of the *Anopheles gambiae* complex. Med Vet Entomol 4:367–373

Pasqualini R, Chamone DF, Brentani RR (1989) Determination of the putative binding site for fibronectin on platelet glycoprotein IIb-IIIa complex through a hydropathic complementarity approach. J Biol Chem 264:14566–14570

Pharmacia (1979) Affinity Chromatography, Principles and Methods. Pharmacia, Uppsala

Plückthun A (1990) Antibodies from *Escherichia coli*. Nature 347:497–498

Prestwich GD (1991) Photoaffinity labeling and biochemical characterization of binding proteins for pheromones, juvenile hormones and peptides. Insect Biochem 21:27–40

Prestwich GD, Boehm MF, Eng W-S, Kulcsár P, Maldonado N, Robles S, Sinha U, Wawrzeńczyk C (1988) New radioligands for characterization of JH and IGR receptors: synthesis and *in vitro* assay. In Sehnal F, Zabza A, Denlinger DL (eds) Endocrinological Frontiers in Physiological Insect Ecology, Wroclaw Technical University Press, Wroclaw, pp 964–973

Prestwich GD, Eng W-S, Boehm MF (1987a) Radioligands for identification of receptors for juvenile hormones and JH analogs. UCLA Symp Mol Cell Biol New Ser 66:279–288

Prestwich GD, Eng W-S, Boehm MF, Wawrzeńczyk C (1987b) High specific activity tritium- and iodine-labeled JH homologs and analogs for receptor binding studies. Insect Biochem 17:1033–1037

Prestwich GD, Koeppe JK, Kovalick GE, Brown JJ, Chang ES, Singh AK (1985) Experimental techniques for photoaffinity labeling of juvenile hormone binding proteins of insects with epoxyfarnesyl diazoacetate. Meth Enzymol 111:509–530

Prestwich GD, Kovalick GE, Koeppe JF (1982) Photoaffinity labeling of juvenile hormone binding proteins in the cockroach *Leucophaea maderae*. Biochem Biophys Res Commun 107:966–973

Prestwich GD, Singh AK, Carvalho JF, Koeppe JK, Kovalick GE (1984) Photoaffinity labels for insect juvenile hormone binding proteins. Synthesis and evaluation *in vitro*. Tetrahedron 40:529–537

Prestwich GD, Wawrzeńczyk C (1985) High specific activity enantiomerically enriched juvenile hormones: synthesis and binding assay. Proc Natl Acad Sci USA 82:5290–5294

Ray A, Memmel NA, Orchekowski RP, Kumaran AK (1987) Isolation of two cDNA clones coding for larval hemolymph proteins of *Galleria mellonella*. Insect Biochem 17:603–617

Raymond M, Callaghan A, Fort P, Pasteur N (1991) Worldwide migration of amplified insecticide resistance genes in mosquitoes. Nature 350:151–153

Redeuilh G, Secco C, Baulieu EE (1990) Biotinylestradiol for purification of estrogen receptor. Meth Enzymol 184:292–300

Renz M, Kurz C (1984) A colorimetric method for DNA hybridization. Nucl Acids Res 12:3435–3444

Rosenbaum LC, Nilaver G, Hagman HM, Neuwelt EA (1989) Detection of low-molecular-weight polypeptides on nitrocellulose with monoclonal antibodies. Analyt Biochem 183:250–257

Sambrook J, Fritsch EF, Maniatis T (1989) Molecular cloning: A laboratory manual. Cold Spring Harbor Laboratory Press, Cold Spring Harbor

Schick MR, Kennedy RC (1989) Production and characterization of anti-idiotypic antibody reagents. Meth Enzymol 178:36–48

Schmidt DJ, Clarkson CE, Swanson TA, Egger ML, Carlson RE, Van Emon JM, Karu AE (1990) Monoclonal antibodies for immunoassay of avermectins. J Agric Food Chem 38:1763–1770

Schneider LE, Taghert PH (1988) Isolation and characterization of a *Drosophila* gene that encodes multiple neuropeptides related to Phe-Met-Arg-Phe-NH_2 (FMRFamide). Proc Natl Acad Sci USA 85:1993–1997

Scott JK, Smith GP (1990) Searching for peptide ligands with an epitope library. Science 249:386–390

Seed B, Aruffo A (1987) Molecular cloning of the CD2 antigen, the T-cell erythrocyte receptor, by a rapid immunoselection procedure. Proc Natl Acad Sci USA 84:3365–3369

Service MW (1985) *Anopheles gambiae*: Africa's principal malaria vector, 1902–1984. Bull Entomol Soc Amer 31:8–12

Shieh B-H, Stamnes MA, Seavello S, Harris GL, Zuker CS (1989) The *ninaA* gene required for visual transduction in *Drosophila* encodes a homologue of cyclosporin A-binding protein. Nature 338:67–70

Smith DP, Shieh B-H, Zuker CS (1990) Isolation and structure of an arrestin gene from *Drosophila*. Proc Natl Acad Sci USA 87:1003–1007

Souliotis VL, Dimitriadis GJ (1989) Identification and molecular analysis of the cuticle protein genes of *Dacus oleae*. Insect Biochem 19:499–507

Soutar AK, Wade DP (1990) Ligand blotting. In Creighton TE (ed) Protein Function: A Practical Approach. IRL Press, Oxford, pp 55–76

Stanker L, Bigbee C, Van Emon J, Watkins B, Jensen RH, Morris C, Vanderlaan M (1989) An immunoassay for pyrethroids: Detection of permethrin in meat. J Agric Food Chem 37:834–839

Stupp H-P, Peter MG (1984) Juvenile hormone III as a natural ligand for photoaffinity labeling of JH-binding proteins. Z Naturforsch 39C:1145–1149

Stupp H-P, Peter MG (1987) Synthesis of aryl azide analogs of insect juvenile hormones as reagents for the photoaffinity labeling of juvenile hormone binding proteins. Liebigs Ann Chem 1987:327–331

Sundseth SS, Kennel SJ, Waters LC (1989) Monoclonal antibodies to resistance-related forms of cytochrome P450 in *Drosophila melanogaster*. Pestic Biochem Physiol 33:176–188

Tam JP (1987) Solid-phase synthesis of type-α transforming growth factor. Meth Enzymol 146:127–143

Timblin C, Battey J, Kuehl WM (1990) Application for PCR technology to subtractive cDNA cloning: identification of genes expressed specifically in murine plasmacytoma cells. Nucl Acids Res 18:1587–1593

Tovey E (1990) Protein Blotting. Today's Life Science 2:56–67

Triglia T, Argyropoulos VP, Davidson BE, Kemp DJ (1990) Colorimetric detection of PCR products using the DNA-binding protein Tyr-R. Nucl Acids Res 18:1080

Trowell SC, Daly J (1990) A *Heliothis* identification kit. Fifth Australian Cotton Conference, Broadbeach, Queensland, ACGRA

Trowell SC, Healy MJ, East PD, Dumancic MM, Campbell PM, Myers MA, Oakeshott JG (1990) A molecular biological approach to the dipteran juvenile hormone system: Implications for pest control. In Casida J (ed) Pesticides and Alternatives: International Conference on Innovative Chemical and Biological Approaches to Pest Control. Elsevier North Holland Biomedical Press, Amsterdam, pp 259–270

Turkova J (1978) Affinity Chromatography. Elsevier Scientific, Amsterdam

Uhler MD, McKnight GS (1987) Expression of cDNAs for two isoforms of the catalytic subunit of cAMP-dependent protein kinase. J Biol Chem 262: 15202–15207

Vandekerckhove J, Bauw G, Puype G, Van Damme J, Van Montagu M (1985) Protein-blotting on polybrene-coated glass-fiber sheets. Eur J Biochem 152: 9–19

Váradi A, Gilmore-Heber M, Benz EJ (1989) Amplification of the phosphorylation site-ATP-binding site cDNA fragment of the Na^+,K^+-ATPase and the Ca^{2+}-ATPase of *Drosophila melanogaster* by polymerase chain reaction. FEBS Letters 258:203–207

Vaux D, Tooze J, Fuller S (1990) Identification by anti-idiotype antibodies of an intracellular membrane protein that recognizes a mammalian endoplasmic reticulum retention signal. Nature 345:495–502

Walker F, Burgess AW (1985) Specific binding of radioiodinated granulocyte-macrophage colony-stimulating factor to hemopoietic cells. EMBO J 4: 933–939

Ward LD, Reid GE, Moritz RL, Simpson RJ (1990) Strategies for internal amino acid sequence analysis of proteins separated by polyacrylamide gel electrophoresis. J Chromatog 519:199–216

Ward SE, Güssow D, Griffiths AD, Jones PT, Winter G (1989) Binding activities of a repertoire of single immunoglobulin variable domains secreted from *Escherichia coli*. Nature 341:544–546

Weber K, Kuter DJ (1971) Reversible denaturation of enzymes by sodium dodecyl sulfate. J Biol Chem 246:4504–4509

Welsh J, McClelland M (1990) Fingerprinting genomes using PCR with arbitrary primers. Nucl Acids Res 18:7213–7218

Williams JGK, Kubelik AR, Livak KJ, Rafalski JA, Tingey SV (1990) DNA polymorphisms amplified by arbitrary primers are useful as genetic markers. Nucl Acids Res 18:6531–6535

Winter G, Milstein C (1991) Man-made antibodies. Nature 349:293–299

Wiśniewski JR (1989) Identification of juvenile-hormone-binding proteins on blotted electropherograms using tritiated juvenile hormones. Experientia 45:1124–1128

Witz P, Amlaiky N, Plassat J-L, Maroteaux L, Borrelli E, Hen R (1990) Cloning and characterization of a *Drosophila* serotonin receptor that activates adenylate cyclase. Proc Natl Acad Sci USA 87:8940–8944

Yamada T, Takeuchi Y, Komori N, Kobayashi H, Sakai Y, Hotta Y, Matsumoto H (1990) A 49-kilodalton phosphoprotein in the *Drosophila* photoreceptor is an arrestin homolog. Science 248:483–486

Zuker CS, Cowman AF, Rubin GM (1985) Isolation and structure of a rhodopsin gene from *D. melanogaster*. Cell 40:851–858

Note added in proof

Readers are directed to a recent review (Williams et al. 1992) of subtractive immunization techniques for raising monoclonal antibodies to rare antigens, which compares a number of regimes including those that use cylophosphamide. Williams CV, Stechmann CL, McLoon SC (1992) Subtractive immunization techniques for the production of monoclonal antibodies to rare antigens. Biotechniques 12: 842–846

Chapter 12

Prospects and Possibilities for Gene Transfer Techniques in Insects

David A. O'Brochta and Alfred M. Handler

1. Introduction

The ability to analyze and manipulate organisms by classical genetic means has proven to be one of the most powerful tools for studying basic biological questions as well as for addressing important problems in agriculture and medicine. Today we have available a new set of genetic tools, those of molecular genetics, which allow researchers to more incisively address both basic and applied biological questions. Nevertheless, the full potential of these techniques has only been realized in those species whose genes are amenable to manipulation in vitro and efficient transfer back into their genome. Gene-transfer, or transformation, is now possible as a routine technique in a wide range of eukaryotes including mammalian and plant species. In insects, however, efficient gene transfer is presently limited to the genus *Drosophila*. This has resulted in a vast wealth of information regarding the genetics, development, physiology, and biochemistry of *Drosophila* species, much of which can probably be extended to other insect species. Nevertheless, the more applied benefits of molecular genetics, in particular the regulation of insect reproduction and behavior, remain to be attained in nondrosophilid insects. In addition, gene transformation has the potential to provide novel ways of elucidating genetic mechanisms and regulation in those insect species which have thus far not been amenable to classical genetic techniques. The basic biological information attained in this way would be likely, in turn, to greatly improve the efficacy of more conventional means of insect control.

In this chapter we will address the problem of insect gene transformation and discuss the efforts that we and others have made toward achieving this goal. The development of germline transformation methods for *D. melanogaster* leads to the anticipation of similar methods in other insect species, yet such methods have not been developed for nondrosophilids. However, information from *Drosophila* and other genetically well-defined systems has provided avenues of investigation, which may make germline transformation possible or facilitate the development of more straightforward somatic transformation (transient expression) techniques to attain similar goals. More than simply reviewing this subject, we hope that the overview presented here will help researchers confronting this problem make a proper evaluation of the approaches available.

1.1. Gene Transfer as a Tool for Fundamental Studies

One fundamental application of gene transformation in nondrosophilid insects would be the elucidation of the mechanisms that regulate gene activity and the physical mapping of genes in these insects, which to varying extents are a necessary prelude to most field applications. While conceptually this is a relatively straightforward process whereby a cloned gene is mutated in vitro, reintroduced into the genome and assessed using any one of many methods, technically these experiments are far from trivial.

Equally powerful is the ability of gene transformation methods to facilitate the identification and isolation of genes. This can be accomplished in a number of ways. Gene vectors, by virtue of their ability to integrate into chromosomes, are mutagenic (see Lambert et al. 1988). Genes identified by the technique of transposon mutagenesis (ffrench-Constant et al. this volume) can be isolated by using the transposon sequence as a molecular tag. Transposon tagging has been used to isolate dozens of genes from *D. melanogaster* (Kidwell 1987), and similar applications using mammalian retroviral gene vectors have been reported (Soriano et al. 1989). Genes can also be identified by their ability to complement a mutant phenotype (mutant rescue; Tearle et al. this volume). Currently, gene transformation methods permit the screening of relatively small regions of the *Drosophila* genome (10–100 kb) by mutant rescue. Efficient methods for introducing DNA and selecting for transformants would permite entire genomic libraries to be screened using mutant rescue methods (Pirrotta 1987).

Recently, a method based on transformation has been described to identify and isolate genes from *Drosophila* and mice without relying on either the complementation (mutant rescue) or induction (transposon tagging) of a mutant phenotype. This method (enhancer trapping) relies on the ability of regulating sequences (enhancers) within the genome to influence the expression of vector-encoded sequences inserted nearby. Specific enhancers involved in a particular process are identified by virtue of their influence on the cell or tissue specific expression of a vector-encoded reporter gene such as β-galactosidase (O'Kane and Gehring 1987; Allen et al. 1988; Bellen et al. 1989; Gossler et al. 1989; Wilson et al. 1989). Isolating DNA flanking the enhancer can result in the identification of the gene usually influenced by the enhancer.

Finally, gene transformation methods potentially can be used to replace a chromosomal homolog with a vector-residing alternative allele. Although gene replacement has been reported in a variety of systems (Hinnen et al. 1978; Miller et al. 1985; de Lozanne and Spudich 1987; Thomas and Capecchi 1987; Pazkowski et al. 1988), success has resulted largely from the existence of efficient and powerful selection schemes that can be imposed on large numbers of cells in vitro. The prospects for gene replacement in insects are real, although certain obstacles other than the development of gene vectors need to be overcome. Most notable is the need for insect cell lines that are analogous to mouse embryonic stem cell lines and allow gene transformation in vitro followed by the incorporation of transformed cells into the germline by injection of embryos. The clonal restriction of insect germ cells established during early embryogenesis will make this a difficult problem to solve (Technau and Campos-Ortega 1986).

1.2. Field Applications of Gene Transfer Technology

Some important basic information relevant to non-drosophilid insects can be extrapolated from gene transformation research in a variety of organisms, but notably from one insect system, *D. melanogaster*. Since this insect is of very limited economic importance, neither gene transformation nor recombinant DNA techniques in general have been pursued to regulate this species' reproduction or behavior outside the laboratory. Thus, the field applications of gene transformation remain highly prospective, though our knowledge of *Drosophila* does suggest reasonable approaches towards insect management. Field applications of gene transformation which may be of immediate use include: genetic sexing and sterilization for sterile male release programs; introduction of genetic markers to monitor gene flow, insect migration and dispersal patterns; conferring insecticide resistance to beneficial insects; and increasing insect susceptibility to control agents already in use (i.e., chemical insecticides, *Bacillus thuringiensis*, growth regulators, etc.). Notable among these examples is genetic sexing and sterilization of non-drosophilids since extensive classical genetic studies are presently under way worldwide in this area; a molecular genetic approach in this area would be highly efficient, and the possibility of containing genetically altered insects in the laboratory would facilitate the rapid implementation of such techniques. The general approach for constructing genetic sexing strains that permit male selection is to isolate a wild type gene or mutant allele, which can be either selected for in males or selected against in females. A classical genetic approach would be to isolate a translocation linking a gene conferring a dominant selectable phenotype to the Y chromosome. Such sexing strains have proven to be highly susceptible to breakdown (especially under large population conditions) due to chromosome instability, recombination, and mutant reversions (Foster et al. 1980; Rossler 1985; Hooper et al. 1987). Germline transformation would theoretically minimize such complications by creating small gene-fusions linking a sex-specific promoter to a gene encoding a selectable gene product. Genetic damage resulting from transformation would be negligible (or could be selected against) compared to mutation induction and chromosome rearrangements. Breakdown of the strains due to recombination or reversion could be minimized by multiple insertions.

Of particular interest are schemes which, although highly prospective for non-drosophilid insects, would allow both genetic sexing and male sterilization to occur in *D. melanogaster*. For example, manipulation of the recently cloned *transformer-2* (*tra-2*) sex-determination gene (Amrein et al. 1988; Belote this volume) could result in a strain of flies capable, under certain conditions, of producing only sterile males. Chromosomal females (XX; AA) homozygous for mutant *tra-2* alleles develop as morphological males capable of mating but are sterile. Chromosomal males (XY; AA) carrying the same mutations are morphologically normal and are also sterile. Thus, a strain of flies having *tra-2* function under conditional regulation could result in fertile parents giving rise to only sterile male progeny (Belote and Baker 1982). The possibility of such a highly efficient scheme working in non-drosophilids would depend upon recombinant DNA and transformation techniques to identify and isolate sex-determination genes having analogous functions to *tra-2*, as well as creating and testing appropriate gene-fusions.

1.3. Historical Perspective

The first reported attempts of whole animal, DNA-mediated transformation of insects appeared approximately 25 years ago when Caspari and Nawa (1965) reported the somatic transformation of *Ephestia*. In these experiments total DNA from wild-type moths was injected directly into the haemocoel of last instar larvae of animals homozygous for a wing scale coloration mutation and resulted in the recovery of adults with wild-type wing scales. Similar experiments were subsequently reported using the eye color mutation *red* (Nawa and Yamada 1968). Although a putative transformant was recovered, the unusual pattern of inheritance of the transformation marker was not consistent with a simple integration into the chromosome. Initial attempts to transform *Drosophila* by treating mutant embryos with wild-type DNA also yielded putative transformants which displayed unusual non-Mendelian inheritance of the transformation marker (Fox and Yoon 1966; 1970; Lui et al. 1979). The results in *Ephestia* and *Drosophila* were more consistent with the episomal transmission of marker DNA than with an integration event.

The first report of germline transformation consistent with the integration of a marker gene was by Germeraad (1976). *Drosophila* embryos homozygous for a mutant allele of the *vermilion* eye color gene were injected with wild-type DNA and lead to the recovery of progeny with a wild-type eye phenotype. Although never confirmed, the wild-type transformants obtained by Germeraad (1976) may have actually been germline revertants of the transposable element (412)-induced *vermilion* allele used in these experiments.

2. Gene Transfer with the *Drosophila* P Element

2.1. In *D. melanogaster*

Fully 10% of the genome of *D. melanogaster* is composed of a unique class of DNA sequences called transposable elements (Young 1979; Young and Schwartz 1981). One of the most active and well characterized families is the P element family (for a review, see Engels 1989). P elements were discovered by virtue of their involvement in the phenomenon known as hybrid dysgenesis, a syndrome including elevated mutation rates, chromosome rearrangements and sterility. Complete P elements are 2.9 kb in length with 31 bp terminal inverted repeats. They contain a single transcription unit composed of four open reading frames encoding a protein (P transposase) required for P excision and transposition (Figure 12.1). Although implied by the name, P transposase has no known enzymatic activity in vivo, and its precise role in P mobilization is unknown. No other factors are known to be required for P mobilization, although the involvement of another protein has been inferred (Rio and Rubin 1988).

The movement of P is highly regulated. Only particular strains of *D. melanogaster* can support P mobility, and extensive genetic evidence indicates that a maternally inherited cellular condition known as a P cytotype is responsible for repressing P movement in some strains. The basis of cytotype repression is not fully understood. In those strains with a nonrepressive cytotype, termed the M cytotype, P mobility is confined to

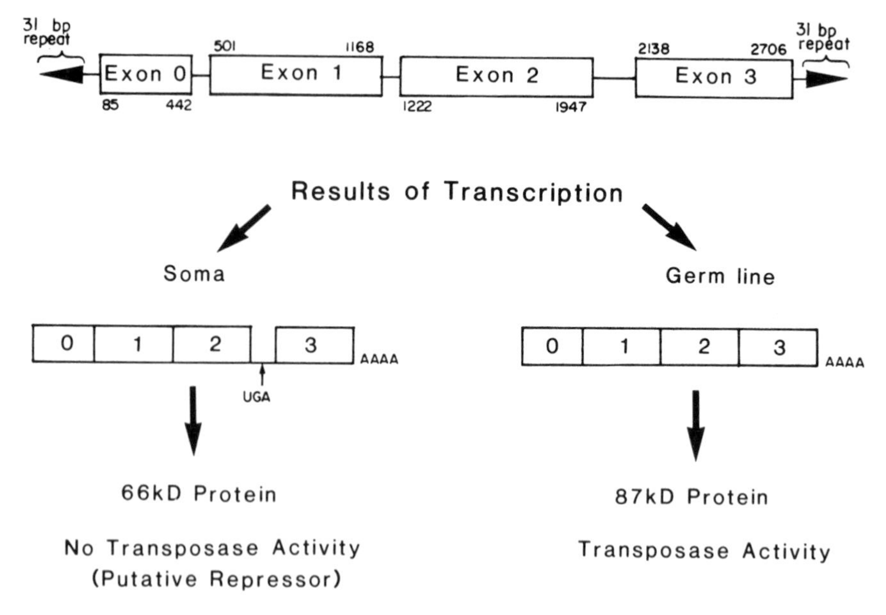

Figure 12.1. Structure of the complete *Drosophila melanogaster* P element. Transcription occurs in somatic and germ cells, however, complete transcript processing leading to functional transposase only occurs in germ cells.

the germline. This restriction results from the germline-specific processing of the third intron from transposase transcripts (Figure 12.1).

The ability to mobilize P elements by expressing transposase in *trans* facilitated their use as efficient gene vectors (Rubin and Spradling 1982; Spradling and Rubin 1982). The generation of transgenic *Drosophila* has become, in the course of a few years, a routine laboratory procedure. Specific details about the physical operations involved in producing transformants, such as the preparation of eggs, needles, and injection procedures, have been described (Spradling 1986) and the general strategy is outlined in Figure 12.2. The general efficiency of this method is difficult to estimate since it is dependent on the host, the size of the vector, concentration of injected DNA, and the way in which the embryos are handled. Nevertheless, Pirrotta (1988) estimated that, following careful injection, 80% of the embryos can be expected to hatch, with 40–50% reaching adulthood. A variable percentage will be sterile as a result of damage or abnormalities induced in the germline by injection. Usually 10–60% of the fertile G_0 animals will give rise to at least one transformed progeny. Very large vectors such as the cosmid vectors, described by Pirrotta (1988), usually result in reduced transformation efficiencies.

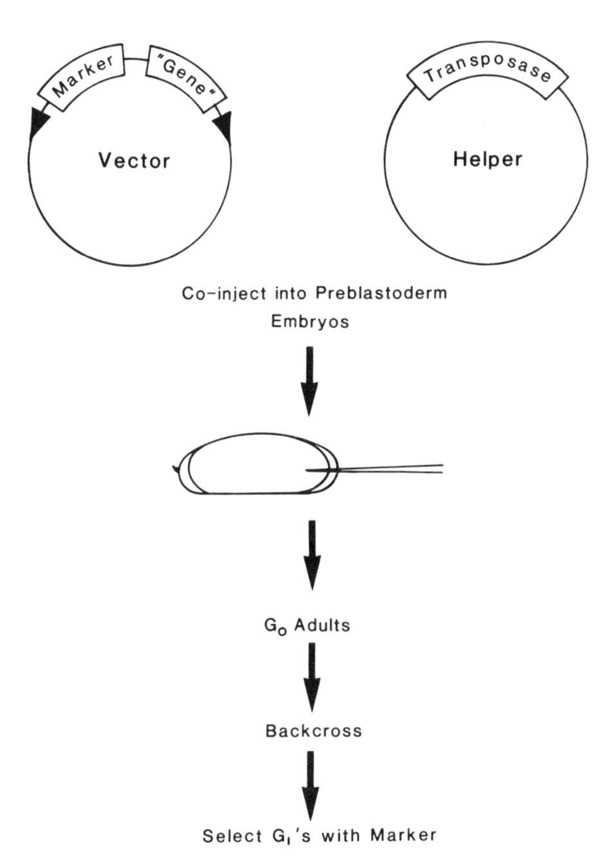

Figure 12.2. Overview of transformation in *Drosophila*. Vectors typically consist of P termini with flanking DNA required in *cis* for transposition, as well as a marker gene, the expression of which results in an identifiable phenotype, and the gene of interest. Vector-containing plasmids are coinjected into preblastoderm embryos with a "helper" plasmid encoding functional transposase. The P sequences contained on the helper are incapable of transposing. Upon cellularization, injected plasmid DNA is incorporated into the presumptive germ cells. Transposase-mediated transposition can occur from plasmids to chromosomes. Chromosomes within the germ cells acquiring a P insertion are identified in the following generation as progeny with a newly acquired phenotype.

2.2. In Other Insects

With the development of gene vectors based on the P element, the prospects for genetically manipulating non-drosophilid insects also improved, and initial results from drosophilid species other than *D. melanogaster* reinforced that optimism. Furthermore, the successful application of the neomycin analog, G418, as a selection agent (Rio and

Rubin 1985) and the elucidation of the basis of tissue specificity of P movement (Laski et al. 1986) encouraged those desiring to transform genetically less well defined insects. Here we review the efforts to use P as a vector in species other than *D. melanogaster* and discuss what is known about the ability of P to function in these other systems.

2.2.1. Drosophilids

The phylogenetic distribution of P elements within the family Drosophilidae is rather unusual. P is found in many species within the subspecies Sophophora but not in some of the closest relatives to *D. melanogaster*, such as *D. simulans* or *D. mauritiana* (Brookfield et al. 1984; Lansman et al. 1985; Daniels and Strausbaugh 1986). There have been reports of sequences homologous to P in species outside the family Drosophilidae, but the precise nature of these sequences is unknown (Anxolabehere and Periquet 1987). Based on the discontinuous distribution of P, both within the species *D. melanogaster* and the subgenus Sophophora, it has been suggested that P was introduced into this species within the last 100 years by the interspecific transfer from a species within the *willistoni* group (Engels 1989).

Scavarda and Hartl (1984) and Daniels et al. (1985) demonstrated that, in the presence of transposase, P elements can be mobilized in the germline of *D. simulans*. Transformation frequencies comparable to those observed in *D. melanogaster* were reported, as well as subsequent increases in P element copy number. Transformed *D. simulans* strains developed characteristics similar to *D. melanogaster* P strains, i.e., a P cytotype and the ability to induce hybrid dysgenesis when crossed to strains of *D. simulans* devoid of P elements (Daniels et al. 1985). The mobilization of P also appears to be confined to the germline of *D. simulans* (DA O'Brochta unpublished observation). The behavior of P in *D. simulans* appears to be quite similar to that seen in *D. melanogaster*.

The more distantly related species, *D. hawaiiensis*, has also been shown to support P transposition. Brennan et al. (1984) coinjected the plasmids pπ25.1 and pASX2 and pASX4 into the preblastoderm embryos of *D. hawaiiensis*. The latter two plasmids contain a 5.2 kb fragment carrying the alcohol dehydrogenase gene of *D. affinidysjuncta* inserted into a nonautonomous P element. Although relying on Southern blot hybridization to detect transformants, Brennan et al. (1984) found two G_0 adults of the 34 screened which produced transformed progeny. Continued transposition of P over successive generations was reported, as was the development of female sterility in those strains that contained the most P elements. Only integrations of pπ25.1 were detected, suggesting that either *trans*-mobilization of P is not possible in this species or that P element mobility is subject to more stringent size limitations.

2.2.2. *Non-Drosophilid Insects*

The demonstration of P transposition in species of *Drosophila* other than *D. melanogaster* suggested that P might be useful in non-drosophilids as well. The development of P vectors carrying the neomycin phosphotransferase (NPT) gene potentially allowed genticin (G418) resistance to be used as a means of selecting transformed individuals.

Recently, transformation studies involving three species of mosquitos were reported in which P element vectors harboring the neomycin resistance gene were apparently integrated into the host's genome, resulting in germline transformation. Miller et al. (1986) reported injecting 2,279 *Anopheles gambiae* eggs with pUChsπΔ2-3 and pUChsneo of which 14% survived to adulthood (145 females and 175 males). The plasmid pUChsneo contains a copy of the NPT gene from the *E. coli* transposon Tn5 under the control of the *D. melanogaster* heat shock promoter (Pirrotta 1988), while pUChsπΔ2-3 contains the transposase gene lacking the third intron and under heat shock promoter control (Laski et al. 1986). The hsp70-NPT portion of pUChsneo resides between the termini of a nonautonomous P element. G_0 adults were mass mated *inter se* and the progeny selected on G418. Of the 9283 larvae selected, 55 survived to adulthood and 36 were subject to Southern analysis. Thirteen animals were found with an integrated copy of pUChsneo, and all yielded identical patterns of hybridization following Southern analysis, indicating that a single premeiotic integration event occurred. The data were consistent with a single recombination event between the host's genome and the *white* gene sequences contained on pUChsneo. This integration event was not the result of a P element mediated transposition and was stable over at least four generations.

More recently, McGrane et al. (1988) reported the recovery of a germline integration of P in the mosquito *Aedes triseriatus*. Like Miller et al. (1987) they used the plasmids pUChsneo and pUChsπΔ2-3. Of the 949 embryos injected, 64 hatched and 57 survived to adulthood. These G_0 adults (23 males and 34 females) were backcrossed to uninjected colony mosquitos and their progeny subjected to G418 selection. Fourteen of the 3,800 progeny selected survived and were again backcrossed to uninjected colony mosquitos. Two hundred and fifty-one of the 10,610 G_2 progeny screened survived to adulthood of which 132 were tested by Southern analysis. Eight yielded patterns of hybridization indicating the presence of sequences homologous to pUChsneo, however, only two different integration patterns were detected. Neither of the integration patterns were consistent with a P element-mediated transposition event, nor were they consistent with a simple recombination event between the vector and the host's genome as reported by Miller et al. (1987). No subsequent G_3 survivors of G418 selection showed any evidence of integrated DNA following Southern analysis, and the precise nature of the putative integration events is unknown.

Identical experiments were conducted with *Aedes aegypti* (Morris et al. 1989). Seventy-one surviving G_0 adults were backcrossed to noninjected individuals. G_1 progeny were selected for resistance to G418 and survivors mated individually to noninjected individuals. The number of surviving G_1's was not reported, however, only one had evidence of sequences homologous to pUChsneo, based on Southern analysis. While the authors suggested that this may have been a P element-mediated transposition, the results of their Southern analysis were not completely consistent with this interpretation. Furthermore, plasmid rescue experiments failed to recover the pUC8 replicon from these putative transformants.

Attempts to transform other species of insects have been made, and all have been less successful than the experiments with mosquitos described above. Efforts by our laboratory and others have failed thus far to recover a P element-mediated integration event in tephritid fruit flies. In the medfly, *Ceratitis capitata*, transformation studies by several laboratories, involving injection of many thousands of embryos, have failed to yield a genetic transformant. McInnis et al. (1988) injected over 15,000 embryos, which yielded 353 G_0 adults, from which 81 matings resulted in progeny surviving G418 selection. Succeeding generations showed variable but decreased resistance to G418. It was concluded (although not specified) that transformation had not occurred, and the observed G418 resistance was either inherent in the tested population or resulted from the extrachromosomal expression of the vector-encoded NPT gene. In a similar study Robinson et al. (1988) reported the injection of 6250 embryos of which 8.1% reached adulthood. In this experiment G_0 adults were inbred, and although a few larvae survived G418 selection, resistance was variable in succeeding generations, and P element integration was not evident by Southern blot analysis. Similarly, Walker (1989) reported that attempts to transform the germline of *Locusta migratoria* with P elements failed to yield a P-mediated integration event.

The attempts to transform the germline of the various insect species described above, although largely unsuccessful, have allowed us to recognize some of the limitations of attempting to assess the mobility properties of P elements in heterologous systems using P element vectors carrying dominant selectable markers. The major limitation is that P mobility can only be detected following transposition, expression of the dominant marker, and the appearance of a new phenotype. It cannot, however, be assumed that P element vectors will have the same mobility properties as unmodified P elements, or that P elements in heterologous environments will have the same structural requirements as P elements in *Drosophila*. For example, the frequency of transposition of P elements in *D. melanogaster* appears to be related to the size of the element and the amount of DNA that the element contains. Robertson et al. (1988) observed fewer transpositions of large elements (e.g., $P[w^+]$) compared to smaller elements ($P[ry^+]$), and the transposition rate of marked P elements ($P[w^+]$ and $P[ry^+]$) was significantly lower than naturally occurring P elements. Finally, Spradling (1986) has shown that small P elements transform embryos more readily than large ones. How these size limitations, defined in *Drosophila*, will apply in other species is not known.

In germline transformation experiments using vectors carrying a genetic marker, transpositions are recognized following the expression of the marker gene and the acquisition of a new phenotype. The transformation experiments described employed the neomycin phosphotransferase gene under the control of the *Drosophila* heat shock promoter as a transformation marker. The ability of neomycin phosphotransferase to confer resistance to the neomycin analog G418 in species other than *D. melanogaster* is largely undetermined. In *D. melanogaster* its use is problematic because of the leakiness of the sensitivity phenotype, the high variability in sensitivity from strain to strain and the frequent occurrence of nontransformed escapers. Similar problems were encountered in using G418 to select for transformed mosquitos. Only the *Anopheles gambiae* transformant was shown to contain and express the NPT gene. The presence and expression of the NPT gene in the putative *Aedes triseriatus* and *Aedes aegypti* transformants were not reported. Both species produced many nontransformed escapers under the selection used, and from the data reported it is not clear if G418 selection even enriched for putative transformants in these species. Similar problems with G418 in tephritids have been encountered. Taken together the data allow us to conclude that the vector system employed (pUChsneo: pUChsπΔ2-3) may not be useful in non-drosophilids. However, these data do not allow us to draw conclusions about the mobility properties of P elements in heterologous systems. The failure to recover P-mediated transformants may have resulted from the absence of P mobility, the inability to confer whole animal resistance to G418, different size and structural constraints on P elements, or the unavailability of P element target sequences in the host genome.

2.3. In Non-Insect Systems

The successful use of P elements as gene vectors and mutagenic agents in *Drosophila* has resulted in an interest in their use not only in other insect systems but in plant and animal systems as well. P elements have been integrated into the germline of mice (Khillan et al. 1985) and murine cell lines (Clough et al. 1985) but in neither case was P transposition involved. Rio et al. (1988) reported the mobilization of P elements in cultured monkey kidney cells stably transformed with a P transposase gene lacking the third intron, and under the promoter control of the simian virus (SV40) early promoter. After transfecting these cell lines with P excision indicator plasmids (pISP or pISP-2) excision products were recovered. (See the following section for a more complete description of the excision assay.) In light of the rather extensive efforts to observe P mobility in heterologous systems in vivo (see also the following section), these in vitro results remain enigmatic. It is known that the mobility properties of transposable elements in cell lines do not necessarily resemble those observed in vivo, and therefore caution is required in interpreting results from in vitro experiments. For example, the *D. melanogaster* retrovirus-like element *copia* shows little evidence of movement in vivo except under extremely stressed conditions (Junakovic et al. 1986) or in particular genetic backgrounds (Pasyukova et al. 1988; Mevel-Ninio et al. 1989), yet is readily mobilized in cell lines (Potter et al. 1979). As suggested by Rubin (1983), the control of transposition may be altered in cells during adaptation to, or growth in, cell culture. The mobilization of P in mammalian cell lines, therefore, may be of limited significance beyond suggesting that other *Drosophila*-specific factors are not necessarily involved in the process.

The P element system has also been analyzed to a limited extent in plants. Martinez-Zapata et al. (1987a; 1987b) introduced a non-autonomous P element and hs$\pi\Delta$2-3 into tobacco and petunia plants using an *Agrobacterium tumefaciens* Ti plasmid as the vector. They were unable to detect any evidence of P excision or transposition in plants regenerated from leaf discs of transformed plants, however, transposase gene expression from hs$\pi\Delta$2-3 was detected. Most of the P-homologous transcription products in these transgenic plants were prematurely truncated (1.5 kb), although full length transcripts (2.5 kb) were detected.

P elements have been introduced into yeast, and although they were not mobile, the expression of a transposase cDNA in the repair deficient mutant *rad 52* resulted in a lethal phenotype (Rio et al. 1988). *rad 52* mutants are unable to repair double stranded breaks in DNA, suggesting that P transposase may have double stranded DNA break activity. However, expression of the cDNA used in these experiments in *D. melanogaster* does not result in the mobilization of P (DA O'Brochta unpublished observation), and while double stranded DNA break activity is a presumed function of P transposase, it may not be involved in normal in vivo activity.

Notwithstanding the unusual results obtained with monkey kidney cell lines, efforts to mobilize P in noninsect systems have yielded results similar to those obtained in nondrosophilid insects. These results support the general conclusion that P mobility is phylogenetically restricted. Direct in vivo analysis of P mobility in nondrosophilid insects described in the following section further supports this conclusion.

2.4. Evaluating P Element Mobility

Here we review what is known about the ability of P elements to function in non-drosophilid insects, independent of their ability to act as a gene vectors. Addressing this issue is critical since currently available P vectors will, apparently, have to be modified for use in non-drosophilids. To avoid the limitations of transformation experiments discussed previously, methods were required that allow the assessment of P mobility directly.

2.4.1. P Element Excision Assays

As with other transposable elements, the transposition and insertion of P into a gene often results in partial or complete elimination of gene activity and function. Reversion of P insertional mutations by the excision of P from the gene can restore the original gene activity and is therefore a convenient way in which to monitor P movement directly. Rio et al. (1986) constructed plasmids (pISP and pISP-2) which permitted the detection of certain P element excisions from a plasmid-residing gene, β-galactosidase. A small nonautonomous P element surrounded by *D. melanogaster white* gene sequences was inserted into the region coding for the lacZ alpha peptide in pUC8. This was done in such a way that the lacZ alpha peptide, which is required for β-galactosidase activity in the appropriate host, was nonfunctional. In the presence of functional transposase and any other factors which may be required, the P element can excise, resulting in the restoration of lacZ alpha peptide expression and β-galactosidase activity. While Rio et al. (1986) used this assay to monitor transposase activity in cell lines, we have modified this assay so P mobility can be directly assessed in the somatic cells of insect embryos. This has enabled us to directly address the question of P mobility in non-drosophilids (O'Brochta and Handler 1988).

The excision assay developed by Rio et al. (1986) only permitted the detection of P excision events which resulted in the reversion of the LacZ phenotype. Genetic evidence from *Drosophila* indicates that there is another class of excision event involving the excision of P without resulting in a phenotypic reversion. In *D. melanogaster* this type of excision can be quite frequent (Searles et al. 1986; Tsubota and Schedl 1986). Therefore, the excision assay just described allows one to witness only a fraction and probably a minority of P excision events. Furthermore, we do not know how or if the proportions of "precise" excisions (phenotype reversions) will change in heterologous systems. To overcome these limitations we constructed a P excision indicator plasmid (pπstrep^{s}) which permitted all excision events to be detected. This was accomplished by inserting the *E. coli* ribosomal protein gene, S12, into a nonessential site of the P element contained in pISP-2. Certain mutations in the chromosomal S12 gene result in streptomycin resistance in *E. coli* and expression of the wild-type S12 gene in such mutants *in trans* restores streptomycin sensitivity (Dean 1981). Therefore, expression of S12 in the appropriate host results in a dominant lethal phenotype in the presence of streptomycin. By incorporating this streptomycin sensitivity gene into the pISP-2 excision indicator plasmid, one can witness all excision events and simultaneously determine the rate of "precise" and "imprecise" excisions by monitoring the LacZ phenotype (Figure 12.3).

Figure 12.4 shows a cumulative summary of P mobility data collected in various species of drosophilids and non-drosophilids using pISP and pπstrep^s. P element mobility was readily detected in all drosophilids tested, including species in the genera *Chymomyza* and *Zaprionus*. In M strains of *D. melanogaster* an excision frequency of approximately 1–2 per 10^3 was seen using pISP-2, whereas a somewhat higher frequency was observed using pπstrep^s (5×10^{-3}). The ratio of P excisions resulting in LacZ reversions ("precise") to those which did not ("imprecise") was approximately 1:2. Within the family Drosophilidae, species most distantly related to *D. melanogaster*, i.e., *Chymomyza* and *Zaprionus*, were the least capable of supporting P mobility. For example, *Chymomyza* supported excision of P from pISP-2 at a rate of 0.2×10^{-3}, while using the total excision assay the rate was about 0.5×10^{-3}.

Sequence analysis of excision products recovered from *Chymomyza proncemis* and *D. melanogaster* indicated that most of the excision reaction products contain both copies of the duplicated eight base pair target site as well as sequences from the P termini (Figure 12.5). Excisions resulting in the perfect restoration of the original target site were rare. In *Chymomyza* many of the excision products had at least one breakpoint in sequences flanking P and the 8 bp target sequence. Occasionally, breakpoints were found in flanking vector sequences approximately 50 bp or more from the usual sites of transposase activity. Thus the activity of transposase and its interaction with P in *Chymomyza*, although similar to that seen in *D. melanogaster*, was unique. This suggests that ancillary factors are required for P movement and that phylogenetic differences may result in reduced and altered transposase activity.

The analysis of P mobility in species outside the family Drosophilidae has, in most cases, failed to detect any evidence of P excision. In those cases where putative excision products have been recovered, their frequency was not above the background level of spontaneous, transposase-independent excisions that were occasionally observed in these species. In addition to the species reported in Figure 12.4, P mobility has been tested using the P excision assay in *Borborillus frigipennis* (family: Sphaeroceridae), *Musca domestica* (family: Muscidae; DA O'Brochta unpublished observations) and *Lucilia cuprina* (family: Calliphoridae; PW Atkinson pers comm), and in all cases P mobility was not detected. Taken together, these data indicate that P elements cannot be mobilized

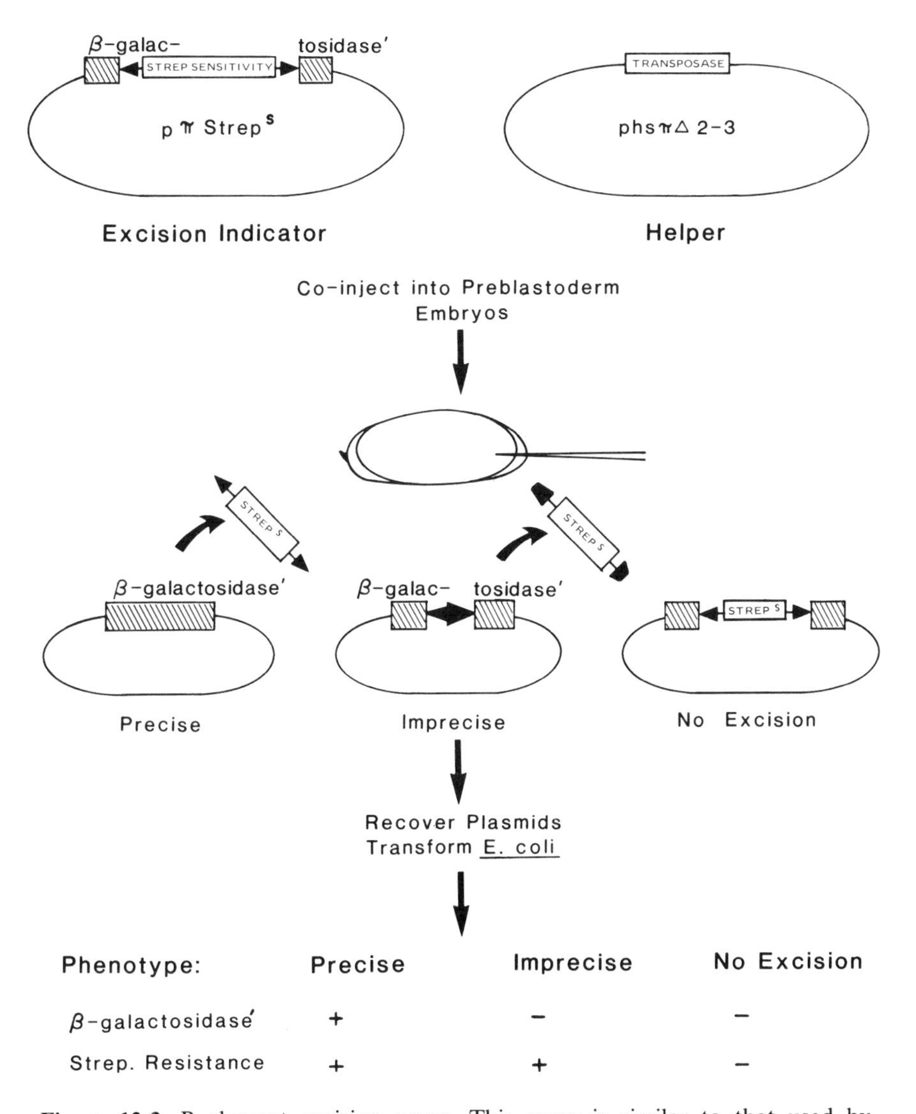

Phenotype:	Precise	Imprecise	No Excision
β-galactosidase'	+	−	−
Strep. Resistance	+	+	−

Figure 12.3. P element excision assay. This assay is similar to that used by O'Brochta and Handler (1988), however, it does not suffer from some of the same limitations. By incorporating the *E. coli* streptomycin sensitivity gene into the original excision indicator plasmid (pISP-2; Rio et al. 1986), all excision events can be detected and recovered. The excision indicator plasmid and a transposase "helper" plasmid are coinjected into preblastoderm embryos. The embryos are allowed to develop, and plasmids are recovered approximately 24 hours post-injection. Recovered plasmids are introduced into the appropriate *E. coli* host, and transformants containing excision products are identified. Because the transposase gene and excision indicator are on independently selectable replicons, excision indicator plasmids can be screened exclusively.

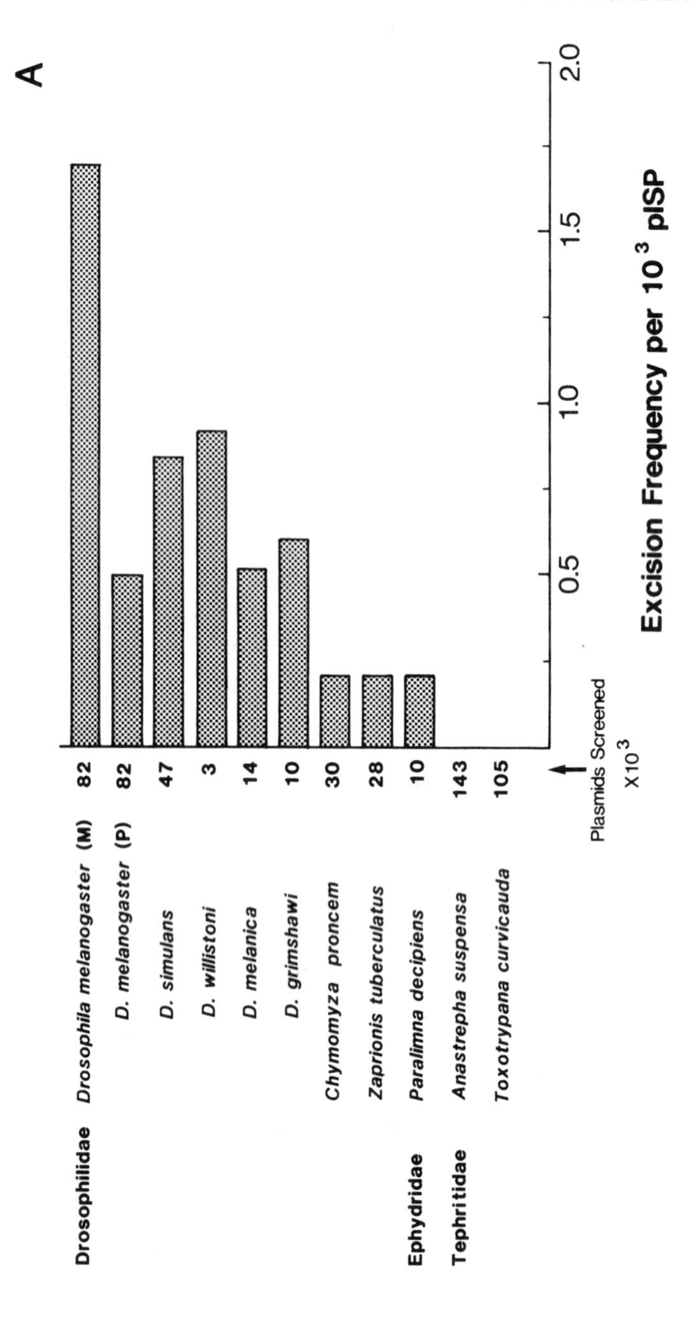
A
Drosophilidae
Ephydridae
Tephritidae
Drosophila melanogaster (M)
D. melanogaster (P)
D. simulans
D. willistoni
D. melanica
D. grimshawi
Chymomyza proncem
Zaprionis tuberculatus
Paralimna decipiens
Anastrepha suspensa
Toxotrypana curvicauda
82
82
47
3
14
10
30
28
10
143
105
Plasmids Screened
X10^3
0.5
1.0
1.5
2.0
Excision Frequency per 10^3 pISP

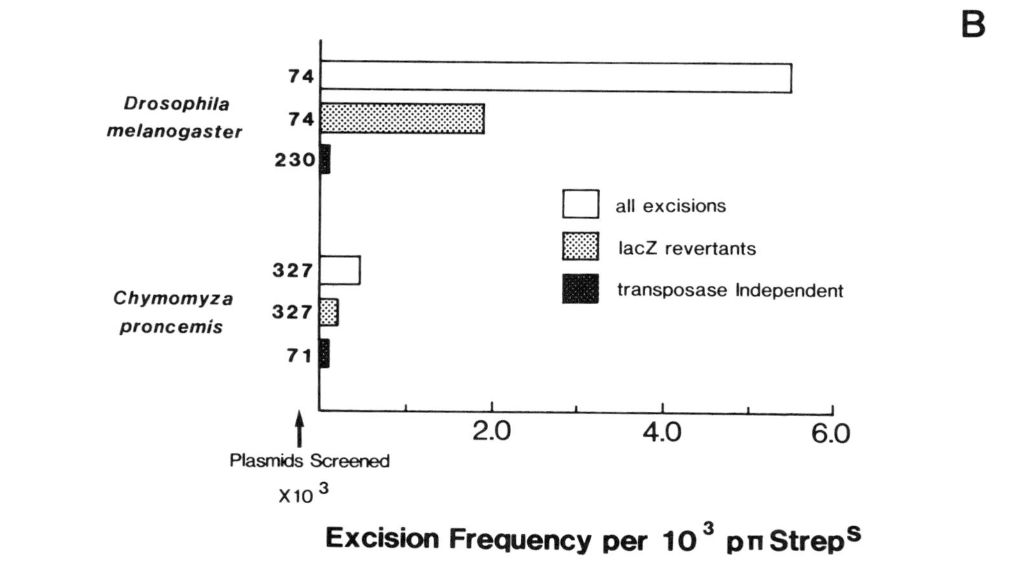

Figure 12.4. **A)** P excision using pISP. P mobility was tested using the original P excision indicator plasmid pISP (Rio et al. 1986). Data for Drosophilidae and Tephritidae are from O'Brochta and Handler (1988). Excision products in these experiments were recognized as LacZ$^+$ revertants. No excision was detected in A. *suspensa* or *T. curvicauda*. **B)** P excision using pπstrep^s. P mobility was monitored in *D. melanogaster* and *C. proncemis* using the "total" excision assay. The excision assay was conducted as described in Figure 3.

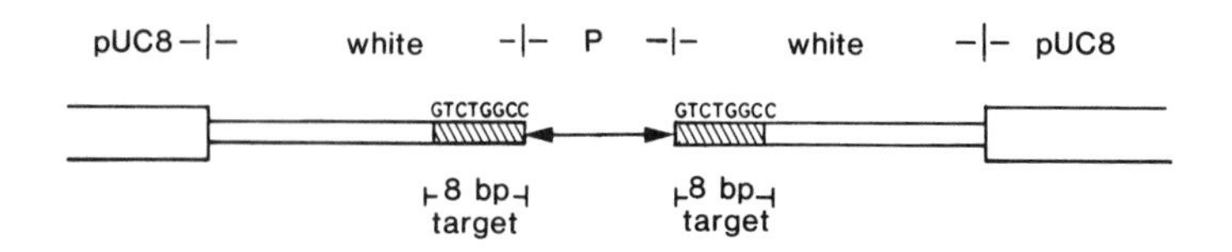

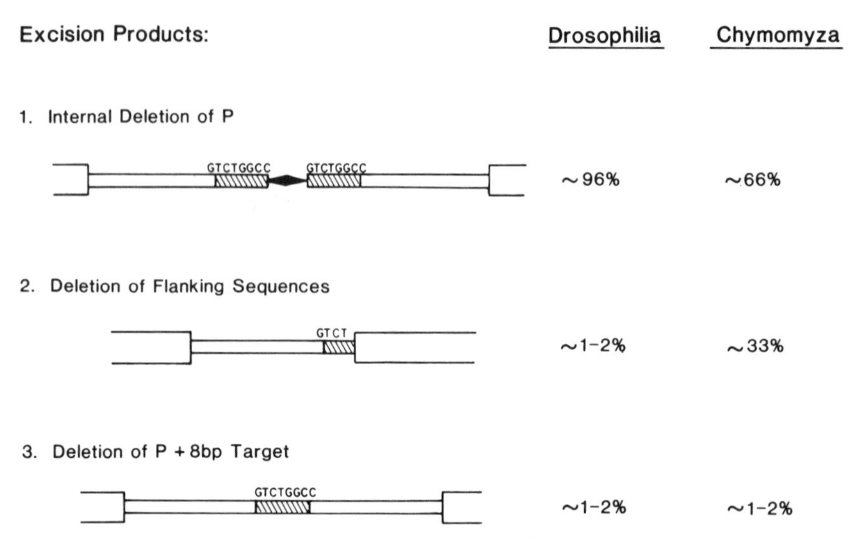

Figure 12.5. Structure of P element excision products. Excision products recovered using the "total" excision assay were sequenced. The products (empty donor sites) were classified into one of three classes. The distribution of excision products was different in the species compared. A higher percentage of the products recovered from *C. proncemis* compared to *D. melanogaster* had break-points in sequences flanking P. Precise excision of P, restoring the original target sequence, was rare.

in species outside the family Drosophilidae under the conditions described. Even within the family, species most unrelated to *D. melanogaster* are the least capable of supporting P excision. Therefore, there appears to be a phylogenetic restriction on P mobility.

2.5. Relevance of Somatic P Element Excision Assays

Most of the plasmids injected into preblastoderm embryos are eventually incorporated into the somatic cells of the developing embryo. Although there is nothing to prevent their incorporation into the presumptive germ cells, the plasmids contained within these cells contribute insignificantly to the total because the germline represents such a small fraction of all cells at the cellular blastoderm stage (DA O'Brochta unpublished observation).

Given a source of functional transposase in the soma and germline, there is no evidence indicating that either the properties of P transposase or the mechanisms regulating P mobility are unique to germ cells. Germ cells and somatic cells can support P excision and P transposition (Laski et al. 1986), and the rate of P excision in the soma is comparable to that measured in the germline, indicating similar transposase activity. For example, the germline rate of P excision from the *white* allele $w^{hd80kl7}$ is approximately 10^{-3} (Rubin et al. 1982), while the somatic excision rate based on excision of P from pISP-2 is also about 10^{-3} (O'Brochta and Handler 1988). The P element and donor site present on pISP-2 are identical to those present in $w^{hd80kl7}$. Finally, cytotype repression is a property of somatic cells as well as germ cells, indicating that the regulation of P in these two cell types is similar (Robertson et al. 1988).

2.6. Transposase Gene Expression in Heterologous Systems

Attempts to mobilize P elements in heterologous systems have relied largely on the use of a transposase gene lacking the third intron and under the transcriptional control of the *D. melanogaster* heat shock promoter (hsp70). Although this promoter has been shown to function and be heat inducible in a variety of animal and plant systems (Durbin and Fallon 1985; Berger et al. 1985; Martinez-Zapata et al. 1987a; 1987b), the heat shock mediated transcription of the transposase gene has not always led to full length, 2.8 kb transcripts. Martinez-Zapata (1987a; 1987b) reported the presence of an abundant 1.5 kb transposase transcript in tobacco plants containing a genomic copy of hs$\pi\Delta$2-3. Sl nuclease protection analysis of the 3′ end of these truncated transcripts indicated that transcription terminated just beyond the second intron. Truncated messages have also been detected in the embryos of tephritid fruit flies and *D. melanogaster* in which hs$\pi\Delta$2-3 has been transiently expressed (O'Brochta and Handler 1988). The significance of these observations is not entirely clear, although inefficient transcription of the transposase gene in heterologous systems may limit the available functional transposase. This could result in an elimination or reduction in P mobility in these systems, or in protein products which interfere with the functioning of complete transposase, thereby acting as repressors.

The P element excision assays described above provide efficient means of analyzing P mobility and with slight modifications can be used to identify and analyze P repressors. In addition to injecting the excision indicator plasmids (pISP-2 or ppstreps) and a transposase source, a third plasmid encoding a putative repressor can also be injected and expressed.

Repressors can be detected by a reduction in the P excision frequency from the indicator plasmids. Using this method we have shown that the somatic expression of an unmodified transposase transcription unit in *D. melanogaster* results in the repression of P mobility (DAO'B, SP Gomez and AMH unpublished observations). Expression of an unmodified transposase transcription unit in the soma results in the production of incompletely processed transcripts due to the absence of intron III splicing (Figure 1). These transcripts encode a 66kD protein product. Other deletion derivatives of the P element have also been implicated in producing repressors, however our understanding of the range of deletion derivatives of P capable of producing repressors is incomplete (Black et al. 1987; Nitasaka et al. 1987; Engels 1989). Therefore, truncated transcription products detected in tephritid embryos transiently expressing hspL2–3 and in transgenic tobacco plants may have resulted in the production of P mobility repressors.

Improper or incomplete processing of transposase transcripts in heterologous systems may also have resulted in the failure to mobilize P. Because P demonstrates strict cell-type specific regulation in *D. melanogaster*, it has been presumed that this may result in the failure of P to be mobilized in non-drosophilids. Therefore, efforts to transform non-drosophilids have involved the use of a modified transposase gene lacking the third intron (hspΔ23). However, the processing of introns I and II, while not regulated tissue specifically in drosophilids, may be species limited. In their analysis of truncated transposase transcripts in tobacco, Martinez-Zapata et al.(1987a; 1987b) also showed that there was a significant accumulation of unprocessed transcripts (with respect to introns I and II). Recently, we have found that splicing of intron I either does not occur or occurs inefficiently in the soma of tephritid fruitflies. This was shown using a splicing reporter plasmid constructed by fusing the *E. coli* β-galactosidase gene in-frame to the second exon of the P transposase gene. This fusion gene was under the control of the hsp70 heat shock promoter and resulted in β-galactosidase expression only following the accurate splicing of exons 0 and 1. Transient expression of this reporter plasmid in *D. melanogaster* embryos resulted in sufficient β-galactosidase activity to be detected by staining whole embryos, while similar experiments with embryos of the tephritid fruitfly, *Anastrepha suspensa*, did not result in detectable expression of β-galactosidase. Direct expression of the β-galactosidase gene in *A. suspensa* embryos did, however, result in sufficient activity to be detected in stained embryos. These results suggest that in tephritid fruitflies intron I processing either does not occur, occurs inefficiently or occurs inaccurately.

The results from tephritid fruitflies and transgenic tobacco plants expressing hsπΔ2-3 suggest that P transcript processing may be inefficient in heterologous systems and that tests of P mobility relying on intron I and II-containing transposase genes may not have been conducted in the presence of functional transposase. A P transposase cDNA under the promoter control of the heat shock promoter might ameliorate this problem, however the transposase cDNA currently available is nonfunctional, based on its failure to catalyze P mobility in *D. melanogaster*. The lack of function of the transposase cDNA may be caused by mutations induced during cloning, or might reflect a requirement for message processing in the biogenesis of stable transposase message.

3. Beyond P: Other Problems and Possibilities

The previous section has shown that the mobility of P is currently restricted to drosophilids, but that recent results suggest modifications of P which may relieve this restriction. While the analysis of P in other insects should therefore be continued, alternate systems should also be considered. In this section we show how efforts to transform non-drosophilids with P elements have also revealed other aspects of the problem as important as vector development, namely efficient DNA delivery systems and reliable genetic markers. Finally, while considerable emphasis has been placed on germline transformation, alternatives such as somatic transformation or transient expression should be considered.

3.1. DNA Delivery Systems

Random integration of DNA introduced into insect eggs has been reported at a frequency of approximately 1% or less (Miller et al. 1987; McGrane et al. 1988; Morris et al. 1989), and given a more efficient means of delivering DNA, this could become a useful method for achieving germline transformation. For many higher dipterans the direct injection of eggs is straightforward and survival can be high. Nevertheless, the need to inject individual eggs is limiting. Recently, a number of "mass injection" methods have been described which have been applied to noninsect systems, and there is some evidence that they may be applicable to insect systems as well. Klein et al. (1987) constructed a device that is capable of propelling DNA-coated microprojectiles at target tissues or cells. Under appropriate conditions some of these microprojectiles can enter the target cells, thereby delivering the encoating DNA. This system has been used to introduce DNA into a variety of cells and organelles, such as the epidermal cells of *Allium cepa* (Klein et al. 1987), mitochondria (Johnston et al. 1988), chloroplasts (Boynton et al. 1988), and limited success has been achieved with insect systems (Carlson et al. 1989). A variation of this method that does not rely on a "particle gun" to produce high speed microprojectiles involves the vortexing of glass beads in a

DNA solution. Costanzo and Fox (1988) recently reported the successful transformation of yeast using this method, and although less efficient than more commonly used transformation procedures, vortexing was considerably more convenient. Preliminary evidence indicates that vortexing insect embryos in a DNA solution containing silicon carbide powder, or "whiskers," can result in the introduction of DNA (A Cockburn pers comm).

Short pulses of high voltage have been used to permeablize a variety of prokaryotic and eukaryotic cells and have been used to introduce DNA into cells (Shigekawa and Dower 1988). Although this technique, termed electroporation, has been used to transform individual cells, its application to dechorionated insect eggs has not been reported. Electroporation usually results in a large amount of cell death and would be expected to kill at least some of the treated embryos, but given the ability to treat a large number of eggs, high mortality may not be a limitation.

The direct injection of DNA into whole animals may also hold some potential as a means of achieving germline transformation, and more likely as a means of achieving somatic transformation of larval and adult tissues. Benvenisty and Reshef (1986) reported the expression of genes in vivo following the intraperitoneal injection of calcium phosphate-precipitated DNA into newborn rats. Similar results have also been reported following direct injection of calcium phosphate precipitated DNA into the liver and spleen of mice (Dubensky et al. 1984) and the intravenous injection of rats with DNA-containing liposomes (Nicolau et al. 1983). These types of experiments might be useful for introducing DNA into the soma or possibly the germline of insects. The results of Caspari and Nawa (1965), discussed earlier, indicate that this may be the case.

To facilitate targeting of particular tissues, DNA can be conjugated to carrier molecules, thereby promoting the receptor-mediated uptake of the DNA (Wu and Wu 1987; 1988). DNA injected into female insects might be targeted to the oocytes by conjugation to yolk proteins, and other cell-specific carriers might be used to target other tissues.

DNA may also be selectively targeted to the nuclei of cells by conjugation with nuclear specific proteins. For example, Kaneda et al. (1989) reported the increased expression of DNA co-introduced with nonhistone chromosomal proteins by liposome fusion into cultured cells. Similar results were obtained with hepatocytes following intravenous injection of liposomes loaded with nuclear protein and DNA (Nicolau et al. 1983). Minden et al. (1989) have shown that histones injected into preblastoderm embryos of *D. melanogaster* become selectively associated with nuclei. These proteins might be used to increase the amount of DNA that reaches the nucleus during more conventional embryo injection experiments.

Prolonging the persistence of DNA introduced into insect cells could further facilitate the recovery of integration events or the utility of transient expression systems. The introduction of DNA sequences that promote the replication of plasmids introduced into eukaryotic cells is one way this might be accomplished. Eukaryotic origins of replication or autonomously replicating sequences (ARS) have been isolated from yeast (Stinchcomb et al. 1979; Struhl et al. 1979; Stinchcomb et al. 1980; Tschumper and Carbon 1980), and sequences from other organisms, including *D. melanogaster*, which act as ARS elements in yeast, have also been identified (Marunouchi and Hosoya 1984; Danilevskaya et al. 1984; Gragerov et al. 1988). However, there is little evidence that these yeast-like ARS elements act as origins of replication in the organism from which they were isolated. ARS elements have recently been isolated from *Ustilago maydis* (Tsukuda et al. 1988) and primates (Frappier and Zannis-Hadjopoulos 1987), and extrachromosomal replication of *copia*-based vectors has been reported in cultured *Drosophila* cells (Sinclair et al. 1983). The ability of these sequences to promote replication or otherwise prolong the persistence of plasmids injected into insect embryos has not been reported.

3.2. Genetic Markers for Transformation

In order to assess the efficiency and utility of any gene transfer method, one must be able to recognize animals possessing a chromosomal integration of the vector. The direct detection of integrated DNA by genomic Southern blot analysis of putative transformants has been used to detect P-mediated transgenic *D. hawaiiensis* (Brennan et al. 1984) and *D. simulans* (Daniels et al. 1985). However, because of the limited number of animals that can be screened, this method will only be useful for detecting relatively frequent events. Therefore, more overt phenotypic markers are needed if less frequent events are to be detected. Positive selection for germline transformants permits a greater number of individuals to be screened than would otherwise be possible using visible markers. The ability to express the prokaryotic gene neomycin phosphotransferase (NPT) in *D. melanogaster* and to confer resistance to the antibiotic G418 has allowed it to be used as a positive selection marker in some transformation experiments (Rio and Rubin 1985; Steller and Pirrotta 1985; Miller et al. 1987). While this selection can be useful, frequent problems have been encountered, such as variability in G418 sensitivity and leakiness in the resistance phenotype, indicating that alternative markers are needed (McGrane et al. 1988; McInnis et al. 1988; Morris et al. 1989; and see Section 2.2.2 above).

A number of genes have been isolated which have been used as positive selection markers in cell lines but have not been tested in whole animals. For example, hygromycin resistance has been used in a variety of eukaryotic cell systems, including mammalian (Margolskee et al. 1988) and insect cells (J Carlson and B Beatty pers comm). Overexpression of dihydrofolate reductase or the mammalian multiple drug resistance gene in insects might result in whole animal resistance to methotrexate and certain plant alkaloids (such as colchicine, vinblastine, doxorubium hydrochloride and actinomycin D), respectively (Gros et al. 1986). While these are potential alternatives to G418 resistance, similar selection problems may also be encountered. Alternatively, genes which can confer resistance to certain insecticides may also be useful as gene transformation markers, although the use of such markers may be limited.

The expression of genes that confer new visible phenotypes, while not permitting the positive selection of transformants, would allow the recognition of transgenic animals within a population and may not be subject to the variances of chemical selection. The introduction and expression of the lacZ gene from *E. coli* in *D. melanogaster* results in high levels of β-galactosidase activity, detectable following a simple staining procedure (Lis et al. 1983). Although *D. melanogaster* has an endogenous β-galactosidase gene, it is developmentally and tissue specifically expressed (Knipple and MacIntyre 1984) and can be distinguished from *E. coli* β-galactosidase by conducting the assays at the appropriate pH (Thummel et al. 1988). Furthermore, β-galactosidase expression in transgenic *D. melanogaster* can be detected in isolated halteres and antennae, thereby allowing the nondestructive testing of individual flies (DA O'Brochta unpublished observation).

Finally, complementation of host mutations with either the cloned homologous wild type gene or a complementary nonhomologous gene from another organism could be used to identify transgenic animals. Recently, experiments conducted in the Australian sheep blowfly *Lucilia cuprina* have suggested that the *vermilion* gene of *D. melanogaster* can complement certain eye color mutations in this species (PW Atkinson pers comm). The recent isolation of the *white* and *topaz* genes from *L. cuprina* using *white* and *scarlet* gene sequences from *D. melanogaster* as probes, respectively, should facilitate the development of germline transformation markers for this species in the future (AJ Howells pers comm).

3.3. Alternative Vectors

Future efforts to develop insect gene vectors should focus not only on P elements but also on alternative systems such as other known mobile genetic elements or gene vectors. In addition the isolation of new mobile genetic elements from the species of interest should be considered. Here we briefly discuss alternative mobile genetic elements we feel hold some potential for use in non-drosophilid insects and some strategies for isolating new elements from non-drosophilid insects.

Mobile genetic elements to be considered for development into insect gene vectors should show some evidence of unrestricted mobility, as reflected in the absence of tissue specificity, a broad phylogenetic distribution or demonstrated mobility in a broad range of species. A number of elements have been isolated meeting all or some of these criteria.

3.3.1. Hobo

The *hobo* element was discovered serendipitously during the course of a molecular characterization of the *Drosophila* glue protein gene (McGinnis et al. 1983). The element, although not as well characterized as the P element, shares many similarities to P. Complete or autonomous *hobo* elements are thought to be 3.0 kb in length, including 12 bp terminal inverted repeats. Like P, *hobo* results in a 8 bp duplication of the target sequence upon integration. The *hobo* element contains codon usages similar to P elements, however, no sequence similarity exists between the two elements (Blackman and Gelbart 1989 for a review). Recently Blackman et al. (1989) have used *hobo* to mediate germline transformation in *Drosophila*. A 3.0 kb *hobo* element containing the wild type allele of the *rosy* gene of *D. melanogaster* was cloned into pUC18. This plasmid, H[(ry^+)har1], was analogous to the P vector Carnegie 20 (Rubin and Spradling 1983; Blackman et al. 1989). Successful transformation of ry^- mutant hosts using H[(ry^+)har1] was accomplished using two strategies. First, a cross was made resulting in "*hobo* dysgenic" embryos which were injected with H[(ry^+)har1]. Because complete *hobo* elements were present within the embryonic genome, the vector was mobilized, resulting in 28% of the fertile G_0 individuals producing transformed progeny. A second set of experiments using a "helper" plasmid containing a complete *hobo* element also resulted in successful transformation. In all cases vector integration appeared to be *hobo*- mediated, as indicated by the fact that *hobo* sequences were integrated without flanking sequences. As with P-mediated transformation, only a single *hobo* element inserted at a given site and the sites of integration were scattered throughout the genome. Attempts to use *hobo* as a vector in *Anopheles gambiae* have yielded results similar to those obtained using P elements (L Miller pers comm). The *hobo* system therefore appears to provide a second independent system for introducing DNA into *Drosophila* genomes, however, if it is to be used in non-drosophilids it will require modification.

3.3.2. Mariner

Mariner is a recently discovered transposable element from *D. mauritiana* which differs from known *Drosophila* transposable elements in a number of ways (Hartl 1989 for a review). First, it is only 1.3 kb in length, making it one of the smallest eukaryotic elements identified. It contains a single, long uninterrupted open reading frame, capable of encoding a protein containing 345 amino acids. This element has been found only in species within the melanogaster subgroup, although not in *D. melanogaster* (Jacobson et al. 1986). Perhaps most unusual is the somatic expression of *mariner* and subsequent somatic mosaicism resulting from *mariner* excision (Jacobson and Hartl 1985; Haymer and Marsh 1986). In *D. mauritiana* the rate of somatic mosaicism is approximately the same as germinal mutations (10^{-3}) (Jacobson and Hartl 1985; Haymer and Marsh 1986). In certain strains of *D. mauritiana*, however, the rate of somatic mosaicism can be extremely high, resulting from the presence of a particular *mariner* allele (*Mos*) (Bryan et al. 1987; Bryan and Hartl 1988; Medhora et al. 1988). When introduced into either *D. simulans* or *D. melanogaster* by P element mediated transformation, *Mos* is capable of destabilizing *mariner* elements previously introduced into these species. Thus the *mariner* system resembles the P and *hobo* systems in that *trans*-mobilization of nonautonomous elements is possible (D Hartl pers comm). The absence of introns within the transcription unit and the generally less restricted mobility properties of *mariner* make it a good candidate for a phylogenetically unrestricted insect gene vector.

3.3.3. Tc

Transposable elements have been isolated and described from a number of species of nematodes, most notably *Caenorhabditis elegans*. The most thoroughly studied element from nematodes is *Tc1* (Moerman and Waterston 1989). This element shares certain general structural features with P, *hobo* and *mariner* elements, such as a relatively small size (1.6 kb), terminal inverted repeats (54 bp), a large open reading frame which can encode a basic protein containing 273 amino acids, and a duplicated target sequence. These elements can transpose and excise at high rates under appropriate conditions, and mobility is not confined to the germline. *Tc1* also has a high degree of sequence similarity to the *HB* family of transposable elements found in *D. melanogaster* (Henikoff and Plasterik 1988). *HB* elements are approximately 1.6 kb in length and have 30 bp terminal inverted repeats showing some similarity to the terminal inverted repeats of *Tc1* (Brierley and Potter 1985). Furthermore, the open reading frames found in *Tc1* and *HB* share about 30% amino acid similarity. It is not known if this widespread distribution of Tc1-like sequences reflects a long evolutionary history or horizontal transmission. The relatively unrestricted mobility within *C. elegans* and the similarity of *Tc1* to insect transposable elements suggests that its mobility might be worth testing in insects.

3.3.4. *Ac*

Of all the eukaryotic transposable elements identified, the *Ac* element from maize has displayed the most phylogenetically unrestricted mobility (Federoff 1989 for a review). *Ac* elements introduced into tobacco, tomato, carrots and *Arabidopsis thaliana* by *Agrobacterium* Ti plasmid-mediated transformation have been shown to be mobile (Baker et al. 1986; Van Sluys et al. 1987; Knapp et al. 1988; Yoder et al. 1988). These results are significant because they indicate either that the *Ac* element is truly autonomous, encoding all of the factors necessary for mobility, or that required non-*Ac*-encoded factors are present even in distantly related dicots.

3.3.5. *FLP Recombinase*

Although all of the systems discussed above are based on transposable elements, other mobile DNA systems, such as the FLP-recombinase system of yeast, may be useful for insect transformation (Cox 1989 for a review). The yeast *Saccharomyces cerevisiae* contains a 2 μm plasmid, present at 60–100 copies per cell. This plasmid is approximately 6.3 kb and contains a pair of inverted repeats (FRTs) of 599 bp, separating regions of 2,346 and 2,774 bp. The 599 bp inverted repeats are the sites of a sequence-specific recombination event catalyzed by the plasmid-encoded protein, FLP recombinase. Normally, recombination between the repeats results in inversion of the unique sequences and plasmid amplification, although intermolecular recombination can also occur. Unlike other site-specific recombination systems, the FLP system requires no proteins other than the recombinase itself; divalent cations and high energy cofactors are not required. Its simplicity is further illustrated by the observation that it functions in *E. coli*. More recently, Golic and Lindquist (1989) have shown that expression of FLP recombinase in *D. melanogaster* results in the excision of a copy of the *white* gene flanked by FRT sites oriented as direct repeats. This reaction is the equivalent of an intramolecular recombination event in the original plasmid, but there is also some evidence that intermolecular recombination between a previously excised FRT-*white* extrachromosomal element and FRT site in the genome may occur.

If the FLP-recombinase system functions in other insects, it might be possible to use this system to promote recombination-mediated plasmid integration. Subsequent to the incorporation of an FRT site into the genome of interest (perhaps by random integration), it could be used as a landing site for subsequent intermolecular recombination events with plasmids containing the gene of interest, a transformation marker and a single FRT site.

3.4. Isolation of New Transposable Elements

Few eukaryotic transposable elements have shown any evidence of phylogenetically unrestricted mobility, and since modification of existing systems to achieve this end remains uncertain, vector development may require the isolation of transposable elements from the insects of interest. Given the relative abundance of middle repetitive DNA in animal genomes, the difficulty will not be in isolating DNA sequences with the structural characteristics of transposable elements but in finding elements that have mobility properties conducive to their development into gene vectors. For this reason we believe methods which bias the search toward actively mobile elements are preferred over, for example, random cloning of middle repetitive DNA or the screening of individuals and populations for restriction fragment length polymorphisms at cloned loci.

Relying on the mutagenic properties of transposons is perhaps the method of choice for isolating an element that will have potential as a gene vector, since the element is isolated by virtue of its mobility. This might be accomplished by screening for mutations at previously identified and cloned loci. Many of the mutations in *D. melanogaster* originally identified as spontaneous mutations have been shown to result from transposon insertions (Green 1988). This approach will not be generally applicable to non-drosophilids because of the difficulty in conducting genetic experiments with many of these systems.

Another possible method for isolating transposable elements from non-drosophilids involves the use of nuclear polyhedrosis viruses (NPVs) (see Vlak this volume). The genomes of these viruses can function as targets for insertion of lepidopteran transposable elements during their passage through lepidopteran cell lines (Miller and Miller 1982; Fraser et al. 1983; 1985; Fraser 1986; Carstens 1987). The passage of exogenous genomes through insect cells may therefore be a general method for capturing transposable elements. Because NPVs are largely limited to Lepidoptera, they may not be useful in capturing transposons from other orders. Alternatively, plasmid molecules introduced into and maintained in insect cells could also be useful as transposon targets.

3.5. Somatic Transformation

This review has focused on efforts to develop a germline transformation system for insects, yet this might not be necessary for the analysis of some gene systems. The ability to introduce plasmids into insect cells by injecting preblastoderm embryos allows plasmids to be used as minichromosomes for as long as they persist in the cells. While the longevity of injected plasmids varies from insect to insect, alternative DNA delivery systems and methods for targeting or protecting the plasmid DNA could expand the utility of this method.

As a consequence of their ability to infect a variety of lepidopteran larval tissues (Adams et al. 1977; Tinsley and Harrap 1977; Smith-Johannsen et al. 1986), NPVs hold some potential as transient expression vectors that might be useful in the analysis of insect genes in vivo. Recently Iatrou and Meidinger (1989) described the construction of a number of mutations in *Bombyx mori* NPV in which the activity of the polyhedron gene promoter was altered or eliminated. These mutant viruses lacking polyhedron gene promoter activity have nevertheless retained the ability to infect cells, replicate and assemble. They can therefore be used as transient expression vectors, both in *Bombyx mori* cell cultures and in pupae (K Iatrou pers comm). For example, K Iatrou (pers comm) has shown that the chorion genes HcA/HcB.12 under the control of their endogenous promoters can be introduced into the follicle epithelium of *Bombyx mori* pupae using recombinant BmNPV and be correctly expressed in that tissue. However, several caveats should be noted. First, other tissues also become infected with the recombinant viruses, and randomly initiated chorion transcripts in other tissues were detected, possibly due to read-through from viral promoters or the influence of viral transcription factors. Without eliminating this background expression, tissue-specific gene regulation will remain ambiguous. Second, NPV infection results in the death of the animal within approximately five days, making the utility of this method somewhat limited in Lepidoptera. Conditional viral mutations that do not result in the death of the host would extend the utility of this method considerably. These limitations notwithstanding, the study of genes under the control of their endogenous promoters using NPV vectors may currently be the best alternative to germline transformation for the in vivo analysis of insect genes expressed relatively late in development.

The use of NPV transient expression vectors as described by Iatrou may be more useful in nonsusceptible species. Sherman and McIntosh (1979) showed that the NPV of *Autogropha californica* (AcNPV) was capable of entering and replicating in *A. aegypti* cell lines without killing the cells, and Carbonell et al. (1985) showed that AcNPV could not only infect *D. melanogaster* cell lines but also the midgut cells of *A. aegypti*, where an endogenous reporter gene was expressed. Because nonsusceptible hosts do not succumb to AcNPV infection but apparently can support virus entry and perhaps even limited replication, they may be more amenable to the use of recombinant NPV transient expression systems. This possibility should be more thoroughly explored.

4. Concluding Remarks

The attainment of a highly efficient germline transformation system for non-drosophilid insects remains a high priority if the full potential of molecular genetic techniques are to be realized for both basic and applied studies. In terms of transposon-based gene vector systems, the *Drosophila* P element system, while not functional in nondrosophilids in its present state, still holds the greatest potential to be rendered functional after modification. Although it may eventually be necessary to develop gene vectors for particular insect species or families, the identification and isolation of transposons for this may not be trivial. Thus, it may be worthwhile initially to consider transposons already isolated having unrestricted mobility properties, including those from heterologous systems such as other insects, and even plants and nematodes. Gene transfer via random integration also remains a viable possibility that has not been adequately evaluated, and certainly for at least some insects such as mosquitos, the limiting factor for this method may only be an efficient means of mass introduction of DNA into eggs.

Beyond developing a means to integrate DNA into an insect host genome, there are ancillary problems affecting not only the successful creation and selection of gene transformants but their subsequent utility in analysis and field applications. Foremost is the development of selectable markers for gene integration. As for gene transfer methods, selectable markers may be subject to species-specific constraints that must be properly evaluated. Although initially encouraged by results in *Drosophila*, the use of neomycin resistance may have served to waste much effort to transform various insects, the medfly being at least one. Selectable markers must be evaluated for their expression and detection, possibly requiring transient expression techniques. Once transformation is achieved it will be necessary to distinguish the activity of the introduced gene from endogenous homologous genes. For genetically uncharacterized species this will be problematic, although development of gene replacement or directed mutagenesis methodologies may provide an eventual answer.

In lieu of germline transformation, gene expression may be analyzed by gene transfer into heterologous systems, such as *Drosophila*, or homologous transient expression. While these methods have inherent limitations, in many instances they are quite adequate, providing a means of initially testing gene constructs that would eventually require stable integration. As noted, transient expression techniques are also critical to a complete evaluation of DNA delivery methods, vector mobility, and selectable markers.

The focus on gene-transfer has been generally limited to tephritid and mosquito species, however, it may be assumed that techniques eventually developed for these species will not be completely universal due to any one of the limitations we have discussed. It is therefore important that research groups interested in the use of germline transformation in diverse insect species take an active part in its achievement. We hope that this discussion has provided some stimulation towards that end.

Acknowledgments. We would like to acknowledge those who contributed unpublished results and the U.S. Department of Agriculture for supporting this work.

References

Adams JR, Goodwin RH, Wilcon TA (1977) Electron microscopic investigations on invasion and replication of insect baculoviruses in vivo and in vitro. Biol Cell 28:261–268

Allen ND, Cran DG, Barton SC, Hettle S, Reik W, Surani MA (1988) Transgenes as probes for active chromosomal domains in mouse development. Nature 333:852–855

Amrein H, Gorman M, Nothiger R (1988) The sex-determining gene *tra*-2 of *Drosophila* encodes a putative RNA binding protein. Cell 55:1025–1035

Anxolabehere D, Periquet G (1987) P-homologous sequences in diptera are not restricted to the Drosophilidae family. Genet Iber 39:211–222

Baker B, Schell J, Loerz H, Fedoroff N (1986) Transposition of the maize controlling element *Activator* in tobacco. Proc Natl Acad Sci USA 83: 4844–4848

Bellen HJ, O'Kane CJ, Wilson C, Grossniklaus U, Pearson RK, Gehring WJ (1989) P-element-mediated enhancer detection: a versatile method to study development in *Drosophila*. Genes Devel 3:1288–1300

Belote JM, Baker BS (1982) Sex determination in *Drosophila melanogaster*: analysis of transformer-2, a sex-determining locus. Proc Natl Acad Sci USA 79:1568–1572

Benvenisty N, Reshef L (1986) Direct introduction of genes into rats and expression of the genes. Proc Natl Acad Sci USA 83:9551–9555

Berg GM, Berg DE, Grocsman EA (1989) Transposable elements and the genetic engineering of bacteria. In Berg DE, Howe MM (eds) Mobile DNA. American Society for Microbiology, Washington DC, pp 879–925

Berger EM, Torrey D, Morganelli C (1986) Natural and synthetic heat shock protein gene promoters assayed in *Drosophila* cells. Somat Cell Mol Genet 12:433–440

Black DMN, Jackson MS, Kidwell MG, Dover GA (1987) KP elements repress P induced hybrid dysgenesis in *D. melanogaster*. EMBO J 6:4125–4135

Blackman RK, Gelbart WM (1989) The transposable element *hobo* of *Drosophila melanogaster*. In Berg DE, Howe MM (eds) Mobile DNA. American Society for Microbiology, Washington DC, pp 523–525

Blackman RK, Grimaila MM, Koehler MD, Gelbart WM (1987) Mobilization of *hobo* elements residing within the decapentaplegic gene complex: suggestion of a new hybrid dysgenesis system in *Drosophila melanogaster*. Cell 49: 497–505

Blackman RK, Macy M, Koehler D, Grimaila R, Gelbart WM (1989) Identification of a fully-functional *hobo* transposable element and its use for germline transformation of *Drosophila*. EMBO J 8:211–217

Boynton JE, Gillham NW, Harris EH, Hosler JB, Johnson AM, Jones AR, Randolph-Anderson BL, Robertson D, Klein TM, Shark KB, Sanford JC (1988) Chloroplast transformation in *Chlamydomonas* with high velocity microprojectiles. Science 240:1534–1537

Brennan MD, Rowan RG, Dickinson WJ (1984) Introduction of a functional P element into the germ line of *Drosophila hawaiiensis*. Cell 38:147–151

Brierley HL, Potter SS (1985) Distinct characteristics of loop sequences of two *Drosophila* foldback transposable elements. Nucl Acids Res 13:485–500

Brookfield JFY, Montgomery E, Langley C (1984) Apparent absence of transposable elements related to the P elements of *D. melanogaster* in other species of *Drosophila*. Nature 310:330–332

Bryan GJ, Hartl DL (1988) Maternally inherited transposon excision in *Drosophila simulans*. Science 240:215–217

Bryan GJ, Jacobson JW, Hartl DL (1987) Heritable somatic excision of a *Drosophila* transposon. Science 235:1636–1638

Carbonell LF, Klowden MJ, Miller LK (1985) Baculovirus-mediated expression of bacterial genes in dipteran and mammalian cells. J Virol 56:153–160

Carlson DA, Cockburn AF, Tarrent CA (1989) Advances in insertion of material into insect eggs via a particle shotgun technique. In Borovsky D, Spielman A (eds) Host Regulated Developmental Mechanisms in Vector Arthropods. University of Florida IFAS Press, Florida, pp 248–252

Carstens EB (1987) Identification and nucleotide sequence of the regions of *Autographa californica* nuclear polyhedrosis virus genome carrying insertion elements derived from *Spodoptera frugiperda*. Virology 161:8–17

Caspari E, Nawa S (1965) A method to demonstrate transformation in *Ephestia*. Z Naturforsch 206:281–284

Clough DW, Lepinske HM, Davidson RL, Storti RV (1985) *Drosophila* P element-enhanced transfection in mammalian cells. Mol Cell Biol 5:898–901

Costanzo MC, Fox TD (1988) Transformation of yeast by agitation with glass beads. Genetics 120:667–670

Cox MM (1989) DNA inversion in the 2 μm plasmid of *Saccharomyces cerevisiae*. In Berg DE, Howe MM (eds) Mobile DNA. American Society for Microbiology, Washington DC, pp 661–670

Daniels SB, Strausbaugh LD (1986) The distribution of P element sequences in *Drosophila*: the *willistoni* and *saltans* species groups. J Mol Evol 23:138–148

Daniels SB, Strausbaugh LD, Armstrong RA (1985) Molecular analysis of P element behavior in *Drosophila simulans* transformants. Mol Gen Genet 200:258–265

Danilevskaya ON, Kurenova EV, Leibovitch BA, Shevelev AY, Bass IA, Khesin RB (1984) Telomeres and P-elements of *Drosophila melanogaster* contain sequences that replicate autonomously in *Saccharomyces cerevisiae*. Mol Gen Genet 197:342–344

Dean D (1981) A plasmid cloning vector for the direct selection of strains carrying recombinant plasmids. Gene 15:99–102

de Lozanne A, Spudich JA (1987) Disruption of the Dictyostelium myosin heavy chain gene by homologous recombination. Science 236:1086–1091

Dubensky TW, Campbell BA, Villarreal LP (1984) Direct transfection of viral and plasmid DNA into the liver or spleen of mice. Proc Natl Acad Sci USA 81:7529–7533

Durbin JE, Fallon AM (1985) Transient expression of the chloramphenicol acetyltransferase gene in cultured mosquito cells. Gene 36:173–178

Engels WR (1989) P elements in *Drosophila melanogaster*. In Berg DE, Howe MM (eds) Mobile DNA. American Society for Microbiology, Washington DC, pp 437–484

Federoff N (1989) Maize transposable elements. In Berg DE, Howe MM (eds) Mobile DNA. American Society for Microbiology, Washington DC, pp 375–411

Foster GG, Maddern RH, Mills AT (1980) Genetic instability in mass-rearing colonies of a sex-linked translocation strain of *Lucilia cuprina* (Wiedemann) (Diptera: Calliphoridae) during a field trial of genetic control. Theor Appl Genet 58:164–175

Fox AS, Yoon SB (1966) Specific genetic effects of DNA in *Drosophila melanogaster*. Genetics 53:897–911

Fox AS, Yoon SB (1970) DNA-induced transformation in *Drosophila*: Locus specificity and the establishment of transformed stocks. Proc Natl Acad Sci USA 67:1608–1615

Frappier L, Zannis-Hadjopoulos M (1987) Autonomous replication of plasmids bearing monkey DNA origin-enriched sequences. Proc Natl Acad Sci USA 84:6668–6672

Fraser MJ (1986) Transposon-mediated mutagenesis of baculoviruses: transposon shuttling and implications for speciation. Ann Entomol Soc Amer 79: 773–783

Fraser MJ, Brusca JS, Smith GE, Summers MD (1985) Transposon-mediated mutagenesis of a baculovirus. Virology 145:356–361

Fraser MJ, Smith GE, Summers MD (1983) Acquisition of host cell DNA sequences by baculoviruses: relationship between host DNA insertions and FP mutants of *Autographa californica* and *Galleria mellonella* nuclear polyhedrosis viruses. J Virol 47:287–300

Germeraad S (1976) Genetic transformation in *Drosophila* by microinjection of DNA. Nature 262:229–231

Golic KG, Lindquist S (1989) The FLP recombinase of yeast catalyzes site specific recombination in the *Drosophila* genome. Cell 59:499–509

Gossler A, Joyner AL, Rossant J, Skames WC (1989) Mouse embryonic stem cells and reporter constructs to detect developmentally regulated genes. Science 244:463–465

Gragerov AI, Danilevskaya ON, Didichenki SA, Kaverina EN (1988) An ARS element from *Drosophila melanogaster* telomeres contains the yeast ARS core and bent replication enhancer. Nucl Acids Res 16:1169–1180

Green MM (1988) Mobile DNA elements and spontaneous gene mutation. In Lambert ME, McDonald JF, Weinstein IB (eds) Eukaryotic Transposable

Elements as Mutagenic Agents. Banbury Report 30, Cold Spring Harbor Laboratory, New York, pp 41–50

Gros P, Neria YB, Croop JM, Housman DE (1986) Isolation and expression of a complementary DNA that confers multiple drug resistance. Nature 323: 728–731

Hartl DL (1989) Transposable element *mariner* in *Drosophila* species. In Berg DE, Howe MM (eds) Mobile DNA. American Society for Microbiology, Washington DC, pp 531–536

Haymer DS, Marsh JL (1986) Germ line and somatic instability of a *white* mutation in *Drosophila mauritiana* due to a transposable element. Devel Genet 6:281–291

Henikoff S, Plasterik RHA (1988) Related transposons in *C. elegans* and *D. melanogaster*. Nucl Acids Res 16:6234

Hinnen A, Hicks JB, Fing GR (1978) Transformation of yeast. Proc Natl Acad Sci USA 75:1929–1933

Hooper GHS, Robinson AS, Marchand RP (1987) Behavior of a genetic sexing strain of Mediterranean fruit fly, *Ceratitis capitata*, during large scale rearing. In Economopoulos AP (ed) Fruit Flies. Elsevier, Amsterdam, pp 349–362

Iatrou K, Meidinger RG (1989) *Bombyx mori* nuclear polyhedrosis virus-based vectors for expressing passenger genes in silkmoth cells under viral or cellular promoter control. Gene 75:59–71

Jacobson JW, Hartl DL (1985) Coupled instability of two X-linked genes in *Drosophila mauritiana*: germinal and somatic instability. Genetics 111:57–65

Jacobson JW, Medhora MM, Hartl DL (1986) Molecular structure of a somatically unstable transposable element in *Drosophila*. Proc Natl Acad Sci USA 83:8684–8688

Johnston SA, Anziano PQ, Shark K, Sanford JC, Butlow RA (1988) Mitochondrial transformation in yeasts by bombardment with microprojectiles. Science 240:1538–1541

Junakovic N, DiFranco C, Barsanti P, Palumbo G (1986) Transposition of *copia*-like elements can be induced by heat shock. J Mol Evol 24:89–93

Kaneda Y, Iwai K, Uchida T (1989) Increased expression of DNA cointroduced with nuclear protein in adult rat liver. Science 243:375–378

Khillan JS, Overbeek PA, Westphall H (1985) *Drosophila* P element integration in the mouse. Devel Biol 109:247–250

Kidwell MG (1987) A survey of success rates using P element mutagenesis in *Drosophila melanogaster*. Dros Inform Serv 66:81–86

Klein TM, Wolf ED, Wu R, Sanford JC (1987) High-velocity microprojectiles for delivering nucleic acids into living cells. Nature 327:70–73

Knapp S, Coupland G, Uhrig H, Starlinger P, Salamini F (1988) Transposition of the maize transposable element *Ac* in *Solanum tuberosum*. Mol Gen Genet 213:285–290

Knipple DC, MacIntyre RJ (1984) Cytogenetic mapping and isolation of mutations of the β-gal-1 locus of *Drosophila melanogaster*. Mol Gen Genet 198: 75–83

Lambert ME, McDonald JF, Weinstein IB (eds) (1988) Eukaryotic transposable elements as mutagenic agents. Banbury Report 30, Cold Spring Harbor Laboratory, New York, 345 pp

Lansman RA, Stacey SN, Grigliatti TA, Brock HW (1985) Sequences homologous to the P mobile element of *Drosophila melanogaster* are widely distributed in the subgenus Sophophora. Nature 318:561–563

Laski FA, Rio DC, Rubin GM (1986) Tissue specificity of *Drosophila* P element transposition is regulated at the level of mRNA splicing. Cell 44:7–19

Lavitrano M, Camiaioni A, Fazio VM, Dolci S, Farace M, Spadafora C (1989) Sperm cells as vectors for introducing foreign DNA into eggs: genetic transformation of mice. Cell 57:717–723

Leopold P, Vailly J, Cuzin F, Rassoulzadegan M (1987) Germ line maintenance of plasmids in transgenic mice. Cell 51:885–886

Lis JT, Simon JA, Sutton CA (1983) New heat shock puffs and β-galactosidase activity resulting from transformation of *Drosophila* with an hsp70-LacZ hybrid gene. Cell 35:403–410

Liu CP, Kreber RA, Valencia JI (1979) Effects on eye color mediated by DNA injection into *Drosophila* embryos. Mol Gen Genet 172:203–210

Margolskee RF, Kavathas P, Berg P (1988) Epstein-Barr virus shuttle vector for stable episomal replication of cDNA expression libraries in human cells. Mol Cell Biol 8:2839–2847

Martinez-Zapata JM, Finkelstein R, Somerville CR (1987a) Introduction of the P-element from *Drosophila* into tobacco. In McIntosh L, Key J (eds) Plant Gene Systems and Their Biology, UCLA Symposia on Molecular and Cellular Biolgy, Vol 62. AR Liss, New York, pp 311–320

Martinez-Zapata JM, Finkelstein R, Somerville CR (1987b) *Drosophila* P-element transcripts are incorrectly processed in tobacco. Plant Mol Biol 11:601–607

Marunouchi T, Hosoya H (1984) Isolation of an autonomously replicating sequence (ARS) from satellite DNA of *Drosophila melanogaster*. Mol Gen Genet 196:258–265

McGinnis W, Shermoen AW, Beckendorf SK (1983) A transposable element inserted just 5′ to a *Drosophila* glue protein gene alters gene expression and chromatin structure. Cell 34:75–84

McGrane V, Carlson JO, Miller BR, Beaty BJ (1988) Microinjection of DNA into *Aedes triseriatus* ova and detection of integration. Amer J Trop Med Hyg 39:502–510

McInnis DO, Tam SYT, Grace CR, Heilman LJ, Courtright JB, Kumaran AK (1988) The Mediterranean fruit fly: progress in developing a genetic sexing strain using genetic engineering methodology. In Modern Insect Control: Nuclear Techniques and Biotechnology. International Atomic Energy Agency, Vienna, pp 251–256

Medhora MM, MacPeek AH, Hartl DL (1988) Excision of the *Drosophila* transposable element mariner: identification and characterization of the *Mos* factor. EMBO J 7:2185–2189

Mevel-Ninio M, Mariol M, Gans M (1989) Mobilization of the *gypsy* and *copia* retrotransposons in *Drosophila melanogaster* induces reversion of the ovo^D dominant female-sterile mutations: molecular analysis of revertant alleles. EMBO J 8:1549–1558

Miller BL, Miller KY, Roberti KA, Timberlake WE (1985) Direct and indirect gene replacement in *Aspergillus nidulans*. Mol Cell Biol 5:1718–1721

Miller DW, Miller LK (1982) A virus mutant with an insertion of a *copia*-like transposable element. Nature 299:562–564

Miller LH, Sakai RK, Romans P, Gwad RW, Kantoff P, Coon HG (1987) Stable integration and expression of a bacterial gene in the mosquito *Anopheles gambiae*. Science 237:779–781

Minden JS, Agard DA, Sedat JW, Aberts BM (1989) Direct cell lineage analysis in *Drosophila melanogaster* by time-lapse, three dimensional optical microscopy of living embryos. J Cell Biol 109:505–516

Moerman DG, Waterston RH (1989) Mobile elements in *Caenorhabditis elegans* and other nematodes. In Berg DE, Howe MM (eds) Mobile DNA. American Society for Microbiology, Washington DC, pp 537–556

Morris AC, Eggelston P, Crampton JM (1989) Genetic transformation of the mosquito *Aedes aegypti* by micro-injection of DNA. Med Vet Entomol 3: 1–7

Nawa S, Yamada S (1968) Hereditary change in *Ephestia* after treatment with DNA. Genetics 58:573–584

Nicolau C, LePape A, Soriano P, Fargette F, Juhel M (1983) In vivo expression of rat insulin after intravenous administration of the liposome-entrapped gene for rat insulin I. Proc Natl Acad Sci USA 80:1068–1072

Nitasaka E, Mukai T, Yamazaki T (1987) Repressor of P elements in *Drosophila melanogaster*: cytotype determination by a defective P element with only open reading frames O through 2. Proc Natl Acad Sci USA 84:7605–7608

O'Brochta DA, Handler AM (1988) Mobility of P elements in drosophilids and nondrosophilids. Proc Natl Acad Sci USA 85:6052–6056

O'Kane CJ, Gehring WJ (1987) Detection in situ of genomic regulatory elements in *Drosophila*. Proc Natl Acad Sci USA 84:9123–9127

Pasyukova EG, Belyaeva ES, Ilyinskaya LE, Bvozdev VA (1988) Outcross-dependent transpositions of *copia*-like mobile genetic elements in chromosomes of an inbred *Drosophila melanogaster* stock. Mol Gen Genet 212: 281–286

Paszkowski J, Baur M, Bogucki A, Potrykus I (1988) Gene targeting in plants. EMBO J 9:4021–4026

Pirrotta V (1988) Vectors for P-mediated transformation in *Drosophila*. In Rodriguez RL, Denhardt DT (eds) Vectors—A Survey of Molecular Cloning Vectors and their Uses. Butterworths, Boston, pp 437–456

Potter SS, Brorein WJ, Dunsmuir P, Rubin GM (1979) Transposition of elements of *412, copia*, and 297 dispersed repeated gene families in *Drosophila*. Cell 17:415–427

Rio DC, Barnes G, Laski FA, Rine J, Rubin GM (1988) Evidence of *Drosophila* P element transposase activity in mammalian cells and yeast. J Mol Biol 200:411–415

Rio DC, Laski FA, Rubin GM (1986) Identification and immunochemical analysis of biologically active *Drosophila* P element transposase. Cell 44:21–32

Rio DC, Rubin GM (1985) Transformation of cultured *Drosophila melanogaster* cells with a dominant selectable marker. Mol Cell Biol 5:1833–1838

Rio DC, Rubin GM (1988) Identification and purification of a *Drosophila* protein that binds to the terminal 31-base pair inverted repeats of the P transposable element. Proc Natl Acad Sci USA 85:8929–8933

Robertson HM, Preston CR, Phillis RW, Johnson-Schlitz DM, Benz WK, Engels WR (1988) A stable genomic source of P element transposase in *Drosophila melanogaster*. Genetics 118:461–470

Robinson AS, Savakis C, Louis C (1988) Status of molecular genetic studies in the medfly *Ceratitis capitata*, in relation to genetic sexing. In Modern Insect Control: Nuclear Techniques and Biotechnology. International Atomic Energy Agency, Vienna, pp 241–250

Rössler Y (1985) Effect of genetic recombination in males of the Mediterranean fruit fly (Ditera: Tephritidae) on the integrity of "genetic sexing" strains produced for sterile-insect releases. Ann Entomol Soc Amer 78:265–270

Rubin GM (1983) Dispersed repetitive DNAs in *Drosophila*. In: Shapiro JA (ed) Mobile Genetic Elements. Academic Press, London, pp 329–361

Rubin GM, Kidwell MG, Bingham PM (1982) The molecular basis of P-M hybrid dysgenesis: the nature of induced mutations. Cell 29:987–994

Rubin GM, Spradling AC (1982) Genetic transformation of *Drosophila* with transposable element vectors. Science 218:348–353

Rubin GM, Spradling AC (1983) Vectors for P element-mediated gene transfer in *Drosophila*. Nucl Acids Res 11:6341–6351

Scavarda NJ, Hartl DL (1984) Interspecific DNA transformation in *Drosophila*. Proc Natl Acad Sci USA 81:7615–7619

Searles LL, Greenleaf AL, Kemp WE, Voelker RA (1986) Sites of P element insertion and structures of P element deletions in the 5′ region of *Drosophila melanogaster Rp II 215*. Mol Cell Biol 61:3312–3319

Sherman KE, McIntosh AH (1979) Baculovirus replication in a mosquito (Dipteran) cell line. Infect Immun 26:232–234

Shigekawa K, Dower WJ (1988) Electroporation of eukaryotes and prokaryotes: a general approach to the introduction of macromolecules into cells. Biotechniques 6:742–751

Sinclair JH, Sang JH, Burke JF, Ish-Horowicz D (1983) Extrachromosomal replication of *copia*-based vectors in cultured *Drosophila* cells. Nature 306: 198–200

Smith-Johannsen H, Witkiewica H, Iatrou K (1986) Infection of silkmoth follicular cells with *Bombyx mori* nuclear polyhedrosis virus. J Invert Pathol 48:74–84

Soriano P, Gridley T, Jaenisch R (1989) Retroviral tagging in mammalian development and genetics. In Berg DE, Howe MM (eds) Mobile DNA. American Society for Microbiology, Washington DC, pp 927–937

Spradling AC (1986) P element-mediated transformation. In Roberts DB (ed) *Drosophila*: A Practical Approach. IRL Press, Oxford, pp 175–197

Spradling AC, Rubin GM (1982) Transposition of cloned P elements into *Drosophila* germ line chromosomes. Science 218:341–347

Steller H, Pirrotta V (1986) P transposons controlled by the heat shock promoter. Mol Cell Biol 6:1640–1649

Stinchcomb DH, Struhl K, Davis RW (1979) Isolation and characterization of a yeast chromosomal replicator. Nature 282:39–43

Stinchcomb DH, Thomas M, Kelly J, Selker E, Davis RW (1980) Eukaryotic DNA segments capable of autonomous replication in yeast. Proc Natl Acad Sci USA 77:2651–2655

Struhl K, Stinchcomb D, Scherer S, Davis RW (1979) High-frequency transformation of yeast: Autonomous replication of hybrid DNA molecules. Proc Natl Acad Sci USA 76:1035–1039

Technau GM, Campos-Ortega JA (1986) Lineage analysis of transplanted individual cells in embryos of *Drosophila melanogaster*. III. Commitment and proliferative abilities of pole cells and midgut progenitors. Wilhelm Roux Arch Devel Biol 195:495–498

Thomas KR, Capecchi MR (1987) Site-direct mutagenesis by gene targeting in mouse embryos-derived stem cells. Cell 51:503–512

Thummel CS, Boulet AM, Lipshitz HD (1988) Vectors for *Drosophila* P-element-mediated transformation and tissue culture transfection. Gene 74:445–446

Tinsley TW, Harrap KA (1977) Viruses of invertebrates. In Fraenkel-Conrad H, Wagner RR (eds) Comprehensive Virology Vol 12. Plenum, New York, pp 1–101

Tschumper G, Carbon J (1980) Sequence of a yeast DNA fragment containing a chromosomal replicator and the TRPl gene. Gene 10:157–166

Tsubota S, Schedl P (1986) Hybrid dysgenesis-induced revertants of insertions of the 5′ end of the *rudimentary* gene in *Drosophila melanogaster*: transposon-induced control mutations. Genetics 114:165–182

Tsukuda T, Carleton S, Fotheringham S, Holloman WK (1988) Isolation and characterization of an autonomously replicating sequence from *Ustilago maydis*. Mol Cell Biol 8:3703–3709

Van Sluys MA, Tempe J, Fedoroff N (1987) Studies on the introduction and mobility of the maize *Activator* element in *Arabidopsis thaliana* and *Daucus carota*. EMBO J 6:3881–3889

Walker VK (1989) Gene transfer in insects. Adv Cell Culture 7:87–124

Wilson C, Pearson RK, Bellen HJ, O'Kane CJ, Grossniklaus U, Gehring WJ (1989) P-element-mediated enhancer detection: an efficient method for isolating and characterizing developmentally regulated genes in *Drosophila*. Genes Devel 3:1301–1313

Wu GY, Wu CH (1987) Receptor-mediated in vitro gene transformation by a soluble DNA carrier system. J Biol Chem 262:4429–4432

Wu GY, Wu CH (1988) Receptor-mediated gene delivery and expression in vivo. J Biol Chem 263:14621–14624

Yoder JI, Palys J, Albert K, Lassner M (1988) *Ac* transposition in transgenic tomato plants. Mol Gen Genet 213:291–296

Young MW (1979) Middle repetitive DNA: a fluid component of the *Drosophila* genome. Proc Natl Acad Sci USA 76:6274–6278

Young MW, Schwartz HE (1981) Nomadic gene families in *Drosophila*. Cold Spring Harbor Symp Quant Biol 45:629–640

Index

A

J

K

L

M

R

S

T